Thin-Layer Chromatography

CHROMATOGRAPHIC SCIENCE SERIES

A Series of Monographs

Editor: JACK CAZES
Cherry Hill, New Jersey

1. Dynamics of Chromatography, *J. Calvin Giddings*
2. Gas Chromatographic Analysis of Drugs and Pesticides, *Benjamin J. Gudzinowicz*
3. Principles of Adsorption Chromatography: The Separation of Nonionic Organic Compounds, *Lloyd R. Snyder*
4. Multicomponent Chromatography: Theory of Interference, *Friedrich Helfferich and Gerhard Klein*
5. Quantitative Analysis by Gas Chromatography, *Josef Novák*
6. High-Speed Liquid Chromatography, *Peter M. Rajcsanyi and Elisabeth Rajcsanyi*
7. Fundamentals of Integrated GC-MS (in three parts), *Benjamin J. Gudzinowicz, Michael J. Gudzinowicz, and Horace F. Martin*
8. Liquid Chromatography of Polymers and Related Materials, *Jack Cazes*
9. GLC and HPLC Determination of Therapeutic Agents (in three parts), *Part 1 edited by Kiyoshi Tsuji and Walter Morozowich, Parts 2 and 3 edited by Kiyoshi Tsuji*
10. Biological/Biomedical Applications of Liquid Chromatography, *edited by Gerald L. Hawk*
11. Chromatography in Petroleum Analysis, *edited by Klaus H. Altgelt and T. H. Gouw*
12. Biological/Biomedical Applications of Liquid Chromatography II, *edited by Gerald L. Hawk*
13. Liquid Chromatography of Polymers and Related Materials II, *edited by Jack Cazes and Xavier Delamare*
14. Introduction to Analytical Gas Chromatography: History, Principles, and Practice, *John A. Perry*
15. Applications of Glass Capillary Gas Chromatography, *edited by Walter G. Jennings*
16. Steroid Analysis by HPLC: Recent Applications, *edited by Marie P. Kautsky*
17. Thin-Layer Chromatography: Techniques and Applications, *Bernard Fried and Joseph Sherma*
18. Biological/Biomedical Applications of Liquid Chromatography III, *edited by Gerald L. Hawk*
19. Liquid Chromatography of Polymers and Related Materials III, *edited by Jack Cazes*
20. Biological/Biomedical Applications of Liquid Chromatography, *edited by Gerald L. Hawk*
21. Chromatographic Separation and Extraction with Foamed Plastics and Rubbers, *G. J. Moody and J. D. R. Thomas*
22. Analytical Pyrolysis: A Comprehensive Guide, *William J. Irwin*

ADDITIONAL VOLUMES IN PREPARATION

Thin-Layer Chromatography

Fouth Edition, Revised and Expanded

Bernard Fried

Department of Biology
Lafayette College
Easton, Pennsylvania

Joseph Sherma

Department of Chemistry
Lafayette College
Easton, Pennsylvania

MARCEL DEKKER, INC.

NEW YORK · BASEL

ISBN: 0-8247-0222-0

This book is printed on acid-free paper.

Headquarters
Marcel Dekker, Inc.
270 Madison Avenue, New York, NY 10016
tel: 212-696-9000; fax: 212-685-4540

Eastern Hemisphere Distribution
Marcel Dekker AG
Hutgasse 4, Postfach 812, CH-4001 Basel, Switzerland
tel: 44-61-261-8482; fax: 44-61-261-8896

World Wide Web
http://www.dekker.com

The publisher offers discounts on this book when ordered in bulk quantities. For more information, write to Special Sales/Professional Marketing at the headquarters address above.

Current printing (last digit):
10 9 8 7 6 5 4 3

PRINTED IN THE UNITED STATES OF AMERICA

Preface to the Fourth Edition

This new edition of our book has been written to update and expand the topics in the third edition. It provides extensive coverage of qualitative and quantitative thin-layer chromatography (TLC) and high-performance thin-layer chromatography (HPTLC) and should appeal to a wide range of scientists including biologists, chemists, biochemists, biotechnologists, and medical technologists.

We have maintained the same overall organization of the earlier editions, namely, a series of initial chapters on theory and practice (Part I) and a second section of chapters covering applications to important compound classes (Part II). As in the previous editions, practical rather than theoretical aspects of TLC have been emphasized. The book is still intended to serve as an introductory primer. For more detailed coverage of all aspects of TLC, the reader is referred to the 1996 *Handbook of Thin-Layer Chromatography: Second Edition, Revised and Expanded*, edited by J. Sherma and B. Fried (Marcel Dekker, Inc., 1996).

This edition updates and extends references to the literature. The TLC literature is voluminous and references in this book are necessarily selective. Citations are mainly to the 1985–1997 literature, with retention of some important earlier references.

The chapter titles in Part I are unchanged, but some subjects have been reorganized within the chapters and there is a general updating of material and elimination of a number of obsolete instruments. Included among the topics for which new sections or increased coverage is provided are the following: sample preparation by supercritical fluid chromatography (Chapter 3); mobile phases for ion pair and chiral separations (Chapter 4); robotics and automation (Chapter 7);

TLC coupled with spectrometry and other analytical methods, video documentation, and computer imaging (Chapter 9); quantification by video densitometry (Chapter 10); validation of quantitative results (Chapter 11); and in situ instrumental measurement of radioactive zones (Chapter 13). The fourth edition continues to provide extensive coverage of sample preparation in Chapter 3. We believe that this coverage is unique and differs from that found in other available treatises on TLC. Because of the primary importance of commercial precoated plates, detailed instructions for preparing layers were removed (Chapter 3), and description of documentation by contact printing has been eliminated (Chapter 9).

Part II is concerned with the applications of TLC to various compound classes. It presents a brief and selective overview of the recent literature and various detailed qualitative and quantitative densitometric experiments on organic dyes, lipids, amino acids, carbohydrates, natural pigments, vitamins, nucleic acid derivatives, steroids and terpenoids, pharmaceuticals, organic acids, antibiotics, and insecticides. This edition presents new experiments on organic dyes, lipids, and antibiotics, and suggestions for new experiments on lipids, pharmaceuticals, organic acids, and insecticides based on the "TLC Applications" catalog of Macherey Nagel GmbH & Co. KG, Postfach 101352, D-522313 Dueren, Germany. This section of the book should be useful to students and faculty at academic institutions and should also provide useful experimental procedures for the practicing chromatographer. The suggestions for new experiments in this edition should also provide information to chromatographers seeking advanced treatment on applications of TLC to various compounds.

Most of the suggestions by reviewers of the third edition have been incorporated into this revision. New line drawings of chromatograms have been added. Some experiments that required the use of plant and animal material that is not easily available were deleted. Corrections were made in Chapter 21 to eliminate minor errors and use of outdated terminology related to steroids. Comments on this edition as well as notification of errors and suggestions for improvements in future editions will be appreciated.

Bernard Fried
Joseph Sherma

Preface to the First Edition

The purpose of this book is to acquaint the reader with the principles, practices, and applications of thin-layer chromatography (TLC). The book is divided into two parts, the first of which is concerned mainly with general practices of TLC and the second with applications based on compound types. Although we feel this book is a valuable source of practical information for any scientist currently using or contemplating the use of TLC, it was written with special consideration for those without extensive training in analytical chemistry. This includes the majority of biology students and graduate biologists, for whom we feel this book will be especially valuable. After reading our book these people should be better equipped to examine critically the more comprehensive texts already published on TLC.

The authors of this book have had considerable experience in both qualitative and quantitative TLC. One author (J.S.), an analytical chemist, has had widespread experience in the separation sciences. The other author (B.F.), an experimental parasitologist, has recently become involved in TLC analyses of hydrophilic and lipophilic excretions of parasitic worms. Both authors have had considerable experience teaching TLC at Lafayette College and at short courses sponsored by Kontes Co., Inc., Vineland, New Jersey, and the Center for Professional Advancement, East Brunswick, New Jersey. Additionally, the authors have advised scientists from academia, government, and industrial research on principles, practices, and applications of TLC.

At Lafayette, which is an undergraduate teaching institution, we have advised colleagues on TLC. For instance, we helped botanists with TLC procedures for the separation of plant pigments and zoologists with separation problems related to hydrophilic and lipophilic substances from animal tissues and fluids. Several of our colleagues have recently introduced one or more TLC experiments in their courses. The aim of this book, in part, is to aid teachers in introducing TLC experiments in courses. The second part presents various detailed TLC experiments that can usually be performed in one or several 3-hr lab periods. Most experiments utilize inexpensive materials that are readily available.

Our experiments are based on TLC separations on silica and cellulose sorbents. Chapter 3, however, discusses less widely used sorbents such as alumina, Sephadex, ion-exchange materials, polyamides, and others. Since TLC is used mainly as a separation tool for small-molecular-weight, nonionized, lipophilic and hyrophilic substances, our detailed experiments emphasize TLC on silica and cellulose.

We have referred to numerous commercial products in this book. These references are based mainly on our combined experiences with these products and are not meant to imply that any product is better than others not mentioned.

Throughout this book the following conventions have been adopted: the terms ''solvent'' and ''mobile phase'' are used more or less interchangeably; ''spot'' and ''zone'' are used interchangeably with no designation of a certain shape; mobile phases are designated in volume proportions unless otherwise stated.

Bernard Fried
Joseph Sherma

Contents

1

Introduction and History

I. INTRODUCTION TO THIN-LAYER CHROMATOGRAPHY

Chromatography encompasses a diverse but related group of methods that permit the separation, isolation, identification, and quantification of components in a mixture. All of the methods involve application of the sample in a narrow (small) initial zone to a stationary phase. Components of the mixture are carried through the stationary phase by the flow of a mobile phase. Separations are based on differences in migration rates among sample components during a multistage distribution process.

Thin-layer chromatography (TLC) is a mode of liquid chromatography in which the sample is applied as a small spot or streak to the origin of a thin sorbent layer supported on a glass, plastic, or metal plate. The mobile phase moves through the stationary phase by capillary action, sometimes assisted by gravity or pressure. TLC separations take place in the "open" layer (or "open bed"), with each component having the same total migration time but different migration distances. The mobile phase in TLC consists of a single solvent or a mixture of organic and/or aqueous solvents. Numerous sorbents have been used, including silica gel, cellulose, alumina, polyamides, ion exchangers, and chemically bonded polar and nonpolar phases, coated onto a suitable support (e.g., glass plate or polyester or aluminum sheet). The wide variety of available mobile and stationary phases provides considerable versatility in choosing a system for resolution of virtually all types of compounds.

The basic procedure for classical TLC consists of the following steps:

1. The sample solution is applied to the plate origin as a spot or zone.
2. The sample solvent is allowed to evaporate from the plate.
3. The plate is placed in a closed chamber containing a shallow pool of mobile phase on the bottom.
4. The mobile phase rises by capillary action through the applied spot.
5. Development is continued until the solvent front is about 10–15 cm beyond the origin.
6. The plate is removed from the chamber and the solvent front is marked.
7. Mobile phase is removed from the plate by drying in air or applying heat.
8. If the compounds are not naturally colored or fluorescent, a detection reagent is applied to visualize the zones.
9. The positions of the zones are used for qualitative identification of compounds and the size and/or intensity of the zones for quantification.

TLC can be used both as an analytical and a preparative technique. Although most of this book is devoted to TLC as an analytical procedure, Chapter 12 discusses preparative-scale TLC. TLC provides for separations in the milligram (mg, 10^{-3} g), microgram (μg, 10^{-6} g), nanogram (ng, 10^{-9} g), and picogram (pg, 10^{-12} g) range. Separated substances that are tentatively identified by TLC can be eluted for further characterization by other microanalytical techniques, such as gas chromatography (GC); high-performance liquid chromatography (HPLC); visible, ultraviolet (UV), infrared (IR), nuclear magnetic resonance (NMR), and mass spectrometry (MS); and electroanalysis. Eluted substances can also be quantified by procedures such as these, but in situ densitometry is the most convenient, accurate, and precise approach for quantitative TLC (Chapter 10).

II. HISTORY OF TLC

The history of TLC has been reviewed by Stahl (1969), Heftmann (1975), Kirchner (1975, 1978, 1980), Pelick et al. (1966), Jork and Wimmer (1986), Wintermeyer (1989), and Berezkin (1995). Sakodynskii (1992) discussed the contributions of M. Tswett, the inventor of chromatography (Strain and Sherma, 1967), to the development of planar chromatography. TLC is a relatively new discipline, and most chromatography historians date the advent of modern TLC from 1958. The Pelick et al. (1966) review contains a tabulation of significant early developments in TLC, and Pelick et al. (1966) and Berezkin (1995) provide translations of the classical studies of Izmailov and Schraiber and of Stahl. In 1938, Izmailov and Shraiber separated certain medicinal compounds on unbound alumina spread on glass plates. Because the solvent was dropped on a glass plate containing the sample and sorbent, their procedure was called drop chromatography. In 1948,

Meinhard and Hall used a binder to adhere alumina to microscope slides; these plates were used for the separation of certain inorganic ions using drop chromatography (Kirchner, 1978). From 1950 to 1954, Kirchner and colleagues at the U.S. Department of Agriculture developed TLC essentially as we know it today. These workers used sorbents bound to glass plates with the aid of a binder, and plates were developed according to conventional ascending development procedures used in paper chromatography. Kirchner used the term "chromatostrips" for his discovery.

The term "thin-layer chromatography" was coined by Egon Stahl in Germany in the late 1950s. Stahl's greatest contribution to the field was the standardization of materials, procedures, and nomenclature and the description of selective solvent systems for the resolution of many important compound classes. His first TLC laboratory manual (Stahl, 1962) popularized TLC, and he elicited the aid of various chemical manufacturers in offering standard materials for TLC. Stahl (1979, 1983) has provided an introspective view of his contributions to the field over a period of 25 years.

Early workers in TLC had no choice but to make their own plates, by procedures that were often messy and arduous and resulted in many rejects. A major breakthrough in this field came in the early 1960s with the commercial availability of convenient precoated plates. This innovation was discussed by Przybylowicz et al. (1965). Today, few workers prepare their own plates in the laboratory unless they require a special type of layer that is not commercially available. Quantitative TLC was first described by Kirchner et al. (1954), for the determination of biphenyl in citrus fruits and products.

III. LITERATURE OF TLC

The majority of TLC papers are published in chromatography journals, most notably *Journal of Planar Chromatography—Modern TLC, Journal of Liquid Chromatography & Related Technologies, Journal of Chromatography parts A and B, Journal of Chromatographic Science, Chromatographia, Acta Chromatographica, and Biomedical Chromatography.* Application papers are also published in a variety of analytical and disciplinary journals, including *Indian Drugs, Analytical Chemistry, Fresenius' Journal of Analytical Chemistry, Analytical Biochemistry, Journal of Analytical Toxicology, Journal of Pharmaceutical and Biomedical Analysis, Journal of AOAC International, Analusis, Analytica Chimica Acta, Analyst, Journal of Forensic Science, Comparative Biochemistry and Physiology,* and *Journal of Parasitology.* The *CRC Handbook of Chromatography* series, introduced in 1972 under the joint editorship of Gunter Zweig and Joseph Sherma and continued from 1991 by the latter, began with publication of two general volumes and included 27 additional volumes on carbohydrates, drugs, amino acids and amines, polymers, phenols and organic acids, terpenoids, pesticides, lipids, steroids, peptides, hydrocarbons, pigments, polycyclic aromatic hydrocarbons, and inorganics when it ceased publication in 1994. Included in each volume are R_F data (Chapter 9), detection reagents, and sample preparation procedures for the various compound classes. TLC is reviewed biennially in the Fundamental Reviews Issue of *Analytical Chemistry*. These reviews contain important books and papers on the principles, techniques, and applications of TLC. The last such review was by Sherma (1998).

IV. COMPARISON OF TLC TO OTHER CHROMATOGRAPHIC METHODS

TLC has been regarded traditionally as a simple, rapid, and inexpensive method for the separation, tentative identification, and visual semiquantification of a wide variety of substances. From a methodologic standpoint, conventional TLC is the simplest chromatographic technique, and basic supplies and apparatus can be obtained for a cost of several hundred dollars.

Despite the widespread use of TLC worldwide over the past 40 years, the technique often has not been considered to be highly efficient or quantitative. It is now realized that modern high-performance or high-efficiency TLC rivals HPLC and GC in its ability to resolve complex mixtures and to provide analyte quantification. High-performance TLC (HPTLC) has come about as a result of improvements in the quality of sorbents and consistency of plate manufacture; use of optimized techniques and equipment for sample application, plate development, detection reagent application, and densitometric scanning; and a greater understanding of chromatographic theory as it applies to TLC. Quantitative HPTLC requires a densitometer and other more or less automated instruments that cannot be fairly characterized as being "inexpensive." However, whether carried out in a less expensive, more manual fashion or with a higher-priced instrumentalized system, HPTLC can have significant advantages compared to other chromatographic methods, such as relative simplicity, high sample throughput, and exceptional versatility.

Many of the principles and techniques of TLC and paper chromatography (PC) are very closely related. Both TLC and PC are open systems in which multiple samples are applied to the planar stationary phase concurrently, the samples are developed by the mobile phase, and solutes are detected statically after removal of the mobile phase. TLC offers superior resolution, speed, and sensitivity because of the more favorable properties of sorbent layers compared to those of cellulose paper. In relation to TLC, the use of PC has been declining for many years. The major applications of PC are for separations of hydrophilic organics and inorganics.

In contrast to TLC, GC and HPLC are closed-column systems in which "plugs" of single samples are sequentially introduced into the moving fluid stream, and solutes are eluted from the column by the mobile phase and are detected in a dynamic, time-dependent fashion in the presence of the phase. Although the mechanisms governing separations in HPLC and TLC are identical, differences in practical aspects lead to the following comparisons of the performance of TLC and HPLC, which should be considered as mutually complementary and competitive techniques:

1. In terms of total available theoretical plates, HPLC is more efficient than HPTLC. A typical HPLC column of 10–25 cm length packed with 3–5 μm particles will provide ~10,000–20,000 theoretical plates. The number of theoretical plates in HPTLC is usually below 5000 for the normal 3–6 cm development.

2. The terms "separation number" (Kaiser, 1977) and "spot capacity" (Guiochon et al. 1982; Guiochon and Siouffi, 1982) are both defined as the maximum number of substances that can be completely separated (resolution = 1) on a plate or column. The separation number for linear HPTLC is 10 to 20 and for HPLC is 20 to 40. Guiochon et al. (1982) have shown that the separation number is raised to 100 to 250 for two-dimensional TLC (Chapter 7), and theoretical calculations indicate that under forced flow conditions (Chapter 7), it should be relatively easy to generate spot capacities well in excess of 500.

4. HPLC is more readily automated than is TLC. In TLC, automated instruments are available for most individual steps (e.g., sample application, development, detection, scanning); however, the plate must be manually handled between steps. Until a robotic, fully automated TLC system is perfected, HPLC will continue to be less labor intensive.

5. Because TLC is open-bed development chromatography, the mobile phase velocity is determined by capillary forces and is not controlled. (An exception is overpressured TLC; Chapter 7.) The mobile phase velocity in closed HPLC is controlled and easily adjusted. In TLC, the mobile phase usually encounters a dry layer at the start of the separation, while the column is in equilibrium with the mobile phase when the sample is applied in HPLC.

6. The simultaneous analysis of samples in TLC leads to higher sample throughput (lower analysis time) and less cost per analysis. Up to 72 samples can be separated on a single plate with convergent horizontal development from both ends. Even with total automation, HPLC cannot compete with TLC in terms of the number of samples processed in a given time period.

7. The ability to process samples and standards simultaneously under the same conditions lends itself to statistical improvements in data handling and better analytical precision and accuracy. There is generally less need for internal standards in quantitative TLC than in GC or HPLC.

8. The development and detection steps are much more flexible and versatile in TLC than in HPLC. Many more solvents can be used for preparation of mobile phases because the phase is completely evaporated before detection and

the plate is used only once. Therefore, the UV-absorbing properties, purity, or acid and base properties of the mobile phase are not as critical as with HPLC. Mobile phases that do not wet the stationary phase can be used in HPLC because the external pressure will serve to transport the solvent, whereas the capillary forces in TLC will not.

9.   Multicomponent mobile phases may separate (demix) during migration through the layer. In HPLC, well-defined gradients are applied to an equilibrated column to improve separations. System equilibration time in TLC is small, and small to large (for gradient elution) in HPLC. Mobile phase vapors can preload the layer during equilibration and affect the resultant separation.

10.   The choice of the solvent in which the sample is dissolved is not as critical in TLC because it is removed by evaporation before development of the layer. In HPLC, the sample solvent chosen must be compatible in terms of composition and strength with the column and mobile phase.

11.   The availability of continuous, multiple, circular, and anticircular development methods and the great variety of available mobile phases allow the TLC system to be optimized for separation of only the compounds of interest from mixtures containing solutes of varying polarities. The rest of the sample can be left at the origin or moved near the solvent front, away from the center region of maximum resolution. This leads to a considerable savings in time compared to HPLC, in which the most strongly sorbed materials have to be completely eluted for each sample.

12.   The variety of commercially available stationary phases is higher in TLC than in HPLC, and the layers can be easily impregnated with selective reagents to further improve selectivity. Optimization of the separation is rapid and cost-efficient because of the ability to easily change the stationary and mobile phase.

13.   Many different detection methods, including inspection under short- and long-wave UV light and sequential application of a series of compatible reagents, can be employed on each layer.

14.   The layer acts as a storage medium for the analytical information contained in the chromatogram. TLC quantification can be carried out in various ways because of the absence of time constraints in development chromatography. Multiple scanning of the same separation at different optimal wavelengths and use of a variety of techniques to enhance sensitivity and selectivity are possible. Multiple development can be applied to separate certain solutes in sequence, with scanning after each step. Spots can be measured both before and after in situ reaction with a derivatizing reagent. Complete visible, UV, or fluorescence spectra can be obtained in situ for each separated zone.

15.   Samples for TLC often require less cleanup because plates are not reused. The presence of strongly sorbed impurities, even solid particles in samples, is of no concern. These materials can build up on an HPLC column and

destroy its performance. In TLC, every sample is separated on fresh stationary phase, without carryover or cross-contamination. Therefore, TLC can require fewer cleanup steps during sample preparation, saving both time and expense.

16. On a TLC plate, the entire sample is available for separation and visual detection. There are no problems with unrecognized loss of peaks or unexpected appearance of peaks from previous samples as in HPLC. Zone identification is facilitated in TLC by the visual nature of the detection using colors and shades and many different reagents and temperatures, and inspection in daylight and under short- and long-wavelength lamps. Separated substances can be subjected to subsequent analytical procedures (e.g., coupled TLC-UV-Vis, TLC-MS, TLC-FTIR) at a later time.

17. Detection limits are approximately the same for TLC and HPLC, ranging typically from picogram to microgram levels.

18. Highly retained substances (low R_F) in TLC form the tightest zones and are detected with the highest sensitivity. In HPLC and GC, highly retained substances (high k') form the widest peaks and are most poorly resolved and detected.

19. Solvent use is much lower for TLC than for HPLC, both on an absolute and, especially, on a per-sample basis, leading to reduced operating and disposal costs and safety concerns.

20. Conventional TLC is inexpensive compared to HPLC, but the cost of computer-controlled instruments can increase the cost of performing modern quantitative HPTLC to an amount comparable to HPLC instrumentation.

21. Under certain conditions, TLC serves well as a pilot procedure for HPLC (Chapter 6).

Comparisons of TLC and HPLC are described in publications by Costanzo (1984), Fenimore and Davis (1981), Borman (1982), Maugh (1982), Coddens et al. (1983), Geiss (1987), Jork and Wimmer (1986), and Sherma (1991).

REFERENCES

Berezkin, V. (1995). The discovery of thin layer chromatography. *J. Planar Chromatogr.—Mod. TLC 8*: 401–405.

Borman, S. A. (1982). HPTLC: taking off. *Anal. Chem. 54*: 790A–794A.

Coddens, M. E., Butler, H. T., Schuette, S. A., and Poole, C. F. (1983). Quantitation in high performance TLC. *LC Magazine, Liquid Chromatography and HPLC 1*: 282–289.

Costanzo, S. J. (1984). High performance thin-layer chromatography. *J. Chem. Educ. 61*: 1015–1018.

Fenimore, D. C., and Davis, C. M. (1981). High performance thin-layer chromatography. *Anal. Chem. 53*: 252A–266A.

Geiss, F. (1987). *Fundamentals of Thin Layer Chromatography*. Alfred Huthig Verlag, Heidelberg, pp. 5–8.

Guiochon, G., and Siouffi, A. M. (1982). Study of the performance of TLC-spot capacity in TLC. *J. Chromatogr. 245*: 1–20.

Guiochon, G., Connord, M.-F., Siouffi, A., and Zakaria, M. (1982). Spot capacity in two-dimensional TLC. *J. Chromatogr. 250*: 1–20.

Heftmann, E. (1975). History of chromatography. In *Chromatography—A Laboratory Handbook of Chromatographic and Electrophoretic Methods*, 3rd ed., E. Heftmann (Ed.). Van Nostrand Reinhold, New York, pp. 1–13.

Jork, H., and Wimmer, H. (1986). Thin-layer chromatography—history and introduction. In *TLC Report: A Collection of Papers*. GIT Verlag, Darmstadt.

Kaiser, R. E. (1977). Simplified theory of TLC. In *HPTLC—High Performance Thin Layer Chromatography*, A. Zlatkis and R. E. Kaiser (Eds.). Elsevier, New York, and Institute of Chromatography, Bad Durkheim, Germany, pp. 15–38.

Kirchner, J. G. (1975). Modern techniques in TLC. *J. Chromatogr. Sci. 13*: 558–563.

Kirchner, J. G. (1978). *Thin Layer Chromatography*, 2nd ed. Wiley, New York.

Kirchner, J. G. (1980). History of TLC. In *Thin Layer Chromatography—Quantitative Environmental and Clinical Applications*, J. C. Touchstone and D. Rogers (Eds.). Wiley-Interscience, New York, pp. 1–6.

Kirchner, J. G., Miller, J. M., and Rice, R. G. (1954). Quantitative determination of diphenyl in citrus fruits and fruit products by means of chromatostrips. *J. Agric. Food Chem. 2*: 1031–1033.

Maugh, T. H., II. (1982). TLC: the overlooked alternative. *Science 216*: 161–163.

Pelick, N., Bolliger, H. R., and Mangold, H. K. (1966). The history of thin-layer chromatography. In *Advances in Chromatography*, J. C. Giddings and R. A. Keller (Eds.), Vol. 3. Marcel Dekker, New York, pp. 85–118.

Przybylowicz, E. P., Staudenmayer, W. J., Perry, E. S., Baitsholts, A. D., and Tischer, T. N. (1965). Precoated sheets for thin layer chromatography. *J. Chromatogr. 20*: 506–513.

Sakodynskii, K. I. (1992). The contribution of M. S. Tswett to the development of planar chromatography. *J. Planar Chromatogr.—Mod. TLC 5*: 210–211.

Sherma, J. (1991). Comparison of thin layer chromatography with liquid chromatography. *J. Assoc. Off. Anal. Chem. 74*: 435–437.

Sherma, J. (1998). Planar chromatography. *Anal. Chem. 70*: 7R–26R.

Stahl, E. (1962). *Dunnschlicht-Chromatographie, ein Laboratoriumshandbuch*. Springer-Verlag, Berlin.

Stahl, E. (1969). The historical development of the method. In *Thin Layer Chromatography*, 2nd ed., E. Stahl (Ed.). Springer-Verlag, Berlin, pp. 1–6.

Stahl, E. (1979). Twenty years of thin-layer chromatography. A report on work with observations and future prospects. *J. Chromatogr. 165*: 59–73.

Stahl, E. (1983). Twenty-five years of thin layer chromatography—an intermediate balance. *Angew. Chem. 95*: 515–524.

Strain, H. H., and Sherma, J. (1967). Michael Tswett's contributions to sixty years of chromatography. *J. Chem. Educ. 44*: 235–237.

Wintermeyer, U. (1989). *The Root of Chromatography: Historical Outline of the Beginning to Thin Layer Chromatography*. GIT Verlag, Darmstadt.

2

Mechanism and Theory

Thin-layer chromatography is a subdivision of liquid chromatography, in which the mobile phase is a liquid and the stationary phase is situated as a thin layer on the surface of a flat plate. TLC is sometimes grouped with paper chromatography under the term "planar liquid chromatography" because of the flat geometry of the paper or layer stationary phases.

The developing solvent or the mobile phase is the transport medium for the solutes to be separated as it migrates through the stationary phase by capillary forces. The movement of substances during TLC is the result of two opposing forces: the driving force of the mobile phase and the resistive or retarding action of the sorbent. The driving force tends to move the substances from the origin in the direction of the mobile phase flow. The resistive action impedes the movement of the substances by dragging them out of the flowing phase back onto the sorbent. Each molecule alternates between a sorbed and unsorbed condition, following a stop-and-go path through the sorbent. Although the zone moves constantly ahead, only a fraction of the molecules in the zone is moving at any one time. At the end of development, each zone has migrated a certain mean distance and has spread owing to the fluctuations in the movement of individual molecules in the zone due to factors such as particle size and uniformity in the layer. The distance traveled by the center of each solute zone in a given time is the resultant of the driving and resistive forces. Substances that move slowly are attracted more strongly to the layer, whereas those that move quickly spend a smaller fraction of their time in the layer because of less affinity for it and more solubility in the mobile phase.

The ability to achieve differential migration (i.e., resolution or separations) among the mixture components is the result of the selectivity, efficiency, and capacity of the chromatographic system (the chromatographic system is defined as the combination of the layer, the mobile phase, and the solutes). The *flow* of the mobile phase is nonselective in that it affects all unsorbed solutes equally. As part of the chromatographic system, however, the mobile phase is selective (in liquid but not in gas chromatography) because it helps determine the relative sorbability of the solutes. The resistive action of the layer (i.e., adsorption, partition, size exclusion, ion exchange, or a combination) is also a selective force. Put another way, all eluted or nonsorbed components or solutes spend equal time in the mobile phase. If there is differential migration along the layer, it is because the solutes spend different amounts of time on the sorbent, as determined by the interactions of the chromatographic system.

The degree of attraction of a solute for the stationary phase is described by the equation

$$K = \frac{C_s}{C_m}$$

where K is called the distribution or partition coefficient, C_s is the equilibrium concentration of the solute in the stationary phase, and C_m is its equilibrium concentration in the mobile phase. The simple linear relationship between C_s and C_m described in this equation applies only at the low solute concentrations normally encountered in chromatography (i.e., microgram levels and below); at higher concentrations, a plot of C_s (y-axis) versus C_m (x-axis), which is termed a sorption isotherm, ceases to be linear and flattens out as the stationary becomes saturated with solute. A solute with a large K value has great affinity for the stationary phase and will travel slowly through the layer (low R_F value). Separation is easiest for two solutes with widely differing K values.

I. CLASSIFICATIONS OF TLC

Depending on its nature, the layer promotes the separation of molecules by (1) physical sorption of solutes from solution onto the surface-active groups of the layer particles (adsorption) (Scott and Kucera, 1979), (2) dissolving of solutes into a stationary liquid held on the layer (partition), (3) attraction of ions to sites of opposite charge on the layer (ion exchange), or (4) retention or rejection of solutes on the basis of molecular size and/or shape (size-exclusion or gel-permeation TLC). The boundary between adsorption and partition is quite obscure because both can involve the same types of physical forces, that is, permanent and induced dipole–dipole forces and hydrogen bonding. In addition, actual separations often involve a combination of these four basic mechanisms. Therefore, the use of the terms "sorbent" and "sorption" has become common when it is not intended to designate any definite type of operable interaction.

In direct- or normal-phase TLC, the layer (e.g., silica or alumina) is more polar than the mobile phase; polar solutes are bound most strongly near the origin. In reversed-phase TLC (RP-TLC), the layer (e.g., silica gel impregnated with mineral oil or chemically bonded with C_2–C_{18} hydrocarbons) is less polar than the mobile phase; nonpolar solutes are most strongly retained. In normal-phase TLC, the strongest (most eluting) mobile phases are the most polar (more aqueous, less organic), whereas in RP-TLC, less polar (less aqueous, more organic) solvents are the strongest eluters. Silica gel chemically bonded with cyano, DIOL, or amino groups can be used in adsorption, RP, and ion exchange separation modes depending on the composition of the mobile phase.

Adsorption TLC is very sensitive to differences in configuration that affect the free energy of adsorption onto the layer surface. Adsorption TLC is, therefore, well suited to the separation of compounds differing in polarity and structural isomers. Partition chromatography is capable of separation based on slight differences in solubility, as would occur in the separation of homologs (Karger et al., 1973).

TLC can also be classified according to the method of mobile phase flow through the stationary phase. In addition to capillary flow in TLC or HPTLC, solvent can be forced through the layer (forced-flow thin-layer or planar chromatography) by the action of external pressure (overpressured layer chromatography and high-pressure planar chromatography), vacuum, electroosmosis (electrochromatography), or centrifugal force (centrifugal or rotational planar chromatography) (Poole and Poole, 1995). Forced-flow methods allow the use of optimized mobile phase velocity, leading to essentially constant efficiency and resolution over the whole separation distance (Nyiredy, 1992).

Another classification of TLC is off-line versus on-line (Nyiredy, 1992). Classical TLC is off-line in that the different steps are performed separately. Overpressured and rotational preparative planar chromatography can be performed on-line by use of an injector, flow-through detector, and fraction collector.

II. CRITERIA OF CHROMATOGRAPHIC PERFORMANCE AND THEIR RELATIONSHIP TO TLC

The main purpose of all chromatographic techniques, including TLC, is to separate or resolve components of mixtures. Resolution (R) of two chromatographic zones is defined as the distance between zone centers (d) divided by the average of the widths (W) of the zones:

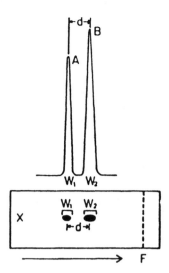

FIGURE 2.1   Chromatographic resolution determined from spots or densitometer scan as the ratio of the separation of zone centers to the average zone width. X = origin and F = solvent front of the TLC plate; the arrow shows the direction of solvent development. The peaks represent the scan of the separated zones.

$$R = \frac{d}{(W_1 + W_2)/2}$$

The values for this equation can be obtained by measuring the actual zones on the layer with a fine millimeter ruler or a densitometric scan of the zones (Figure 2.1). The same is true for calculation of N (Figure 2.2). Use of a scan rather than the zones themselves leads to lower values for R and N because a scanner will broaden the zone dimensions and diminish apparent resolution and efficiency. According to this equation, resolution can be improved by moving the zone centers farther apart (increasing system selectivity) and/or decreasing the zone widths (increasing system efficiency). Resolution is considered to be acceptable when it is equal to, or greater than, 1.0 (baseline separation = 1.5). On a theoretical basis, it is practically impossible to resolve substances with ΔR_F smaller than 0.05 (Guiochon et al., 1979) (see Chapter 9, Section I for definition of R_F).

Efficiency is measured by calculation of the theoretical plate number N or plate height H from the equations

$$N = 16 \left(\frac{X}{W}\right)^2, \quad H = \frac{W^2}{16X}$$

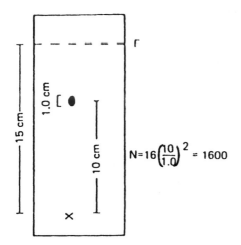

FIGURE 2.2 Calculation of plate number in TLC. X = origin, F = solvent front.

where X is the distance (mm) from the origin to the center of a given zone and W is the width of the zone (the base width of the denistometric scan). The factor 16 in this equation is a carryover from early distillation theory, the use of which relates the plates on the layer to the number of discrete equilibrations of the solute between the mobile and stationary phases (Giddings, 1964). Because all substances in a mixture have the same diffusion time but different migration distances in TLC, plate numbers are correlated to the migration distances of the substances (their specific R_F values). The effect of diffusion is the same for all solutes, and zone broadening is a function of the distance of migration. Plate height has been shown to decrease linearly with the reciprocal of the R_F value (Poole and Poole, 1991).

Figure 2.2 shows an example of a TLC system providing 1600 theoretical plates. Conventional TLC is capable of producing up to 600 theoretical plates, whereas HPTLC can produce up to 5000 plates for a 10-cm development (up to 50,000 plates per meter). However, the normal development distance in HPTLC is 3–6 cm. For comparison, a typical 25 cm × 4.6 mm HPLC column with 5-μm particle-size packing will produce 10,000 to 15,000 theoretical plates (40,000 to 60,000 plates per meter). Capillary and megabore HPLC columns can provide as many as 100,000 plates per meter. Packed GC columns provide typically 2500 to 3000 plates per meter (5000 to 6000 total plates for a 2-m column); capillary GC columns have a similar number of plates per meter but can provide up to 150,000 plates for a 50-m column. Typical plate heights are 12 μm in HPTLC and 6–10 μm in HPLC. (See Table 2.1.) Because of the quadratic decrease in mobile phase velocity, solutes are forced to migrate through regions of different

TABLE 2.1 A Comparison of Conventional and High-Performance Thin-Layer Chromatography

Parameter	Conventional TLC	HPTLC
Plate size (cm)	20 × 20	10 × 10
		10 × 20
Particle size (μm)		
Average	20	5–15
Distribution	10–60	Tight
Adsorbent layer thickness (μm)	100–250	200
Plate height (μm)	30	5–20
Total number of usable theoretical plates	< 600	< 5000
Separation number	7–10	10–20
Sample volume (μl)	1–5	0.01–0.2
Starting spot diameter (mm)	3–6	1–2
Diameter of separated spots (mm)	6–15	2–6
Solvent migration distance (cm)	10–15	3–6
Separation time (min)	30–200	3–20
Detection limits (ng)		
Absorption	1–5	0.1–0.5
Fluorescence	0.05–0.1	0.005–0.01
Samples per plate	10	18–36

Source: Adapted from Poole and Poole (1991) and Robards et al. (1994).

local efficiency, and plate heights in TLC and HPTLC must be specified by an average value (Poole and Poole, 1995). Average plate height increases with particle diameter, with increasing diffusion constants of the solute, and, in particular, with increasing development time (Robards et al., 1994).

Separation number (SN) (Kaiser, 1977; Geiss, 1987; Guiochon and Siouffi, 1982; Guiochon et al., 1982), or spot capacity, indicates the maximum number of spots than can be separated with $R = 1$ over a given separation distance, between $R_F = 0$ and $R_F = 1$. It is determined experimentally using the equation

$$ SN = \frac{Z}{b_0 + b_1 - 1} $$

where Z is the migration distance from the origin to the solvent front, b_0 is the width for a spot with $R_F = 0$, and b_1 is the width of a spot with $R_F = 1$. Widths are measured at half height from a densitogram of a mixture of substances having different R_F values spread over the entire chromatogram. As shown in Table 2.1, a separation number of 10–20 is typical in capillary-controlled HPTLC whereas for forced-flow TLC, a capacity of 40–80 is possible. Theoretical calculations indicate a spot capacity of 100–400 for two-dimensional TLC with capillary flow and 500–3000 with forced flow (Poole and Poole, 1995).

An alternative equation for resolution is

$$R = \frac{1}{4}\left(\frac{\alpha - 1}{\alpha}\right)(N)^{1/2}\left(\frac{k'}{1 + k'}\right)$$

This equation shows that resolution is a function of three factors, namely, selectivity (α), number of theoretical plates (N), and capacity factor (k'). As discussed above, selectively relates to the ability to separate zone centers (difference in R_F values), whereas the number of theoretical plates measures zone spreading throughout the chromatographic system. The capacity factor describes retention of a component by the stationary phase. In HPLC, k' is stated in terms of column volumes, and values of 1 to 10 are normal.

In linear TLC,

$$k' = \frac{1 - R_F}{R_F} \text{ and } R_F = \frac{1}{k' + 1}$$

Based on this definition, the R_F value of the zone in Figure 2.2 is 10/15 = 0.67. The k' values of 1 to 10 correspond to R_F values of 0.09 to 0.5 by the above equation. Most often, k' is controlled by varying the mobile phase strength, although the stationary phase can also be changed. For maximum resolution, it is clear that the values of α and N should be large, and k' should be within its optimum range. For optimum resolution in circular and anticircular TLC (Chapter 7), the preferred approximate values of k' are 100 and 1, respectively (Poole and Schuette, 1984). These correspond to optimum R_F values of 0.009 and 0.5 for these techniques. Thus, circular development provides good separations of compounds with low-R_F values and anticircular development for high-R_F compounds.

The resolution equation for TLC (Poole and Poole, 1991) related to the HPLC equation above is

$$R = \left(\frac{(n_1 R_{F2})^{1/2}}{4}\right)\left[\frac{k'_1}{k'_2} - 1\right](1 - R_{F2})$$

where n_1 is the number of theoretical plates passed over by the zones of the two compounds (1 and 2) if they moved to the solvent front (Guiochon et al., 1979), k'_1 and k'_2 are the respective capacity factors of compounds 1 and 2 (compound 2 is the compound with the higher R_F value), and R_{F2} is the R_F value of compound 2. The terms in this equation define system efficiency, selectivity, and capacity for resolution by a single development under capillary-flow-controlled conditions. The optimum R_F value for maximum resolution is ~ 0.3 for the faster moving zone, but resolution decreases by no more than 8% of maximum for R_F values in the range 0.2–0.5 (Poole and Poole, 1991).

The effect of three independent spreading mechanisms on efficiency in column chromatography (gas and liquid) is expressed by the Van Deemter equation, the classical form of which is

$$H = A + \frac{B}{v} + C_v$$

In this equation, H is the height equivalent to a theoretical plate (the column length divided by N). A low value of H indicates high efficiency. A is the term involving eddy diffusion, which results from different sample molecules experiencing unequal flow velocities (path lengths) as they pass through the column. The B term involves longitudinal molecular diffusion in the mobile phase. The C term expresses nonequilibrium resulting from resistance to mass transfer between the stationary and mobile phases, that is, finite time required for sorption and desorption of the sample. From the Van Deemter equation, it is seen that B is inversely proportional to the mobile phase velocity (v), whereas C is directly proportional. The terms are, therefore, not totally independent.

The classical Van Deemter equation and that modified for planar chromatography by Stewart (1965) indicate some experimental factors that can be important in TLC for high efficiency. The A term is lowered if the layer is composed of small, uniform particles as is the case for commercial HP plates. Small particle size also lowers C by increasing the opportunity for equilibration. In HPTLC, resistance to mass transfer can usually be ignored at the normal mobile phase velocities employed. In conventional TLC with larger-particle sorbents, mass transfer kinetics cannot be ignored (Guiochon et al., 1978). Under capillary-flow-controlled conditions, zone broadening is largely controlled by longitudinal molecular diffusion, and the useful development distance for a separation is defined by the range of mobile phase velocities that are sufficient to minimize zone broadening (Poole and Poole, 1995; Guiochon et al., 1978). The A (packing) term controls average plate height for coarse particles below a mobile phase migration distance of 10 cm (Robards et al., 1994).

An exact model of TLC is impossible to define because of the complexities of the TLC system (Poole and Poole, 1991; Poole, 1988; Guiochon and Siouffi, 1978a; Belenkii et al., 1990). The mobile phase, the stationary phase, and the vapor phase inside the chamber are not well defined and are changing during chromatographic development. The degree of activation of the spotted stationary phase when placed in the chamber is impossible to control exactly, and during development, the layer adsorbs the vapor phase in front of the migrating solvent. Different components of the mobile phase are adsorbed at different rates. Evaporation of the solvent within the tank (Stewart and Gierke, 1970; Stewart and Wendel, 1975) and the temperature gradient generated between the front and bulk solvent as the solvent moves along the initially dry bed (Miller, 1975) are major variables in TLC. The different amounts of evaporation and rates of equilibration (Viricel and Gonnet, 1975) in various types of development chambers (Chapter 7) are the basis for the different resolutions obtained in these chambers (Jaenchen, 1964; Dallas, 1965). The speed of mobile phase migration is dependent on the distance of development, the type and size of the sorbent particles, the type of chamber, and other experimental variables (Geiss, 1987, 1988; Guiochon and Siouffi, 1978b; Guiochon et al., 1980; Halpaap et al., 1980; Kalasz, 1984; Gimpelson and Berezkin, 1988; Poole, 1989). Concentration gradients (multiple fronts) are produced as a multicomponent mobile phase containing solvents of different strength migrates through the layer (solvent demixing) (Geiss, 1987, 1988; Zlatanov et al., 1986), and volume gradients (Robards et al., 1994) can result because of nonuniform penetration of the sorbent layer even with single-component mobile phases. Greater control of these types of parameters is achieved in a low-volume horizontal development chamber and in overpressured TLC (Chapter 7). However, even in these cases, the TLC system is much too complicated for an exact theoretical description.

A study of variation of efficiency as a function of solvent front migration distance found that the average plate height for HPTLC was only slightly more favorable than for TLC ($\sim$55 μm vs 62 μm, respectively), but that the minimum shifted to longer development length for TLC ($\sim$7 cm) compared to HPTLC ($\sim$4 cm). Therefore, when the development distance is optimized, the separation performance of TLC and HPTLC are not very different (Poole and Poole, 1995). Solvent migration is slower in HPTLC than in TLC because the decreased mobile phase velocity experienced with fine-particle HPTLC layers compared to coarse-grained TLC layers (Geiss, 1988) is not completely offset by the increased velocity on thinner HPTLC compared to thicker TLC layers (Poole et al., 1990a, 1990b) and the higher permeability constant of the more uniform, tightly packed HP layers. Despite the slower development time per centimeter, HPTLC is classified as being faster than TLC because the development distances in the former technique are much shorter. For HPTLC layers composed of 5 μm particles, the maximum migration distance used should be 5–6 cm; for longer distances, mobile phase velocity decreases to the point where zone broadening exceeds the rate of zone center separation and resolution is decreased (Poole and Poole, 1995).

The Van Deemter equations also imply the existence of an optimum particle diameter for any given system, which has been estimated to be in the range 1–5 μm by numerous workers. In opposition to this, Guiochon et al. (1978, 1979), Siouffi et al. (1981), and Guiochon and Siouffi (1978a) have compared average plate height versus development length for several different layer particle sizes. It was found that plates with particle sizes less than 10 μm are significantly influenced by diffusion of the solute. As a consequence, the efficiency is decreased with increasing development distance. In contrast, plates with larger particles are less influenced by diffusion of the solute, and resolution is improved with an increase in development distance. This led to the conclusion that easier separations should be carried out on small-particle (HP) layers developed for short distances and more difficult separations on large-particle (conventional) layers developed for longer distances. In addition, it was found by these workers that temperature has little effect on general performance, except through change in R_F and relative retention. Layers with a wider particle diameter range are undesirable because they have the high plate values of the coarse fractions and the low mobile phase velocity of the fine fractions (Robards et al., 1994).

Extra-bed factors can also influence efficiency in TLC. Perhaps the most important of these factors is the influence of the initial spot size on plate height and resolution. It is especially true in high-performance TLC that optimum efficiency cannot be obtained unless the initial zones are as small as possible. When large sample solution volumes must be applied, results can often be improved by predevelopment of the plate for a short distance with a highly eluting solvent after sample application. This solvent will carry the sample forward with the solvent front and concentrate it into a tight zone. The same effect can be accomplished without solvent predevelopment by applying the sample to special plates with a preadsorbent area that provides negligible sample retention with most solvents (Chapter 5).

The speed of solvent front movement in capillary-flow-controlled ascending TLC is given by the equation:

$$(Z_f)^2 = kt$$

where Z_f is the position reached by the solvent front above the solvent entry position and k is the velocity constant (also called velocity coefficient, flow constant, or rate coefficient). k depends on a number of factors such as the permeability and particle diameter of the layer; viscosity, surface tension, and contact angle of the mobile phase; and saturation level of vapor phase in contact with the layer, which is affected by chamber type (Guiochon and Siouffi, 1978b; Guiochon et al., 1980; Poole and Poole, 1995). Solvent velocity is not constant throughout the run, cannot be controlled at an optimum value, and will eventually reach zero if the layer is sufficiently long. With some hydrophobic chemically bonded reversed-phase layers and solvents of high water content,the solvent does not sufficiently wet the layer and cannot ascend the plate, and chromatography is impossible (Guiochon et al., 1980; Halpaap et al., 1980). For these RP layers, larger particle size and a lower level of surface modification will increase mobile-phase velocity. In some cases, solvents with high water content cause swelling and/or flaking of the precoated RP layer.

The Knox equation, $h = Av^{1/3} + B/v = Cv$, where h is the reduced plate height and v the reduced velocity, has a form similar to the Van Deemter equation. This equation has been used recently in studies of the properties of HPLC columns and HPTLC precoated layers. The conclusions of these studies were that there is little room for improving the properties of commercial precoated layers used for capillary-flow-controlled HPTLC, for which mobile phase velocity is too slow for optimum kinetic performance. Flow resistance characteristics cannot be significantly decreased, and a reduction in layer thickness to 100 μm would increase velocity by a factor of only 1.1–2.5 (Poole and Poole, 1995).

III. FORCED-FLOW DEVELOPMENT

Unlike capillary-controlled TLC development modes, forced-flow development using centrifugal force or pressure allows the mobile phase flow rate through the layer to be optimized (Kalasz, 1984; Tyihak and Mincsovics, 1988). In rotational planar chromatography (RPC), solvent flow through an open layer is controlled by the rotation speed and the supply rate of the mobile phase to the layer (Nyiredy et al., 1989). In overpressured layer chromatography (OPLC), solvent is pumped through a sealed layer at a constant velocity (Kalasz, 1984; Witkieiewicz and Bladek, 1986).

In OPLC, the developing system is essentially an S-chamber that resembles an HPLC column. The distance traveled by the solvent front is proportional to time; development time does not increase with distance according to the quadratic law (Geiss, 1987) as in capillary-flow TLC. Aqueous solvents that do not wet the layer can be used with RP layers.

In OPLC, a special development chamber is used in which the layer is sealed by applying hydraulic pressure to one side of a polymeric membrane in intimate contact with the layer surface, and the mobile phase is forced through the layer by a pump; the edges of the layer must be sealed before use by impregnation with glue (Poole and Poole, 1995). Solvent forced through the dry, sealed layer by pressure displaces air from the layer, producing a secondary (beta) front and a "disturbing zone" in which migrating solutes may be distorted (Tyihak and Mincsovics, 1988; Nyiredy et al., 1987; Velayudhan et al., 1988). Means of reducing or eliminating this disturbing zone include removing air prior to development by predevelopment of the layer or drawing vacuum, or application of back pressure to increase the solubility of air in the mobile phase (Poole and Poole, 1991).

The minimum average plate height under capillary-controlled flow is always greater than with forced-flow development for the same layer (Poole and Poole, 1995). The average plate height is independent of migration distance in OPLC, so the total number of plates and resolution increase fairly linearly with development length. Separation number also increases linearly with front migration, reaching 60 for a 35-cm development compared to 15 for conventional TLC with a development distance of 20 cm (Geiss, 1987; Poole and Poole, 1991; Poole and Poole, 1995). Efficiency is better on layers with small particles than on those with coarse particles because the flow rate is maintained at a constant, optimum value, chosen in relation to the migration distance and minimum average plate height (Witkiewicz and Bladek, 1986; Hauck and Jost, 1983) and does not decrease with migration distance as in capillary-flow TLC. Zone broadening by molecular diffusion is not significant in OPLC when the optimum solvent velocity is employed (Poole and Poole, 1991). The minimum plate height of ~15 μm for OPLC with HPTLC plates and ~35 μm with TLC plates is obtained at a flow of ~1.5 cm/min (Geiss, 1987; Robards et al., 1994). Whereas reducing the contribution from resistance to mass transfer leads to only minor improvements in separation performance in capillary-flow-controlled systems, it is critical in achieving optimum performance and reducing separation time in forced-flow development because spot broadening by diffusional processes plays a minor role at the relatively high linear mobile phase velocities commonly employed (Poole and Poole, 1995; Robards et al., 1994).

Maximum resolution in OPLC is obtained by using the optimum mobile phase velocity, and resolution increases linearly with migration distance (Hauck and Jost, 1983) up to the limit imposed by the length of the plate, inlet pressure, and the available time (Poole and Poole, 1991). OPLC is capable of better resolution than capillary-flow TLC given the same layer/mobile phase/mixture system (Poole and Poole, 1991), unless a solvent mixture is used that demixes in the sandwich-type chamber and ruins the results (Geiss, 1987). Resistance to mass transfer is significantly greater for precoated layers than for columns, but under capillary-controlled-flow conditions the mobile phase velocities are too slow for this factor to contribute significantly to zone broadening. Under forced-flow conditions, mass transfer considerations require lower mobile phase velocities than for columns, thereby decreasing separation speed (Poole and Poole, 1995).

IV. UNIDIMENSIONAL MULTIPLE DEVELOPMENT

Efficiency is improved in capillary-flow-controlled HPTLC with multiple development because of the zone refocusing mechanism. The zone is compressed by the mobile phase front advancing through the bottom of the zone in each development, then it broadens as usual by longitudinal diffusion after the front passes the top of the zone. If the compression and broadening are balanced, the zone will migrate in the form of a band without significant increase in width. Poole and Poole (1995) summarized the optimum conditions for isocratic multiple development TLC as follows: resolution is controlled primarily by the zone center separation because zone center widths are relatively constant after the first development; the maximum zone center separation of two solutes of similar migration properties occurs when the zones have migrated 0.63 of the development distance; the maximum resolution observed is largely independent of the average R_F value of the zones, but if the average value is low a large number of developments will be needed to reach the maximum resolution value; and compounds that are difficult to separate should be developed repeatedly with the most selective mobile phase that will produce low R_F values.

V. DETERMINATION OF LIPOPHILICITY WITH REVERSED-PHASE TLC

Reversed-phase TLC has been widely used in structure–activity relationship (QSAR) studies to determine lipophilicity or hydrophobicity of bioactive compounds, which is generally well correlated with biological activity and can aid in the design of drugs. The method is rapid and simple; only a small amount of the solute is needed, and impurities that will separate during TLC can be tolerated. Silica gel plates pre-impregnated by development with paraffin or methylsilicone or chemically bonded C_{18} plates have been used successfully with aqueous-organic mobile phases. The principles and procedures have been reviewed by Biagi et al. (1994), Cserhati and Forgacs (1996a), and Dross et al. (1994). The method has been applied to compounds such as plant growth–stimulating amido esters of ethanolamine (Gocan et al, 1994), anticancer drugs (Forgacs and Cserhati, 1995), bioactive heterocyclic compounds (Darwish et al, 1994), morphine derivatives (Kalasz et al., 1995), piperazine derivatives (Kastner et al., 1997), and tetrazolium salts (Seidler, 1995).

Readers interested in detailed discussions of the mechanism and theory of TLC are referred to the books by Geiss (1987) and Ceshati and Valko (1994), the book chapters by Kowalska (1996) and Poole and Poole (1991), and the review papers by Poole and Poole (1995) and Cserhati and Forgacs (1996b). The theory of mobile phase optimization is covered in Chapter 6.

REFERENCES

Belenkii, B., Kurenbin, O., Litvinova, L., and Gankina, E. (1990). A new approach to optimization in TLC. *J. Planar Chromatogr.—Mod. TLC 3*: 340–347.

Biagi, G. L., Barbaro, A. M., Sapone, A., Recanatini, M (1994). Determination of lipophilicity by means of reversed phase thin layer chromatography. I. Basic aspects and relationship between slope and intercept of TLC equations. *J. Chromatogr. 662*: 341–361; II. Influence of the organic modifier on the slope of the thin layer chromatographic equation. *J. Chromatogr. 669*: 246–253.

Cserhati, T., and Forgacs, E. (1996a). Introduction to techniques and instrumentation. In *Practical Thin Layer Chromatography—A Multidisciplinary Approach*, B. Fried and J. Sherma (Eds.). CRC Press, Boca Raton, FL, pp. 1–18.

Cserhati, T., and Forgacs, E. (1996b). Use of multivariate mathematical methods for the evaluation of retention data matrixes. *Adv. Chromatogr. (N.Y.) 36:* 1–63.

Cserhati, T., and Valko, K. (1994). *Chromatographic Determination of Molecular Interactions*. CRC Press, Boca Raton, FL.

Dallas, M. S. J. (1965). Reproducible R_F values in thin layer adsorption chromatography. *J. Chromatogr. 17:* 267–277.

Darwish, Y., Cserhati, T., and Forgacs, E. (1994). Reversed phase retention characteristics of some bioactive heterocyclic compounds. *J. Chromatogr. A 668:* 485–494.

Dross, K., Sonntag, C., and Mannhold, R. (1994). Determination of the hydrophobicity parameter R_{Mw} by reversed phase thin layer chromatography. *J. Chromatogr. A 673:* 113–124.

Forgacs, E., and Cserhati, T. (1995). Effect of various organic modifiers on the determination of the hydrophobicity parameters of non-homologous series of anticancer drugs. *J. Chromatogr. A 697:* 59–69.

Geiss, F. (1987). *Fundamentals of Thin Layer Chromatography*. Alfred Huthig Verlag, Heidelberg.

Geiss, F. (1988). The role of the vapor in planar chromatography. *J. Planar Chromatogr.— Mod. TLC 1:* 102–115.

Giddings, J. C. (1964). The theoretical plate as a measure of efficiency. *J. Gas Chromatogr. 2:* 167–169.

Gimpelson, V. G., and Berezkin, V. G. (1988). On the migration of the mobile phase in a thin sorptive layer. *J. Liq. Chromatogr. 11:* 2199–2212.

Gocan, S., Irimie, F., and Cimpan, G. (1994). Prediction of the lipophilicity of some plant growth-stimulating esters of ethanolamine using reversed phase thin layer chromatography. *J. Chromatogr. A 675:* 282–285.

Guiochon, G., and Siouffi, A. (1978a). Study of the performance of TLC. II. Band broadening and plate height equation. *J. Chromatogr. Sci. 16:* 470–481.

Guiochon, G., and Siouffi, A. (1978b). Study of the performance of TLC. III. Flow velocity of the mobile phase. *J. Chromatogr. Sci. 16:* 598–609.

Guiochon, G., and Siouffi, A. M. (1982). Spot capacity in thin layer chromatography. *J. Chromatogr. 245:* 1–20.

Guiochon, G., Siouffi, A., Engelhardt, H., and Halasz, I. (1978). Study of the performance of TLC. I. A phenomenological approach. *J. Chromatogr. Sci. 16:* 152–157.

Guiochon, G., Bressolle, F., and Siouffi, A. (1979). Study of the performance of TLC. IV. Optimization of experimental conditions. *J. Chromatogr. Sci. 17:* 368–386.

Guiochon, G., Korose, G., and Siouffi, A. (1980). Study of the performance of TLC. V. Flow rate in reversed phase plates. *J. Chromatogr. Sci. 18:* 324–329.

Guiochon, G., Gonnord, M. F., Siouffi, A., and Zakaria, M. (1982). Spot capacity in two-dimensional thin layer chromatography. *J. Chromatogr. 250:* 1–20.

Halpaap, H., Krebs, K.-F., and Hauck, H. E. (1980). Thin layer chromatographic and high performance thin layer chromatographic ready-for-use layers with lipophilic modifications. *J. High Resolut. Chromatogr. Chromatogr. Commun. 3:* 215–240.

Hauck, H.-E., and Jost, W. (1983). Investigations and results obtained with overpressured layer chromatography. *J. Chromatogr. 262:* 113–120.

Jaenchen, D. (1964). Uber die verwendung extrem kleinvolumiger trennkammern in der dunnschicht-chromatographie. *J. Chromatogr. 14:* 261–264.

Kaiser, R. E. (1977). Simplified theory of TLC. In *High Performance Thin Layer Chromatography*, A. Zlatkis and R. E. Kaiser (Eds.). Journal of Chromatography Library, Vol. 9. Elsevier, Amsterdam, pp. 15–38.

Kalasz, H. (1984). Processes dominating forced flow thin layer chromatography. *Chromatographia 18*: 628–632.

Kalasz, H., Kerecsen, L., Csermely, T., and Goetz, H. (1995). TLC investigation of some morphine derivatives. *J. Planar Chromatogr.—Mod. TLC 8*: 17–22.

Kastner, B. L., Snyder, L. R. and Horvath, C. (1973). *An Introduction to Separation Sciences*. Marcel Dekker, New York, pp. 170–178.

Kastner, P., Kuchar, M., Klimes, J., and Dosedlova, D. (1997). Relationship between structure and reversed phase thin layer chromatographic lipophilicity parameters in a group of piperazine derivatives. *J. Chromatogr. A. 766*: 165–170.

Kowalska, T. (1996). Theory and mechanism of thin layer chromatography. In *Handbook of Thin Layer Chromatography*, 2nd ed., J. Sherma and B. Fried (Eds.). Marcel Dekker, New York, pp. 49–80.

Miller, J. M. (1975). *Separation Methods in Chemical Analysis*. Wiley-Interscience, New York, pp. 267–269.

Nyiredy, Sz. (1992). Planar chromatography. In *Chromatography, Part A: Fundamentals and Techniques*, 5th ed., E. Heftmann (Ed.). Elsevier, Amsterdam, pp. A109–A150.

Nyiredy, Sz., Meszaros, S. Y., Dallenbach-Tolke, K., Nyiredy-Mikita, K., and Sticher, O. (1987). The "disturbing zone" in overpressure layer chromatography (OPLC). *J. High Resolut. Chromatogr. Commun. 10*: 352–356.

Nyiredy, Sz., Botz, L., and Sticher, O. (1989). *J. Planar Chromatogr.—Mod. TLC 2*: 53–61.

Poole, C. F. (1988). Band broadening and the plate height equation in thin layer chromatography. *J. Planar Chromatogr.—Mod. TLC 1*: 373–376.

Poole, C. F. (1989). Solvent migration through porous layers. *J. Planar Chromatogr.—Mod. TLC 2*: 95–98.

Poole, C. F., and Poole, S. W. (1991). *Chromatography Today*. Elsevier, Amsterdam, Chapter 7, pp. 649–743.

Poole, C. F., and Poole, S. K. (1995). Multidimensionality in planar chromatography. *J. Chromatogr. A 703*: 573–612.

Poole, C. F., and Schuette, S. A. (1984). *Contemporary Practice of Chromatography*. Elsevier, New York, Chapter 9, pp. 619–699.

Poole, S. K., Ahmed, H. D., Belay, M. T., Fernando, W. P. N., and Poole, C. F. (1990a). The influence of layer thickness on chromatographic and optical properties of high performance silica gel thin layer chromatographic plates. *J. Planar Chromatogr.—Mod. TLC 3*: 133–140.

Poole, S. K., Ahmed, H. D., and Poole, C. F. (1990b). Evaluation of high performance thin layer chromatographic plates prepared from 3 μm silica gel. *J. Planar Chromatogr.—Mod. TLC 3*: 277–279.

Robards, K., Haddad, P. R., and Jackson, P. E. (1994). *Principles and Practice of Modern Chromatographic Methods*. Academic Press, San Diego, CA, pp. 179–226.

Scott, R. P. W., and Kucera, P. (1979). Solute-solvent interactions on the surface of silica gel. II. *J. Chromatogr. 171*: 37–48.

Seidler, E. (1995). Experimental estimation of hydrophobic parameters of tetrazolium salts. *Acta Histochem.* *97*: 295–299.

Siouffi, A., Bressolle, F., and Guiochon, G. (1981). Optimization in TLC—some practical considerations. *J. Chromatogr.* *209*: 129–147.

Stewart, G. H. (1965). The stationary phase in paper chromatography. *Adv. Chromatogr.* *1*: 93–112.

Stewart, G. H., and Gierke, T. D. (1970). Evaporation in thin layer chromatography and paper chromatography. *J. Chromatogr. Sci.* *8*: 129–133.

Stewart, G. H., and Wendel, C. T. (1975). Evaporative thin layer chromatography, solvent flow, and zone migration. *J. Chromatogr. Sci.* *13*: 105–109.

Tyihak, E., and Mincsovics, E. (1988). Forced flow planar liquid chromatographic techniques. *J. Planar Chromatogr.—Mod. TLC 1*: 6–19.

Velayudhan, A., Lillig, B., and Horvath, C. (1988). Analysis of multiple front formation in the wetting of thin layer plates. *J. Chromatogr.* *435*: 397–416.

Viricel, M., and Gonnet, C. (1975). Influence of the activity of the adsorbent in thin layer chromatography. *J. Chromatogr.* *115*: 41–55.

Witkiewicz, Z., and Bladek, J. (1986). Overpressured thin layer chromatography. *J. Chromatogr.* *373*: 111–140.

Zlatanov, L., Gonnet, C., and Marichy, M. (1986). Demixing effects in thin layer chromatography with NH_2-modified silica gel precoated plates. *Chromatographia 21*: 331–334.

3

Sorbents, Layers, and Precoated Plates

I. STRUCTURE AND PROPERTIES OF COATING MATERIALS

A. Silica Gel

Silica gel, the sorbent with undoubtedly the widest range of TLC applications, is an amorphous, porous adsorbent variously referred to as silica, silicic acid, or porous glass. Silica gels used in traditional column and thin-layer chromatography are similar, except that the particle size used in the latter is much finer. Particle sizes for TLC are typically comparable to that used for modern high-performance column liquid chromatography.

Silica gel is prepared by spontaneous polymerization and dehydration of aqueous silicic acid, which is generated by adding mineral acid to a solution of sodium silicate (Szepesi and Nyiredy, 1996). The framework of silica gel can be considered as follows (Figure 3.1). The OH groups attached to the silicon atoms are reactive and mainly account for the adsorptive properties of silica gel. There are several different types of active sites, each varying in strength of adsorption. Substances are adsorbed on silica via hydrogen bonding and induced dipole–dipole interactions, and the surface hydroxyls serve as hydrogen donors (Snyder, 1975). The surface pH of silica ranges from mildly acidic to neutral, ~ 5–7.

According to Felton (1979), the most nearly "ideal" condition for general TLC silica gel separations is considered to be at the level of 11–12% water (by weight). This amount of water leaves available the most plentiful "middle-activ-

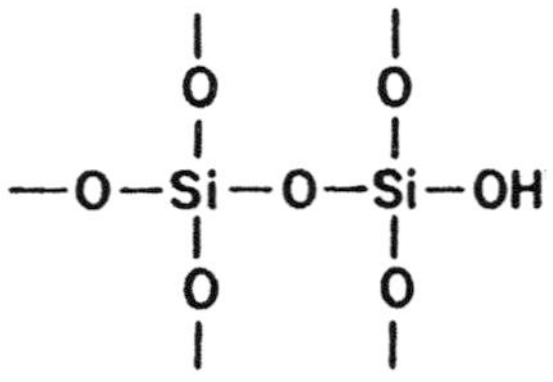

$$-O-\underset{\underset{|}{O}}{\overset{\overset{|}{O}}{\underset{|}{Si}}}-O-\underset{\underset{|}{O}}{\overset{\overset{|}{O}}{\underset{|}{Si}}}-OH$$

FIGURE 3.1   Framework of silica gel.

ity'' silica gel adsorption sites, whereas the most active sites are attached to water. A level of 11–12% water is achieved when silica gel is at equilibrium with air having a relative humidity of 50% at 20°C, conditions that are approximated in many modern analytical laboratories. Heat activation is, therefore, often not required prior to TLC. If precoated plates have been exposed to high humidity, it is good practice to heat them at 70–80°C for 30 min and then allow them to cool in an open, clean environment to return to the 11–12% water level. The same procedure should be followed for homemade plates. Some precoated layers containing organic polymer binder rather than gypsum are prepared with a certain amount of polymer adsorbed on active silica gel sites to simulate an 11–12% water content. Such plates will not generally require heat activation. Heating gypsum-bound layers above 130°C will cause $CaSO_4$ to lose its water of hydration and no longer provide any binding action. Some sources claim that the most firmly bound water cannot be removed except by heating in excess of 400°C, at which temperature the actual physical and chemical structures of silica gel are modified and relatively nonpolar siloxane groups (Figure 3.2) are formed. Others caution that a maximum of 200°C must be observed to avoid sintering and other changes in silica such as splitting of water from SiOH groups. In addition to its use as a selective adsorbent in its activated form, silica gel serves for the chromatography of more polar solutes after partial deactivation, as a liquid–liquid partition medium when fully deactivated with water or supporting a stationary organic phase, or as a molecular sieve when pore size is suitable.

The activation of TLC plates is a subject open to dispute. Hahn-Deinstrop (1993) recommends heating silica gel and alumina layers at 120°C for 30 min

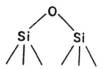

FIGURE 3.2   Structure of the siloxane group.

to remove surface water physically adsorbed from the atmosphere but not chemisorbed surface water, even though later she states that an equilibrium, depending on the relative humidity, is set up within a few minutes between the laboratory atmosphere and the sorbent. Others suggest that heat activation of precoated TLC plates is essentially useless under normal conditions (see Chapter 11). Heat conditioning is usually carried out in a conventional oven, but use of a microwave oven reduces the required time (Khan and Williams, 1993). Hahn-Deinstrop (1993) also recommends prewashing adsorbent layers by development or dipping (once or twice between 1 and 7 minutes) prior to sample application and conditioning with a defined vapor phase before chromatography.

The preparation conditions determine the special properties of silica gel. Physicochemical properties such as hardness and polarity of silica gel are related to particle size (μm), size distribution, shape, surface area (m^2/g), pore system [size (A) and distribution], and presence or absence of additives—that is, binders, contaminants, or indicators. The surface area of silica gel adsorbent for TLC is typically 300 to 600 m^2/g, pore volume is ~0.75 ml/g, and pore diameters range from 40 to 80 A (most often 60 A or 6 nm). Adsorbent with larger surface area (smaller particle size) will generally give better resolution but a slower development time. The usual binder for commercial TLC adsorbent powders is 5–20% gypsum (silica gel G). Precoated plates usually have organic binders such as polyesters or polyvinyl alcohol.

During manufacture, various inorganic and organic contaminants may be introduced into TLC silica gel. During slurry preparation or handling of a finished plate, additional contaminants may inadvertently be added to the silica gel. Slurries or plates can be washed in chloroform–methanol or the mobile phase to help reduce contaminants. Silica gel is usually prepared as an aqueous slurry approximately 2 parts water to 1 part powder, prior to application to a glass plate (see Section II). When properly purified, silica gel has little or no tendency to catalyze the reactivity of labile substances.

B. Alumina

Various hydrated aluminum hydroxides serve as starting material for TLC alumina. By a series of nonuniform thermal dehydration processes, a variety of aluminas are obtained, and the ones most suitable for TLC are the crystalline modifications of χ-Al$_2$O$_3$ and γ-Al$_2$O$_3$ (Rossler, 1969; Snyder, 1975). The physicochemical properties or the exact nature of adsorption sites of alumina are not well understood. Snyder (1975) has suggested that exposed Al atoms, strained Al–O bonds, and perhaps other cationic sites serve as adsorption sites, whereas, unlike silica, surface hydroxyl groups are probably not important. Acids are probably retained by interaction with basic sites such as surface oxide ions. Gasparic and Churacek (1978) report that every Al atom is surrounded by six atoms of oxygen on which molecules of water are bound by hydrogen bonds. After heating, alumina is alkaline (aqueous slurry = pH 9) and contains Na^+ ions that may be bound to acidic substances during TLC.

Physical properties of TLC alumina are somewhat similar to TLC silica in terms of particle size, surface area, and average pore diameter. Aluminas for chromatography have surface areas between 50 and 350 m^2/g, average pore diameters.between 20 and 150 A, and specific pore volumes between 0.1 and 0.4 ml/ g (Rossler, 1969; Hahn-Deinstrop, 1992; Robards et al., 1994).

The optimum activation temperature for alumina is not known with certainty. Heating for 30 min at 75–110°C generally produces an alumina with activity suitable for adsorption TLC. Alternatively, the most active preparation (activity I on the Brockmann scale) can be obtained by heating commercial alumina to 400–450°C, followed by the gradual addition of defined amounts of water to achieve a lower activity (III–V) for TLC. The percentage of water ranges from 0% for activity I to 15% for activity V. Test dyes are used to evaluate the activity of alumina layers as follows:

| | R_F | | | |
| | Brockmann activity | | | |
Dye	I	III	IV	V
Azobenzene	0.59	0.74	0.85	0.95
p-Methoxyazobenzene	0.16	0.49	0.69	0.89
Sudan yellow	0.01	0.25	0.57	0.78
Sudan red	0.00	0.10	0.33	0.56
p-Aminoazobenzene	0.00	0.03	0.08	0.19

As for silica, a variety of alumina powders are available commercially, and layers can be prepared as aqueous slurries following manufacturers' instructions. Commercial powders are available in different degrees of activity according to the Brockmann scale and can be obtained as acidic (pH 4.0–4.5), neutral (7.0–8.0), or basic (9.0–10.0). They are also available with or without a gypsum binder or fluorescent indicator. Commercially precoated alumina plates are usually received with approximately 6–8% water content (activity III–IV). The variations in pH, specific surface area, and pore size among various TLC aluminas lead to differences in separation properties, and the conditions used to obtain certain results must be carefully documented.

Like silica gel, alumina mainly separates components of the sample according to polarity, with hydrogen bonds or dipole forces being the primary mechanism of interaction. The selectivity of alumina in TLC adsorption chromatography is similar to that of silica gel and, therefore, this sorbent is most useful for the separation of neutral and acidic lipophilic substances. Acidic alumina strongly attracts basic compounds, whereas basic alumina attracts acidic compounds. Aromatic compounds are more strongly retained on alumina than on silica gel.

A drawback of alumina is that it can catalyze certain labile substances. Aromatic hydrocarbons are more strongly retained on alumina than silica gel. Alumina has been used to separate compounds such as fat-soluble vitamins, alkaloids, certain antibiotics, and chlorinated hydrocarbons. The TLC analysis of inorganic and organometallic compounds on alumina over 60 years was reviewed (Ahmad, 1996).

C. Cellulose

Cellulose is a highly polymerized polysaccharide characterized by the cellobiose unit (Figure 3.3). The presence of a large number of free OH groups in cellulose permits hydrogen bonding with low-molecular-weight liquids such as water and alcohols. Cellulose is, therefore, ideally suited for the separation of hydrophilic substances such as carbohydrates and amino acids, mostly by the mechanism of normal-phase partition chromatography.

TLC cellulose powders and precoated plates are of two types, native fibrous and microcrystalline. An example of the former is Macherey-Nagel (MN) 300/301, which has a degree of polymerization of 400–500 and 2–20 μm long fibers, and of the latter "Avicel," prepared in the form of cellulose crystallites with a mean degree of polymerization between 40 and 200 by hydrolysis of high purity cellulose with hydrochloric acid. For details of degree of polymerization, particle size, and commercial preparation of these types, see Wollenweber (1969). The separation behaviors of fibrous and microcrystalline cellulose will vary because they have different particle (fiber) size, surface characteristics, degree of polycondensation, and swelling behavior. Either type of cellulose may be superior for a particular separation.

Preparation of cellulose sorbent in the laboratory involves making a 15–35% aqueous slurry of the powder, followed by brief homogenization. The coated plates should be air dried rather than oven heated. Because of the particular nature of cellulose powder, a binder is not necessary. However, impurities and additives

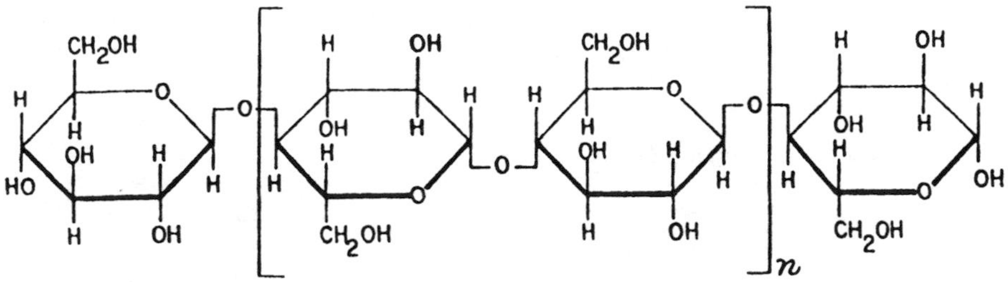

FIGURE 3.3 Fraction of a cellulose molecule.

may be present in the cellulose powder and could affect separations. Corrosive visualization agents must be avoided in cellulose TLC.

Acetylated cellulose plates, prepared by esterification of cellulose with acetic acid or acetic anhydride, are used for reversed-phase TLC. Plates are available with varying degrees of acetylation ranging from 10 to 40%; the polarity of the layer decreases with increasing acetylation. Enantiomeric compounds were resolved by RP-TLC on microcrystalline triacetyl cellulose using 2-propanol-water (6:4) as the mobile phase (Lepri, 1995).

D. Polyamide

Separations on polyamide layers are affected mainly by virtue of hydrogen bonding between the solutes and the functional amide groups of the sorbent (Endres, 1969). Although a variety of supports are available, the ones most frequently used are ε-polycaprolactam (= polyamide 6 = nylon 6) and 11-aminoundecanoic acid (= polyamide 11). The lipophilic properties of these materials differ, as do their chromatographic properties. Layers are coated on glass plates or plastic films from a slurry consisting of 20 g of sorbent in 100 ml of 75% formic acid (Wang and Weinstein, 1972). Polyamide 6.6 and acetylated polyamide are also used. Precoated polyamide layers can be obtained commercially. As with cellulose, polyamide layers are not activated before use. Mobile phases must be capable of displacing the hydrogen bonds formed between the solutes and the polyamide molecules.

Although corrosive visualization agents cannot be used in polyamide TLC, visualization is made possible by the inherent fluorescent properties of the support, and iodine can also be applied. Ninhydrin detection is possible on mixed polyamide–silica gel layers but not on pure polyamide. Because minimal diffusion occurs on polyamide, spots are sharp and compact, as is apparent in the numerous excellent chromatograms pictured by Wang and Weinstein (1972). As stated by these authors, "any organic compound which can interact through a hydrogen bond with the polyamide support is a candidate for polyamide chromatography." Polyamide TLC has been used for the separation of phenols, carboxylic and sulfonic acids, amino acid derivatives, steroids, quinones, aromatic nitro compounds, N-terminal groups in polypeptide chains, acid dyes, alkaloids, heterocyclic nitrogen compounds, nucleotides, bile pigments, and carbamate and urea pesticides.

E. Size-Exclusion Gels

Size exclusion gels are prepared from cross-linked dextran (Sephadex, Pharmacia) and poly(acrylamide) (Bio-Gel P, Bio-Rad) for the separation of hydrophilic biopolymers such as peptides, proteins, and nucleic acids based on size by gel filtration TLC. The mechanism of separation is partition chromatography governed by size exclusion in swollen gels containing pores of carefully controlled dimensions. Layers range in thickness from 0.4 to 1.0 mm, with 0.60 mm (600 μm) being most common. In addition to Sephadex and other Sephadex-like materials (Hung et al., 1977), porous glass beads (Waksmundzki et al., 1979), Styragels (Shibukawa et al., 1978), and polystyrene-divinylbenzene copolymers (Pietrzyk et al., 1979) have been used for TLC separations based on size differences. Sephadex ion exchangers (Pharmacia) have been used for the TLC separation of charged high-molecular-weight biological compounds (Steverle, 1966) and for determination of protein molecular weights (Hauck and Mack, 1996). Sephadex used for TLC must be the superfine grade (10–40 μm), with fractionation ranges for dextrans of 100–5,000 (G-25) to 5,000–200,000 (G-200) molecular weight (daltons) (Hauck and Mack, 1996). Precoated plates are not available; gels must be totally swollen and layers prepared in the laboratory.

F. Ion Exchangers

Three types of ion exchanges are available commercially as powders for preparing layers and/or as precoated plates. Layers coated with a mixture of silica and a cation- or anion-exchange resin with an inert binder (Ionex, Macherey-Nagel) are suitable for the separation of amino acids, nucleic acid hydrozylates, amino sugars, aminocarboxylic acids, antibiotics, inorganic phosphates, cations, and other compounds with ionic groups.

Polyethyleneimine (PEI) and diethylaminoethyl (DEAE) celluloses are strong base anion exchangers that are used mostly for separations of amino acids, peptides, enzymes, nucleic acid constituents (nucleotides, nucleosides), and other high molecular weight bioorganic compounds. The former is prepared as a complex of cellulose and PEI, whereas the latter is a chemically modified cellulose containing a diethylaminoethyl group bonded to cellulose via an ether group. Mixed layers of DEAE cellulose and unmodified HR cellulose or Avicel cellulose are also available, which provide reduced ion-exchange capacity but higher chromatographic mobility. Other ion-exchange layers available as precoated layers include Poly-P (polyphosphate) impregnated cellulose cation exchanger and the following surface-modified materials: AE (aminoethyl) and ECTEOLA (reaction product of epichlorhydrin, triethanolamine, and alkaline cellulose) anion exchangers and P (phosphorylated) and CM (carboxymethyl) cation exchangers (Robards et al., 1994).

Homemade plates are also used for ion-exchange TLC. For example, separation of platinum from 27 other cations was achieved using 0.5 M ammonium hydroxide as mobile phase with cerium (III) silicate layers (Husain et al., 1996).

G. Kieselguhr

Kieselguhr for TLC is refined from fossilized diatoms (diatomaceous earth), and the main constituent ($\sim$90%) is SiO_2. A variety of other inorganics are present in kieselguhr, and the pore size of this sorbent is quite variable. Relative to most TLC silica gels, the surface of kieselguhr is only slightly active, and the pore volume is relatively large. Typically, the surface area of kieselguhr for TLC is 1–5 m^2/g, the average pore diameter 10^3–10^4, and specific pore volume 1–3 ml/ g. These characteristics permit use of this neutral sorbent for separation of sugars, amino acids, and other similarly polar substances; as an inert support for partition chromatography; and as a preadsorbent. The most frequently used kieselguhr (kieselguhr G) contains approximately 15% gypsum binder and is prepared as an aqueous slurry, 1 part sorbent to 2 parts water. For further details, see Rossler (1969). Precoated aluminum-backed layers are available commercially. Synthetic silica 50,000 is also used widely as a support for partition TLC; it has chromatographic properties similar to kieselguhr but greater chemical purity because it is synthetic rather than a natural siliceous algae.

H. Miscellaneous Inorganic Sorbents

Various other inorganic sorbents are used occasionally in TLC for rather specific separation problems. These sorbents are generally unavailable as precoated plates and must be prepared from aqueous or alcoholic slurries. For further details, see Rossler (1969). Magnesium oxide layers were developed with petroleum ether (30–50°C)–benzene (3:1) for the separation and quantification of alpha- and beta-carotene in lettuce and snails (Drescher et al., 1993). Zinc carbonate with a starch binder has been used to separate aldehydes, ketones, and other carbonyl groups (Rossler, 1969). Magnesium silicate (talc or Florisil) prepared as an alcoholic slurry has been used to separate pesticides (Getz and Wheeler, 1968), fatty acids, and lanatosides (Rossler, 1969). Charcoal has been combined with silica gel for ketone separations (Rossler, 1969). Two forms of charcoal are available, polar and nonpolar, and each type has very different capabilities. Charcoal has had very limited use in TLC, partly because of the difficulty of zone detection on the layer. For a discussion of the adsorptive properties of charcoal, see Snyder (1975).

I. Miscellaneous Organic Sorbents

Of all the organic sorbents, cellulose and its derivatives are used most frequently in TLC. Some lesser-used organics include starch and sucrose (Wollenweber, 1969). Starch is hydrophilic, generally prepared as an aqueous slurry, and has been used for the separation of hydrophilic substances. Starch has also been used as a support in reversed-phase TLC (Davidek, 1964). Sucrose containing a small amount of starch has been used for the separation of chloroplast pigments (Colman and Vishniac, 1964). Because of its solubility in water, sucrose is prepared as an alcoholic slurry (Wollenweber, 1969).

Chitin is a natural celluloselike biopolymer consisting predominantly of unbranched chains of *N*-acetyl-D-glucosamine residues, but the presence of amine and amide groups makes the selectivity of chitin different from cellulose. Laboratory-prepared layers of chitin, its deacetylated derivative chitosan, and chitin modified with metal cations have been used to separate a variety of organic and inorganic compounds, including amino acids (Malinowska and Rozylo, 1991; Rozylo et al., 1989).

Organic sorbents have the disadvantage of charring with corrosive detection reagents.

J. Mixed Layers

Commercial mixed layers available from various manufacturers include aluminum oxide/acetylated cellulose (normal- and reversed-phase chromatography), cellulose/silica (normal-phase chromatography), and kieselguhr/silica gel (normal-phase chromatography, reduced adsorption capacity compared to silica). Magnesium in aluminum alloys was determined using handmade plates with a mixture of cellulose and the strong acid cation exchanger Amberlite IRP-69 in the H^+ form and 0.5–2 M HCl or HNO_3 mobile phase (Petrovic et al., 1993). Preparation of mechanically blended, mixed C_{18}/CN and C_8/diol stationary phases for one- and two-dimensional TLC of polycyclic nuclear hydrocarbons with normal- and reversed-phase mechanisms was described (Hajouj et al., 1995 and 1996).

K. Impregnated Layers

Acid, alkaline, or buffered layers can be prepared by using a suitable aqueous solution of an acid, base, or salt mixture instead of water to prepare the adsorbent slurry used to cast the layer. Impregnation can also be carried out on a manually or commercially precoated layer, usually kieselguhr, cellulose, or silica gel, by a variety of methods. In the irrigation or predevelopment method, the dry plate is placed in a TLC tank containing the impregnation solution, for example, 5 ml of paraffin oil in 95 ml of low-boiling petroleum ether, for preparation of a lipophilic reversed-phase layer. The solution is allowed to ascend to the top of the plate, followed by air drying for 5–10 min before spot application.

Another method for pretreatment of coated plates is by immersion in a solution of the impregnating agent in a readily volatile solvent contained in a chromatography developing tank, a dip tank (Chapter 8), or a tray. After immersion for several seconds, the excess solution is allowed to drain, and the volatile solvent is evaporated before use. Predevelopment requires more time than immersion, but its application has been recommended because the resulting coverage is claimed to be more uniform (Cserhati and Forgacs, 1996). Additional methods for impregnation include equilibration in a closed vessel with the vapors of the stationary liquid, or spraying with the stationary phase alone or a solution of it in a volatile solvent. A comparative study of methods for impregnation of silica gel layers with ion-pairing reagents (TMA, TBA, CTMA, TOMA) found overdevelopment in a normal unsaturated chamber for 18 h gave a smoother gradient than other types of development, but dipping using an immersion device resulted in almost uniform reagent distribution on the layer (Kovacs-Hadady and Varga, 1995).

The range of applications of both TLC and high-performance TLC is considerably expanded through the use of different impregnation agents added to layers in various concentrations (Halpaap and Ripphahn, 1977). Stable hydrophilic stationary phases are formed by treatment with agents such as formamide, dimethyl formamide, dimethyl sulfoxide, poly(ethylene glycol), or various buffers. Lipophilic stationary phases for reversed-phase TLC of homologs or very polar compounds that bind too strongly or streak on silica gel are obtained by impregnation with liquid paraffin, undecane, mineral oils, silicone oils, or ethyl oleate. These nonpolar layers strongly attract nonpolar molecules and polar molecules move most readily, the reverse order compared to a polar silica gel layer. Impregnation with specific reagents aids the separation of certain types of compounds, such as silver nitrate (Li et al., 1995) for unsaturated compounds based on the number and configuration of double bonds, boric acid for polar lipids with vicinal hydroxy groups, sodium borate for sugars, sodium bisulfite for carbonyl compounds, trinitrobenzene or picric acid for polynuclear compounds, tributylamine or tributylphosphate or salicylic acid for metal ions, and tricaprylammonium chloride for various bioactive compounds such as drugs and antibiotics. Layers impregnated with transition metal ions were used to separate sulfonamides (Bhushan and Ali, 1995), antihistamines (Bhushan and Joshi, 1996), and sugars (Bhushan and Kaur, 1997), and with EDTA to resolve some cephalosporins (Bhushan and Parshad, 1996a). Combinations of two different impregnations are readily made on the same plate. For example, two-dimensional fatty acid separations are carried out on a strip impregnated with silver nitrate followed by a second run at right angles on a portion of the layer impregnated with paraffin. Analtech offers precoated silica gel layers impregnated with 1% potassium oxalate, 5–7% magnesium acetate, $0.1N$ sodium hydroxide, 5 to 20% silver nitrate, and ammonium sulfate. The first four plates improve separations of certain compounds, whereas the last produces sulfuric acid upon heating for charring detection of organic compounds.

Analtech supplies precoated plates for reversed-phase TLC consisting of a support layer impregnated with a long-chain hydrocarbon that gives separations similar to those obtained on a C_{18}-bonded layer (Section III.C). Mobile phases can be totally aqueous (100% water) or aqueous-organic mixtures, and they need not be presaturated with the hydrocarbon phase. These layers are less expensive than bonded RP layers.

A massive review containing 650 references, covering the procedures and applications of TLC and HPTLC on layers impregnated with organic stationary phases, has been written by Gasparic (1992). In addition to reviewing analytical aspects of impregnated-layer TLC, lipophilicity determination by TLC (see Chapter 2, Section IV) is described and more than 100 references on this topic are tabulated in the review. Reversed-phase partition TLC has been shown to be a simple and fast alternative to liquid–liquid partition for determining lipophilicity through determination of R_M values (Chapter 9), which are related to the logarithm of the partition coefficient of the chromatographed compound between the stationary and mobile phases. A silica gel layer impregnated with silicone oil and developed with an aqueous acetone mobile phase has been used for a wide variety of drugs and is a typical reversed-phase system applied for these studies. Other RP systems include undecane or paraffin oil as the stationary phase, cellulose or kieselguhr as the support, and a variety of buffered and unbuffered polar aqueous mobile phases. Normal-phase systems have also been used, for example, silica gel or cellulose impregnated with formamide and developed with heptane or heptane–diethylamine for local anesthetics. Thermogravimetry was found to have some advantages over spectrometric methods for studying the distribution of different impregnation reagents on silica gel layers (Kovacs-Hadady and Balazs, 1993).

II. DEVICES AND METHODS FOR LAYER PREPARATION

A variety of procedures are available for the manual preparation of thin layers from sorbents that are manufactured especially for TLC by various firms, including spreading, pouring, spraying, or dipping a slurry of sorbent on a backing material, usually a glass plate. Manually prepared layers should have a uniform, bright, translucent appearance when viewed in incident and transmitted light after drying, and the layer must form a good bond with the support. Standard TLC plates are 20 × 20 or 20 × 10 cm, and the thickness of the dried layer for analytical purposes is 0.1–0.3 mm, most commonly 0.2–0.25 mm. Unquestionably, when plates are prepared manually, the most uniform layers are produced with the aid of various commercial spreaders. This is also by far the most convenient method when a larger number of plates is to be prepared.

It is impossible to achieve the exacting conditions employed by plate manufacturers when preparing homemade plates, and whenever possible commercial precoated TLC and HPTLC plates should be used for best results. Several reasons that might lead to the preparation of homemade plates in the laboratory include (1) use of a very large number of plates and the ability of personnel to successfully prepare layers of adequate quality on reused glass at lower cost than that for which this number of precoated plates could be purchased; (2) the need of a special sorbent or combination layer not offered commercially, such as magnesium oxide for the separation of carotene isomers; (3) the need for a backing material not offered commercially, such as borosilicate glass for heating to a very high temperature for charring techniques; and (4) the desire to include in the sorbent formulation a special material to facilitate separation or detection, such as a mixture of three fluorescent additives that give colored spots of PTH-amino acids under 250–400-nm ultraviolet light (Nakamura et al., 1979).

Previous editions of this book contained discussions of three commercial layer-spreading devices, preparation of gradient layers (Stahl and Mueller, 1980), and manual layer preparation by pouring, spraying, and dipping. These will not be repeated here.

Manufacturers supplying TLC sorbents will provide instructions for making slurries, air drying times, and oven activation conditions in order to hand-prepare TLC plates using a suitable apparatus such as the Camag automatic or hand-operated plate coater or Desaga 120131/12305. These devices prepare 20 × 20 cm plates with layer thicknesses up to 2 mm.

III. PRECOATED LAYERS

A. General Properties

Currently, silica gel, alumina, cellulose, kieselguhr, polyamide, ion-exchange, and a variety of bonded-phase precoated plates are commercially available. Kieselguhr, polyamide, and ion exchangers are available only in TLC grade, whereas silica gel, alumina, cellulose, and bonded phases are available as HPTLC plates and are most suitable for quantification. Layers are sold with three types of backings or supports: glass, plastic (polyethylene terephthalate), and heavy aluminum foil. Glass is by far the most popular backing. It is inert, its rigidity allows layers to be coated with more uniform thickness, and its transparency facilitates transmission scanning for quantification of zones.

Most precoated analytical TLC and HPTLC layers on all three backings are between 100 and 250 μm (0.1–0.25 mm) in thickness. Preparative layers have thicknesses between 0.5 and 2 mm to provide greater sample capacity. Thicknesses are optimized by manufacturers for each of their particular sorbent/binder/backing combinations. The usual plate sizes are 10 × 10 cm, 10 × 20 cm, and 20 × 20 cm. Some companies will also prepare layers of other sorbents on a special-order basis, and some sell scored glass plates that can be snapped or broken cleanly into smaller, more convenient sizes. Nonscored glass plates can be easily and cleanly cut to smaller sizes by use of one of the diamond-tipped glass cutters or plate-cutting machines (Hahn-Deinstrop, 1993) sold by a number of companies. Charring procedures with the use of corrosive reagents and high temperature are best done on layers with glass supports. However, charring at less than 130°C is possible on plastic-backed layers. For ease of handling, cutting (see Hahn-Deinstrop, 1993, for the correct procedure), documenting, and storing chromatograms, plastic and aluminum are preferred to glass. The variety of sorbents available on aluminum or plastic is less than on glass, and plastic- and aluminum-coated foils may curl irreversibly on heating and are unsuitable for use with certain solvents, such as those containing strong acids and bases.

Most precoated plates contain some type of binder to impart mechanical strength, durability, and abrasion resistance to the layer. Binders that are used include calcium sulfate (gypsum), silicic acid, starch, and a variety of polymeric organic compounds (polymethacrylates, polyols, polycarboxylates, etc.) in amounts of 0.1–10%. Organic binders result in layers that are especially stable to handling and shipping. It has been determined that the binder resides largely within the pores of the layer material but does not specifically block the pore entrances (Poole and Poole, 1995). The presence of the binder may modify the chromatographic behavior of the original sorbent.

Precoated plates can also contain a fluorescent compound to facilitate detection by fluorescence quenching. Manganese-activated zinc silicate, which produces yellow-green fluorescence when irradiated with 254-nm UV light, is often used. At the levels employed, these phosphor indicators do not normally alter separations.

For in situ densitometric quantification, the use of highly uniform precoated layers on glass is virtually mandatory for adequate reproducibility and accuracy. Plates divided into a series of parallel lanes or channels (usually ~1 cm in width) restrict lateral zone diffusion and facilitate manual sample application and densitometric scanning (Figure 3.4).

Flexible sheets with the backing material combined directly with the sorbent are available as Empore sheets containing plain or bonded silica mixed with microfibrils of poly (tetrafluoroethylene) (PTFE) in an approximate 9:1 weight ratio. They have not been widely used, apparently because of inferior separation performance and slow mobile phase migration velocity using capillary flow. They have advantages of easier sample spotting and sample recovery from the developed layer, and may find most application in overpressured layer chromatography (Poole and Poole, 1990; Robards et al., 1994). Glass has been used for TLC in three different forms: as porous glass sheets that can be regenerated by washing with strong acids for repeated use (Yoshioka et al., 1990), as sintered glass plates (Okumura, 1980), and as thin, porous sintered glass rods with embedded silica gel or alumina (Mukherjee, 1996). A new plate with a layer closed by a film transparent to UV radiation has been described, and its characteristics studied;

FIGURE 3.4 Commercial precoated, glass-backed 20 × 20-cm channeled pre-adsorbent silica gel plate. (Photograph courtesy of Analtech.)

it was used with one end attached to a vacuum source and the other to a mobile-phase source to shorten development (Berezkin and Buzaev, 1997).

A comparison of different precoated plates was given by Brinkman and de Vries (1985).

B. High-Performance Silica Gel and Cellulose

High-performance (HP) precoated silica gel plates are available from several manufacturers in sizes ranging from 5 × 5 to 20 × 20 cm, usually with an organic binder. The basic difference in TLC and HPTLC plates is the particle sizes of the sorbents. For precoated TLC plates, the average particle size is ~11 μm with a 3–18 μm distribution; for HPTLC plates, the average particle size is ~5–6 μm with a 4–7 μm distribution. The smaller particle size and distribution allow HPTLC plates to provide higher efficiency and resolution and to be used with shorter development distances. See Chapter 2 for a complete description of the advantages of HPTLC compared to TLC.

Precoated cellulose HPTLC layers (0.1 mm thickness) are commercially available and have been shown to be more efficient than conventional cellulose plates and to have shorter development times (Brinkman et al., 1980).

Layers of wide-pore (5000-nm) silica gel have been reported (Hauck and Halpaap, 1980) to have separation properties similar to cellulose, with the advantages of resisting swelling in organic solvents and allowing the use of charring detection reagents.

C. Chemically Bonded Plates

Chemically bonded C_{18} plates (Beesley, 1982; Musial and Sherma, 1998) consist of a layer produced by reacting octadecyl monochlorosilane with some of the surface silanol groups of silica gel. Other chemically bonded phases are produced by use of organosilane reagents having other functional groups. The C_{18} nonpolar silanized silica gel layers are suitable for reversed-phase separations of compounds that have similar polarity but differ in lipophilicity, such as a series of homologs with a single functional group, by use of a hydrophilic mobile phase. The advantage of this chemically bonded medium compared to conventional impregnated nonpolar phases (Section I.K) is that the phase is permanently bonded and cannot be stripped off during development, and no pains need to be taken to saturate the mobile phase with the stationary liquid to attempt to prevent this loss.

Mobile phases used with these plates are usually simple, two-component mixtures of water and an organic solvent (most commonly methanol, acetonitrile, or tetrahydrofuran) that are easily optimized by adjusting the proportions. The more nonpolar the sample, the higher the proportion of the organic solvent used. Conversely, as the polarity of the solute increases, the water content of the mobile phase is raised. Development times vary with percentage of water in the mobile phase (Jost and Hauck, 1987). For many bonded RP plates, decreased water content causes development times to drop rapidly. For other RP layers and some polar-bonded phases, the development time increases and then decreases with decreasing water content, reaching a maximum value in the range of 30–50% water, the exact value depending on the particular layer and solvent modifier. Whatman plates can be used with high percentages of water in the mobile phase, but when the mobile phase is greater than 40% aqueous, salt solution (i.e., $0.5M$ NaCl) must be substituted for water to maintain the stability of the layer (Brinkman and deVries, 1985). NaCl or other neutral salts can influence the retention of solutes (salting-in or salting-out effects), sometimes reducing separation efficiency; the effect of salt is higher when the solute has one or more dissociable polar substructure (Cserhati and Forgacs, 1996). Poole and Poole (1991) reported that the degree of bonding for commercial layers ranges from ~20 to 50% of the available silanol groups, which have a concentration of 8 μmole/m^2 on the silica surface. The 1996 Macherey-Nagel products catalog lists C_{18} plates that are described as being totally (100%) or partially (50% of the reactive groups) bonded, but these numbers probably do not represent the actual percentage of silanol groups reacted. As the percentage of silanol groups reacted increases for

RP layers, the ability to use mobile phases with a high water content decreases. RP layers plates with a ''W'' designation have fewer reacted silanol groups and allow the use of mobile phases containing a high amount of water. See also Chapter 6 for more information on mobile-phase selection.

C_{18} layers have been found to have close correlation with HP liquid chromatography reversed-phase columns (Gonnet and Marichy, 1979), and TLC is used widely to scout appropriate mobile phases for HPLC (see Chapter 6). Very polar substances (basic and acidic pharmaceuticals) can be separated on RP layers by ion pair chromatography.

C_2-, C_8-, and diphenyl-bonded layers are also manufactured and have been compared with C_{18} layers for the TLC of 30 compounds representing seven polar and nonpolar compound types (Heilweil and Rabel, 1985). Retention generally increased with longer bonded chain length, as did development times. Polymeric-bonded phases, made with a polyfunctional silane, were more stable than monomeric or ''brush'' phases. The aliphatic-bonded layers showed similar selectivities with the various compound classes. Diphenyl and C_2 layers had similar selectivities, leading the authors to propose a solvation-layer mechanism for reversed-phase separations.

Precoated HPTLC layers with moderately polar CN (cyano) (Omori et al., 1983) or NH_2 (amino) (Jost and Hauck, 1983) groups bonded to silica gel through a C_3-spacer chain are available commercially. Depending on the choice of mobile phases, they can separate mixtures according to either polarity (by interaction with the functional group; normal-phase TLC) or lipophilic properties (by interaction with the spacer; reversed-phase TLC). Diol layers consist of (CH_2—CH_2—CH_2—O—CH_2—CHOH—CH_2—OH) groups bonded to the silica gel surface. Like unmodified silica gel, the diol layer has active surface hydroxyl groups, but the groups are alcoholic and bonded to silica gel through a spacer, rather than silanol groups directly on the surface. The unique nature of the diol and silica gel functionality leads to differences in chromatographic properties in terms of activity, selectivity, and effect of atmospheric humidity. The diol plate can provide normal-phase separations with solvents such as petroleum ether-acetone and reversed-phase separations with acetone-water (Hauck et al., 1989; Witherow et al., 1990). Cyano layers are usually used as a weak adsorbent for normal-phase TLC, but the organic surface of the layer imparts some lipophilic character so that reversed-phase TLC can also be done. Amino layers act as weak anion exchangers when used with acidified aqueous normal-phase solvents. Hydrophobic and polar interactions are also evident on these layers, depending on the solutes and mobile phases, and sulfonamides have been separated by ion-pairing TLC (Bieganowska and Petruczynik, 1996). Placed on a polarity scale, the chemically bonded phases can be roughly arranged as follows: silica > amino > cyano = diol > C_2 > C_8 > C_{18}. When classifying a layer, the distinction between normal and reversed phase is made by noting the effect of mobile-phase water content on retention; in normal phase, addition of water reduces retention.

Other than the bonded phases already mentioned above, silica layers chemically bonded with ethyl and propanediol are commercially available. The latter is a weaker sorbent than silica gel but has comparable selectivity (Poole and Poole, 1991).

Bonded phases are prepared using a variety of derivatizing agents (mono-, bi-, and trifunctional), silica gels, and reaction conditions, leading to products that vary in important physical properties, such as percentage of silanol groups reacted, carbon loading, extent of silanol deactivation, average particle size, and chromatographic properties. Layers of supposedly the same kind from different manufacturers can yield very dissimilar R_F values, migration sequences, and development times with the same mobile phase.

D. Preadsorbent Layers

Some precoated TLC and HPTLC plates contain silica gel or chemically bonded silica gel with a bottom zone consisting of an inert preadsorbent or concentration zone. After the application of the sample as a spot (Figure 3.4) or streak, even over a large area of the zone, soluble compounds pass through the preadsorbent with the mobile phase front and are automatically concentrated or focused in a narrow band at the junction with the separation sorbent (Figure 3.5). The formation of narrow bands leads to a greater number of theoretical plates and enhanced separations and limits of detection. The interface between the preadsorbent and the separation layer serves as the origin for calculation of R_F values. The preadsorbent acts as a blotter or dispensing area and can accommodate a relatively large sample of crude biological materials such as blood or urine. Certain impurities may be retained in this zone, thereby achieving some sample cleanup, but material soluble in the solvent system, such as salts, may still rise. Preadsorbent plates are particularly advantageous when applying large sample volumes of dilute samples and when sample mixtures contain substances of very different proportional amounts. The final R_F value of the separated zone is independent of the exact position of application on the preadsorbent zone. The preadsorbent area of current commercial TLC and HPTLC plates is a $\sim$23–33-mm strip consisting of diatomaceous earth or a low-surface-area (0.5 m^2/g), large-pore-volume (5000-nm) silica (Halpaap and Krebs, 1977), the former having higher sample capacity. This low-activity, wide-pore silica is useful in the form of a precoated layer for separation of highly polar compounds (Hauck and Halpaap, 1980).

E. Combination (Biphasic) Layers

Combination 20 $\times$ 20-cm partition-adsorption layers are available consisting of a 3.0-cm strip of C_{18} chemically bonded reversed-phase medium adjacent to a

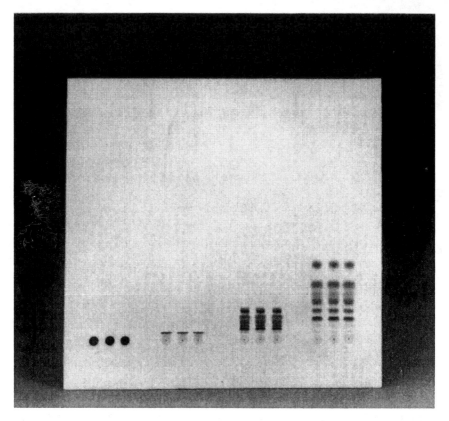

FIGURE 3.5 Sequence of separation on a preadsorbent layer. (Photograph courtesy of Analtech.)

silica gel layer. The sample is spotted and developed on the RP strip, followed by development at right angles on the silica gel. Resolutions of complex mixtures with a wide range of polarities can be improved by this two-dimensional development utilizing two dissimilar, complementary separation mechanisms. Plates are also available having the same configuration except that the 3.0-cm strip is silica gel and the rest of the layer is C_{18}-bonded silica gel. Other available combination layers for separating mixtures by two types of polar adsorption mechanisms include CN/silica gel and NH_2/silica gel.

F. Chiral Phases

Phases that have been used for the TLC separation of enantiomers include cellulose and triacetyl cellulose; chemically bonded layers containing cyclodextrins and their derivatives (cyclodextrins have also been used as mobile-phase additives for enantiomer separations); dinitrobenzoyl amide phases produced by chemically bonding a Pirkle-type selector to an amino-bonded silica gel layer; silica gel layers impregnated with an optically active acid or base such as (+)-ascorbic acid; and the CHIR plate. The latter, which operates by the mechanism of chiral ligand exchange and is the most important phase for separation of enantiomers, consists of a preadsorbent sample application area and a bonded C_{18} layer impregnated with a copper salt and an optically active hydroxyproline derivative, the chiral selector. The usual mobile phases consist of methanol–water–n-propanol mixtures. Water-insoluble beta-cyclodextrin polymer beads show retention characteristics deviating from those of traditional adsorptive and RP sorbents and mainly depending on steric parameters (Cserhati, 1994; Cserhati et al., 1995). Silica gel plates impregnated with erythromycin as the chiral selector were used to separate enantiomeric dansylamino acids (Bhushan and Parshad, 1996b) and with L-arginine for resolution of ibuprofen enantiomers (Bhushan and Parshad, 1996c). A TLC plate made by the molecular imprinting technique provided rapid chiral separation of L-and D-phenylalanine anilide; this was a preliminary report on a novel and promising family of stationary phases based on predetermined selectivity (Kriz et al., 1994).

The overwhelming majority of chiral separations up to now have been carried out by adding various chiral selectors to the mobile phase rather than by using a chiral plate. For example, enantiomers of amino acids were separated by using alpha- and beta-cyclodextrins as mobile phase additives with bonded C_{18} and cellulose layers (Cserhati and Forgacs, 1996).

The methods and applications of enantiomer separations by TLC were reviewed in a paper by Mack and Hauck (1989) and book chapters by Gunther and Moeller (1996) and Grinberg et al. (1990).

G. The Nomenclature of Precoated Plates

The symbols and names given to different precoated plates (and sorbent powders for manual preparation of layers) are given in Table 3.1.

IV. REPRODUCIBILITY

Although Chapter 11 deals with factors affecting the reproducibility of TLC results, it should be emphasized here that the properties of sorbent powders and precoated plates can differ widely from manufacturer to manufacturer. The scientist is strongly advised to standardize on one particular source for a given application and to become thoroughly familiar with the properties of that sorbent. The manufacturer of the plates used in research should be named in publications so that results can be reproduced. TLC plates can absorb water and other vapors that

TABLE 3.1 Nomenclature of TLC Plates

"Sil"	A product composed of silica gel, e.g., Anasil from Analabs
G	Gypsum [CaSO$_4$ · (1/2) H$_2$O] binder ("soft" layers)
S	Starch binder
O	Organic binder, such as polymethacrylate or polycarboxylate; layers are "hard" or abrasion resistant
H or N	No "foreign binder"; products may contain a different form of the adsorbent, e.g., colloidal or hydrated silica gel or colloidal silicic acid, to improve layer stability
HL	Hard or abrasion resistant layer containing an organic binder (Analtech)
GHL	Hard or abrasion-resistant layer containing an inorganic hardener (Analtech)
HR	Specially washed and purified (highly refined)
P	Thicker, preparative layer or material for preparing such layers (for cellulose, see below)
P + CaSO$_4$	Preparative layer containing calcium sulfate binder
F or UV	Added fluorescent indicator (phosphor) such as Mn-activated zinc silicate
254 or 366	Used after F or UV to indicate the excitation wavelength (nm) of the added phosphor. An s after the number designates an acid-stable fluoresecnt indicator
60	Silica gel 60 (Merck) has pore size of 60 Å (10 Å = 1 nm). Other pore-size designations are 40, 80, 100
D	Plates divided into a series of parallel channels or lanes
L	Layer with a preadsorbent sample dispensing area (Whatman)
K	Symbol used in all Whatman products
RP	Reversed-phase layer; RP$_{18}$ or RP-C$_{18}$ would indicate that octadecylsilane groups are chemically bound to silica gel
4, 7, 9	These numbers after the adsorbent name usually indicate pH of a slurry
E, T	Aluminum oxide layers having specific surface areas of 70 m^2/g and 180–200 m^2/g, respectively
MN-300 or -400	Macherey-Nagel proprietary fibrous celluloses
Avicel	American Viscose cellulose (microcrystalline)
CM	Carboxymethyl cellulose
DEAE	Diethylaminoethyl cellulose
Ecteola	Cellulose treated with ethanolamine and epichlorhydrin
PEI	Polyethyleneimine cellulose
P	Phosphorylated cellulose
R	Specially purified
W	Water tolerant, wettable layer
RP	Reversed-phase, RP-8 or RP-18 indicates the carbon number of the bonded hydrocarbon chain

TABLE 3.1 Continued

NH₂	Layer with bonded amino group
CN	Layer with bonded cyano group
DIOL	Layer with diol modification
CHIR	Chiral layer for separating enantiomers
Preadsorbent	Plate contains an area that concentrates the sample before reaching the separation area; also called preconcentration zone
Prescored	Plate can be broken down into smaller plates
Silanized	Silanized silica gel is RP-2 silica gel with a dimethylsilyl modification

may cause anomalous results. If convenient, plates can be protected by storage in a desiccated cupboard. Measures taken for quality assurance of reproducibility in the manufacture of plates conforming with GLP and ISO 9001 standards have been described (Hauck et al., 1993 and 1995).

More detailed information on precoated layers and sorbents is available in book chapters written by Hauck and Moeller (1996) and Gocan (1990), the latter of which contains 810 references, and in reviews by Hauck and Jost (1990a, 1990b) and Hauck and Mack (1990). Hauck and Jost (1990b) tabulated all bulk sorbents and precoated plates commercially available worldwide in 1990.

REFERENCES

Ahmad, J. (1996). Use of alumina as stationary phase for thin layer chromatography of inorganic and organometallic compounds. *J. Planar Chromatogr.—Mod. TLC 9*: 236–246.

Beesley, T.E. (1982). Chemically bonded reverse phase TLC. In *Advances in TLC—Clinical and Environmental Applications*, J. C. Touchstone (Ed.). Wiley-Interscience, New York, pp. 1–12.

Berezkin, V.G., and Buzaev, V.V. (1997). New thin layer chromatography plate with a closed layer and details of its application. *J. Chromatogr., A 758*: 125–134.

Bhushan, R., and Ali, I. (1995). TLC separation of sulfonamides on impregnated silica gel layers, and their quantitative estimation by spectroscopy. *J. Planar Chromatogr.—Mod. TLC 8*: 245–247.

Bhushan, R., and Joshi, S. (1996). TLC separation of antihistamines on silica gel plates impregnated with transition metal ions. *J. Planar Chromatogr.—Mod. TLC 9*: 70–72.

Bhushan, R., and Kaur, S. (1997). TLC separation of some common sugars on silica gel plates impregnated with transition metal ions. *Biomed. Chromatogr. 11*: 59–60.

Bhushan, R., and Parshad, V. (1996a). Separation and identification of some cephalosporins on impregnated TLC plates. *Biomed. Chromatogr. 10*: 258–260.

Bhushan, R., and Parshad, V. (1996b). Thin layer chromatographic separation of enantiomeric dansylamino acids using a macrocyclic antibiotic as a chiral selector. *J. Chromatogr., A 736*: 235–238.

Bhushan, R., and Parshad, V. (1996c). Resolution of (+/−)-ibuprofen using L-arginine-impregnated thin layer chromatography. *J. Chromatogr., A 721*: 369–372.

Bieganowska, M. L., and Petruczynik, A. (1996). Thin layer and column chromatography of sulfonamides on aminopropyl silica gel. *Chromatographia 43*: 654–658.

Brinkman, U. A. T., and deVries, G. (1985). TLC on chemically bonded phases—A comparison of precoated plates. In *Techniques and Applications of TLC*, J.C. Touchstone and J. Sherma (Eds.). Wiley-Interscience, New York, pp. 87–107.

Brinkman, U. A. T., deVries, G., and Cuperus, R. (1980). Stationary phases for high performance TLC. *J. Chromatogr. 198*: 421–428.

Colman, B., and Vishniac, W. (1964). Separation of chloroplast pigments on thin layers of sucrose. *Biochim. Biophys. Acta. 82*: 616–618.

Cserhati, T. (1994). Relationship between the physicochemical parameters of 3,5-dinitrobenzoic acid esters and their retention behavior on beta-cyclodextrin support. *Anal. Chim. Acta 292*: 17–22.

Cserhati, T., and Forgacs, E. (1996). Introduction to techniques and instrumentation. In *Practical Thin-Layer Chromatography—A Multidisciplinary Approach*, B. Fried and J. Sherma (Eds.). CRC Press, Boca Raton, FL, pp. 1–18.

Cserhati, T., Fenyvesi, E., Szejtli, J., and Forgacs, E. (1995). Water-insoluble beta-cyclodextrin polymer as thin-layer chromatographic sorbent. *Chim. Oggi 13*: 21–24.

Davidek, J. (1964). Chromatography on thin layers of starch with reversed phases. In *Thin Layer Chromatography*, G. B. Marini-Bettolo (Ed.). Elsevier, New York, pp. 117–118.

Drescher, J. N., Sherma, J., and Fried, B. (1993). Thin layer chromatographic determination of alpha-carotene on magnesium oxide layers. *J. Liq. Chromatogr. 16*: 3557–3561.

Endres, H. (1969). Polyamides as adsorbents. In *Thin Layer Chromatography. A Laboratory Handbook*, 2nd ed., E. Stahl (Ed.). Springer-Verlag, New York, pp. 41–44.

Felton, H. R. (1979). Moisture and silica gel. Technical Report 7905, Analtech, Newark, DE.

Gasparic, J. (1992). Chromatography on thin layers impregnated with organic stationary phases. *Adv. Chromatogr. 31*: 153–252.

Gasparic, J., and Churacek, J. (1978). *Laboratory Handbook of Paper and Thin Layer Chromatography*. Halsted Press, New York, pp. 39–44.

Getz, M. E., and Wheeler, H. G. (1968). Thin layer chromatography of organophosphorus insecticides with several adsorbents and ternary solvent systems. *J. Assoc. Off. Anal. Chem. 51*: 1101–1107.

Gocan, S. (1990). Stationary phases in thin layer chromatography. In *Modern Thin Layer Chromatography*, N. Grinberg (Ed.). Marcel Dekker, New York, pp. 5–137.

Gonnet, C., and Marichy, M. (1979). Reversed phase thin layer chromatography on chemically bonded layers and validity of data transfer to columns. *Analusis 7*: 204–212.

Grinberg, K., Armstrong, D. W., and Han, S. M. (1990). Chiral separations in thin layer chromatography. In *Modern Thin Layer Chromatography*, N. Grinberg (Ed.). Marcel Dekker, New York, pp. 398–427.

Gunther, K., and Moeller, K. (1996). Enantiomeric separations. In *Handbook of Thin Layer Chromatography*, 2nd ed. J. Sherma and B. Fried (Eds.). Marcel Dekker, New York, pp. 621–682.

Hahn-Deinstrop, E. (1992). Stationary phases, sorbents. *J. Planar Chromatogr.—Mod. TLC 5*: 57–61.

Hahn-Deinstrop, E. (1993). TLC plates: initial treatment, prewashing, activation, conditioning. *J. Planar Chromatogr.—Mod. TLC 6*: 313–318.

Hajouj, Z., Thomas, J., and Siouffi, A.M. (1995). Two-dimensional TLC on mechanically blended silica based bonded phases. Evaluation of the behaviors of C_8-diol mixtures using polynuclear aromatic hydrocarbons and comparison with C_{18}-cyano mixtures. *J. Liq. Chromatogr. 18*: 887–894.

Hajouj, Z., Thomas, J., and Siouffi, A.M. (1996). Mechanically blended silica-based bonded phases. Application to the simple modulation of the polarity of the stationary phase to TLC. *J. Liq. Chromatogr. Relat. Technol. 19*: 2419–2422.

Halpaap, H., and Krebs, K.-F. (1977). Thin layer chromatographic and high performance thin layer chromatographic ready-for-use preparations with concentrating zones. *J. Chromatogr. 142*: 823–853.

Halpaap, H., and Ripphahn, J. (1977). Performance, data, and results with various chromatographic systems and various detection patterns in high performance thin layer chromatography. *Chromatographia 10*: 613–623, 643–650.

Hauck, H. E., and Halpaap, H. (1980). HPTLC on inert pre-coated layers of cellulose and silica 50000. *Chromatographia 13*: 538–548.

Hauck, H. E., and Jost, W. (1990a). Application of hydrophilic modified silica gels in TLC. *Am. Lab. 22*(8): 42, 44–45, 47–48.

Hauck, H. E., and Jost, W. (1990b). Sorbent materials and precoated layers in thin layer chromatography. In *Packings and Stationary Phases in Chromatographic Techniques*, K. K. Unger (Ed.). Chromatographic Science Series, Volume 47. Marcel Dekker, New York, pp. 251–330.

Hauck, H. E., and Mack, M. (1990). Precoated plates with concentrating zones: a convenient tool for thin layer chromatography. *LC-GC 8*(2): 88, 92, 94, 96.

Hauck, H. E., and Mack, M. (1996). Sorbents and precoated layers in thin layer chromatography. In *Handbook of Thin Layer Chromatography*, 2nd ed., J. Sherma and B. Fried (Eds.). Marcel Dekker, New York, pp. 101–128.

Hauck, H. E., Junker-Buchheit, A., Wenig, R., and Schoemer, S. (1995). Good laboratory practice (GLP) in chromatography. *LaborPraxis*: 70–72, 109–110.

Hauck, H. E., Junker-Buchheit, A., and Wenig, R. (1993). Reproducibility of silica gel 60 precoated plates in TLC and HPTLC. *GIT Fachz. Lab. 37*: 973–974, 976–977.

Hauck, H. E., Mack, M., Reuke, S., and Herbert, H. (1989). Hydrophilic surface modified HPTLC precoated layers in pesticide analysis. *J. Planar Chromatogr.—Mod. TLC 2*: 268–275.

Heilweil, E., and Rabel, F. M. (1985). Reversed-phase TLC on C2, C8, C18, and diphenyl bonded phases. *J. Chromatogr. Sci. 23*: 101–105.

Hung, C. H., Strickland, D. K., and Hudson, B. G. (1977). Estimation of molecular weights of peptides and glycopeptides by thin layer gel filtration. *Anal. Biochem. 80*: 91–100.

Husain, S.W., Avanes, A., and Ghoulipour, V. (1996). Thin layer chromatography of metal ions on cerium (III) silicate: a new ion-exchanger. *J. Planar Chromatogr.—Mod. TLC 9*: 67–69.

Jost, W., and Hauck, H. E. (1983). High performance thin layer chromatographic precoated plates with amino modification and some applications. *J. Chromatogr. 261*: 235–244.

Jost, W., and Hauck, H. E. (1987). The use of modified silica gels in TLC and HPTLC. *Adv. Chromatogr. 27*: 129–165.

Khan, M. U., and Williams, J.P. (1993). Microwave-mediated methanolysis of lipids and activation of thin layer chromatographic plates. *Lipids 28*: 953–955.

Kovacs-Hadady, K., and Balazs, E. (1993). Thermoanalytical study of the impregnation of silica gel layers. *J. Planar Chromatogr.—Mod. TLC 6*: 463–466.

Kovacs-Hadady, K., and Varga, T. (1995). A systematic study of impregnation of thin layers with different ion-pairing reagents and by different methods. *J. Planar Chromatogr.—Mod. TLC 8*: 292–299.

Kriz, D., Kriz, C. B., Andersson, L. I., and Mosbach, K. H. (1994). Thin layer chromatography based on the molecular imprinting technique. *Anal. Chem. 66*: 2636–2639.

Lepri, L. (1995). Reversed phase planar chromatography of enantiomeric compounds on microcrystalline triacetyl cellulose. *J. Planar Chromatogr.—Mod. TLC 8*: 467–469.

Li, T.-S., Li, J.-T., and Li, H.-Z. (1995). Modified and convenient preparation of silica impregnated with silver nitrate and its application to the separation of steroids and triterpenes. *J. Chromatogr., A 715*: 372–375.

Mack, M., and Hauck, H.E. (1989). Separation of enantiomers by thin layer chromatography. *J. Planar Chromatogr.—Mod. TLC 2*: 190–193.

Malinowska, I., and Rozylo, J. K. (1991). Chitin, chitosan, and their derivatives as stationary phases in thin layer chromatography.*J. Planar Chromatogr.—Mod. TLC 4*: 138–141.

Mukherjee, K. D. (1996). Applications of flame ionization detectors in thin layer chromatography. In *Handbook of Thin Layer Chromatography*, 2nd ed., J. Sherma and B. Fried (Eds.). Marcel Dekker, New York, pp. 361–372.

Musial, B., and Sherma, J. (1998). Determination of the sunscreen 2-ethylhexyl-*p*-methoxycinnamate in cosmetics by reversed phase HPTLC with ultraviolet absorption densitometry on preadsoebent plates. *Acta Chromatogr. 8*: 5–12.

Nakamura, H., Pisano, J.J., and Tamura, Z. (1979). Thin layer chromatography using a layer containing a mixture of fluorescent additives. Application to amino acid phenylthiohydantoins and other derivatives. *J. Chromatogr. 175*: 153–161.

Okumura, T. (1980). Sintered thin layer chromatography. *J. Chromatogr. 184*: 37–78.

Omori, T., Okamoto, M., and Yamada, F. (1983). Cyanoalkyl-bonded HPTLC plates and their chromatographic characteristics. *J. High Resolut. Chromatogr. Chromatogr. Commun. 6*: 47–48.

Petrovic, M., Kastelan-Macan, M., Turina, S., and Ivankovic, V. (1993). Quantitative determination of magnesium in aluminum alloys by ion exchange TLC. *J. Liq. Chromatogr. 16*: 2673–2684.

Pietrzyk, J., Rotsch, T. D., and Leuthauser, S. W. Ch. (1979). Applications of the porous Amberlite XAD copolymers as stationary phases in TLC. *J. Chromatogr. Sci. 17*: 555–561.

Poole, C. F., and Poole, S. K. (1991). *Chromatography Today*. Elsevier, Amsterdam, Chap. 7, pp. 649–734.

Poole, C. F., and Poole, S. K. (1995). Multidimensionality in planar chromatography. *J. Chromatogr., A 703*: 573–612.

Robards, K., Haddad, P.R., and Jackson, P. E. (1994). *Principles and Practice of Modern Chromatographic Methods*. Academic Press, San Diego, CA, pp. 179–226.

Rossler, H. (1969). Aluminas and other inorganic adsorbents. In *Thin Layer Chromatography. A Laboratory Handbook*, 2nd ed., E. Stahl (Ed.). Springer-Verlag, New York, pp. 23–32.

Rozylo, J. K., Malinowska, I., and Musheghyan, V. (1989). Separation process of amino acids on chitin in TLC.*J. Planar Chromatogr.—Mod. TLC 2*: 374–377.

Shibukawa, M., Oguma, K., and Kuroda, R. (1978). TLC of various β-diketones and their metal complexes on Styragel 60A.*J. Chromatogr. 166*: 245–252.

Snyder, L. R. (1975). Adsorption. In *Chromatography*, 3rd ed., E. Heftmann (Ed.). Reinhold, New York, pp. 46–76.

Stahl, E., and Mueller, J. (1980). New spreader for the preparation of gradient layers in TLC. *J. Chromatogr. 189*: 293–297.

Steverle, H. (1966). Separation and determination of aromatic sulfonic acids by chromatography on Sephadex G25. *Z. Anal. Chem. 220*: 413–420.

Szepesi, G., and Nyiredy, Sz. (1996). Pharmaceuticals and Drugs. In *Handbook of Thin Layer Chromatography*, 2nd ed., J. Sherma and B. Fried (Eds.). Marcel Dekker, New York, pp. 819–876.

Waksmundzki, A., Pryke, P., and Dawidowicz, A. (1979). Thin layer gel permeation chromatography on a porous glass bead support.*J. Chromatogr. 168*: 234–240.

Wang, K.-T., and Weinstein, B. (1972). Thin layer chromatography on polyamide layers. In *Progress in Thin Layer Chromatography and Related Methods*, A. Niederwieser, and G. Pataki (Eds.), Vol. 3. Ann Arbor Science Publishers, Ann Arbor, MI, pp. 177–231.

Witherow, L., Thorp, R. J., Wilson, I. D., and Warrander, A. (1990). Comparison of diol bonded silica gel plates with underivatized silica gel. *J. Planar Chromatogr.—Mod. TLC 3*: 169–172.

Wollenweber, P. (1969). Organic adsorbents. In *Thin Layer Chromatography. A Laboratory Handbook*, 2nd ed., E. Stahl (Ed.). Springer-Verlag, New York, pp. 32–41.

Yoshioka, M., Araki, H., Kobayashi, M., Kaneuchi, F., Seki, M., Miyazaki, T., Utsuki, T., Yaginuma, T., and Kakano, M. (1990). Porous glass sheets for use in thin layer chromatography. *J. Chromatogr. 515*: 205–212.

4

Obtaining Material for TLC and Sample Preparation

I. OBTAINING MATERIAL

A. Human

For teaching purposes, human blood, saliva, urine, or feces are suitable and easily available material for TLC experiments. One- to 10-μl samples of serum or urine can often be spotted directly for various TLC analyses to show the presence of hydrophilic and lipophilic compounds. Saliva can be applied directly to the preadsorbent spotting area of precoated silica gel plates and analyzed by TLC (Touchstone et al., 1982; Zelop et al., 1985). Fecal material is useful to demonstrate bile pigments and various sterols using relatively simple techniques. Serum, urine, and feces are often "cleaned up" (purified) using sample preparation techniques described in Section II of this chapter. Considerable TLC literature is available on natural products from humans [for instance, see Zollner and Wolfram (1969) or Scott (1969)]. Introductory level students are often interested in the results of TLC analyses obtained from human samples. Moreover, abnormal samples of blood or urine may be available from pathology departments at local hospitals. These samples may be useful to detect by TLC such abnormal conditions as elevated lipid levels in the blood (hyperlipemia), or elevated levels in the urine (aminoacidurias) or particular amino acids associated with pathologic conditions. Analyses by simple TLC procedures will usually detect these disorders in abnormal urine and serum samples. An interesting and readily available source of human material is skin surface exudate, which can be obtained from the forehead with the aid of a cotton swab. The swab containing the lipids can be extracted in chloroform-methanol (2:1) and then analyzed by TLC for skin lipids as described by Downing (1979). Jain (1996) has provided useful information for obtaining human samples for use in TLC determinations in clinical chemistry.

B. Warm-Blooded Animals

The above-mentioned samples can be obtained from warm-blooded laboratory animals such as rats, mice, and domestic chickens, which are often maintained in biology departments. Laboratory animals provide a good source of organs that are not readily available from humans. Information is available, particularly on the TLC of lipids, for brain, liver, kidney, and blood of laboratory mice and rats (Kates, 1986; Christie, 1982; Mangold, 1969). Numerous books are available on handling warm-blooded laboratory animals [see Griffith and Farris (1942); Green (1966); Sturkie (1965)]. A simple and useful laboratory manual that describes routine methods for handling common laboratory animals is that of MacInnis and Voge (1970). For routine bleeding purposes, blood may be obtained following venipuncture, ear puncture, cardiac puncture, or after cervical flexure or decapitation procedures. A book that describes methods of obtaining and preparing tissues and fluids from warm-blooded animals is that of Paul (1975). Tissues and fluids to be used for TLC should be obtained from freshly killed and prepared animals. Precautions for the preparation and handling of samples are given in Kates (1986) and Christie (1982). Fried and Haseeb (1996) have provided information for obtaining material for TLC from warm-blooded animals. Methods useful for obtaining skin secretions from warm-blooded vertebrates for use in TLC have been reviewed by Weldon (1996).

C. Cold-Blooded Animals

Cold-blooded animals include fish, frogs, salamanders, snakes, turtles, and invertebrates. Much less TLC work has been done on samples from these animals compared with material from humans and warm-blooded animals. Therefore, cold-blooded vertebrate and invertebrate samples provide a good source of material to initiate student research projects. Most biology departments maintain frogs for teaching exercises in physiology, and tropical fish for display, teaching, or research purposes. Additionally, a variety of living invertebrates, such as crayfish, cockroaches, clams, snails, and planarians, are often maintained in laboratories. Excellent guides for the maintenance, handling, and processing of cold-blooded animals are those of MacInnis and Voge (1970), Welsh et al. (1968), and Needham et al. (1937).

Moreover, animals such as frogs, cockroaches, and snails are usually parasitized with an array of protozoa (single-celled organisms) and helminths (parasitic worms). These parasites can easily be removed from hosts (MacInnis and Voge, 1970) and processed for TLC in a manner similar to that of host tissues or organs. As determined by TLC, the profile of some substances from the host may differ from that of the parasite (Fried and Butler, 1977).

Cold-blooded animals can provide a source of blood or hemolymph (blood of invertebrates is usually referred to as hemolymph) for TLC studies, which can be used to compare work done on human or mammalian blood. Hemolymph from invertebrates can be obtained easily by crushing an invertebrate such as a snail, clam, or insect. More sophisticated methods of obtaining hemolymph can also be used. Hoskin and Hoskin (1977) have described a method for withdrawing hemolymph from the adductor muscle of the clam, *Mercenaria mercenaria*, and then processing the fluid for further TLC studies. Loker and Hertel (1987) described a method for obtaining hemolymph from the heart of *Biomphalaria glabrata* snails. Shetty et al. (1990) used various chromatographic procedures to analyze sterols from cardiac blood samples of the *B. glabrata* snail. Relatively few studies of invertebrate hemolymph constituents by TLC are available. Fried and Sherma (1990) have reviewed the literature on thin-layer chromatographic analysis of lipids in the hemolymph of snails (Gastropoda). TLC studies on the analysis of hemolymph from mosquitoes (Insecta) can provide a useful model for future work (Mack and Vanderberg, 1978). Fried and Haseeb (1996) have provided information for obtaining material for TLC from cold-blooded animals. Methods useful for obtaining skin secretions from cold-blooded vertebrates for use in TLC have been reviewed by Weldon (1996).

D. Microbial Organisms

Organisms such as bacteria, viruses, yeast, protozoa, algae, and fungi are arbitrarily grouped as microbial organisms. Some of these organisms are considered as protists or prokaryotes. For additional classification information on these microbial organisms, see Burdon and Williams (1964). Although microbial organisms can be collected in the wild, it is best to obtain pure species or strains from suppliers. Protozoa and algae are available from Carolina Biological Supply, Wards Natural Science Establishment, and Turtox. An excellent source for bacteria, fungi, and yeasts is the American Type Culture Collection (ATCC). Daggett and Nerad (1992) published a list of sources of strains of microorganisms for teaching and research purposes.

Bacteria, bacteriophage, lyophilized cultures, and fungal cultures are available from Presque Isle Cultures and Carolina Biological Supply Co. Difco microbiological culture media can be purchased through Fisher and VWR. Other microbiological supplies can be purchased from the International Institute of Biochemical and Biomedical Technology, Chicago, Illinois. Knowledge of aseptic procedures and in vitro culture techniques as given in Taylor and Baker (1978) are necessary prior to handling microbial organisms for TLC. Contamination of pure cultures with spurious organisms may invalidate the results of TLC analyses of microorganisms. TLC can be used as a tool for taxonomy, and studies on the chemotaxonomy of certain microbial organisms are available. TLC has been used to differentiate among mycobacterial species (Tsukamura and Mizuno, 1975) species of yeast (Kaneko et al., 1976), and strains of the parasitic protozoan, *Blastocystis hominis* (Keenan et al., 1992).

Fungal cultures that grow as floating mycelial mats on liquid media are easy to filter (Buchner funnel). The resultant filtrate is usually clear and provides synthetic products, that is, auxin along with the enzyme for auxin degradation from *Omphalia flavida* (Sequeria and Steeves, 1954; Ray and Thimann, 1956) and gibberellic acid from *Fusarium moniliforme* (Moore, 1974). These filtrates can be used for subsequent TLC analyses. Many microbes grow suspended in the nutrient medium of choice. The organisms can be filtered or separated from the medium in various ways, such as by low-speed centrifugation. Both the growth medium and the organisms that are collected as a pellet can be prepared for TLC analysis (Clark, 1964). Because of the tough outer coat of many microbes, special equipment, such as homogenizers or sonicators, are employed to free cell contents for further analysis. The use of TLC for studies in bacteriology has been reviewed recently by Maloney (1996).

E. Plant Materials

Plant materials can be collected in the wild, or purchased from local greenhouses or supermarkets. Additionally, biology departments often maintain greenhouses for their individual needs. Examination of a Carolina Biological Supply House catalog will show an array of plant material available from most major groups. Live cultures can also be purchased from Triarch., Inc. A wide variety of plant tissue culture supplies are available through Sigma Co. and Fisher Scientific Co. Plant material, particularly leaves, provide an important starting point for pigment analyses. TLC studies on chloroplast and other pigments are discussed in Strain and Sherma (1969) and Sherma et al. (1992). Leaf material is often used as a starting point for the extraction of phenolic substances. Such substances provide chemotaxonomic information about hybrids and closely related species. For a study on the chemotaxonomy of members of the genus *Prunus*, see Muszynski and Nybom (1969). Materials other than leaves are used for TLC analytical work, and such studies include analyses of roots, stems, seeds, and fruits. For instance, carpel tissue (fruit) of the yellow wax pepper (*Capsicum frutescens*) was analyzed by means of two-dimensional thin-layer chromatography to examine changes in chlorophyll and other pigments during the ripening of the fruit (Schanderl and Lynn, 1966). Tissues of higher plants differ from those of microorganisms in that they contain considerable amounts of phenolic substances, particularly in woody tissues. Plant tissues are often extracted by boiling them in aqueous ethanol (Harborne, 1984). Several groups of chemical compounds are extracted by such treatment, including organic acids, amino acids, photosynthetic pigments, and sugars. The ethanol extract can be further processed for analysis of each of the constituents. Harborne (1984) has provided a general procedure for extracting fresh plant tissues and fractionating them into different classes according to polarity. Some of the extracts can be analyzed further by TLC. Harborne's procedure is shown in Figure 4.1.

Organic acids such as citric, malic, acetic, and oxalic are commonly contained in plant cell vacuoles of fresh leaves and stems. Hot alcohol extracts are filtered and concentrated prior to spotting on TLC plates (Harborne, 1984).

Auxins can be extracted from fresh shoot tips or from fresh roots. Roots

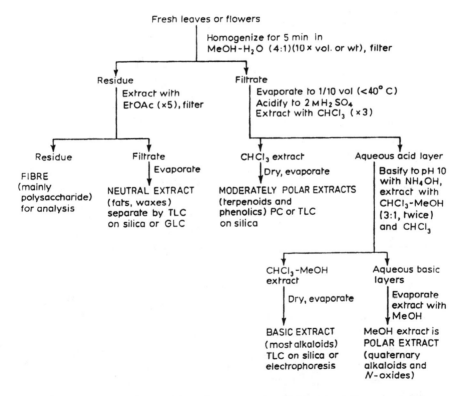

FIGURE 4.1 A general procedure for extracting fresh plant tissues and fractioning them into different classes according to polarity. [From Harborne (1984). Reproduced with permission of Chapman and Hall Ltd.]

washed free of soil and debris are frozen with dry ice and ground in methanol using a mortar and pestle. Particles are removed from the extract by centrifugation or cheesecloth filtration. The extract is concentrated, reconstituted in ether, and applied to TLC plates (Harborne, 1984). Pothier (1996) has recently reviewed the use of TLC in the plant sciences.

F. Concluding Remarks

Materials can be obtained by collecting organisms in the wild, or by purchasing plants and animals from biological suppliers or other dealers. Material usually can be obtained from workers who have published on research done with the desired organism. Therefore, a review of recent literature is a way of determining where an organism may be obtained. Exchange of materials with colleagues is a possibility. For several years, one of us (B. F.) has been sending fresh water snails (shipped on moist filter paper) to colleagues in the United States and abroad. These snails have been used for parasitological and TLC analyses. Fried (1997) and Irwin (1997) have provided useful information about obtaining materials for various research purposes.

Farm ponds, lakes, and rivers provide excellent sites for the collection of aquatic invertebrates and plants. Special permits, federal, state, or local, may be needed to collect vertebrates, and even some invertebrates. Consultation with federal and local game wardens may be necessary prior to collecting.

Mention has already been made of the Carolina Biological Supply House, which maintains and sells a variety of living animals and plants. Additional companies that provide living organisms are Champlain Biological Company and Ann Arbor Biological Center. A good source of marine invertebrates is Woods Hole Biological Company on the east coast of the United States and Sea-Life Supply Company on the west coast. For those near intertidal areas, marine invertebrates can be collected in this habitat.

Various suppliers, such as Grand Island Biological Company (GIBCO) and the United States Biochemical Company (USBC), provide media and tissue culture products that can be analyzed by TLC. Media obtained from suppliers may be supplemented with other products; for example, defined medium may be supplemented with egg yolk as a lipid source (Butler and Fried, 1977). Analysis of the defined medium by TLC before and after supplementation is possible. Some companies, such as Supelco, Applied Science, and Cal-Biochem, supply authentic standards that are suitable for TLC work as received or following dilution with appropriate solvents. Natural products may also be useful for preparing standards. However, the purity of such standards should be suspect. Substances such as lard (a good source of triacylglycerols) or egg yolk (a good source of triacylglycerols, cholesterol, and sterol esters) provide useful and inexpensive sources of crude TLC standards.

II. STORAGE, PRESERVATION, AND FREEZING

Tissues or fluids should be processed for TLC as soon as possible after samples are removed from plants or animals. If this is not possible, samples should be stored overnight in a refrigerator at 2–5°C, or for longer periods in a freezer at −20°C. Preservatives such as formaldehyde or neutral buffered formalin should not be added to samples because they can alter the characteristics of the sample and produce anomalous results during subsequent TLC analysis.

Chemically clean glassware is necessary for handling tissues or fluids prior to TLC to avoid sample contamination. For lipid work, glassware that has already been cleaned in sulfuric acid is often further cleaned in chloroform–methanol (2:1) prior to use. Plastic or metal containers may leach substances into a sample and should be avoided in handling materials prior to and during the TLC process. Glass vials, jars, and bottles are recommended along with aluminum foil or Teflon-lined lids.

Particular caution must be taken with lipid samples because they are easily autooxidized or contaminated. Use of the antioxidant, butylated hydroxytoluene (BHT), is recommended (10 mg of BHT to 100 ml of solvent) as an additive to lipid samples that are to undergo prolonged storage. Whenever possible, lipids should be handled and processed under nitrogen; they should not be stored in a dry state. Lipids are best stored with a small volume of a suitable solvent under nitrogen and maintained in a freezer.

The presence of certain components in a sample following TLC analyses may indicate contamination or degradation of the sample. For instance, significant amounts of methyl esters rarely occur in animal tissues (Lough et al., 1962), and their presence may indicate a preparation artifact. The neutral lipid standard, 18-4A (Nu-Chek), contains methyl oleate in the mixture. When this standard is used along with fluid or tissue samples, the chromatographer can check for the presence of methyl esters in the sample.

Solvents used during sample preparation or extraction may contain contaminants that can alter subsequent TLC results. Solvents can be evaporated to small volumes and then spotted or streaked on a TLC plate to check their purity.

An extract of a blank can be used to determine the validity of a TLC observation. For example, if a saline solution is supplemented with hen's egg yolk, and this mixture is extracted and then spotted on a TLC plate, neutral lipid TLC procedures should detect mainly triacylglycerols and cholesterol. If the saline alone is used as the blank and is treated in an identical manner, the saline should be lipid negative as shown by TLC.

To determine the reliability of TLC, replicate experiments (at least two or more replicates) should be carried out with different representative samples. For each experiment, at least two TLC runs are made. When consistent spots are found, the data can usually be considered representative of the sample.

III. SAMPLE PREPARATION

A. General Considerations

Samples for TLC are dissolved in a suitable solvent prior to application to a TLC sheet or plate. Volumes between 1 and 10 µl are usually applied to the origin. These volumes should contain an amount of compound that can be detected by the method of choice. Sensitivities of detection methods are generally in the range of 0.1–1 µg. If the compound of interest is present in low concentration in a complex sample matrix such as biological materials, extraction, isolation, and concentration steps normally precede TLC. These steps may take considerable time but are essential to the success of the TLC procedure because of the adverse effects of impurities on the chromatography.

As an example, the determination of amino acids in urine or neutral lipids in serum can be approached in a number of different ways, depending on the loading capacity of the layer and the sensitivity of the detection method. Urine, saliva, or serum can sometimes be directly spotted on the layer and compounds of interest separated with no additional cleanup steps. More often, compounds of interest must be extracted from the sample; the extract may then require further cleanup by liquid–liquid partition or column chromatography; the final extract is then concentrated prior to TLC.

The types of sample preparation steps needed prior to TLC are similar to those required for gas chromatography and high-performance column liquid chromatography. These steps may include extraction, solvent partitioning, column chromatography, desalting, deproteinization, centrifugation, evaporation, and dissolving of residues. These techniques are discussed below.

B. Direct Spotting of Samples

If the compound of interest is present in sufficient concentration and purity, it may be possible to spot the sample directly without further extraction or cleanup procedures. The concentration should be such that the normal spotting volume of 1–10 µl for TLC will give a detectable zone. Impurities must not retain the compound of interest at the origin, distort the shape (i.e., tailing), or change the R_F (higher or lower) of the spot. As an example of this approach, Sherma and Gray (1979) determined caffeine in cola, tea, and coffee by direct spotting of these beverages. One to 6 µg of standard caffeine were spotted on a silica gel F layer from a 2-µl Microcap micropipet along with either 10 µl of the cola sample or 5 µl of instant coffee or tea dissolved in water. Development with ethyl acetate-methanol (85:15) separated caffeine as a discrete zone at $R_F = 0.6$. Most of the remainder of sample material stayed as residue at the origin. Determination of the caffeine involved densitometric scanning in the fluorescence quench mode (Chapter 10).

C. Spotting on Plates with a Preadsorbent Zone

In place of conventional thin layers, layers with a lower spotting and cleanup strip to retain impurities and an upper adsorbent area of silica gel for the analytical separation can be used. Samples are applied directly to the plate, or after a crude cleanup procedure. For example, Linear K (Whatman) preadsorbent silica gel plates incorporate a 3-cm-wide diatomaceous earth (kieselguhr) strip in the spotting area. These plates allow for fast, diffuse application of relatively large volumes of crude sample material. As the solvent passes over the preadsorbent, material to be separated is moved to the solvent front and applied to the origin of the silica gel as a concentrated band. For these layers to be effective, the preadsorbent area must be nonadsorbent for the compounds to be separated.

D. In Situ Extraction on a Conventional Thin Layer

A similar effect can be achieved after crude samples (e.g., urine or serum) are spotted on conventional layers and then predeveloped for a distance of 1–2 cm with a polar solvent. After drying, the extracted and concentrated substance is then developed a second time for a distance of 10 cm with a suitable but less polar solvent. Interfering materials (e.g., proteins) are left behind at the original origin.

   An example of the in situ extraction procedure used to determine lipid profiles in rat plasma and liver homogenate was reported by Kupke and Zeugner (1978). One-half microliter of plasma from capillary blood or liver homogenate was applied to a 10×20-cm HP silica gel layer. Samples were applied to the origin over a 15-μl spot of absolute methanol, and each spot was covered immediately with a few microliters of additional methanol. After drying in a stream of cold air, the plate was developed with chloroform–methanol–water (65:30:5) twice, each time for a distance of 3.7 cm, and then with hexane-diethyl ether–acetic acid (80:20:1.5) to within 1 cm of the top of the plate. Fluorescent spots were then produced by treating the plates with $(NH_4)HCO_3$ followed by heating for 10 min at 150°C. The lipid spots obtained were sharp and reproducible.

E. Dissolving and Simple Extraction Procedures

When the compound of interest is a major constituent of the sample, simply dissolving the sample in an appropriate amount of a suitable solvent followed by direct spotting of a portion of the sample may be adequate for TLC. As an example of this approach, Sherma and Beim (1978) separated caffeine in APC tablets on C_{18} reversed-phase layers containing a fluorescent indicator. The tablet was extracted in 25 ml of chloroform–methanol (1:1), and 1-μl portions of the extract were spotted on the plate along with caffeine standards. After development with methanol–0.5M NaCl (1:1), aspirin, phenacetin, and caffeine were resolved; the compound of major interest, caffeine, had an R_F of 0.5 and was quantified by in situ densitometry. Although the filler in the APC tablet did not dissolve, this did not cause problems during TLC analysis.

Simple extractions can be used to separate a single tissue sample into major chemical fractions (Graff, 1964). Starting with 1 g or less of a fresh biological tissue sample, 10 ml of 70% ethanol should be added and the sample allowed to sit overnight; the supernatant fluid should be decanted and saved for analyses of free pool compounds such as amino acids, organic acids, and simple carbohydrates. The tissue should then be homogenized in 10 ml of 70% ethanol and centrifuged. The precipitate will contain lipids, polysaccharides, proteins, and nucleic acids. The supernatant should be decanted and combined with the original free pool material. Additional cleanup may be needed, that is, liquid–liquid partition, column chromatography, desalting, to separate the various free pool constituents prior to TLC.

F. Complex Extraction Procedures—Partition of Extracts

Extracts are often too impure for direct spotting and require partitioning with an immiscible solvent or solvents as a cleanup step. The principle behind such differential partitioning is to leave impurities behind in one solvent while extracting the compound of interest into another. This method of cleanup may require considerable knowledge of wet chemistry, trial and error, and a good knowledge of the constituents of the sample. A review of various liquid–liquid partitioning systems has been presented by Sherma (1980).

As an example of a complex extraction procedure, the use of chloroform–methanol (2:1) extraction and the Folch wash will be given. This classical technique was first described by Folch et al. (1957). There are now numerous modifications of this original procedure as explained in Christie (1982) and in Phillips and Privett (1979). In this procedure, a tissue sample of 1 g or less or a fluid sample of 1 ml or less is extracted in 20 ml of chloroform–methanol (2:1). If protein is present in the sample, it will precipitate and should be removed by centrifugation or filtration through glass wool. The residue should be rinsed with additional chloroform–methanol (2:1) and the supernatant combined with that from the original extraction. To achieve a partitioning effect, either distilled water or a dilute salt solution, that is, 0.88% KCl, is added at a volume of 25% of the supernatant. The addition of the aqueous phase, the so-called Folch wash, allows for a biphasic separation in which most hydrophilic materials, that is, amino acids, simple carbohydrates, salts, organic acids, are in the top layer (mainly a methanolic–water layer) and most lipophilic material, that is, simple and complex lipids, are in the bottom or lipophilic layer (the main organic solvent constituent of this layer is chloroform). The two layers can be separated in a separatory funnel, by decantation, or by pipeting off the top from the bottom layer. As mentioned in Section III.E of this chapter, additional cleanup of both hydrophilic and lipophilic layers may be necessary prior to TLC.

G. Column Chromatography

The use of column chromatography can be advantageous for sample cleanup prior to TLC. The use of a Bio-Rad 11A8 (Bio-Rad Laboratories) ion-retardation resin combines desalting and removal of peptides from urine samples prior to amino acid TLC. Urine is passed into the column and eluted with distilled water. Interfering salts and peptides are retained on the column (Heathcote, 1979). Columns packed with silica gel or Florisil (Floridin) can be used for cleanup of lipid samples with minimal autooxidation effects on the sample. For instance, if complex lipids (phospholipids) are to be separated from simple lipids (neutral lipids), use of a silicic acid or Florisil column as an adsorbent may be helpful. After the lipid sample is applied to the column, elution with chloroform will separate the simple from the complex lipids. The complex lipids may then be recovered by elution with methanol. Column chromatography was used to clean up the lipid extract of the parasitic nematode, *Onchocerca gibsoni*, prior to TLC analysis of the worm glycolipids (Maloney and Semprevivo, 1991). Because phospholipids interfered with the analysis of glycolipids, it was necessary to isolate the glycolipids from the other lipid classes. The glycolipids were acetylated prior to isolation by Florisil column chromatography. The purified glycolipids were then deacylated and analyzed by high-performance thin-layer chromatography (HPTLC).

H. Solid-Phase Extraction (SPE)

Solid-phase extraction is a relatively new sample preparation method based on chemically modified silica gel or other sorbent type packed into a small plastic disposable cartridge or column. The cartridges are usually connected to the end of a syringe and samples passed through by pressure, and the columns and disks are normally used in a vacuum manifold. Sorbents with various functional groups can be obtained in different configurations and sizes from commercial suppliers, which can supply procedures for analyses and bibliographies of applications.

SPE is applied to liquid samples. If the sample contains components that can interfere with SPE, preliminary steps may be necessary, such as filtration to eliminate particulate matter; liquid–liquid extraction to remove oils, fats, or lipids; ion exchange, use of a desalting column, dialysis, or passage through a nonpolar sorbent to remove inorganic salts; removal of proteins by pH modification, denaturing with chaotropic agents or organic solvents, precipitation with acid, adding a compound that competes for binding sites, or use of restricted-access media (Boos and Rudolphi, 1997); or dilution of viscous samples with a compatible solvent.

The sorbent in the cartridge or column and the analyte eluent are selected with consideration for type of sample, the retention characteristics of the analyte(s), and the composition of the sample matrix. The following mechanisms and phases are used:

1. Reversed phase (C_{18}, C_8, C_4, C_2, phenyl, diphenyl, cyclohexyl, cyanopropyl) for nonpolar analytes (alkyl, aromatic) in polar solutions (water, aqueous buffers), with high to medium polarity eluents (methanol, acetonitrile, water)
2. Normal phase (silica, Florisil, alumina adsorbents and bonded phase such as diol, cyano, amino, diamino) for polar analytes in nonpolar solvents (hexane, toluene, dichloromethane), with low to medium polarity eluents (hexane, dichloromethane, ethyl acetate, acetone, acetonitrile)
3. Cation exchange (sulfonic acid, carboxylic acid) for compounds with positively charged functional groups (amines) in aqueous, low ionic strength solution, with buffer eluents (acetate, citrate, phosphate)
4. Anion exchange (quaternary amine, DEAE, amino) for compounds with negatively charged functional groups (organic acids) in aqueous, low ionic strength solutions, with buffer eluents (phosphate, acetate).

In addition to these four sorbent types, other bonded phases (e.g., sulfonylpropyl, diethylaminopropyl, N-propylethylaminediamine), polymers (styrene divinylbenzene, polyvinylpyrrolidine), chelating resins, and specialty phases for specific compounds, such as columns for analyzing drugs of abuse in urine, pesticides according to standard EPA methods, or aldehydes and ketones from air, are commercially available. Reversed-phase sorbents have been prepared with different chemistries to provide unique selectivities—for example, light-loaded, non-endcapped C_{18}, monofunctional or trifunctional C_{18}, and polar, non-endcapped C_8.

The selected sorbent is conditioned with a solvent of the same type as the sample loading solvent, that is, a nonpolar solvent (e.g., hexane, dichloromethane) for adsorbents and normal-phase-bonded phases, a polar solvent (acetonitrile, methanol, or THF followed by water or buffer) for reversed-phase packings, or a low ionic strength buffer (~0.01 M) for ion exchangers). Excess conditioning solvent is removed but the sorbent is not allowed to dry out. The sample is loaded in a weak solvent, and interferences are eluted with a weak eluent, followed by the analytes with a strong eluent. For reversed-phase sorbents, polar solvents are weak, nonpolar solvents are strong, and the most polar analytes elute first; for adsorbents and normal-phase-bonded sorbents, the polarity and elution orders are reversed. The analytes are collected in the smallest possible eluent volume, which is evaporated to concentrate the solution before TLC, if necessary. This strategy, in which the analyte is retained and the matrix interferences pass through unretained, is generally chosen when the analytes are present at low levels, or multiple components of widely differing polarities need to be isolated by selective elution. In some cases, conditions may be chosen to retain matrix components while the analytes pass through the sorbent without retention; this strategy is usually chosen when the analyte is present in high concentration. Typical flow rates are 1–10 ml/min for normal and reversed phase and 1–2 ml/min for ion exchange.

SPE is primarily used to extract and concentrate analytes from dilute solution (trace enrichment), or to clean up solvent extracts (Lautie and Stankovic, 1996). Other secondary uses are to exchange phases to eliminate emulsions, carry out solid-phase derivatizations, and store samples taken remotely for transport to the laboratory (Majors, 1997). Much smaller volumes of solvent are required for SPE compared to conventional liquid–liquid extraction in a separatory funnel or cleanup using large chromatography columns, saving the cost of purchase and disposal of solvents. SPE columns are usually eluted with relatively little solvent, which minimizes evaporation required to concentrate samples prior to TLC. A number of eluents with increasing strength may be used to isolate compounds of different polarities in separate fractions; also, multiple columns of different types can be connected in series to optimize cleanup and fractionation. Other advantages of SPE compared to conventional liquid–liquid extraction include ease of operation, speed, the ability to produce cleaner extracts, and possibility of automation using flow-processing manifolds or robotics (Robards et al., 1994). See Sherma (1996) for a more detailed description of SPE systems for different classes of samples and analytes. The Guide to Sample Preparation, published by *LC-GC Magazine* in 1997, is an excellent source of information on sample preparation strategies, including SPE, for chromatographic analysis.

Figure 4.2 shows a commercial cartridge SPE system. Cartridges are available in various sizes and shapes from different manufacturers containing sorbent weights of 35 mg to 10 g, with 100 and 500 mg probably being most commonly used. The disposable cartridge, packed with silica gel, C_{18} bonded silica gel (reversed phase), or other sorbent, is fitted to a syringe with the plunger removed. Sample is poured into the barrel of the syringe, and the plunger is replaced and pressed to force the sample through the cartridge under pressure. The process is repeated with an appropriate solvent or solvents to purify and elute the extracted solutes. On silica gel, the sample is applied and nonpolar impurities are eluted with a nonpolar solvent. The solute of interest is then removed with a more polar solvent, just polar enough for its elution. Very polar impurities remain on the cartridge. On a C_{18} cartridge, polar impurities are eluted with a polar solvent such as water and then the analyte is removed by a solvent sufficiently nonpolar for its elution. Very nonpolar impurities remain on the cartridge. On ion exchange cartridges, the most weakly ionized components elute first; to elute retained com-

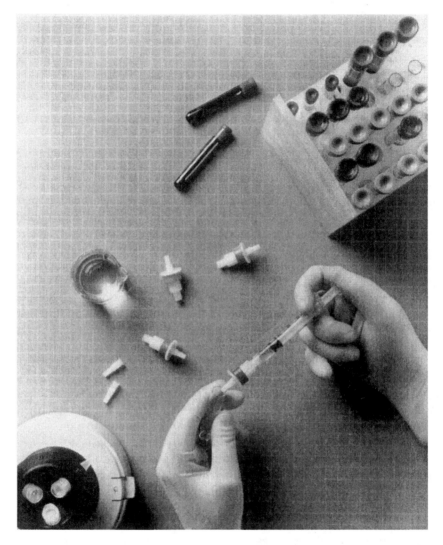

FIGURE 4.2   A Sep-Pak cartridge in use. The cartridge is fitted on the Luer tip of a syringe barrel. (Photograph courtesy of Millipore, Milford, MA.)

ponents, ionic strength is increased or pH is changed (increased for anion exchange or decreased for cation exchange).

Figure 4.3 shows a vacuum manifold used to simultaneously process up to 12 SPE columns; a 24-port version is also available. Columns are available in a variety of sizes, e.g., with column volumes ranging from 1 to 6 ml and

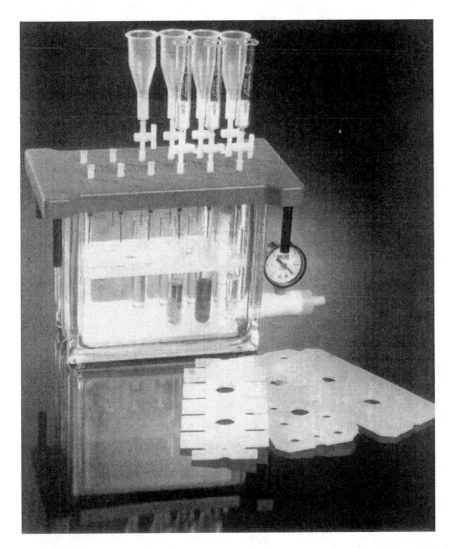

FIGURE 4.3 The BAKER spe-12G Column Processor (Solid Phase Extraction) system. (Photograph courtesy of J. T. Baker, Phillipsburg, NJ.)

sorbent weights from 100 to 1000 mg. Solvents and samples are aspirated through the columns by means of the vacuum manifold, and sample eluates are collected in tubes placed in a removable rack inside the manifold. Collected eluates may be analyzed directly for qualitative work, or evaporated and reconstituted in a known volume of solvent for quantitative analysis.

The packing and use of SPE columns in the laboratory were described in earlier editions of this book. This material will not be repeated here because almost all analysts now make use of commercial columns, which are more convenient and reproducible. There may be a small number of applications for which the previously described homemade columns could have selectivity advantages over available commercial columns.

The newest form of SPE makes use of extraction disks, which are especially useful for rapid, large volume extractions of water samples. High flow rates (e.g., 200 ml/min) are possible compared to SPE columns or cartridges, and clogging by samples containing suspended solids is minimized. Applications of extraction disks have included analyses of drinking water, process streams, ground water, and solid waste. Different types of disks are commercially available, including sorbent particles (10 μm compared to ~40 μm for SPE cartridges and columns) embedded in an expanded Teflon network (Empore) or fiberglass matrix (Supelco), or a bed of micro-particles of sorbent sandwiched between two borosilicate glass fiber filter mats with graded porosity and supported in a cartridge housing (Bakerbond Speedisk). The disks are flat, usually 1 mm or less in thickness, with diameters ranging from 4 to 96 mm. Sorbents supplied in the form of disks include C_{18}, C_8, strong acid anion exchanger, divinylbenzene, and specialty products such as Bakerbond PolarPlus, for slightly polar and nonpolar hydrocarbons from oil and grease. Figure 4.4 shows the J.T. Baker Diskmate II Rotary Extraction Station, which is a vacuum manifold system for simultaneous processing of as many as six samples of up to one liter volume.

SPE combined HPTLC has been especially widely used for determinations of multi-pesticide residues in water, most often with use of a C_{18} or aminopropyl extraction column and automated multiple development (AMD) of an HP silica gel plate (Pfaab and Jork, 1994; Stan and Butz, 1995; Butz and Stan, 1995; Mouratidis and Thier, 1995). In other applications of SPE-TLC, pentachlorophenol and cymiazole were determined in water and honey (Sherma and McGinnis, 1995), N-methylcarbamate insecticides in water (McGinnis and Sherma, 1994), atrazine and simazine in drinking and surface waters (Zahradnickova et al., 1994), metribuzin and metabolites in soil and water (Johnson and Pepperman, 1995), pesticides in contaminated soil (Bladek et al., 1996), and pentachlorophenol in leather goods (Fischer et al., 1996). Most applications of SPE with TLC have employed columns or cartridges rather than disks. The use of cartridges with a syringe has been found to be more difficult than columns in a manifold for quantitative TLC analyses involving accurately measured volumes of samples and eluents.

King et al. (1992) described a unique combination extraction/sample application procedure in which a disk was placed into a hole in a TLC sheet to determine the presence of cannabinoid (THC) metabolite in human urine. Using this procedure, hydrolyzed urine samples were aspirated directly through a small,

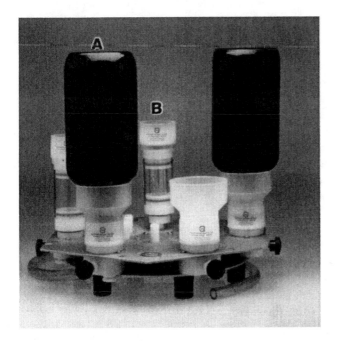

Figure **4.4** DISKMATE II Rotary Extraction Station configured to illustrate sample application (A) and elution (B) steps: (A) a 185-ml reservoir bottle is fitted into a BAKERBOND Speedisk extraction disk with a 50-mm diameter × 1 mm high sorbent bed for sample application; (B) the disk is attached to a collection chamber for the elution step. (Photograph courtesy of J. T. Baker, Phillipsburg, NJ.)

porous disk composed of bonded silica impregnated within a glass fiber matrix. The THC metabolite was retained by the bonded silica adsorbent through the process of solid-phase extraction. The disk was inserted into a hole in the silica gel/glass fiber TLC plate for subsequent TLC development and detection. Greater future use of SPE disks for preparing samples for TLC analysis is inevitable.

I. Supercritical Fluid Extraction (SFE)

Supercritical fluid extraction (SFE) has been used increasingly since the mid-1980s for extracting components of interest from solid samples prior to chromatographic analyses. Liquids can be placed in an extraction thimble and extracted successfully, but procedures are more difficult (Majors, 1997). Extractions with a supercritical fluid are generally faster compared to conventional extraction with liquid solvents (e.g., Soxhlet or ultrasonic); other advantages include reductions in operating costs, solvent usage and waste, time, labor, space, and glassware compared with many traditional extraction approaches. Supercritical carbon dioxide, which has been the most widely used extraction fluid, extracts nonpolar and moderately polar compounds efficiently, but polar compounds are not fully recovered. Methanol has been added to carbon dioxide as a modifier to allow extraction of more polar analytes than carbon dioxide alone at reasonable pressures. Variables that must be optimized for successful SFE include the choice of fluid, pressure, temperature, time, and sample size; fluid density, which is related to solvating power, can be controlled by changing extraction temperature and pressure.

SFE has been carried out in static (sample and fluid sealed in an extraction vessel), dynamic (fluid flows through the vessel), off-line (analytes are collected for later analysis), or on-line (analytes are directly transferred to a coupled instrument, such as a gas chromatography) modes. The principle of the dynamic, off-line method is as follows: sample is placed in a flow-through container and supercritical fluid is passed through the sample; after depressurization, extracted analyte is collected in solvent or trapped on an adsorbent, followed by desorption by rinsing with solvent (Majors, 1997). In some cases, such as analysis of fatty foods for pesticides, SFE results in recovery of both analytes and unwanted contaminants, and column chromatography steps must be employed for sample cleanup. For a discussion of the modes and equipment used in SFE, see Robards (1994). Disadvantages of SFE include the expense and limitations of currently available commercial instruments and lengthy method development time due to the large number of experimental parameters.

In general, SFE has been most widely applied in environmental (e.g., PAHs and pesticides in soils and sludges), food (cholesterol in fats), natural product (medicinal extraction), pharmaceutical (drug materials in tablets and tissues), and polymer (additive extraction) sample preparation. Particular applications of SFE combined with TLC include the determination of multi-pesticide residues in soil using carbon dioxide with methanol modifier (Koeber and Niessner, 1996), aloin and aloe-emodin in aloe leaf products (Kiridena et al., 1995), and the automated determination of hydroperoxides in combustion aerosols (Esser and Klockow, 1994). In the latter analysis, a dichloromethane solution containing the hydroperoxides, with 0.01 M citric acid in methanol as modifier, was loaded on glass fiber filters. The hydroperoxides were extracted with CO_2 (20 MPa), equivalent to 60 ml/min of gaseous CO_2, at 35°C for 2 min in the static mode, followed by 20 min in the dynamic mode. Samples from the cell impinged onto silica gel plates via a fused silica constrictor held at 70°C within a stainless steel jacket. Temelli (1992) used supercritical fluid extraction to extract lipids from canola flakes. Phospholipids used as emulsifiers were sparingly soluble in supercritical $CO_2(SC\text{-}CO_2)$ but could be recovered with the addition of ethanol as an entrainer. A temperature of 70°C and a pressure of 62 MPa plus 5% ethanol gave maximal yield of phospholipids. TLC was used to show the presence of phospholipids in the SC-CO$_2$–ethanol extracts.

J. Derivatization

The preparation of derivatives for TLC has been reviewed by Gasparic (1961) and Edwards (1980). Derivatization has been used in the TLC of amino acids where certain amino acids are easily converted into the so-called DANS forms (*d*imethyl *a*mino *n*aphthalene *s*ulfonyl) (Pataki and Niederweiser, 1967). These authors have studied a variety of amino acid derivatives on silica gel. The use of both pre- and post-chromatographic derivatization in TLC applications in food analysis has been discussed (Shalaby, 1996). For a recent update on the use of derivatization in TLC, see Cserhati and Forgacs (1996).

Edwards (1980) illustrated derivatizing reagents in TLC and discussed those reagents that are most useful. He suggested that derivatization techniques can be applied to a wide range of analytical problems including the analysis of amino acids, steroids, drugs, and environmental pollutants. He noted that although derivatization reagents are used mainly after TLC by spraying techniques, there are times when it is advantageous to carry out the derivatization step prior to TLC.

Derivatives are often used when a volatile compound can be changed into a nonvolatile compound, or when the detection sensitivity of the derivative is greater than that of the original compound. Derivatization is advantageous as a cleanup procedure when it is easier to clean up the derivative than the parent compound.

Derivatives are often used in lipid chemistry to prepare fatty acid methyl esters needed to determine the fatty acid composition of lipids. Fried et al. (1992) published a technique on the transesterification of 500-mg samples of snail tissue. Because the technique is applicable to other biological tissues and fluids, it is presented herein. Lipids were extracted from snail bodies with 10 ml of chloroform–methanol (2:1); the extracts were filtered through a plug of glass wool contained in a Pasteur pipet, and nonlipid contaminants were removed by extraction with 8 ml of Folch wash (0.88% aqueous KCl). The lipid-containing lower phase was separated and evaporated just to dryness under a stream of nitrogen at room temperature. The total lipid sample was dissolved in about 30 ml of methanol, and 0.5–1.0 ml of concentrated sulfuric acid was added. The mixture was refluxed for 1 h; the formed fatty acid methyl esters were extracted with 30–40 ml of petroleum ether and the extract dried over anhydrous sodium sulfate. The fatty acid methyl esters were concentrated on a Rotoevaporator at 40°C and the volume reduced to 1 ml. The fatty acid methyl esters can be separated by TLC on silica gel impregnated with silver nitrate (Christie, 1982).

K. Desalting Procedures

Various techniques have been used to eliminate salt, particularly from urine, serum, and tissue culture media. Salts in these fluids may cause streaking, trailing,
or unclear, nonresolved spots during the TLC of amino acids, carbohydrates, and
other hydrophilic substances. For a good review of desalting procedures, see
Heathcote et al. (1971). A simple desalting procedure suitable for 0.1–0.2 ml of
urine, serum, and saline solutions is as follows: A sample is dried under air at
45°C and then desalted by extraction in 1 ml of 0.5% HCl in 95% ethanol for
24 h. The extract is then evaporated to dryness and the residue dissolved in 100
µl of 0.5% HCl in 95% ethanol prior to spotting for TLC (Bailey and Fried,
1977).

   The ion-retardation resin Bio-Rad 11A8 (Bio-Rad Laboratories) is often
used to desalt samples prior to amino acid TLC (Heathcote, 1979). A list of
suggested ion-exchange resins for desalting blood and urine samples prior to
carbohydrate TLC is presented in Scott (1969). A desalting procedure used prior
to amino acid TLC that does not involve resin columns, but instead uses acidified
butanone, is given in Heathcote (1979). A convenient method for desalting samples prior to carbohydrate TLC is by preferential extraction of carbohydrates with
pyridine as described in Scott (1969).

L. Deproteinization

Where protein may interfere with subsequent TLC analyses, it should be removed
(deproteinization procedures). A suitable procedure for an approximate 50-µl
sample of serum involves addition of 100 µl of methanol to precipitate the protein, followed by shaking and centrifugation of the mixture to obtain a clear
supernatant. This technique has been used to deproteinize biological fluids prior
to the determination of drugs in these fluids (Touchstone and Dobbins, 1979).
Other commonly used techniques to precipitate protein in either serum, urine, or
tissue homogenates involve the addition of either trichloroacetic acid, perchloric
acid, or sulfosalicylic acid followed by centrifugation, and then removal of the
supernatant. The supernatant may require further cleanup or be used directly for
TLC. Further information on deproteinization can be found in Jain (1996).

M. Evaporation of Solutions

Sample preparation procedures usually include steps to concentrate or evaporate
samples to or near dryness. Care must be taken to concentrate the sample without
losing or degrading the compound of interest. A common method of evaporating
sample volumes in excess of approximately 25 ml is by using a rotary evaporator
with an attached round-bottom flask. For descriptions of roto-evaporators, see
Kates (1986). For sample volumes of less than 25 ml, simple nitrogen blow-

down procedures are adequate. A suitable nitrogen blow-down setup for handling numerous samples is described in Kates (1986). A simple procedure for nitrogen blow-down involves attaching a shell vial containing the sample to a ring stand in a fume hood. The vial is placed in a beaker of water maintained on a hot plate set at no higher than 45°C. A disposable pipet attached via rubber tubing to a nitrogen tank is inserted in the shell vial. Nitrogen gas bubbled slowly into the vial will effect evaporation of the solution. A similar setup can be used with compressed air instead of nitrogen if oxidation of the sample is not a problem during the evaporation procedure.

N. Dissolving of Evaporated Residues

Following evaporation of solutions, the residue is dissolved (reconstituted) in a known volume of an appropriate solvent. All or a portion of the reconstituted solution can be spotted on the TLC plate. The least polar solvent in which the compound of interest is soluble should be used for reconstitution. In this way, some additional purification can be achieved if some highly polar impurities are left undissolved in the residue. Very polar solvents are difficult to remove from the sorbent during application of the sample. If a small amount of solvent is retained on the sorbent after sample application, it can adversely affect the separation by altering the size or shape of the spot, interfering with the migration of the compound of interest, or altering the R_F value. Also, caution must be used in drying the sample because excessive hot air applied to the origin could alter or otherwise decompose labile substances on the surface of an active sorbent.

REFERENCES

Bailey, R. S., Jr., and Fried, B. (1977). Thin layer chromatographic analyses of amino acids in *Echinostoma revolutum* (Trematoda) adults. *Int. J. Parasitol.* 7: 497–499.

Bladek, J., Rostkowski, A., and Miszczak, M. (1996). Application of instrumental thin layer chromatography and solid phase extraction to the analysis of pesticide residues in grossly contaminated samples of soil. *J. Chromatogr., A 754*: 273–278.

Boos, K.-S., and Rudolphi, A. (1997). The use of restricted access media in HPLC, Part I—Classification and review. *LC–GC 15*: 602, 604, 606–611.

Burdon, K. L., and Williams, R. P. (1964). *Microbiology.* Macmillan, New York.

Butler, M. S., and Fried, B. (1977). Histochemical and thin layer chromatographic analyses of neutral lipids in *Echinostoma revolutum* metacercariae cultured in vitro. *J. Parasitol. 63*: 1041–1045.

Butz, S., and Stan, H.-J. (1995). Screening of 265 pesticides in water by thin layer chromatography with automated multiple development. *Anal. Chem. 67*: 620–630.

Christie, W. W. (1982). *Lipid Analysis—Isolation, Separation, Identification and Structural Analysis of Lipids*, 2nd ed. Pergamon Press, Oxford.

Clark, J. M., Jr. (1964). *Experimental Biochemistry.* Freeman, San Francisco.

Cserhati, T., and Forgacs, E. (1996). Introduction to techniques and instrumentation. In *Practical Thin-Layer Chromatography—A Multidisciplinary Approach*. B. Fried and J. Sherma (Eds.). CRC Press, Boca Raton, Fl., pp. 1–18.

Daggett, P. M., and Nerad, T. A. (1992). Sources of strains for research and teaching. In *Protocols in Protozoology*, J. J. Lee and A. T. Soldo (Eds.). Allen Press, Lawrence, KS, pp. E1.1–E1.4.

Downing, D. T. (1979). Lipids. In *Densitometry in Thin Layer Chromatography—Practice and Applications*, J. C. Touchstone and J. Sherma (Eds.). Wiley, New York, pp. 367–391.

Edwards, D. J. (1980). Derivative formation for TLC. In *Thin Layer Chromatography: Quantitative Environmental and Clinical Applications*, J. C. Touchstone (Ed.). Wiley, New York, pp. 51–79.

Esser, G., and Klockow, D. (1994). Detection of hydroperoxides in combustion aerosols by supercritical fluid extraction coupled to thin layer chromatography. *Mikrochim. Acta 113*: 373–379.

Fischer, W., Bund, O., and Hauck, H. E. (1996). Thin layer chromatographic analysis of phenols on TLC aluminum sheet RP-18F_{254}. *Fresenius' J. Anal. Chem. 354*: 889–891.

Folch, J., Lees, M., and Sloane-Stanley, G. H. (1957). A simple method for the isolation and purification of total lipids from animal tissue. *J. Biol. Chem. 226*: 497–509.

Fried, B. (1997). An overview of the biology of trematodes. In *Advances in Trematode Biology*. B. Fried and T. K. Graczyk (Eds.). CRC Press, Boca Raton, Fl., pp. 1–30.

Fried, B., and Butler, M. S. (1977). Histochemical and thin layer chromatographic analyses of neutral lipids in metacercarial and adult *Cotylurus* sp. (Trematoda: Strigeidae). *J. Parasitol. 63*: 831–834.

Fried, B., and Haseeb, M. A. (1996). Thin-layer chromatography in Parasitology. In *Practical Thin-Layer Chromatography—A Multidisciplinary Approach*. B. Fried and J. Sherma (Eds.). CRC Press, Boca Raton, Fl., pp. 51–70.

Fried, B., and Sherma, J. (1990). Thin layer chromatography of lipids found in snails (Gastropoda: Mollusca). *J. Planar Chromatogr.—Mod. TLC 3*: 290–299.

Fried, B., Rao Sundar, K., and Sherma, J. (1992). Fatty acid composition of *Biomphalaria glabrata* (Gastropoda Planorbidae) fed hen's egg yolk versus leaf lettuce. *Comp. Biochem. Physiol. 101A*: 351–352.

Gasparic, J. (1961). Paper chromatography of organic substances after their conversion into derivatives. *Chem. Listy. 55*: 1439–1443.

Graff, D. J. (1964). Metabolism of C^{14}-glucose by *Moniliformes dubius* (Acanthocephala). *J. Parasitol. 51*: 72–75.

Green, E. L. (Ed.). (1966). *Biology of the Laboratory Mouse*, 2nd ed. McGraw-Hill, New York.

Griffith, J. Q., and Farris, E. J. (Eds.). (1942). *The Rat in Laboratory Investigation*. Lippincott, Philadelphia.

Harborne, J. B. (1984). *Phytochemical Methods—A Guide To Modern Techniques of Plant Analysis*, 2nd ed., Chapman and Hall, London.

Heathcote, J. G. (1979). Amino acids. In *Densitometry in Thin Layer Chromatography—Practice and Applications*, J. C. Touchstone and J. Sherma (Eds.). Wiley, New York, pp. 153–179.

Heathcote, J. G., Davies, D. M., and Haworth, C. (1971). An improved technique for the analysis of amino acids and related compounds on thin layers of cellulose. V. The quantitative determination of urea in urine. *J. Chromatogr. 60*: 103–109.

Hoskin, G. P., and Hoskin, S. P. (1977). Partial characterization of the hemolymph lipids of *Mercenaria mercenaria* (Mollusca: Bivalvia) by thin-layer chromatography and analysis of serum fatty acids during starvation. *Biol. Bull. 152*: 373–381.

Irwin, S. W. B. (1997). Excystation and cultivation of trematodes. In *Advances in Trematode Biology*. B. Fried and T. K. Graczyk (Eds.). CRC Press, Boca Raton, Fl., pp. 57–86.

Jain, R. (1996). Thin-layer chromatography in clinical chemistry. In *Practical Thin-Layer Chromatography—A Multidisciplinary Approach*. B. Fried and J. Sherma (Eds.). CRC Press, Boca Raton, Fl., pp. 131–152.

Johnson, R. M., and Pepperman, A. B. (1995). Analysis of metribuzin and associated metabolites in soil and water samples by solid phase extraction and reversed phase thin layer chromatography. *J. Liq. Chromatogr. 18*: 739–753.

Kaneko, H., Hosohara, M., Tanaka, M., and Itoh, T. (1976). Lipid composition of 30 species of yeast. *Lipids 11*: 837–840.

Kates, M. (1986). Techniques of lipidology—Isolation, analysis and identification of lipids. In *Laboratory Techniques in Biochemistry and Molecular Biology*, 2nd ed., R. H. Burdon and P. H. van Knippenberg (Eds.). Elsevier, Amsterdam, pp. 1–464.

Keenan, T. W., Huang, C. M., and Zierdt, C. H. (1992). Comparative analysis of lipid composition in axenic strains of *Blastocystis hominis*. *Comp. Biochem. Physiol. 102B*: 611–615.

King, D. L., Fukui, P., Schultheis, S. K., and Donnell, C. M. (1992). Evaluation of a thin-layer chromatographic method for the detection of cannabinoid (THC) metabolite using a microcolumn/disk extraction technique. In *Recent Developments in Therapeutic Drug Monitoring and Clinical Toxicology*, I. Sunshine (Ed.). Marcel Dekker, New York, pp. 521–525.

Kiridena, W., Poole, S. K., Miller, K. G., and Poole, C. F. (1995). Determination of aloin and aloe-emodin in aloe consumable products by thin layer chromatography. *J. Planar Chromatogr.—Mod. TLC 8*: 416–419.

Koeber, R., and Niessner, R. (1996). Screening of pesticide-contaminated soil by supercritical fluid extraction and high performance thin layer chromatography with automated multiple development. *Fresenius' J. Anal. Chem. 354*: 464–469.

Kupke, I. R., and Zeugner, S. (1978). Quantitative high-performance thin-layer chromatography of lipids in plasma and liver homogenates after direct application of 0.5 µl samples to the silica gel layer. *J. Chromatogr. 146*: 261–271.

Lautie, J.-P., and Stankovic, V. (1996). Automated multiple development TLC of phenylurea herbicides in plants. *J. Planar Chromatogr.—Mod. TLC 9*: 113–115.

Loker, E. S., and Hertel, L. A. (1987). Alterations in *Biomphalaria glabrata* plasma induced by infection with the digenetic trematode *Echinostoma paraensei*. *J. Parasitol. 73*: 503–513.

Lough, A. K., Felinski, L., and Garton, G. A. (1962). The production of methyl esters of fatty acids as artifacts during the extraction or storage of tissue lipids in the presence of methanol. *J. Lipid Res. 3*: 478–479.

MacInnis, A. J., and Voge, M. (1970). *Experiments and Techniques in Parasitology*. Freeman, San Francisco.

Mack, S. R., and Vanderberg, J. P. (1978). Hemolymph of *Anopheles stephensi* from noninfected and *Plasmodium berghei*-infected mosquitoes. 1. Collection procedure and physical characteristics. *J. Parasitol. 64*: 918–923.

Majors, R. E. (1997). Sample preparation in analytical chemistry (organic analysis). In *Handbook of Instrumental Techniques for Analytical Chemistry*, F. Settle (Ed.). Prentice Hall PTR, Upper Saddle River, NJ, pp. 17–53.

Maloney, M. D. (1996). Thin-layer chromatography in bacteriology. In *Practical Thin-Layer Chromatography—A Multidisciplinary Approach*. B. Fried and J. Sherma (Eds.). CRC Press, Boca Raton, Fl., pp. 19–31.

Maloney, M. D., and Semprevivo, L. H. (1991). Thin-layer and liquid column chromatographic analyses of the lipids of adult *Onchocerca gibsoni. Parasitol. Res. 77*: 294–300.

Mangold, H. K. (1969). Aliphatic lipids. In *Thin Layer Chromatography*, 2nd ed., E. Stahl (Ed.). Springer-Verlag, New York, pp. 361–421.

McGinnis, S. C., and Sherma, J. (1994). Determination of carbamate insecticides in water by C-18 solid phase extraction and quantitative HPTLC. *J. Liq. Chromatogr. 17*: 151–156.

Mouratidis, S., and Thier, H.-P. (1995). Solid phase extraction for the confirmation of results in polar pesticides residue analysis by HPTLC. *Z. Lebensm.-Unters.-Forsch. 201*: 327–330.

Moore, T. C. (1974). *Research Experiences in Plant Physiology—A Laboratory Manual*. Springer-Verlag, New York.

Muszynski, S., and Nybom, N. (1969). The relationships between *Prunus avium, Prunus mahaleb* and *Prunus fontanesiana*, studied by means of thin layer chromatography. *Acta Soc. Bot. Polon. 38*: 507–515.

Needham, J. G., Galtsoff, P. S., Lutz, F. E., and Welch, P. S. (1937). *Culture Methods for Invertebrate Animals*. Dover, New York.

Pataki, G., and Niederwieser, A. (1967). Thin-layer chromatography of nucleic acid bases, nucleosides, nucleotides and related compounds. IV. Separation on PEI-cellulose layers using gradient elution and direct fluorometry of spots. *J. Chromatogr. 29*: 133–141.

Paul, J. (1975). *Cell and Tissue Culture*, 5th ed. Churchill Livingstone, Edinburgh.

Pfaab, G., and Jork, H. (1994). Application of AMD for the determination of pesticides in drinking water. Part 3. Solid phase extraction influencing factors. *Acta Hydrochim. Hydrobiol. 22*: 216–223.

Phillips, F., and Privett, O. S. (1979). A simplified procedure for the quantified extraction of lipids from brain tissue. *Lipids 14*: 590–595.

Pothier, J. (1996). Thin-layer chromatography in plant sciences. In *Practical Thin-Layer Chromatography—A Multidisciplinary Approach*. B. Fried and J. Sherma (Eds.). CRC Press, Boca Raton, Fl., pp. 33–49.

Ray, P. M., and Thimann, K. V. (1956). The destruction of indole-acetic acid. *Arch. Biochem. Biophys. 64*: 175–192.

Robards, K., Haddad, P. R., and Jackson, P. E. (1994). *Principles and Practice of Modern Chromatographic Methods*, Academic Press, San Diego, CA, pp. 407–455.

Schanderl, S. H., and Lynn, D. Y. C. (1966). Changes in chlorophylls and spectrally related pigments during ripening of *Capsicum frutescens. J. Food Sci. 31*: 141–145.

Schwartz, D. P. (1985). Quantitative colorimetric method for sulfamethazine in swine feeds. *J. Assoc. Off. Anal. Chem. 68*: 214–217.

Scott, R. M. (1969). *Clinical Analysis by Thin-Layer Chromatography Techniques.* Ann Arbor Science Publishers, Ann Arbor, MI.

Sequeria, L., and Steeves, T. A. (1954). Auxin inactivation and its relation to leaf drop caused by the fungus *Omphalia flavida. Plant Physiol. 29*: 11–16.

Shalaby, A. R. (1996). Thin-layer chromatography in food analysis. In *Practical Thin-Layer Chromatography—A Multidisciplinary Approach.* B. Fried and J. Sherma (Eds.). CRC Press, Boca Raton, Fl., pp. 169–192.

Sherma, J. (1980). Sample preparation for quantitative TLC. In *Thin Layer Chromatography—Quantitative Clinical and Environmental Applications*, J. C. Touchstone and D. Rogers (Eds.). Wiley-Interscience, New York, pp. 17–35.

Sherma, J. (1996). Basic techniques, materials and apparatus. In *Handbook of Thin Layer Chromatography*, 2nd ed., J. Sherma and B. Fried (Eds.). Marcel Dekker, New York, pp. 3–47.

Sherma, J., and Beim, M. (1978). Determination of caffeine in APC tablets by densitometry on chemically bonded C_{18} reversed phase layers. *HRC&CC J. High Resolut. Chromatogr. Chromatogr. Commun. 1*: 309–310.

Sherma, J., and Gray, D. (1979). Quantitation of caffeine in beverages by TLC densitometry with direct spotting of samples. *Am. Lab. 11*: 21–24.

Sherma, J., and McGinnis, S. C. (1995). Determination of pentachlorophenol and cymiazole in water and honey by C-18 solid phase extraction and quantitative HPTLC. *J. Liq. Chromatogr. 18*: 755–761.

Sherma, J., O'Hea, C. M., and Fried, B. (1992). Separation, identification, and quantification of chloroplast pigments by HPTLC with scanning densitometry. *J. Planar Chromatogr.—Mod. TLC 5*: 343–349.

Shetty, P. H., Fried, B., and Sherma, J. (1990). Sterols in the plasma and digestive gland-gonad complex of *Biomphalaria glabrata* snails, fed lettuce versus hen's egg yolk, as determined by GLC. *Comp. Biochem. Physiol. 96B*: 791–794.

Stan, H.-J., and Butz, S. (1995). Multimethod applying AMD-TLC analysis to drinking water. *Chem. Plant Prot. 12*: 197–216.

Strain, H. H., and Sherma, J. (1969). Modifications of solution chromatography illustrated with chloroplast pigments. *J. Chem. Educ. 46*: 476–483.

Sturkie, P. D. (1965). *Avian Physiology*, 2nd ed. Cornell University Press, Ithaca, NY.

Taylor, A. E. R., and Baker, J. R. (1978). *Methods of Cultivating Parasites in Vitro.* Academic Press, London.

Temelli, F. (1992). Extraction of triglycerides and phospholipids from canola with supercritical carbon dioxide and ethanol. *J. Food Sci. 57*: 440–442.

Touchstone, J. C., and Dobbins, M. F. (1979). Steroids. In *Densitometry in Thin Layer Chromatography—Practice and Applications*, J. C. Touchstone and J. Sherma (Eds.). Wiley, New York, pp. 633–659.

Touchstone, J. C., Hansen, G. J., Zelop, C. M., and Sherma, J. (1982). Quantitation of cholesterol in biological fluids by TLC with densitometry. In *Advances in Thin*

Layer Chromatography—Clinical and Environmental Applications, J. C. Touchstone (Ed.). Wiley, New York, pp. 219–228.

Tsukamura, M., and Mizuno, S. (1975). Differentiation among mycobacterial species by
thin-layer chromatography. *Int. J. System. Bacteriol. 25*: 271–280.

Weldon, P. J. (1996). Thin-layer chromatography of the skin secretions of vertebrates. In
Practical Thin-Layer Chromatography—A Multidisciplinary Approach. B. Fried
and J. Sherma (Eds.). CRC Press, Boca Raton, Fl., pp. 105–130.

Welsh, J. H., Smith, R. I., and Kammer, A. E. (1968). *Laboratory Exercises in Invertebrate
Physiology*, 3rd ed. Burgess, Minneapolis, MN.

Zahradnickova, H., Simek, P., Horicova, P., and Triska, J. (1994). Determination of atrazine and simazine in drinking and surface waters by solid phase extraction and high
performance thin layer chromatography. *J. Chromatogr., A 688*: 383–389.

Zelop, C. M., Koska, L. H., Sherma, J., and Touchstone, J. C. (1985). Quantification of
phospholipids and lipids in saliva and blood by thin layer chromatography with
densitometry. In *Techniques and Applications of Thin Layer Chromatography*,
J. C. Touchstone and J. Sherma (Eds.). Wiley, New York, pp. 291–304.

Zollner, N., and Wolfram, G. (1969). TLC in clinical diagnosis. In *Thin Layer Chromatography*, 2nd ed., E. Stahl (Ed.). Springer-Verlag, New York, pp. 1–6.

5

Application of Samples

I. PREPARATION OF SAMPLES FOR APPLICATION

Prior to TLC, samples are dissolved in a suitable solvent and then applied, usually in 1–10-μl volumes, to the origins of a TLC sheet or plate (0.1–1 μl for HPTLC). When the concentration of the solute is known—for instance, when standards are used—the solution applied is usually at a concentration of 0.01–1 μg/μl. When using a complex mixture such as a biological sample, the concentration of the solute of interest is often not known. If the amount of sample applied is too small, the compound of interest may not be detected. Conversely, if too much sample is applied, the material may not be adsorbed throughout the whole thickness of the layer but only on the surface, leading to sample overloading and streaked or trailing zones. Therefore, the compounds of interest may not be resolved following the TLC process. Trial and error is needed when applying samples to a TLC plate to determine the optimal concentration for spotting, which is usually governed by the method of detection employed. Typically, detection methods have sensitivities in the microgram or nanogram range, but detection of picogram amounts is sometimes possible (see Table 2.1).

The selection of a solvent for application of the sample can be a critical factor in achieving reproducible chromatography with distortion-free zones. In general, the application solvent should be a good solvent for the sample and should be as volatile and weak as possible. For silica gel TLC, a weak solvent is nonpolar; for reversed-phase TLC, it is polar. High volatility promotes solvent elimination from the initial zone which, if not complete, could cause distorted zones and altered R_F values, but if volatility is too great, the solvent will undergo partial evaporation before application from the capillary. Highly viscous samples may not fill capillary pipets or empty from them completely. Too strong a solvent can cause preliminary circular chromatographic development of the sample at the origin, especially when sample increments are spotted on top of each other, leading to distorted zones. It is also important, particularly with reversed-phase layers, that the application solvent wet the sorbent so that the sample penetrates the layer.

II. PREPARATION OF THE PLATE PRIOR TO SPOTTING

The top of the plate should be marked lightly with a pencil, and this should be the only area handled to avoid placing fingerprints on the plate. The use of disposable plastic gloves during the handling of plates is a good idea to avoid introducing extraneous substances onto the active surface of the sorbent. Prewashing is carried out to the top of the plate in chloroform–methanol (1:1) or in the development solvent prior to use. This will clean up extraneous material contained in the layer. This procedure is particularly recommended when compounds of interest may migrate at or near the solvent front during the development process, and for quantitative procedures where high contrast between the zone and background is important for precise and sensitive scanning. Plates that are not precleaned may contain considerable extraneous material at or near the region of the solvent front.

Precoated plates exposed to high humidity or kept on hand for long periods may be activated by placing them in an oven at 70–110°C for 30 min prior to spotting. Trial and error will be necessary to determine if prewashing and activation procedures are needed for a particular separation (see Chapter 11).

The origin line is usually located 1.5–2.5 cm from the bottom of a plate (1.0 cm for HPTLC). The distance between the the mobile-phase entry position and the origin line has a considerable influence on the efficiency of the chromatography, with 10 mm being optimum for TLC and 6 mm for HPTLC (Cserhati and Forgacs, 1996). The origin line along with the solvent front line can be marked using a soft pencil with the aid of a template or spotting guide available from several manufacturers; care must be taken to avoid damaging the layer. Spots should be applied in a straight horizontal line so that R_F values are comparable. The solvent front is marked as a line parallel to the origin and from 10 to 15 cm from the origin (5–7 cm for HPTLC). A 10-cm solvent front distance is convenient when R_F values are to be calculated.

III. SAMPLE APPLICATION FOR TLC AND HPTLC

A. General Considerations

Sample application is a critical step for obtaining good resolution and quantification in TLC (Kaiser, 1988). The sample should be completely transferred from the applicator to the layer and form a compact initial zone, and the application procedure must not damage the layer. It is important to rinse the application device thoroughly with pure solvent between the various samples and standards to avoid cross-contamination, and to then rinse with the solution to be spotted to avoid dilution with solvent.

Manual spotting methods, which involve the use of a variety of applicators, are most frequently employed in qualitative TLC. Care should be taken not to dig the applicator into the surface of the sorbent so as to avoid abnormally shaped spots following the development process.

The application of small initial zones during the spotting procedure will produce compact spots following the separation procedure. The application of large spots to the origins of a TLC plate will result in diffuse, nonresolvable spots following development. Spot size is minimized by applying the sample volume in small increments on top of each other, with complete drying of the solvent after each application. Successive spotting requires additional time and increases chances for damage to the layer from gouging with the applicator and spot deformation during chromatography. However, application of large volumes in a smaller zone can greatly enhance resolution.

Initial zones must be completely dry prior to development. Solvent evaporation may be aided by use of a hairdryer, stream of nitrogen, or drying in a fume hood or oven at low temperature. Nitrogen is recommended to dry labile compounds that might decompose if subjected to a stream of air. Heat should be avoided for drying spots unless it is known that decomposition will not occur. Most workers rely on visual observation of solvent evaporation to indicate that drying is complete and development can be started. However, many solvents, such as ethyl acetate, acetone, and alcohols, require drying times that are longer than visual evaporation times. Therefore, to achieve reproducible results, it is important to standardize drying conditions of samples before development. Typical conditions are air drying in a fume hood for 10 min, or blowing with hot air or cold air or nitrogen.

One of the great advantages of TLC is that many samples can be applied on the origins of a 20 × 20-cm plate or sheet. If spot size is kept to a minimum, a plate may accommodate up to 19 samples. In practice, spot size for qualitative TLC is usually from 3 to 6 mm in diameter (1–2 mm in HPTLC). A distance of 1–2 cm should be maintained between spots applied to the origin line (5 mm for HPTLC), and placement should be no closer than 1.5–2 cm from the side of the plate to avoid the classical "edge effect" (Chapter 7).

Some chromatographers prefer to apply bands or streaks rather than spots to the origins. In many cases, the application of bands rather than spots will lead to smaller broadening in the direction of development, better zone resolution, and improved instrumental quantification. Overloading the layer, which can lead to trailing, can be minimized by applying bands. A disadvantage of band application is that fewer samples or standards can be applied per plate. For quantitative TLC, short bands matched to the width of the scanner light beam and the pattern of automatic track changes are desirable. These are facilitated by the use of the automated Linomat applicator or preadsorbent plates. In preparative-layer chromatography, the streak may cover a distance of 10–20 cm. Figure 12.1 illustrates an applicator for preparative-layer chromatography.

B. Manual Sample Application

A variety of manual spotters are available that are useful for applying 1–100 µl of sample to a plate. The simplest of these devices are wooden applicator sticks, attenuated Pasteur pipets, and capillary tubes. For more precise delivery of sample, micropipets or microsyringes calibrated in microliters are recommended.

Of these applicators, Drummond Microcap micropipets are probably the most often used. Microcaps (Figure 3.4) are available in a wide range of sizes (from 0.1 to 200 µl). They are disposable, inexpensive, simple to use, and very precise in sample application (Emanuel, 1973). The glass capillary tube is placed through the septum of the support tube, filled with solution, held vertically, and gently touched to the layer. The solution is drawn out of the pipet by the layer contact, forming a round spot. If the solution is not expelled completely, the hole on top of the rubber bulb is covered and the bulb gently squeezed. Each capillary should be used only once to avoid contamination of the following samples. The capillaries fill quickly when dipped into solutions with organic solvents; with aqueous solutions, filling will be much slower and it may be necessary to draw the solution up with the aid of the rubber cap.

Drummond also manufactures a series of glass tube/plunger-type digital microdispensers for application of variable volumes in the 2–1000-µl range. These pipets (Figure 5.1) are especially useful for spotting preadsorbent plates for quantitative TLC.

As stated in Chapter 2, one of the major requirements for HPTLC is that the initial zones be as small as possible (1–2 mm or less) and uniform in size and shape to achieve the maximum possible system efficiency. This has led to the application of submicroliter volumes of highly concentrated solutions for HPTLC. Disposable glass capillaries with 100-, 200-, and 500-nl volumes are available (Analtech) and can be used for manual spotting as described above for microliter microcaps. However, because of their small size, they are more difficult to use, even with the special capillary holder designed for them.

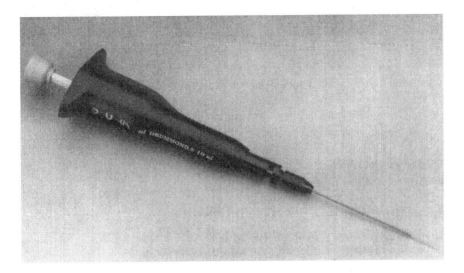

FIGURE 5.1 A 10-µl digital microdispenser. (Photograph courtesy of Drummond Scientific Company.)

TLC plates with preadsorbent spotting area were described in Chapter 3. These plates automatically produce thin streaks of sample at the sorbent–preadsorbent junction (Figure 3.5), leading to increased efficiency and detection sensitivity (Rabel and Palmer, 1992). No special skills are required for sample application with respect to geometry, positioning, or spreading of the initial zones in the preadsorbent area, leading to a considerable saving of time. Possible decomposition or irreversible adsorption of substances on the active silica gel is avoided because the sample contacts this portion of the layer only after being dissolved in the mobile phase during passage through the preadsorbent area. This is somewhat equivalent to the application of the sample in the mobile phase as normally carried out in column chromatography. Preadsorbent plates are especially advantageous for applying large amounts of sample, very dilute sample solutions, or crude (e.g., biological) samples, and plates with channels or lanes have been widely used for quantitative analysis by scanning densitometry, especially when expensive spotting instruments are not available. Although errors are possible if the sample is not uniformly distributed across the band that is formed at the preadsorbent–layer interface, many excellent results have been reported (e.g., Sherma and Incarvito, 1997). Typical sample volumes are 1–25 µl for TLC. Standardization in quantitative analysis can be carried out by spotting variable volumes of a single solution rather than constant volumes of a series of standards with increasing concentration, as is required on non-preadsorbent plates.

The use of HP layers with preadsorbent spotting area allows application of larger samples, typically 1–10 μl instead of the usual 10–200 nl spotted for HPTLC on nonpreadsorbent layers. The sample collects at the interface as a narrow line that provides the effect of having applied a low-volume sample in a small area. Halpaap and Krebs (1977) demonstrated that for levels of samples between 0.01 and 50 μg, HP preadsorbent plates gave separation numbers that were up to six times higher than on comparable plates without preadsorbent.

A similar concentrating effect is obtained by spotting large samples (1–10 μl) on nonpreadsorbent HPTLC plates and focusing them into a line with a short, preliminary development with a strong solvent. The plate is thoroughly dried prior to the analytical development.

C. Instruments for Sample Application

Although of limited value in qualitative analytical TLC, a variety of automatic or semiautomatic spot and streak applicators can be purchased. These applicators are used most frequently for quantitative TLC and HPTLC, and for preparative TLC. Jaenchen (1996) has reviewed instruments for application of samples for TLC and HPTLC.

Figure 5.2 illustrates the Nanomat III fixed-volume spot applicator. The

FIGURE 5.2 Camag Nanomat III with universal capillary holder. (Photograph courtesy of Camag.)

Nanomat is used with 0.5–5-µl disposable glass micropipets loaded into a metal universal capillary holder using the Capillary Dispenser System. The pipet of choice is filled by dipping it into the sample solution; the Nanomat applicator head is moved to the selected position, and the metal holder with the filled pipet is placed against the applicator head, which is magnetic so as to hold the pipet in position; the applicator head is pushed down against a spring, the pipet touches the layer surface with a constant contact pressure that is determined by the resting frictional force against the permanent magnet, and the sample discharges; after the head is released and rises, the capillary is refilled or replaced, the applicator head is moved to the next application position, and the process is repeated. Controlled positioning facilitates repetitive delivery of larger samples, whereas regulated pressure eliminates layer damage that is virtually inevitable with hand spotting. The volume precision of the Nanomat is <1% when used properly.

Figure 5.3 shows the microprocessor-controlled Camag Linomat IV, which applies samples in the form of narrow bands by a spray-on technique in which nitrogen carrier gas atomizes the sample from a syringe onto the plate, which is moving back and forth under the atomizer. Band length is selectable between 0 (−spot) and 190 mm. Although sample volumes of 1–100 µl can be applied by the Linomat, the sample size most used is 5–20 µl for TLC and HPTLC plates.

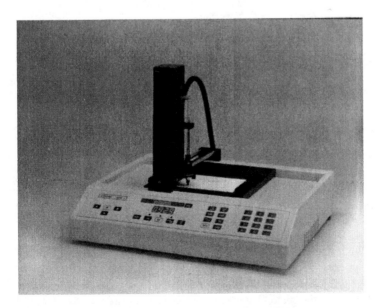

FIGURE 5.3 Camag Linomat IV. (Photograph courtesy of Camag.)

A typical spotting pattern is 10-mm bands, 5 mm apart, for TLC and 6-mm bands, 4 mm apart, for HPTLC. For multilevel calibration in quantitative analysis, different volumes of a single standard can be applied instead of the more laborious procedure of applying identical volumes of different concentrations. The standard addition or spiking technique can be carried out by overspraying spotted unknown samples with different amounts of a standard solution, and, in some cases, prechromatographic derivatization can be carried out in situ by overspraying the applied samples with reagent solution. In addition to the improved resolution realized by application of samples as narrow bands, an advantage of use of the Linomat, because of uniform mass distribution over the full length of the bands, is the ability to scan developed zones by the aliquot technique, which employs a slit of one-half to two-thirds of the length of the band originally applied. This scanning method is claimed (Jaenchen, 1996) to reduce systematic measuring errors in quantitative densitometric evaluation to a significant degree and ensure maximum accuracy. For preparative separations, up to 500 µl can be applied as a narrow band across the entire width of the plate. The Linomat IV-Y is a modified instrument that applies large-volume samples in the shape of rectangles, which are then focused into narrow bands by development for a short distance with a strong solvent. This approach can be useful for samples whose volumes overload and wash away the layer or when matrix concentration is high, leading to deposits that can cause tailing. An inert gas blanket accessory is available for the Linomat IV to protect sensitive samples from air oxidation.

The Camag Automatic TLC Sampler III (Figure 5.4) provides completely automatic, programmable application of 10 nl to 50 µl from septum-covered sample vials held in a 16-position rack. Filling, spotting fixed volumes at a controlled rate, and rinsing of the steel capillary, as well as positioning of the spots for unidirectional linear development, linear development from both sides of a plate, circular development, or anticircular development, are all PC-controlled. Spots are applied by contact transfer and bands by spraying. The delivery capillary moves to selected positions at a lateral speed of 25 cm/sec, and sample dispensing is selectable between 10 and 1000 nl/sec.

The AS-30 (Desaga) is another commercial microprocessor-controlled instrument for application of spots or bands (Figure 5.5). A stream of gas carries the sample from the cannula tip onto the TLC or HPTLC plate, preventing damage to the layer and allowing the application tower to be moved during sample ejection. During the filling process, the dosing syringe is positioned over a tray, which collects rinsing and flushing solvent and excess sample. The sample is injected into the body of the syringe through a lateral opening and, after filling, a stepping motor moves the piston downward to close the fill-port. A second stepping motor moves the tower across the plate. The microprocessor controls the two stepping

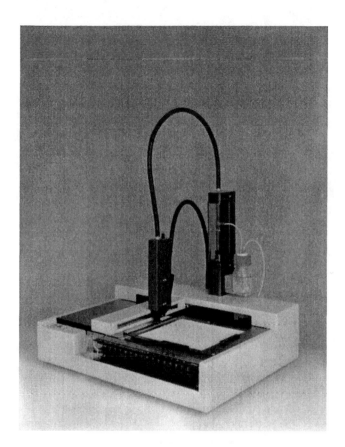

FIGURE 5.4 Camag Automatic TLC Sampler III. (Photograph courtesy of Camag.)

motors and the gas valve; all parameters for the application of up to 30 samples or standards are entered via the keyboard, and 10 different methods can be stored in the computer memory.

These two PC-controlled applicators are closest in function to the automatic injectors available for column HPLC. Combination with a densitometer that is operated by the same PC provides a quantitative system that can collect and document data according to modern GMP/GLP principles (Zieloff, 1995). Applicators of this type would appear to be be an integral part of fully automated, robotic TLC systems, which are under development (Delvorde and Postaire, 1995; Buhlmann et al., 1994; Postaire et al., 1996).

FIGURE 5.5   Desaga AS 30 TLC-Applicator. (Photograph courtesy of Desaga GmbH.)

REFERENCES

Buhlmann, R., Carmona, J., Donzel, A., Donzel, N., and Gil, J. (1994). An open software environment to optimize the productivity of robotized laboratories. *J. Chromatogr. Sci. 32*: 243–248.

Cserhati, T., and Forgacs, E. (1996). Introduction to techniques and instrumentation. In *Practical Thin Layer Chromatography—A Multidisciplinary Approach*, B. Fried and J. Sherma (Eds.). CRC Press, Boca Raton, FL, pp. 1–31.

Delvorde, P., and Postaire, E. (1995). Automation and robotics in planar chromatography. *J. Planar Chromatogr.—Mod. TLC 6*: 289–293.

Emanuel, C. F. (1973). Delivery precision of micro-pipettes. *Anal. Chem. 45*: 1568–1569.

Halpaap, H., and Krebs, K.-F. (1977). Thin layer chromatographic and high performance thin layer chromatographic ready-for-use preparations with concentrating zones. *J. Chromatogr. 142*: 823–853.

Jaenchen, D. (1996). Instrumental thin layer chromatography. In *Handbook of Thin Layer Chromatography*, 2nd ed., J. Sherma and B. Fried (Eds.). Marcel Dekker, New York, pp. 129–148.

Kaiser, R. (1988). Scope and limitations of modern planar chromatography—sampling. *J. Planar Chromatogr.—Mod. TLC 1*: 182–187.

Postaire, E. P. R., Delvorde, P., and Sarbach, C. (1996). Automation and robotics in planar chromatography. In *Handbook of Chromatography*, 2nd ed. J. Sherma and B. Fried (Eds.). Marcel Dekker, Inc., New York, pp. 373–385.

Rabel, F., and Palmer, K. (1992). Advantages of using thin-layer plates with concentrating zones. *Am. Lab. 24*(17): 20BB.

Sherma, J., and Incarvito, C. (1997). Analysis of tablets and caplets containing ketoprofen by normal- and reversed-phase HPTLC with ultraviolet absorption densitometry on preadsorbent plates. *Acta Chromatogr. 7*: 124–128.

Zieloff, K. (1995). Automation in thin layer chromatography and good laboratory practice (GLP). *Oesterr. Chem. Z. 96*: 120–122.

6

Solvent Systems

I. CHOICE OF THE MOBILE PHASE

In thin-layer chromatography, the mobile phase is often called the solvent system, developing solvent, or, simply, the solvent. It has been recommended (Robards et al., 1994) that these terms should not be used interchangeably because the solvent (the liquid in the developing chamber) differs in composition from the mobile phase actually passing through the layer during development. However, this distinction in terms will not be made in this book.

The mobile phase is generally chosen by controlled trial and error based on the analyst's experience and search of the literature. It is usually possible to find a layer–solvent combination already reported in the literature for compounds of interest, or at least very similar compounds. The handbooks by Sherma and Zweig (1972–1989), Sherma (1991–1993), and Sherma and Fried (1996), as well as the chapters on specific compounds in Part II of this book, are especially recommended. For many classes of compounds, general, standardized solvent systems have evolved with time, such as those listed for drug analysis by Sherma (1997). Plate manufacturers usually can supply bibliographies of separations that have been accomplished with their products. As part of the overall chromatographic system, the mobile phase should be chosen to match the nature of the analytes and sorbent layer being used. Because the mobile phase competes with the chromatographed substances for sorbent sites, polar substances will require a polar solvent to cause migration on a silica gel or alumina adsorbent layer. A stronger solvent would increase R_F values (often with an accompanying decrease in resolution) and in the case of normal-phase TLC would be more polar. In reversed-phase TLC, nonpolar substances are strongly attracted to the layer, and nonpolar mobile phases are required to effect migration. Stronger solvents are more nonpolar.

Mobile phases should be prepared from the purest grade of solvent available. ACS reagent-grade solvents are recommended for qualitative TLC and specially purified chromatography grades for quantitative TLC because nonvolatile impurities remain sorbed to the layer and may cause irregular baselines in scanning densitometry (Poole et al., 1989). Stabilizers and antioxidants, such as ethanol, chloroform, ethers, tetrahydrofuran, and dioxane, which are added to some solvents, may affect separations, and their presence must be considered. If possible, mixtures of more than three or four components should be avoided because of problems associated with reproducible preparation. Mobile-phase proportions are usually designated in parts by volume so that the sum is 100, for example, toluene–chloroform–20% aqueous ammonium hydroxide (50:40:10 v/v). If one solvent is strong and one is weak in a binary mixture, the percentage of the strong solvent will usually be small, 0.05–5%. If both solvents are quite weak, the stronger solvent will be present in higher concentration, e.g., 20–50%, to achieve the required overall strength and selectivity.

Quantitative techniques and volumetric apparatus such as pipets, volumetric flasks, and graduated cylinders should always be used in preparing mobile phases. Solvent compounds should be measured individually and then mixed thoroughly in a storage vessel before being placed in the development chamber. A multisolvent mobile phase should not be used more than once because its composition will change during a run due to differential degrees of evaporation and adsorption by the layer of its components or chemical reaction between the components, so that results will not be consistent if the solvent is reused. It is also important to remember that the solvent mixture that is prepared and poured in the chamber is not usually equivalent to the effective mobile phase causing the separation because of solvent demixing and the effects of evaporation and condensation (Bauer et al., 1991).

For the separation of complex mixtures containing a large number of components, systematic, computer-assisted mobile-phase optimization methods are preferable to unguided trial and error. The systematic schemes that have been used in TLC are mostly adapted from HPLC, but their use for TLC must take into account differences in the processes, such as changes in the bulk mobile phase during TLC development caused by demixing. Optimization strategies have included a structural approach that assumes selectivity and solvent strength are independent variables (Geiss, 1987), window diagrams (Nurok et al., 1982a; Wang and Wang, 1990), overlapping resolution maps (Nurok et al., 1982b; Wang and Wang, 1990; Issaq et al., 1981), simplex methods (Bayne and Ma, 1987; Turina, 1986; De Speigeleer et al., 1987), pattern-recognition procedures (De

Spiegeleer et al., 1987; De Spiegeleer and De Moerloose, 1988), a graphical method (Matyska and Soczewinski, 1993), numerical taxonomy and information content derived from Shannon's equation (Medic-Saric et al., 1996), and the PRISMA system (Nyiredy et al., 1988, 1989, 1991; Nyiredy and Fater, 1995; Dallenbach-Toelke et al., 1986) (see Section I.D). All of these optimization procedures involve the use of some form of statistical design to select a series of solvents for evaluation or to indicate the best system by comparing the results obtained from an arbitrarily selected group of solvents (Poole and Poole, 1991).

Different approaches for systematic optimization of the solvent system for one- and two-dimensional planar chromatography have been described in detail by Kowalska (1996), De Spiegeleer (1991), Geiss (1987), Wang (1996), Sarbu and Haiduc (1993), and Nurok (1988, 1989). Computer-assisted methods were described for optimizing mobile phases for stepwise gradient HPTLC (Wang et al., 1994a; Matysik and Soczewinski, 1996; Soczewinski and Swieboda, 1995; Wang et al., 1996) and for stepwise gradients with one- and two-dimensional and multiple development TLC (Markowski, 1996; Markowski and Czapinska, 1995; Yan et al., 1996). Optimization methods were applied to forced-flow as well as capillary-flow TLC; for example, a statistical method was described for optimizing mobile phases for one- and two-dimensional overpressured layer chromatography (Nurok et al., 1997). Szepesi and Nyiredy (1996) have summarized the possibilities of transferring optimized mobile phases between different planar chromatographic methods, as well as HPLC (see Section III) and fully on-line forced-flow planar chromatography techniques. The optimization methods used in TLC have been compared and critically evaluated (Cavalli et al., 1993; Wang and Yan, 1996; Prus and Kowalska, 1995). Despite extensive research on their development, these optimization methods have not been used widely for routine, practical TLC up to this time.

A. The Elutropic Series

Solvents can be grouped for adsorption chromatography into a so-called elutropic series according to their elution strength. As seen in Table 6.1, the solvent strength, as measured by the semiempirical parameter ε^0, increases with the polarity of the solvent (ability to form hydrogen bonds). Although the values in the table are for alumina, they can be used interchangeably with reasonable reliability for silica gel as well, although there are variations in relative strength among various adsorbents. For example, more basic solvents are generally stronger when used with acidic silica or alumina, whereas acidic solvents are stronger with amino-bonded silica as the adsorbent (Robards et al., 1994). In addition to elution strength considerations, solvents with low boiling points (to facilitate evaporation from the layer), low viscosity (to decrease development time and spot diffusion), low toxicity and flammability, nonreactivity toward the layer and sample compo-

TABLE 6.1   Solvent Strength Data on Alumina Adsorbent

Solvent	ε^0	Solvent	ε^0
Fluoroalkanes	−0.25	Methyl chloride	0.42
n-Pentane	0.00	Ethylene dichloride	0.44
Isooctane	0.01	Methyl ethyl ketone	0.51
Petroleum ether	0.01	1-Nitropropane	0.53
n-Decane	0.04	Triethylamine	0.54
Cyclohexane	0.04	Acetone	0.56
Cyclopentane	0.05	Dioxane	0.56
1-Pentene	0.08	Tetrahydrofuran	0.57
Carbon disulfide	0.15	Ethyl acetate	0.58
Carbon tetracholoride	0.18	Methyl acetate	0.60
Xylene	0.26	Diethylamine	0.63
i-Propyl ether	0.28	Nitromethane	0.64
i-Propyl chloride	0.29	Acetonitrile	0.65
Toluene	0.29	Pyridine	0.71
n-Propyl chloride	0.30	Dimethyl sulfoxide	0.75
Benzene	0.32	i- or n-Propanol	0.82
Ethyl bromide	0.35	Ethanol	0.88
Ethyl sulfide	0.38	Methanol	0.95
Chloroform	0.40	Ethylene glycol	1.1

Source: Data from Snyder (1968).

nents, low cost, and suitable solubility characteristics are preferred for preparing mobile phases.

To prepare TLC mobile phases, solvents from the elutropic series are blended into binary or ternary mixtures of the correct strength. In most cases, the strength of a solvent mixture will be intermediate between the strengths of the two (or more) components of the mixture. In addition to the proper strength, the mobile phase must also provide adequate selectivity to achieve the required separation.

B. Designing a Separation on Silica Gel

Silica gel is by far the most widely used layer for normal-phase TLC, but others include alumina and polar-bonded silica phases such as cyano, amino, and diol. Mobile phases for these normal-phase layers are usually a single nonpolar organic solvent or a mixture of a nonpolar solvent with a polar modifier to control solvent strength and selectivity (Robards et al., 1994).

Silica gel has proven to be a very versatile layer material because it provides a moderate degree of adsorption for a wide range of solute types in a great variety of mobile phases. To design a separation on silica gel, it is recommended to spot the mixture on microscope slides and to develop the slides in small beakers or jars with single solvents of increasing strength from the elutropic series. For example, hexane, carbon tetrachloride, benzene (or toluene), chloroform, diethyl ether, and methanol can be chosen. The opportunity can also be taken to try out different visualization reagents. The solubility of the compound in each solvent can provide a valuable preliminary clue as to the required solvent strength for the mobile phase; all solutes should be moderately soluble in the mobile phase. Instead of microscope slides, radial development (Chapter 7) can be used for the preliminary scouting. Screening of mobile phases can also be performed using the Vario KS chamber (Chapter 7), which allows a number of different solvents to simultaneously develop chromatograms in individual channels on a single TLC plate.

The solvent that moves the zones near the center of the plate ($R_F = 0.2$ to 0.8) has the correct strength. If the resolution is not adequate with any of the single solvents, mixtures of these or other solvents with different selectivity properties should be prepared with approximately the same solvent strength as the single solvent giving the best range of R_F values. As a general rule for choosing a replacement solvent to enhance selectivity, a second solvent from the same selectivity group will normally not enhance the separation but a solvent with similar strength from a different selectivity group may do so. The greatest improvement in selectivity is likely if the substituted solvent is from a widely separated group in the selectivity triangle (Snyder, 1974, 1978). The change in strength as a function of the volume percent of the more polar component in a binary solvent is not a linear function. A large increase in solvent strength is produced by a small increase in concentration of the polar solvent if it is present in a small concentration, and the increase in strength is less rapid as the concentration of the polar component increases (Robards et al., 1994). Data such as those shown in Tables 6.2 and 6.3 and Figure 6.1 can aid in the preparation of these solvent blends.

The advantage of solvent mixtures is that resolution or selectivity can be improved if solvent–solute and/or solvent–layer interactions change as the solvent composition is varied. The solvents diethyl ether–hexane (1:1), chloroform–hexane (34:66), chloroform–methylene chloride–hexane (17:23:60), and diethyl ether–chloroform–methylene chloride–hexane (17:11:15:57) all have the same strength and would give a similar R_F range for a given solute mixture, but the selectivities would be different. Addition of a small amount ($\sim 1\%$) of a certain modifier to a mobile phase often makes a significant difference in the selectivity of the system. This is especially true for the addition of acid (acetic acid), base (ammonia), or a buffer.

TABLE 6.2   Solvent Strengths for Mobile Phases of Polar Solvents in Pentane

ε^0	Solvent A: Solvent B:	Pentane							
		CS_2	i-PrCl[a]	Benzene	Ethyl ether	$CHCl_3$	CH_2Cl_2	Acetone	Methyl acetate
0.00		0	0	0	0	0	0	0	0
0.05		18	8	3.5	4	2	1.5		
0.10		48	19	8	9	5	4	1.5	
0.15		100	34	16	15	9	8	3.5	2
0.20			52	28	25	15	13	6	3.5
0.25			77	49	38	25	22	9	5
0.30				83	55	40	34	13	8
0.35					81	65	54	19	13
0.40						100	84	28	19
0.45								42	29
0.50								61	44
0.55								92	65
0.80									100

Note: Values refer to volume percent of solvent B in solvent A (pentane) for the given ε^0 value. These data are for alumina as sorbent. For silica gel, the order will be the same, but the actual ε^0 value will be slightly different.
[a]i-Propyl chloride.
Source: Reprinted from Snyder (1968), by courtesy of Marcel Dekker, Inc.

C.   Designing a Partition Separation

In reversed-phase TLC, mobile phases are generally water mixed with an organic modifier. Solvent strength increases with increasing concentration and decreasing polarity of the modifier. For separations on reversed-phase chemically bonded C_{18} layers, the empirical solvent strength parameter S can be used as a measure of solvent strength (Snyder et al., 1979).

Solvent	S
Water	0
Methanol	3.0
Acetonitrile	3.1
Dioxane	3.1
Ethanol	3.6
Isopropanol	4.2
Tetrahydrofuran	4.4

TABLE 6.3 Solvent Strengths for Mobile Phases of Polar Solvents in Benzene and Pentane

	Solvent A:	Pentane	Benzene					
ε^0	Solvent B:	Diethyl-amine	Acetone	Methyl acetate	Diethyl-amine	Aceto-nitrile	*i*-PrOH[a]	MeOH[b]
0.30		2.5						
0.35		5	6	4	2			
0.40		8	18	12	7			
0.45		13	36	24	14	1.5		
0.50		22	60	42	26	6		
0.55		38	93	66	45	14	4	
0.60		73		100	77	36	7	
0.65						100	12	
0.70							21	
0.75							40	4
0.80							75	8
0.85								18
0.90								44
0.95								100

Note: Values refer to volume percent of solvent B in solvent A (benzene or pentane) for the given ε^0 value. These data are for alumina as sorbent. For silica gel, the order will be similar but ε^0 will be different. In most cases, toluene can be substituted for benzene and hexane for pentane without significantly changing the results.
[a] Isopropanol.
[b] Methanol.
Source: Reprinted from Snyder (1968), by courtesy of Marcel Dekker, Inc.

For mixtures, the solvent strength S is the arithmetic average of S values for the pure components, weighted according to volume fraction vf:

$$S = vf_a S_a + vf_b S_b + \ldots$$

For example, the S value for 40% methanol–60% water would be

$$0.40 \times 3.0 + 0.60 \times 0 = 1.2$$

This equation can be used when preparing solvent mixtures with different selectivities at a constant, optimum strength.

In conventional partition systems, a stationary polar or nonpolar liquid is supported by an inert (ideally) cellulose, kieselguhr, or silica gel layer. The mobile phase and the stationary liquid phase must be as immiscible as possible, and the mobile phase should be previously saturated with the stationary liquid so the latter will not be stripped from the layer during development. The resolution of

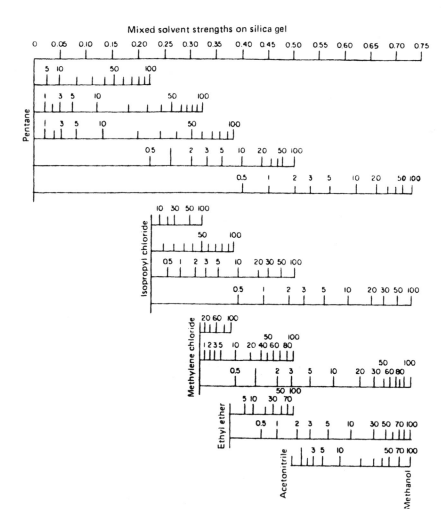

FIGURE 6.1 Mixed solvent strengths on silica gel. The first five lines give the percentages of isopropyl chloride, methylene chloride, ethyl ether, acetonitrile, and methanol, respectively, required to be mixed with pentane to yield the mixed solvent strengths listed along the top. The more polar solvent referred to in each case is determined by looking vertically below the "100" entry at the end of each line. Lower lines present data for the other solvents (along the Y-axis) mixed with more polar components, each of which is again read below the "100" at the end of each line. [From Touchstone (1992), by permission of John Wiley & Sons, Inc., New York.]

solutes is controlled by changing the mobile phase by trial and error or based on consultation of the literature. Many partition systems for resolution of polar (normal phase) and nonpolar (reversed phase) solutes are described in books on paper chromatography [e.g., Sherma and Zweig (1971)]. The majority of these separations are probably transferable to cellulose layers with little or no variations in parameters.

Solvent mixtures containing greater than 35–60% water are impossible to use with some chemically bonded reversed-phase layers, the exact amount depending on the brand and type of plate and the chain length of the bonded moiety. Addition of salt to the water may allow mobile phases with higher water content to be used.

D. Examples of General Solvent Systems for Normal- and Reversed-Phase TLC; Systematic Solvent Optimization

Benzene (or toluene) has proven to be a good solvent for many separations on silica gel. If R_F values are too high, 5 or 10% of hexane is added to reduce polarity. If benzene is too weak, 5 to 10% of methanol can be added. A percentage of acetic acid or ammonia (or pyridine) is added, respectively, to keep acidic and basic solutes nonionized so as to prevent tailing of zones.

A solvent system that has proven versatile for the separation of more or less polar solutes in an unsaturated N-tank (Chapter 7) consists of hydrocarbon plus acetone plus chloroform. A typical example is hexane–acetone–chloroform (65:30:5). Pentane, isooctane, or toluene are other suitable hydrocarbons to replace hexane to improve selectivity. For very polar solutes, a small percentage of methanol is included, and a little acid or base is added to control ionization of acidic or basic solutes. The effect of the chloroform appears to be control of the evaporation rates of the various solvent components during the development–equilibrium process.

The following series of solvent blends provides a regular increase in ε^0 from 0 to 0.95. A variety of these can be used instead of single solvents of the elutropic series for initial screening of optimum solvent strength:

100% pentane ($\varepsilon^0 = 0$)
3.5–83% benzene in pentane ($\varepsilon^0 = 0.05$–0.30)
0.5–100% acetonitrile in toluene ($\varepsilon^0 = 0.35$–0.65)
4–100% methanol in toluene ($\varepsilon^0 = 0.70$–0.95)

To modify selectivity after locating the optimum ε^0, the most polar component is substituted in whole or part by one just above or just below in the elutropic series. For example, acetonitrile–toluene (80:20) can be changed to acetonitrile–nitromethane–toluene (40:40:20).

An optimization procedure based on the Geiss (1987) structural approach uses a Vario KS chamber (Chapter 7) with three strong solvents, methyl-*t*-butyl ether, acetonitrile, and methanol, that are diluted with a weak solvent such as 1,2-dichloroethane to produce a series of mobile phases ranging in ε^0 values from 0.0 to 0.70 in 0.05 increments. Once the appropriate solvent strength is determined, the separation is fine-tuned by blending solvent mixtures of this strength but with different selectivity (Szepesi and Nyiredy, 1996).

A systematic approach for mobile-phase design and optimization for reversed- and normal-phase TLC was adapted from HPLC practice by Sherma and Charvat (1983) and Heilweil (1985). Using the Snyder (1978) solvent classification triangle, three solvents with diverse selectivity parameters based on their dipoles and their properties as proton donors and proton acceptors are chosen. For normal-phase TLC, the solvents are diethyl ether, methylene chloride, and chloroform, with hexane (solvent strength = 0) as the strength-adjusting solvent, or *t*-butylether, methylene chloride, and acetonitrile with hexane. For reversed-phase TLC, the solvents are methanol, acetonitrile, and tetrahydrofuran, with water used for strength adjustment. Other solvents can be substituted for those listed as long as they are from diverse solvent groups located near the corners of the selectivity triangle and have good mutual solubility characteristics.

The selection approach is as follows for RP-TLC on a C_{18} layer. Trial-and-error is used to find the optimum proportions for a binary mixture of methanol and water (R_F values 0.2 to 0.8 for the components of interest). Using the equation in Section I.C, the solvent strength (S) of this mixture is calculated, and this value is used to calculate the compositions of binary mixtures of the other two solvents with water, three ternary mixtures involving the three solvents with water, and a four-component mixture. These seven experiments, involving all possible selectivity-determining interactions, will provide the best separations of the solutes obtainable on the C_{18} layer. Usually, all of the components of interest will be separated by use of one or two of the calculated systems. If no single system or combination is adequate for the required separation, a different layer (e.g., silica gel, C_8, cyano, phenyl) should be tried. The same seven-experiment, four-solvent approach is used for a normal-phase layer, starting with chloroform–hexane binary mixtures to find to optimum strength. It may be necessary to add a small proportion of acid or base to the mobile phases chosen by these schemes to control the ionization of acidic or basic solutes.

The PRISMA system for mobile-phase optimization is a more elaborate, structured trial-and-error version of the normal-phase and reversed-phase procedures described above. The PRISMA system, which is the most widely used of the systematic optimization methods, involves selection of the stationary phase, individual solvents, and vapor phase; optimal combination of the solvents by means of the PRISMA model; and selection of the appropriate development mode. With silica gel, 10 mobile phases representing the Snyder (1978) selectivity groups are first tested in unsaturated chambers. If necessary to obtain R_F values between 0.2 and 0.8, solvent strength is decreased by adding hexane or increased by adding water or acetic acid. The solvents used are diethyl ether (selectivity group I), isopropanol and ethanol (II), THF (III), acetic acid (IV), methylene chloride (V), ethyl acetate and dioxane (VI), toluene (VII), and chloroform (VIII). A similar procedure is used for RP-TLC with water-soluble solvents and water as the strength-reducing solvent. The two to five originally tested solvents that give the best resolutions (or others from the respective selectivity groups that are tested) are selected for further optimization by construction of the PRISMA model. Modifiers such as acids, bases, or ion-pairing reagents may also be added, usually in a low and constant amount. The PRISMA model is a three-dimensional geometrical design that correlates strength and selectivity of the solvent system. The model has three parts, one for optimizing the separation of polar compounds, one for nonpolar compounds, and the third representing the modifiers. The system can be used for analytical and preparative NP- and RP-planar chromatography in capillary- and forced-flow modes, such as OPLC (Mazurek and Witkiewicz, 1991). Details are beyond the scope of this book, but Nyiredy et al. (1988), Nyiredy (1992), and Szepesi and Nyiredy (1996) present guidelines to the PRISMA system that are easy to follow.

E. Ion-Exchange Separations

Mobile phases for separations of inorganic ions or ionized organic compounds on ion-exchange layers are usually composed of buffer solutions with a controlled ionic strength or simple salt solutions. An organic solvent can be added to either of these types of mobile phases to superimpose a solvent-partitioning effect in addition to the ion-exchange mechanism. Systematic procedures or theory for selection of solvents for ion-exchange TLC are not well advanced, and it is recommended that the reader consult the literature [e.g., Sherma and Zweig (1972–1989), Sherma (1991–1993), Sherma and Fried (1996)] for appropriate chromatographic systems for any particular separation.

F. Ion-Pair and Chiral Separations

As an example of separations in ion-association systems, a series of alkaloids was separated by ion-pair reversed-phase TLC using silanized silica gel either alone or containing a counterion as the sorbent, with phosphate buffer–organic modifier mixtures containing low concentrations of anionic ion-pairing reagent as the mobile phase (Bieganowska and Petruczynik, 1994).

As mentioned in Chapter 3, most enantiomeric separations have been performed on layers developed with a mobile phase containing a chiral selector rather than on a chiral layer. For example, chiral separations of amino acids and derivatives were achieved on C_{18}-bonded silica gel (Lepri et al., 1994) and on cellulose (Xuan and Lederer, 1994; Huang et al., 1996) using cyclodextrin-containing mobile phases. TLC separations of amino acid derivatives and racemic drugs were reported using the macrocyclic antibiotic vancomycin as a chiral mobile-phase additive (Armstrong and Zhou, 1994).

II. TROUBLESHOOTING SEPARATION PROBLEMS

A. Sample Streaking

Streaked or trailing zones can be caused by sample overloading or highly active adsorption sites on the layer surface. Reducing the sample size is the obvious answer to the first problem. Layer deactivation can be achieved by the addition of traces of acid or base to the mobile phase as described before, or by the deactivation of the layer by treatment with water vapor. If nothing else helps, a less polar derivative of the solute with less tendency to streak on an adsorbent layer can be prepared, or a different type of layer can be employed.

B. Insufficient Separation

If the attempt to find a mobile phase with adequate strength and selectivity on silica gel as outlined above fails, a change in the layer is indicated. This could involve impregnation of silica gel with a selective reagent, the use of another adsorbent, or a change from adsorption to a partition or ion-exchange layer. In all cases, the strength of the mobile phase is always chosen in relation to the sorptivity of the layer and the polarity of the solutes. All of these stationary phases are described in Chapter 3. A change in the development method (Chapter 7) might also be of help.

C. Multiple Solvent Fronts

Multicomponent mobile phases often are separated (''demixed'') as they migrate through the layer because different compounds of the solvent mixture are attracted more-or-less strongly to the sorbent. This leads to a mobile-phase gradient effect along the layer and the appearance of extra fronts behind the bulk mobile-phase front. These fronts can aid separations and should not affect the reproducibility of results if TLC conditions are constant from run to run. Results will not be reproducible, however, unless fresh mobile-phase and the same equilibration conditions are used for each development.

III. SELECTION OF MOBILE PHASES FOR HPLC

TLC has been widely used to scout or pilot mobile phases for HPLC (Jork et al., 1981). Because the vast majority of HPLC separations are performed on reversed-phase C_{18}- or C_8-bonded silica gel columns, comparable bonded-phase layers are usually used. Impregnated layers can also be used if the results are known to be interchangeable. Microscope slide sized plates or standard 20×10- or 20×20-cm plates are spotted with the mixture of interest and developed with series of mobile phases, for example, different proportions of acetonitrile and water. Only isocratic solvents can be tested by TLC, but HPLC gradients can be approximated from the TLC results (Golkiewicz, 1981a). If a 50:50 mixture of acetonitrile and water gives somewhat separated zones near the center of the plate by TLC, this mobile phase should be tried for isocratic HPLC, or a gradient from 25 to 75% acetonitrile in water ($\pm 25\%$ of the best isocratic solvent).

Although theoretical calculations have shown a relationship between the retention behavior of solutes in TLC and HPLC (Rozylo and Janicka, 1991a, 1991b, 1995; Rozylo et al., 1994; Kossoy et al., 1992), results from TLC are not directly related to those by HPLC because the two processes are basically different (Chapter 1) and the layer and column sorbents are usually not identical. Samples are applied to a dry layer in TLC but are injected into a solvent-equilibrated column in HPLC; solutes are developed for a limited distance in TLC but are eluted in HPLC; vapor-phase equilibrium is a factor in TLC but not in HPLC, in which the column and liquid are in complete equilibrium; the pressurized driving force maintains a constant flow rate in HPLC that does not exist in capillary-flow TLC; multicomponent solvents can produce gradients in TLC due to demixing that will not exist in isocratic HPLC with the same solvent; and layers usually contain binders not present in column sorbents. Despite these considerable differences, results from TLC are often transferrable to HPLC columns, both in the reversed phase, bonded phase mode, and for normal-phase liquid–solid chromatography. OPLC, which is the planar chromatography technique most similar to HPLC, is especially useful for transfer of TLC data (Ferenczi-Fodor et al., 1991; Mincsovics et al., 1996).

HPLC distribution coefficients can be approximated from the Geiss and Schlitt transfer equation, $k' = (1/R_F - 1)$ (Geiss, 1987). A good mobile phase for HPLC should give R_F values in the range 0.1–0.7, which translates to k' values between 0.4 and 9. The TLC separation may not appear to be complete with the chosen solvent, especially if R_F values are low, but the HPLC column will generate more theoretical plates and improve the separation. The best chance for successful transfer of data using solvent mixtures occurs when the plate is preloaded with the mobile-phase vapors in an N-chamber (Guinchard et al., 1982; Geiss, 1987) or when TLC is performed in a sandwich chamber (Rozylo and Janicka, 1991a).

Studies of data transfer between TLC and HPLC were reported by Kaiser and Reider (1977), Buglio and Venturella (1982), Golkiewicz (1981b), Hara (1977), Soczewinski and Golkiewicz (1976), Soczewinski et al. (1977), and Bieganowska and Glowniak (1988). Jost et al. (1984) demonstrated a linear relationship between k' derived from TLC and k' from HPLC for polycyclic aromatic hydrocarbons on silica gel and C_{18}-bonded silica gel, fatty acids on C_2-bonded silica gel, stilbestrol derivatives on C_8-bonded silica gel, and methylated phenols on amino-bonded silica gel. For compounds, such as substituted benzoic acids, for which the correlation was not linear, at least retention order could be obtained. Successful application of TLC to acquire information useful for HPLC was reported for pharmaceuticals (Bliesner, 1994), ajmaline stereoisomers (Sosa et al., 1994), coumarins (Vuorela et al., 1994), and furocoumarin isomers (Mincsovics et al., 1996), but correlation of TLC and HPLC retention behavior was poor for pesticides on alumina (Cserhati and Forgacs, 1994) and for antihypoxia drugs in reversed-phase systems (Wallerstein et al., 1993). The theory and techniques for the transfer of TLC separations to columns were reviewed in book chapters by Geiss (1987) and Grinbeg (1990).

TLC is also useful for solvent scouting for removal of strongly sorbed compounds from the top of HPLC columns. TLC has the advantage of allowing detection of all compounds in a sample, including those strongly sorbed at the origin. Solvents that are found to elute these compounds on the TLC plate will also be useful for cleaning them from the column inlet between sample injections.

REFERENCES

Armstrong, D. W., and Zhou, Y. (1994). Use of a macrocyclic antibody as the chiral selector for enantiomeric separations by TLC. *J. Liq. Chromatogr. 17*: 1695–1707.

Bauer, K., Gros, L., and Sauer, W. (1991). *Thin Layer Chromatography—An Introduction*, E. Merck, Darmstadt, Germany.

Bayne, C. K., and Ma, C. Y. (1987). Optimization of solvent composition for high performance thin layer chromatography. *J. Liq. Chromatogr. 10*: 3529–3546.

Bieganowska, M. L., and Glowniak, K. (1988). Retention behavior of some coumarins in normal phase high performance thin layer and column chromatography. *Chromatographia 25*: 111–116.

Bieganowska, M. L., and Petruczynik, A. (1994). Thin layer reversed phase chromatography of some alkaloids in ion-association systems. *Chem. Anal. (Warsaw) 39*: 445–454.

Bliesner, D. M. (1994). Graphical correlation of TLC and HPLC data in the development of pharmaceutical compounds. *J. Planar Chromatogr.—Mod. TLC 7*: 197–201.

Buglio, B., and Venturella, V. S. (1982). Selection of a mobile phase for LSC from TLC data. *J. Chromatogr. Sci. 20*: 165–170.

Cavalli, E., Truong, T. T., Thomassin, M., and Guinchard, C. (1993). Comparison of optimization methods in planar chromatography. *Chromatographia 35*: 102–108.

Cserhati, T., and Forgacs, E. (1994). Relationship between the high performance liquid and thin layer chromatographic retention of non-homologous series of pesticides on an alumina support. *J. Chromatogr., A 668*: 495–500.

Dallenbach-Toelke, K., Nyiredy, Sz., Meier, B., and Sticher, O. (1986). Optimization of overpressured layer chromatography of polar, naturally occurring compounds by the "PRISMA" model. *J. Chromatogr.* *365*: 63–72.

De Spiegeleer, B. M. J. (1991). Optimization. In *Handbook of Thin Layer Chromatography*, J. Sherma and B. Fried (Eds.). Marcel Dekker, New York, pp. 71–85.

De Spiegeleer, B. M. J., and De Moerloose, P. H. M. (1988). High performance thin layer chromatography of seventeen diuretics. *J. Planar Chromatogr.—Mod. TLC 1*: 61–64.

De Spiegeleer, B. M. J., De Moerloose, P. H. M., and Slegers, G. A. S. (1987). Criterion for evaluation and optimization in thin layer chromatography. *Anal. Chem. 59*: 62–64.

Ferenczi-Fodor, K., Mincsovics, E., and Tyihak, E. (1991). Overpressured layer chromatography. In *Handbook of Thin Layer Chromatography*, J. Sherma and B. Fried (Eds.). Marcel Dekker, New York, pp. 155–181.

Geiss, F. (1987). *Fundamentals of Thin Layer Chromatography*. Alfred Huthig Verlag, Heidelberg.

Golkiewicz, W. (1981a). Use of data obtained from RP-18 plates for the prediction of gradient programs in reversed phase liquid chromatography. *Chromatographia 14*: 629–632.

Golkiewicz, W. (1981b). TLC as a pilot technique for optimization of gradient HPLC. 1. Experimental verification of a graphical method for the optimization of stepwise gradient elution. *Chromatographia 14*: 411–414.

Grinberg, N. (1990). *Modern Thin Layer Chromatography*. Marcel Dekker, New York, pp. 453–463.

Guinchard, C., Mason, J. D., Truong, T. T., and Porthault, M. (1982). Research on the best chromatographic system for separation of polar aromatic compounds on silica by different layer preloading for transposition to a column. *J. Liq. Chromatogr. 5*: 1123–1140.

Hara, S. (1977). Use of thin layer chromatographic systems in high performance liquid chromatographic separations. Procedure for systematization and design of the separation process in synthetic chemistry. *J. Chromatogr. 137*: 41–52.

Heilweil, E. (1985). A systematic approach to mobile phase design and optimization for normal and reversed phase TLC. In *Techniques and Applications of TLC*. J. C. Touchstone and J. Sherma (Eds.). Wiley-Interscience, New York, Chap. 3, pp. 37–49.

Huang, M.-B., Li, H.-K., Li, G.-L., Yan, C.-T., and Wang, L.-P. (1996). Planar chromatographic direct separation of some aromatic amino acids and aromatic amino alcohols into enantiomers using cyclodextrin mobile phase additives. *J. Chromatogr., A 742*: 289–294.

Issaq, H. J., Klose, J. R., McNitt, K. L., Haky, J. E., and Muschik, G. M. (1981). Systematic statistical method of solvent selection for optimal separation in liquid chromatography. *J. Liq. Chromatogr. 4*: 2091–2120.

Jork, H., Reh, E., and Wimmer, H. (1981). To what extent is TLC effective as a pilot technique for HPLC? *GIT Fachz. Lab. 25*: 566–570.

Jost, W., Hauck, H. E., and Eisenbeis, F. (1984). Possibilities and limits of transfer of TLC separation systems to HPLC. *Fresenius' Z. Anal. Chem. 318*: 300–301.

Kaiser, R. E., and Rieder, A. (1977). C_8 and C_{18} reversed phase high performance thin layer chromatography on chemically bonded layers for environmental trace analysis and for optimization of high performance liquid chromatography. *J. Chromatogr. 142*: 411–420.

Kossoy, A. D., Risley, D. S., Kleyle, R. M., and Nurock, D. (1992). Novel computational method for the determination of partition coefficients by planar chromatography. *Anal. Chem. 64*: 1345–1349.

Kowalska, T. (1996). Theory and mechanism of thin layer chromatography. In *Handbook of Thin Layer Chromatography*, 2nd ed., J. Sherma and B. Fried (Eds.). Marcel Dekker, New York, pp. 49–80.

Lepri, L., Coas, V., and Desideri, P. (1994). Planar chromatography of optical and structural isomers with eluents containing modified beta-cyclodextrins. *J. Planar Chromatogr.—Mod. TLC 7*: 322–326.

Markowski, W. (1996). Computer-aided optimization of gradient multiple development thin layer chromatography. III. Multistage development over a constant distance. *J. Chromatogr., A 726*: 185–192.

Markowski, W., and Czapinska, K. L. (1995). Computer simulation of the separation in one- and two-dimensional thin layer chromatography by isocratic and stepwise gradient development. *J. Liq. Chromatogr. 18*: 1405–1427.

Matysik, G., and Soczewinski, E. (1996). Computer-aided optimization of stepwise gradient TLC of plant extracts. *J. Planar Chromatogr.—Mod. TLC 9*: 404–412.

Matyska, M., and Soczewinski, E. (1993). Optimization of chromatographic systems in TLC by graphical method and with a computer program. *Chem. Anal. (Warsaw) 38*: 555–563.

Mazurek, M., and Witkiewicz, Z. (1991). The analysis of organophosphorus warfare agents in the presence of pesticides by overpressured thin layer chromatography. *J. Planar Chromatogr.—Mod. TLC 4*: 379–384.

Medic-Saric, M., Males, Z., Stanic, G., and Saric, S. (1996). Evaluation and selection of optimal solvent and solvent combinations in thin layer chromatography of flavonoids and phenolic esters of Zizyphus jujuba Mill. *Croat. Chem. Acta 69*: 1265–1274.

Mincsovics, E., Ferenczi-Fodor, K., and Tyihak, E. (1996). Overpressured layer chromatography. In *Handbook of Thin Layer Chromatography*, 2nd ed., J. Sherma and B. Fried (Eds.). Marcel Dekker, New York, pp. 171–203.

Nurok, D. (1988). Computer aided optimization of separation in planar chromatography. *LC-GC Mag. 6*: 310–322.

Nurok, D. (1989). Strategies for optimizing the mobile phase in planar chromatography. *Chem. Rev. 89*: 363–375.

Nurok, D., Becker, R. M., Richard, M. J., Cunningham, P. D., Gorman, W. B., and Bush, C. L. (1982a). Optimization of binary solvents in thin layer chromatography. *J. High Resolut. Chromatogr. Chromatogr. Commun. 5*: 373–376.

Nurok, D., Becker, R. M., and Sassic, K. A. (1982b). Time optimization in thin layer chromatography. *Anal. Chem. 54*: 1955–1959.

Nurok, D., Kleyle, R. M., McCain, C. L., Risley, D. S., and Ruterbories, K. J. (1997). Statistical method for quantifying mobile phase selectivity in one- and two-dimensional overpressured layer chromatography. *Anal. Chem. 69*: 1398–1405.

Nyiredy, Sz. (1992). Planar Chromatography. In *Chromatography*, 5th ed. E. Heftmann (Ed.). Elsevier, Amsterdam, pp. A109–A150.

Nyiredy, Sz., Dallenbach-Toelke, K., and Sticher, O. (1988). The "PRISMA" optimization system in planar chromatography. *J. Planar Chromatogr.—Mod. TLC 1*: 336–342.

Nyiredy, Sz., Dallenbach-Toelke, K., and Sticher, O. (1989). Correlation and prediction of the *k′* values for mobile phase optimization in HPLC. *J. Liq. Chromatogr. 12*: 95–116.

Nyiredy, Sz., and Fater, Zs. (1995). Automatic mobile phase optimization, using the "PRISMA" model, for the separation of apolar compounds. *J. Planar Chromatogr.—Mod. TLC 8*: 341–345.

Nyiredy, Sz., Wosniok, W., Thiele, H., and Sticher, O. (1991). PRISMA model for computer-aided HPLC mobile phase optimization based on an automatic peak identification approach. *J. Liq. Chromatogr. 14*: 3077–3110.

Poole, C. F., and Poole, S. K. (1991). *Chromatography Today*. Elsevier, Amsterdam, pp. 649–734.

Poole, C. F., Poole, S. K., Dean, T. A., and Chirco, N. M. (1989). Sample requirements for quantitation in thin layer chromatography. *J. Planar Chromatogr.—Mod. TLC 2*: 180–189.

Prus, W., and Kowalska, T. (1995). A comparison of selected methods for optimization of separation selectivity in reversed phase liquid–liquid chromatography. *J. Planar Chromatogr.—Mod. TLC 8*: 288–291.

Robards, K., Haddad, P. R., and Jackson, P. E. (1994). *Principles and Practice of Modern Chromatographic Methods*. Academic Press, San Diego, CA, pp. 179–226.

Rozylo, J., and Janicka, M. (1991a). Some theoretical aspects of the use of TLC as a pilot technique for column liquid chromatography. II: Comparison of Different TLC Techniques as pilot methods for mixture separation by column liquid chromatography. *J. Planar Chromatogr.—Mod. TLC 4*: 241–245.

Rozylo, J., and Janicka, M. (1991b). Thermodynamic description of liquid–solid chromatography process in the optimization of separation conditions of organic compound mixture. *J. Liq. Chromatogr. 14*: 3197–3212.

Rozylo, J., and Janicka, M. (1995). Experimental and theoretical problems in the use of different TLC techniques for prediction of solute retention by the LCC method. *Acta Chromatogr. 4*: 9–32.

Rozylo, J., Janicka, M., and Siembida, R. (1994). Advantages of TLC as a pilot technique for HPLC. *J. Liq. Chromatogr. 17*: 3641–3653.

Sarbu, C., and Haiduc, I. (1993). Optimal choice of solvent systems in bidimensional thin layer chromatography. *Stud. Univ. Babes-Bolyai, Chem. 38*: 49–55.

Sherma, J. (1991–1993). *Handbooks of Chromatography*. CRC Press, Boca Raton, FL. Carbohydrates (Vol. II) (1991); steroids (Vol. II) (1992); aromatic hydrocarbons (1993); lipids (Vol. III) (1993); polymers (Vol. II) (1993).

Sherma, J. (1997). Thin layer chromatography. In *Encyclopedia of Pharmaceutical Technology*, Volume 15, J. Swarbrick and J. C. Boylan (Eds.). Marcel Dekker, New York, pp. 81–106.

Sherma, J., and Charvat, S. (1983). Solvent selection for RP-TLC. *Am. Lab. 14(7)*: 138–144.

Sherma, J., and Fried, F. (Eds.) (1996). *Handbook of Thin Layer Chromatography*, 2nd ed., Marcel Dekker, NY.

Sherma, J., and Zweig, G. (1971). *Paper Chromatography*. Academic Press, New York.

Sherma, J., and Zweig, G. (1972–1989). *Handbooks of Chromatography*. CRC Press, Boca Raton, FL. General (Vols. I and II) (1972); drugs (Vols. I and II) (1981); carbohydrates (Vol. I) (1982); polymers (Vol. I) (1982); phenols and organic acids (Vol. I) (1982); amino acids and amines (Vol. I) (1983); pesticides (Vol. I) (1984); terpenoids (Vol. I) (1984); lipids (Vols. I and II) (1984); steroids (Vol. I) (1986); peptides (Vol. I) (1986); inorganics (Vol. I) (1987); nucleic acids and related compounds (Vols. IA and IB) (1987); plant pigments (Vol. I) (1988); drugs (Vols. III–VI) (1989); amino acids and amines (Vol. II) (1989).

Siouffi, A. M., Guillemonat, A., and Guiochon, G. (1977). An attempt to use HPTLC data for prediction of HPLC retention coefficients. *J. Chromatogr. 137*: 35–40.

Snyder, L. R. (1968). *Principles of Adsorption Chromatography*. Marcel Dekker, New York.

Snyder, L. R. (1974). Classification of the solvent properties of common liquids. *J. Chromatogr. 92*: 223–230.

Snyder, L. R. (1978). Classification of the solvent properties of common liquids. *J. Chromatogr. Sci. 16*: 223–234.

Snyder, L. R., Dolan, J. W., and Gant, J. R. (1979). Gradient elution in high performance liquid chromatography. I. Theoretical basis for reversed-phase systems. *J. Chromatogr. 165*: 3–30.

Soczewinski, E., and Golkiewicz, W. (1976). A graphical method for the determination of retention times in column liquid-solid chromatography from thin layer chromatographic data. *J. Chromatogr. 118*: 91–95.

Soczewinski, E., and Swieboda, R. (1995). Computer-aided optimization of stepwise gradient thin layer chromatography. *Acta Chromatogr. 5*: 25–33.

Soczewinski, E., Dzido, T., Golkiewicz, W., and Gazda, K. (1977). Comparison of high performance liquid chromatographic and thin layer chromatographic data obtained with various types of silica. *J. Chromatogr. 131*: 408–411.

Sosa, M. E., Valdes, J. R., and Martinez, J. A. (1994). Determination of ajmaline stereoisomers by combined high performance liquid chromatography and thin layer chromatography. *J. Chromatogr., A 662*: 251–254.

Szepesi, G., and Nyiredy, S. (1996). Pharmaceuticals and drugs. In *Handbook of Thin Layer Chromatography*, 2nd ed., J. Sherma and B. Fried (Eds.). Marcel Dekker, New York, pp. 819–876.

Touchstone, J. C. (1992). *Practice of Thin Layer Chromatography*, 3rd ed. Wiley-Interscience, New York.

Turina, S. (1986). Optimization in chromatographic analysis. In *Planar Chromatography*, Volume 1, R. E. Kaiser (Ed.). Huthig, Heidelberg, pp. 15–46.

Vuorela, P., Rahko, E.-L., Hiltunen, R., and Vuorela, H. (1994). Overpressured layer chromatography in comparison with thin layer and high performance liquid chromatography for the determination of coumarins with reference to the composition of the mobile phase. *J. Chromatogr., A 670*: 191–198.

Wallerstein, S., Cserhati, T., and Fischer, J. (1993). Determination of the lipophilicity of some anti-hypoxia drugs: Comparison of TLC and HPLC methods. *Chromatographia 35*: 275–280.

Wang, Q.-S. (1996). Optimization. In *Handbook of Thin Layer Chromatography*, 2nd ed., J. Sherma and B. Fried (Eds.). Marcel Dekker, New York, pp. 81–99.

Wang, Q.-S., and Wang, H.-Y. (1990). Computer assisted comprehensive optimization of mobile phase selectivity in HPTLC. *J. Planar Chromatogr.—Mod. TLC 3*: 15–19.

Wang, Q.-S., and Yan, B.-W. (1996). Criteria for comparing and evaluating the optimization of separation in TLC. *J. Planar Chromatogr.—Mod. TLC 9*: 192–196.

Wang, Q.-S., Yan, B.-W., and Zhang, Z.-C. (1994). Computer-assisted optimization of mobile phase composition in stepwise gradient HPTLC. *J. Planar Chromatogr.— Mod. TLC 7*: 229–232.

Wang, Q.-S., Yan, B.-W., and Zhang, L. (1996). Computer-assisted optimization of two-step development high performance TLC. *J. Chromatogr. Sci. 34*: 202–205.

Yan, B.-W., Zhang, L., Wang, Q.-S. (1996). Computer-assisted optimization of mobile phase compositions and development distance selectivity in gradient two-step development high performance thin layer chromatography. *Chin. J. Chem. 14*: 354–358.

Xuan, H. T. K., and Lederer, M. (1994). Adsorption chromatography on cellulose. XI. Chiral separations with aqueous solutions of cyclodextrins as eluents. *J. Chromatogr. 659*: 191–197.

7

Development Techniques

I. INTRODUCTION

The process of separation of the sample mixture by migration of the mobile phase through the layer is known as development. This chapter describes the different modes and apparatus for development of thin layers. The advantages and disadvantages of the various development modes and instruments were recently reviewed (Cserhati and Forgacs, 1996b).

When a mobile phase that has two or more solvents of widely differing polarities is used with an adsorbent layer, solvent demixing will occur. This phenomenon involves separation of the solvent itself by the layer during the development operation because the adsorbent has a stronger affinity for the more polar solvents, leading to the formation of different solvent zones with secondary fronts. The zone near the top of the layer will be richer in the less polar solvent component, whereas that near the bottom will be richer in the more polar components. Solvent demixing can cause problems in thin-layer chromatography but is also the reason for many successful separations that have been achieved in multicomponent mobile phases.

Chamber saturation has a great influence on R_F values and the separation achieved. During development in an unsaturated tank, solvent evaporates from the plate (mainly in the region of the solvent front), so that more solvent is required for a given distance of front migration and R_F values increase. In a vapor-presaturated chamber lined with solvent-soaked filter paper, the inserted plate is preloaded with vapors and requires less solvent for the same distance of front migration, and lower R_F values result (Bauer, 1991). Pre-equilibration of the dry layer with solvent vapors is desirable with some solvents but not others; likewise, the effect of relative humidity can be large with some solvents but is not important with others. In general, the importance of saturated versus unsaturated development is less in RP-TLC than in adsorption TLC (Cserhati and Forgacs, 1996a).

The role of the vapor phase in TLC was reviewed in detail by Geiss (1988), but little attention is generally paid to it in normal practice. Separations obtained in TLC can be significantly affected by the vapor phase, which depends on the type, size, and saturation condition of the chamber. The interactions of the layer, mobile phase, and vapor phase, as well as other factors (Chapter 11), must be controlled to obtain reproducible TLC separations. A procedure was proposed by Nyiredy et al. (1992) for characterization of chamber saturation in FFPC based on the R_F range obtained on development of a marker test dye mixture with dichloromethane. It was concluded that as a general rule, a saturated chamber should be used for samples containing less than seven compounds to be quantified, whereas OPLC in unsaturated chambers with an optimized mobile phase is required for more than seven compounds or difficult separations.

II. MODES OF DEVELOPMENT

A. Ascending

Ascending is by far the most frequent mode of linear development in TLC. Following sample application, a plate is placed in an appropriate tank, jar, or sandwich chamber so that the solvent is below the point of application of the sample. The solvent is then allowed to rise by capillarity, usually to a distance of 10–15 cm above the origin on 20 × 20-cm TLC plates and 3–7 cm on smaller, high-performance TLC plates. The distance to which the solvent will migrate can be marked on the plate in advance, at the time of sample application, or at the end of the development. The result of linear development is usually a series of compact, symmetrical zones of increasing size toward the solvent front. Zones near the solvent front can be elliptical or distorted.

B. Descending

For descending development of TLC plates, the mobile phase must be fed to the top of a vertical inclined plate through a wick arrangement. This more cumbersome arrangement has no significant advantages over ascending development with regard to development time or resolution, and so it is rarely used. Descending development must be used for Sephadex thin-layer gel filtration chromatography (TLG), the apparatus for which was sold by Pharmacia but has been discontinued. Bhushan and Martens (1996) have described the use of TLG for separation of peptides and proteins.

C. Two-Dimensional

Two-dimensional development is used for examination of complex mixtures. Following application of the sample in one corner of a 20 × 20-cm TLC or 10 × 10-cm HPTLC plate, ascending development is carried out for the full length of the plate to achieve maximum resolution. The plate is removed from the chamber and air dried to remove solvent completely. The plate is then rotated at a 90° angle and redeveloped, usually with a second solvent having a different selectivity. The line of partly resolved components from the first development becomes the origin for the second development. Resolving power is nearly the square of that attained in one-dimensional TLC and often exceeds that of column HPLC. Using convergent flow in the HPTLC horizontal development chamber (see Figure 7.6), four samples applied in the corners of a plate can be chromatographed simultaneously. The plate is developed simultaneously from the top and bottom into the center with the same solvent, and then turned 90° and developed from the top and bottom again with a second solvent.

Variations of the two-dimensional (2D) technique include modification of the characteristic properties of the layer between developments by impregnation with a chemical reagent or immiscible solvent (e.g., determination of parabens by NP-TLC on silica gel followed by impregnation with paraffin prior to RP-TLC in the second direction; Dimov and Filcheva, 1996); use of thin-layer electrophoresis in the second direction; or reacting the separated solutes, such as by treatment with a chemical reagent, reactive gas such as bromine, or ultraviolet radiation, before rechromatography with the same or a different solvent in the second direction. The use of bonded plates that separate according to different mechanisms depending on the solvent are especially advantageous for 2D TLC (e.g., amino: ion exchange or adsorption; cyano: normal or reversed phase). Precoated bilayer plates with a strip of chemically bonded C_{18} reversed-phase silica gel joined to the major plain silica gel area (or vice versa) allow combination reversed-phase partition and adsorption two-dimensional TLC, each with its own unique selectivity, to be performed. Concentration of the separated zones at the interface by development through the strip with a strong solvent may be required after the first run. Poole et al. (1989) and Poole and Poole (1995) have reviewed methods for combining different retention mechanisms in 2D TLC.

Quantitative analysis by slit-scanning densitometry is not too successful for 2D TLC because standards can be applied only after the first development and will not have the same zone configuration as the doubly developed analyte zones. Alternatively, standards and samples must be developed and scanned on different plates under conditions that have to be assumed identical. Best results are expected to be obtained in the fluorescence mode, for which zone shape has less effect on scan area, or by use of a flying spot (zigzag) scanner that can represent the chromatogram as a contour map or three-dimensional image (Poole and Poole, 1995). Instead of mechanical slit-scanning, electronic scanning or video densitometry (Chapter 10, section F) has the potential to be more successful for quantification of 2D chromatograms, but acceptance and use of this technique is not yet widespread.

Theoretical studies have shown that for capillary-flow 2D TLC, it should be possible to achieve a spot capacity of 100–250, that it is difficult to reach 400, and nearly impossible to reach 500. For forced-flow development, it is theoretically possible to generate a spot capacity in excess of 500 relatively easily, with an upper boundary of several thousand (Poole and Poole, 1991). An instrument for forced-flow 2D TLC was described (Gonnord and Siouffi, 1990). Several methods for computer simulation of 2D chromatograms employing unidimensional chromatographic data and either a mathematical function for rankings of chromatograms or visual interpretation using contour diagrams have been developed, and their principles and usefulness were reviewed by Poole and Poole (1995).

D. Horizontal

Various chambers, described below, are available for development of a plate in a horizontal position. A continuous supply of solvent must be transferred from a reservoir to the layer by some type of wick or capillary slit arrangement. Especially reproducible R_F values have been reported with horizontal development in these chambers. Development can be carried out from one side or from both sides.

Simple and inexpensive PTFE horizontal S-chambers for 5 × 5 cm or 10 × 10 cm plates are shown in Figure 7.1. The chambers are covered with a 4-mm-thick glass cover, and a glass frit rod is used to transport the mobile phase to the layer. Other commercial chambers for horizontal TLC include the Camag horizontal development chamber (Section IIIC), the Vario-KS chamber (Section IIIF), and the BN-chamber (no longer available commercially). For construction of a chamber suitable for horizontal TLC, see Stahl (1969) and Dzido (1990).

E. Multiple (Manual and Automated)

In multiple (manual) development, following a single development in the ascending mode, the chromatogram is removed from the chamber and dried, usually in air for 5–10 min. The chromatogram is then redeveloped in a fresh batch of the same solvent in the same direction for the same distance. This process, which may be repeated numerous times, increases the resolution of components with R_F values below 0.5. The theory of unidimensional multiple development, including an equation for calculation of the optimum number of developments, was reviewed by Perry et al. (1975). The final R_F value (^nR_F) for a multiple-

FIGURE 7.1 Desaga H-Separating Chambers. (Photograph courtesy of Desaga.)

developed solute can be predicted from the R_F after one development (^1R_F) by the expression

$$^nR_F = 1 - (1 - {}^1R_F)^n$$

where n is the number of developments. Repeated movement of the solvent front through the spots from the rear causes spot reconcentration and compression in the direction of the multiple developments, reducing the broadening that usually occurs during chromatographic migration (Poole et al., 1989).

Variations of the multiple-development technique for difficult separations involve moving the starting position of the development higher (Buncak, 1984) or removing some of the layer below the lowest spot (Poole et al., 1984) between runs. Poole and Poole (1992) found that raising the solvent entry position and simultaneously increasing the development distance in successive developments increased zone separation because migration distance increased while normal band broadening was effectively counteracted. Multiple chromatography with a variable solvent entry position was shown to always yield improved resolution compared with a fixed-entry position (Poole and Poole, 1995).

Other variations have been described by Szepesi and Nyiredy (1996) and Nyiredy (1996). Incremental multiple development (IMD) (Szabady et al., 1995 and 1997) involves rechromatography with the same composition mobile phase for distances that increase, usually by the same amount (linearly). If development occurs in the same direction with the same distance but different mobile phases having distinctive strength and selectivity, the method is termed gradient multiple development (GMD); this method most significantly increases the separation capacity of the system. In bivariate multiple development, the development distance and mobile phase composition are varied simultaneously for successive runs; this method, which is effective for samples of differing polarity, has been used especially for preparative layer chromatography plates.

Multiple development carried out with different solvents in the same direction, each being run the same or different distances (Stahl, 1959), is termed "step-wise development." A less polar phase can be used first, followed by a more polar phase, or vice versa. As an example, nonpolar impurities can be removed from polar solutes by development on silica gel with a nonpolar mobile phase such as toluene for the full length of the plate. This moves the nonpolar materials to the top part of the layer, leaving the polar solutes undisturbed at the origin. After drying, the polar solutes are separated by development with a polar mobile phase.

Multiple development combined with scanning densitometry can provide quantitative analysis of components of mixtures with widely differing R_F values that is optimized in terms of resolution and time. Solvent strength is successively changed to separate compounds of different polarity in the center region of the plate, with scanning after each development. Attempts to resolve the whole mixture in a single run are not needed. Lee et al. (1980) used this approach to separate and quantify five antiarrhythmia drugs in serum using two different mobile phases run for 7 and 13 min, respectively. Thirteen mycotoxins of widely differing polarity were completely separated and analyzed in 1 h by the sequential development and scanning technique, with detection limits in the low nanogram (for absorbance) and picogram (fluorescence) range and relative standard deviations between 0.7 and 2.2%.

Automated multiple development (AMD) is performed using the Camag AMD system (Section III.E.). The combination of AMD with HPTLC plates, automated sample application, and scanning densitometry represents one of the most powerful planar chromatography analytical systems currently available. Gradient elution in HPLC is almost always carried out on RP phases, but AMD allows reproducible gradient elution with silica gel. Multiple development is also carried out with forced solvent flow methods, in which spot capacity increases with the square root of the development distance instead of going through a maximum as with capillary-flow TLC (Nyiredy, 1992).

The theory and practice of multiple development were reviewed by Poole and Poole (1995).

F. Overrun (Continuous)

Like multiple development, continuous development is a method for effectively lengthening the plate to improve the resolution of slowly moving solutes. Using a commercial tank with a slot in the lid or a similar tank that is constructed in the laboratory (Stahl, 1969; Soczewinski, 1986), the plate is allowed to protrude from the top of the tank so that the solvent continuously evaporates. Continuous development can be conveniently carried out in the SB/CD chamber (see Section III).

In continuous development, low-R_F solutes are allowed to move farther than in normal development, increasing the number of theoretical plates available for their resolution. Weaker solvents (i.e., less polar for silica gel) must be used than in noncontinuous development so that high-R_F spots do not run together at the top of the plate. Disadvantages of the method include long development times and broadening of high-R_F zones. Nurok and co-workers studied the theory of continuous TLC (Tecklenburg and Nurok, 1984). They concluded that separations can be predicted (Tecklenburg et al., 1984; Nurok et al., 1984), and that continuous development is always faster than conventional development for a given separation (Nurok et al., 1982).

Evaporative TLC is a technique similar to continuous TLC, except the plate is heated to allow controlled evaporation as development occurs (Stewart and Wendel, 1975). Although improved speed of separation is claimed, the method has seldom been used.

G. Gradient Development

The composition of the mobile phase can be changed continuously during the course of development to improve separation of mixtures with components having a wide polarity range. A chamber for this has been proposed (Wieland and Determann, 1962) in which a polar solvent is pumped through a capillary tube on the side of a circular chamber and is mixed with the initial nonpolar mobile phase already in the chamber. Mixing is carried out by magnetic stirring, and the total volume of mobile phase is controlled by an overflow device.

In addition to continuous, stepwise, or exponential mobile-phase gradients, gradients can be produced in the stationary phase—for example, layers with defined pH regions, different sorbents, or activity variations. Gradients can be applied in a direction other than the development direction, and the gradient arrangement (i.e., parallel, diagonal, orthogonal, antidiagonal, antiparallel) is defined based on the gradient direction and the solvent flow direction (Niederwieser,

1969). Other types of gradients that have been employed include vapor, temperature (Marutoiu and Fabian, 1995), flow rate, particle size, and layer thickness (e.g., preparative taper plates; Chapter 12), as well as combination gradients such as mobile-phase/layer thickness. The history, nomenclature, classification, techniques, and apparatus for gradient development in TLC were reviewed by Golkiewicz (1996) and Grinberg (1990), and a reference book on the subject has been published (Liteanu and Gocan, 1974). The widest application of gradient TLC today is probably the use of stepwise gradients in connection with automated multiple development (Section III.E.). Gradients have been used with analytical and preparative forced-flow layer chromatography (Vajda et al., 1986), and gradient TLC is often combined with densitometric evaluation of chromatograms (Matysik, 1994).

H. Development of Heat-, Light-, and Air-Sensitive Compounds

Some heat-sensitive compounds or unstable compounds such as vitamins require that TLC br carried out at low temperature in a cold room or thermostatically controlled box (Stahl, 1965). Some substances, such as vitamin A compounds (De Leenheer and Lambert, 1996) are readily oxidized when exposed to air and they are very sensitive to light. For this type of compound, TLC is carried out in chambers that are hermetically sealed and wrapped with aluminum foil or used in the dark.

I. Radial (Circular)

Radial or circular development is used infrequently with conventional TLC plates but is of historical significance because it was employed in early spot chromatography procedures.

Mobile phase is slowly applied to the center of a ring of spots in the middle of a horizontal TLC plate. Sample components move outward in the form of concentric arcs under the influence of a negative solvent gradient that causes the rear of each zone to move relatively faster than the front. Zones near the origin (low R_F) remain symmetrical and compact and are especially well separated (Kaiser, 1988). Working with the same mobile phase, the resolution, especially in the low R_F range, is about 4–5 times greater in circular compared to linear development (Szepesi and Nyiredy, 1996). Zones with high R_F values become compressed in the direction of development and elongated in a perpendicular direction (Figure 7.2A). A series of concentric rings will be formed if the mobile phase is applied to a single spot located in the center of the plate, or if sample is injected into the mobile phase entering at the center of the plate. The separation power

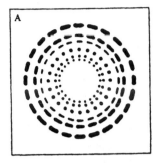
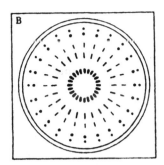

FIGURE 7.2 Circular development with the point of solvent entry at the plate center (A). Anticircular development from the outer circle toward the center (B). [Reprinted with permission of the American Chemical Society from Fenimore and Davis (1981).]

of circular development is optimized if samples are spotted near to the center (Nyiredy and Fater, 1994).

The relation between linear and circular migration is given by $R_{F(circ)}^2 = R_F(\text{lin})$ (Po and Irwin, 1979; Kaiser, 1977). In circular development, the solvent is pulled out into an increasing plate area; therefore, the flow rate remains relatively constant. This is in contrast to linear development, in which the flow decreases with the square of the distance the solvent front travels.

Radial development has been used for quickly selecting a suitable mobile phase for linear or circular TLC (Stahl, 1958) by adding solvents from a pipet onto single spots of mixture placed on microscope slides and observing the optimum pattern of rings. A commercial device for simultaneously testing 16 solvents on a single 20 × 20-cm plate by horizontal radial development is available from Schleicher and Schuell. The mobile phases enter the layer through holes connected to their reservoirs. This Selecta Sol apparatus accomplishes the same results as the pipet/microscope slide approach in a more elegant manner.

Radial chromatography is not as versatile or convenient as normal linear TLC, for example, for comparison of samples with standards, elution of separated zones, and in situ quantification. More elaborate and precise instruments for radial and anticircular chromatography are described in Sections III I, J, K of this chapter.

Circular chromatography can be carried out with forced-flow as well as capillary-flow mobile-phase migration (Botz et al., 1992). No special plate preparation is required for off-line radial OPLC or RPC, but for on-line radial OPLC, a sector must be isolated by scraping the layer and then impregnated with solvent (Nyiredy, 1992).

J.   Anticircular (Antiradial)

The anticircular development mode is the exact opposite of conventional circular development. Mobile phase is applied to the layer along a precise outer circle, from where it flows over the initial zones inward toward the center. Migration in linear and anticircular TLC are related by $R_{F(lin)} = 1 - (1 - R_{F(AC)})^2$ (Kaiser, 1977). Spots near the origin remain compact, whereas those close to the solvent front are narrow and elongated (Figure 7.2B). In comparison to linear development, substances with high R_F values are separated particularly well (Kaiser, 1988). The flow rate of solvent is the fastest with respect to separation distance and is almost constant; a migration distance of 45 mm requires ~4 min. Anticircular development provides the highest sample capacity per plate of any TLC technique, but has the poorest separation performance per unit bed length (Robards et al., 1994).

Antiradial on-line OPLC (Section III.K.) and U-RPC (Section III.L.) separations require specially prepared plates. Radial and antiradial development are combined in the sequential mode of RPC; the radial mode is used for separation of zones, whereas the antiradial mode pushes zones back toward the center of the plate with a strong solvent before drying with nitrogen and the next radial development with another suitable mobile phase (Nyiredy, 1992).

Consult Kaiser (1978) for a description of the theory and techniques of anticircular TLC.

K.   Forced-Flow Planar Chromatography

In forced-flow planar chromatography (FFPC), external pressure of 10–100 bars drives the mobile phase through the sorbent rather than capillary action as in conventional planar chromatography. The mobile-phase flow is constant and continuous and is adjustable to an optimal rate along the entire layer, leading to an extension of the useful separation distance of the layer and higher resolution. A larger number of theoretical plates is available in FFPC, the total number being limited by the available pressure drop and bed length. Separation times are reduced (e.g., 30 cm separation distance in 30 min), and solvents that poorly wet the layer can be used (Robards et al., 1994). The two modes of FFPC are overpressured layer chromatography (OPLC) and rotation (centrifugal) planar chromatography (RPC). OPLC is carried out by forcing solvent into a spotted, dry or solvent-equilibrated layer covered with a membrane under pressure, and the method can be done in on-line or off-line modes (Nyiredy, 1996). OPLC was reviewed by Mincsovics et al. (1996) and used by Tyrpien (1996) to separate hydroxy derivatives of polycyclic aromatic hydrocarbons.

High-pressure planar chromatography (HPPC) (Kaiser and Reider, 1986) is a radial version of OPLC in which a single sample is applied to the center of a 10 × 10-cm HPTLC plate and development is carried out with ~30 bars of external pressure. HPPC is said to be useful both for sample analysis and transfer of solvents to HPLC. Developing times are very short and flow rates optimal. As an example, D,L-amino acids were separated on a Chiralplate into their antipodes within 2–3 min by use of HPPLC, but its application has not been reported to any significant degree.

The mobile phase in RPC is driven by centrifugal force through the samples, up to 70 of which can be applied near the center of the plate for analytical separations. In preparative RPC, one sample is applied in the center as a circle. Again, either off-line or on-line modes are possible, the latter for UV detection of effluents and collection of sample fractions separated in preparative RPC (Nyiredy, 1996). Instruments consist of two parts, one to rotate the plate and the other to feed sample to the center of the rotating layer. Preparative RPC was reviewed by Nyiredy (1996).

Mobile-phase velocity is higher with forced-flow development than in capillary-flow TLC. The actual flow rate is influenced by the type of chamber (rectangular or sandwich, saturated or unsaturated), the pressure and solvent viscosity (OPLC), or the rotational speed (RPC) (Nyiredy et al., 1988a). Nyiredy (1992) discussed the relation among resolution, separation distance, and time for forced-flow planar chromatography compared to capillary flow. It was stated that for separation of nonpolar compounds by FFPC on silica gel, a separation time of ~1–2.5 min over a separation distance of 18 cm can be used without great loss in resolution. By contrast, longer separation times are needed for separation of polar compounds.

III. CHAMBERS AND INSTRUMENTS

Instruments for development of plates were reviewed by Jaenchen (1996).

A. Rectangular Glass Tanks (N-Tanks)

A variety of chambers or tanks are available for ascending development of TLC and HPTLC plates of different sizes (Figure 7.3). A rectangular glass tank, or N-tank, is most frequently used in TLC to develop 20 × 20-cm TLC plates. The most popular model has inner dimensions of 21 × 21 × 9 cm and can accommodate two plates using 50–100 ml of solvent to fill the tank to a depth of 5 mm. The tank is lined on the back wall and sides with thick filter paper (Whatman 3MM) that becomes thoroughly soaked when the solvent is poured in. The tank should stand for 5–10 minutes to allow the inside atmosphere to be saturated with solvent vapor and to obtain equilibrium before the plate is inserted. If the cover is removed and replaced quickly, there is only minimal disturbance of the equilibration inside the chamber. To facilitate saturation, bulldog clips can be

FIGURE 7.3 Glass developing chambers for TLC and HPTLC. (Photograph courtesy of Analatech.)

used to secure the lid to the tank, or a heavy object can be placed on top of the lid. However, it is usually adequate to simply place a fairly thick, flat glass plate on the ground-glass top surface of the tank. Glass is advantageous as tank construction material because of its inertness, ease of cleaning, and transparency, which allows observation of any anomalies in solvent front flow that can adversely alter chromatographic results. The sides of the plate must not contact the filter paper lining the tank or solvent will enter the layer on the side and cause anomalous development. Analtech supplies a TLC solvent front detector that sounds an alarm when development has reached a preselected level.

If two plates are developed simultaneously, care must be taken to lean them against opposite walls so that they are not in contact. This is facilitated by glass block chambers with a ridge across the center to serve as a support for the bottoms of the plates. Upon dipping the plates into the mobile phase, the origins must be above the level of the solvent in the tank so that the sample will not diffuse away. Grease should never be applied to the lid because this can be transferred to the layer and cause anomalous results.

Analtech supplies a "plate conditioning apparatus" that allows one to five 20×20-cm plates to hang in the tank containing the mobile phase so that both the plates and the tank atmosphere can be equilibrated before the plates are lowered into the mobile phase using a bar that protrudes through the chamber cover. In this case, the equilibrated tank does not have to be opened for insertion of the plate.

Some workers prefer to use unsaturated tanks, that is, to pour in the mobile phase and to add the plate and begin development at once. Unsaturated tanks will produce generally higher R_F values because of evaporation of mobile phase from the surface of the layer and subsequent higher mobile-phase flow through the sorbent. Superior resolution with multicomponent solvents has also been reported in some situations. The improvements are caused by a concentration gradient and mostly depend on the differential rates of evaporation of the solvent components and their affinities for the sorbent layer (De Zeeuw, 1968a). However, because evaporation of solvent proceeds more rapidly at the edge of the plate than in the center, a concave solvent front is more likely to form in an unsaturated tank, leading to higher R_F values for a given solute near the edges compared to the center (Stahl, 1959), and poor reproducibility in R_F values from plate to plate.

The twin-trough developing chamber (Camag) is a very useful variation of the N-tank that facilitates layer preconditioning. The tank has a raised glass wedge-shaped ridge down the center that effectively separates the chamber into two separate compartments (Figure 7.4). When used as a normal N-tank, development conditions are identical, except that much less solvent is used (20 ml/20 × 20-cm plate) (Figure 7.5A). Smaller twin-trough chambers are available for 10 × 20-cm and 10 × 10-cm plates that require ~12 ml and 4 ml of solvent, respectively. Equilibration is conveniently achieved by filling the paper-lined right trough with the mobile phase and putting the plate in the empty left trough. After ~10 min, the tank is carefully tipped to transfer the solvent and begin development. It is therefore, not necessary to remove the lid to insert the plate into the tank after preequilibration, as in the case of the conventional N-chamber. Alternatively, one trough can be used for development, while the other trough contains a different solvent or a sulfuric acid–water humidity-control mixture for adsorbent preconditioning (Figure 7.5B). In this case, development starts only when mobile phase is added to the trough holding the plate. For example, mobile phases containing ammonium hydroxide or acetic acid usually demix and form a second (beta) front below the front representing the farthest movement of the solvent (the alpha front). If an ammonia- or acetic-acid containing solvent is instead placed in one of the troughs to impregnate the layer, it may be possible to develop the plate in the other trough with a neutral solvent that will not demix (Jork et al., 1990). The use of solvent vapors to modify adsorbent properties and improve resolutions has been exploited by De Zeeuw (1968b) in "vapor-programmed" TLC (see also the section on the Vario-KS chamber).

FIGURE 7.4 Camag Twin-Trough Chambers for development of 20 × 20-cm plates, shown (front to back) with stainless steel and glass lids and without a lid. (Photograph courtesy of Camag.)

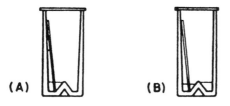

(A)   (B)

FIGURE 7.5 The Camag Twin-Trough Chamber setup for standard development (A) or for layer preconditioning with the mobile-phase vapor or any conditioning liquid or volatile reagent (B). (Reproduced with permission of Camag.)

B. Sandwich Chambers

The sandwich design (S-chamber) has a very thin chamber (<3 mm) to accommodate a single plate or foil and a minimal amount of solvent for ascending chromatography. S-chambers with a plain glass cover plate are "unsaturated," whereas "saturated" conditions are achieved by employing a counterplate coated with sorbent (often cellulose) and soaked with solvent. S-chambers offer two distinct advantages over rectangular tanks, namely, faster gas-phase equilibration and excellent reproducibility (Dallas, 1965), both due to the much smaller internal volume when compared with N-tanks. A sandwich chamber is used when the sorbent layer is not to be preloaded with solvent or when evaporation of solvent from the layer is to be avoided (Bauer et al., 1991). Optimal results in classical TLC/HPTLC are achieved using saturated S-chambers. Unsaturated or saturated S-chambers are best used for optimization of solvent systems for the various modes of OPLC and RPC (Nyiredy, 1992).

A variety of commercial S-chambers are available for 10 × 10-, 10 × 20-, and 20 × 20-cm plates. One type is a sandwich plate cover with permanently fixed glass spacers on two sides. It is 2 cm shorter than a 20 × 20-cm plate and is positioned so that it does not dip into the solvent. Therefore, there is no need to scrape off the layer at the sides. The sandwich composed of the TLC plate and the cover plate, which is open at the top and held together by four stainless-steel clamps, can be developed in a regular N-tank. Other models consist of a cover plate plus a small-volume chamber for vertical development.

C. Horizontal Development Chambers

The Camag Horizontal Developing Chambers (Figure 7.6) accomodate 10 × 10- or 10 × 20-cm HPTLC plates. Development can be carried out from one end to the other, maximizing resolution, or from both ends into the middle, increasing sample capacity. Solvent is carried from reservoirs to the layer by capillary flow by microscope slides (Figure 7.7). Better performance is achieved because the plate is horizontal (solvent flow is not against gravity) and free volume around the plate is minimal (leading to rapid and uniform equilibration). If the tank is perfectly level and development is carried out from both ends of the plate, the solvent fronts will meet exactly in the center and development will stop. Up to 70 spots can be applied 5 mm apart on a 20 × 10 cm plate in this configuration, or 36 seven-mm bands separated by 3 mm. Figure 7.6 shows 15 chromatograms developed from each end of a single layer. As a sandwich chamber, solvents with volatile acids and bases or those with a large concentration of volatile polar solvent should not be used (Poole and Poole, 1991). The chamber can be operated as a normal tank as well as in the sandwich configuration, in which case it is suitable for all types of solvents. Conditioning at a desired relative humidity can

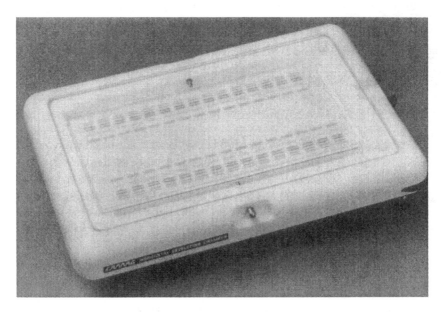

FIGURE 7.6 Camag Horizontal Developing Chamber for 20 × 10-cm plates. (Photograph courtesy of Camag.)

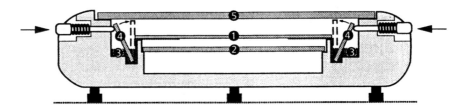

FIGURE 7.7 Principle of operation of the Camag Horizontal Developing Chamber in the sandwich configuration with development from both sides. The HPTLC plate (1) is placed layer-down at a distance of 0.5 mm from the counterplate (2). If the tank configuration is to be used, the counterplate (2) is removed and the recess below is either filled with conditioning liquid or left empty. Narrow troughs (3) hold the mobile phase. Development is started by pushing the levers, which tilt the glass strips (4) inward. Solvent travels to the layer through the resulting capillary slit. Development stops automatically when the solvent fronts meet in the center (development distance, 4.5 cm). The chamber, which is constructed from poly(tetrafluoroethylene), is kept covered with a glass plate (5) at all times. (Courtesy of Camag.)

be established with an appropriate sulfuric acid–water mixture placed in the tray beneath the counterplate.

The horizontal DS chamber has been described for use in isocratic, gradient, and multiple development, as well as simultaneous development using different mobile phases under different conditions (Waksmundzka-Hajnos and Wawrzynowicz, 1994; Matysik, 1994; Matysik and Wojtasik, 1994; Matysik et al., 1994; Golkiewicz, 1996; Matysik and Toczolowski, 1997). A modified version of the DS chamber allows the solvent entry position to be changed by movement of the plate, increasing efficiency in single and multiple development (Dzido et al., 1995).

D. Automatic Chromatogram Development

A microprocessor-controlled instrument is available from Desaga (TLC-MAT) for automatic ascending chromatogram development without supervision (Figure 7.8). The dry plate can be preconditioned in the vapor without contact with the mobile phase. Chromatography takes place on plates with 10 or 20 cm height in a chamber protected from the atmosphere and light; a sensor recognizes the solvent front (programmable from 2–18 cm in steps of 0.1 cm) and reports the exact development time, and solvent vapors are removed by an integrated fan after development. Up to 10 programs can be stored in battery-buffered memory.

Similar Automatic Developing Chambers (ADC) for either 20 × 10 or 20 × 20 cm plates are available from Camag. The ADC was used to separate saturated and unsaturated oligogalacturonic acids by multiple development with 1-propanol–water mixtures on silica gel 60 plates (Dongowski, 1997).

E. Automated Multiple Development System

The first automated multiple development (AMD) technique was termed programmed multiple development (PMD). The commercial device for PMD originally manufactured by Regis is no longer available.

The Camag AMD system (Poole et al., 1989) (Figure 7.9) consists of a developing module and a microprocessor control unit. It is essentially a closed N-chamber with connections for adding and removing developing solvents and vapor phases. Mobile phases are prepared by a pump and gradient mixer from pure solvents contained in six storage bottles. The plate is subjected to multiple (up to 25) linear developments over increasingly longer (~3–5 mm distances) with solvents of decreasing strength (i.e., decreasing polarity for silica gel). The gradient used for AMD (termed a "universal gradient") has been often composed of methanol or acetonitrile as the most polar component; dichloromethane, diisopropyl ether, or *t*-butylmethyl ether as the "basis" or central solvent; and hexane as the nonpolar component. Poole and Poole (1995) have described systematic approaches to mobile-phase selection for AMD. Solvent is removed and the layer

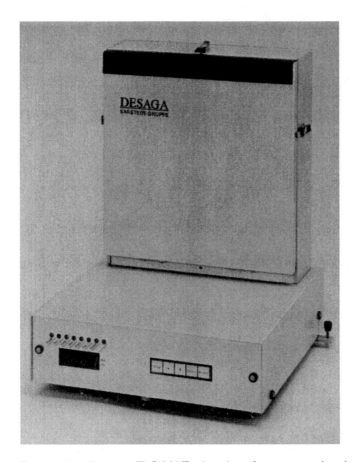

FIGURE 7.8 Desaga TLC-MAT chamber for automatic plate development. (Photograph courtesy of Desaga.)

is dried under vacuum between runs. Before the next development, the plate is reconditioned by pumping a vapor phase from a reservoir into the chamber. The first development with strong solvent focuses the samples without separation, and later developments with the stepwise mobile-phase gradient cause zones to compress into thin bands with excellent resolution. The migration distance of the individual components is largely independent of the sample matrix. All steps are computer controlled and fully automated, and the process is highly reproducible. Gradient elution allows compounds with widely different polarities to be separated in a single chromatogram.

A separation number of 18 was reported for the HPTLC-AMD of addictive

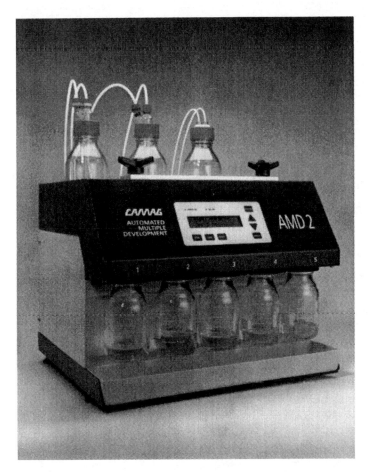

FIGURE 7.9 Camag AMD 2 System. (Photograph courtesy of Camag.)

drugs in a comparative study of different development methods (Kovar, 1997), but Camag claims that up to 40 components can be baseline-resolved in the usable separation distance of 80 mm for AMD on a 10 × 10 cm HPTLC plate, which makes it an attractive alternative to OPLC for separations requiring high separation numbers (Jaenchen, 1996). AMD is not useful for volatile compounds that may be lost or unstable compounds that break down during the repeated drying steps, and solvents that are not completely removed between developments cannot be used.

The techniques and applications of AMD have been reviewed (Poole et al., 1989; Poole and Belay, 1991; Jaenchen, 1991). Figure 7.10 shows the separation

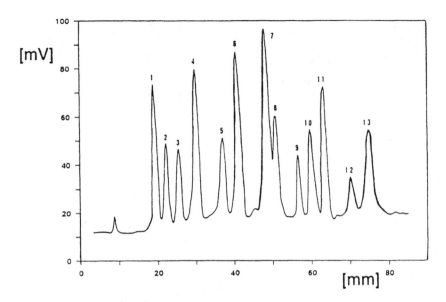

Figure 7.10 Separation of a standard mixture of 13 phenolic compounds by automated multiple development by use of a 20-step universal gradient composed of acetonitrile, methylene chloride, and hexane. The drying time was 4 min, and the gas phase for conditioning the plate between runs was prepared from 10% aqueous formic acid. The phenols were (1) rutin, (2) kaempferol-3-rutinoside, (3) quercetin-3-arabinoside, (4) quercetin-3-galactoside, (5) chlorogenic acid, (6) myricetin, (7) caffeic acid, (8) quercetin, (9) apigenin, (10) ferulic acid, (11) acacetin, (12) flavon, (13) coumarin. [Reprinted from Lodi et al. (1991) with permission of Alfred Huethig Verlag GmbH.]

of 13 phenol standards by AMD-TLC (Lodi et al., 1991). Recent applications of HPTLC-AMD (on silica gel unless otherwise noted) include the analysis of pulp mill bleaching process effluent (methanol-dichloromethane gradient) (Dethlefs and Stan, 1996) and determinations of plant extracts (25-step gradient using methanol, ethyl acetate, toluene, 1,2-dichloroethane, 25% ammonia, and anhydrous formic acid) (Gocan et al., 1996); amines in textiles (Kralecik, 1996); phenylurea herbicides in plants (Lautie and Stankovic, 1996); sugars (diol layers) (Lodi et al., 1997); gangliosides (chloroform–methanol–20 mM aq. CaCl$_2$, 120:85:20, mobile phase, triple development) (Muething and Ziehr, 1996); pesticide residues in foods (Stan and Wippo, 1996); and lipids (25-step gradient based on methanol, diethyl ether, and *n*-hexane) (Zellmer and Lasch, 1997).

F. Vario-KS Chamber

This Vario-KS chamber is a versatile but elaborate sandwich-type horizontal development chamber, designed by Geiss and Schlitt (1968) and manufactured and distributed by Camag, for evaluation of the effects of different solvents, solvent vapors, and relative humidities on TLC separations and for obtaining very reproducible chromatograms. Saturated versus unsaturated conditions can be compared, and developments with solvent or vapor gradients can be carried out. The optimum conditions worked out in the Vario-KS chamber can be easily transferred to the twin-trough tank (Figure 7.4) for more simple, routine operation, or to forced-flow separations of complex mixtures.

In use, a 20 × 20-cm plate is placed layer downward over a glass tray containing 5, 10, or 25 compartments to hold appropriate conditioning solutions. Five samples on a single plate can be developed simultaneously with a variety of mobile phases (Sandroni and Schlitt, 1970), or with the same mobile phase after preconditioning each track with a different sulfuric–water mixture or saturated salt solution to produce a range of activities or with vapors of a different solvent (De Zeeuw, 1970). The design of the chamber ensures that saturating and developing solvents are segregated. The chamber can be used with a heating accessory for continuous development.

The Camag HPTLC Vario system for 10 × 10- or 10 × 20-cm HPTLC plates is shown in Figure 7.11. Development with six different solvents can be performed side by side, and up to six different preequilibration conditions can be tested simultaneously in N-tank or sandwich configuration.

The Vario chambers, which can be used in sandwich or tank configurations, allow selection of the optimal vapor-phase conditions and are best suited for working with the PRISMA optimization system (Chapter 6) (Nyiredy, 1992).

The theory and techniques of vapor programming (De Zeeuw, 1968b) and adsorbent, solvent, and solvent-vapor gradients (Niederwieser, 1969; Snyder and Saunders, 1969; Golkiewicz, 1996) have been studied and reviewed. These techniques are rather complex and have not been widely used in practical TLC.

G. Horizontal BN-Chamber

The horizontal BN-chamber, which is illustrated by Stahl and Mangold (1975), was designed by Brenner and Niederwieser (1961) for the separation of amino acids. It is essentially a sophisticated sandwich chamber useful for horizontal development under controlled conditions. The chamber will not be described because it is no longer commercially available.

H. SB/CD Chamber

Substances with very similar R_F values in a given mobile phase will not separate on a TLC plate. Addition of a nonpolar solvent such as hexane to lower the

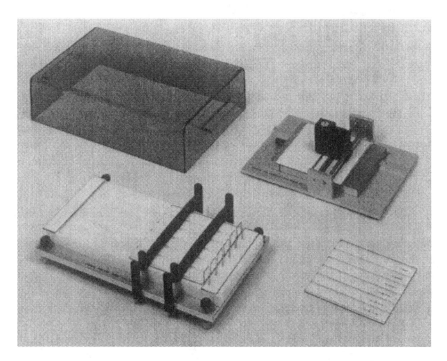

FIGURE 7.11   Camag HPTLC Vario System.

mobile-phase strength will reduce the R_F values, and, in many cases, the reductions will occur at different rates so that selectivity is improved (Perry, 1979). The Regis Short Bed/Continuous Development (SB/CD) chamber (Figure 7.12) is a flat N-chamber that permits plates to be developed continuously over a short distance at constant, high solvent velocities so that slow-moving solutes will leave the origin in a reasonable time to be resolved in these selective, low-strength mobile phases. Because the distance of movement is small, zone diffusion is low, and sensitivity of detection is increased. Four ridges in the chamber base provide five different plate angles and development lengths (Figure 7.12). Saturated conditions can be attained by lining the walls and lid of the chamber with filter paper. The SB/CD chamber has been used for time-optimized TLC separations with binary mobile phases composed of a weak and strong solvent (Tecklenburg et al., 1983). In addition to using the SB/CD chamber for continuous development with a single solvent, it has been employed to carry out multiple development with different solvents, plate lengths, and development times. For example, amino acid PTH-derivatives were separated using five developments with four changes

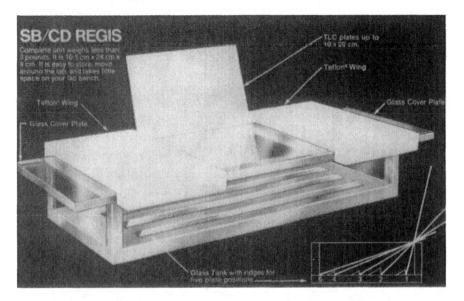

FIGURE 7.12 Regis SB/CD chamber. (Photograph courtesy of Regis.)

in mobile-phase composition (Schuette and Poole, 1982). The total development time was less than 1 h, and the plate was scanned after three of the runs to quantify the resolved acids. Some USP methods for TLC drug analysis specify use of the SB/CD chamber. Details of use and applications of the SB/CD chamber are available from Regis.

I. Circular Chambers

The U-Chamber (Kaiser, 1988) for radial TLC, which was described in the third edition of this book, is no longer available from Camag. Analtech offers a simple, inexpensive stainless steel radial HPTLC chamber with a centered wick for feeding solvent to the layer.

J. Anticircular U-Chamber

Figure 7.13 illustrates the Camag Anticircular U-chamber. samples (24 or 48) are applied in an outer ring with a Nanomat spotter (Figure 5.5) or manually using a positioning block. Solvent is fed from a reservoir to a narrow circular channel in direct contact with the layer outside of the initial zones. Sorbent is scraped away outside of the channel so that solvent can only flow inward toward the center of the plate. Mobile phase is transferred by capillary action and is not

FIGURE 7.13 Camag Anticircular U-Chamber. (Photograph courtesy of Camag.)

externally controlled. Anticircular development has the highest sample capacity and development speed of all TLC modes. Sample resolution and minimum detectable levels are optimum for high R_F values, despite elongation of these zones (Figure 7.2B). Solvent consumption is very low with both the U-chamber and anticircular U-chamber. Anticircular TLC has been used for the simultaneous separation of up to 48 samples in 3–4 min. Consult Kaiser (1978) for a description of the theory and techniques of this TLC variation.

K. Overpressured Layer Chromatography

The technique termed overpressured (or overpressure, overpressurized) layer chromatography is carried out in commercial instruments called Chrompres 10 or

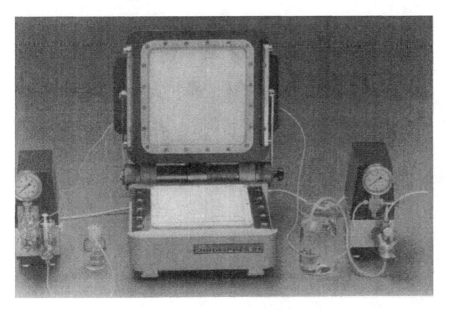

Figure 7.14 Chrompress-25 OPLC system showing the chamber in the center, eluent pump on the left, and water pump on the right. (Photograph courtesy of LABOR Instrument Works, Budapest, Hungary.)

25 (Figure 7.14) (LABOR Instrument Works, Budapest, Hungary), the numbers referring to the maximum external pressure (bar) (Newman, 1985). The chambers have a flexible membrane under external pressure (produced by pumping water to create a cushion) completely covering the surface of the layer to eliminate the vapor phase (essentially an unsaturated S-chamber), and mobile phase is applied at a constant rate by a second pump. The presence of a disturbing zone and multifront effect result from the absence of a vapor phase and must be considered when choosing the mobile phase and performing separations (Szepesi and Nyiredy, 1996). Development can be carried out in one-directional linear (development from bottom to top), two-directional linear (development of two lines of samples from the center toward the ends of the plate), circular (radial), antiradial, and linear two dimensional. Linear development is most common but requires special preparation of the plate, by scraping and application of a sealant to avoid mobile-phase leakage, before it is inserted in the OPLC instrument. A narrow channel is placed before the eluent inlet to achieve linear migration of the front. Longer plates can be used with pressurized solvent flow, allowing separation of more compounds than on shorter plates with capillary-flow development. OPLC

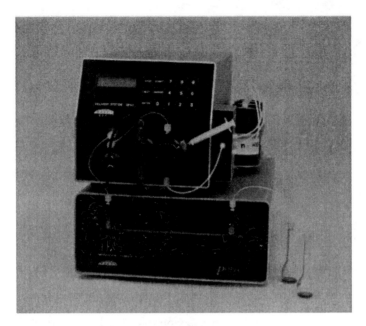

FIGURE 7.15   Personal OPLC BS 50 system including separation chamber for 20 × 20 or 20 × 10 cm plates, holding unit, layer casette, and liquid delivery system. (Photograph courtesy of OPLC-NIT Engineering Co., Ltd., Budapest, Hungary.)

is most closely related to HPLC and is useful for scouting solvents for HPLC. Continuous development with on-line detection by use of an HPLC UV detector is valuable for preparative OPLC. A newer commercial instrument, the Personal OPLC BS 50 (OPLC-NIT Engineering Company, Ltd., Budapest, Hungary) (Figure 7.15), allows automated linear, isocratic, or three-step gradient developments in on- or off-line modes for analytical or preparative separations (Mincsovics et al., 1996). This instrument was used for OPLC separation of ascorbigen and 1'-methylascorbigen on silica gel (Katy et al., 1997). Another instrument was described by Witkiewicz et al. (1993), in which the mobile phase is fed to the layer from below via a syringe pump; gas is used to supply external pressure to the plate. This instrument can be set up quickly and is easy to use (Mincsovics et al., 1996). Readers should consult Mincsovics et al. (1996) and Nyiredy (1992 and 1996) for details of theory, procedures, instrumentation, and applications of OPLC.

An instrument for high-speed circular or anticircular high-pressure planar chromatography on 10 × 10-cm HPTLC plates, the HPPLC 3000 (Institute for

FIGURE 7.16 Chromatotron Model 8924 with solvent delivery pump. The Chromatotron Model 7924 shown schematically in Figure 7.18 differs from the 8924 only in the use of a Teflon funnel, instead of a pump, for solvent supply. (Photograph courtesy of Harrison Research Inc., Palo Alto, CA.)

Chromatography, Bad Durkheim, Germany), is described by Nyiredy (1992), Kaiser and Rieder (1986), and Tyihak and Mincsovics (1988). The mobile phase is supplied by a pump to the layer, which is under very high pressure produced by a hydraulic press. An apparatus for high-pressure TLC up to a final pressure drop of 140 bar based on the principle of OPLC was described by Flodberg and Roeraade (1995).

L. Rotation Planar Chromatography

Chamber types for rotation planar chromatography (RPC) differ according to the vapor space: in N-RPC, the layer rotates in an unsaturated N-chamber; in M- and U-RPC (Nyiredy et al., 1988b), the chamber is the saturated (M) and unsaturated (U) sandwich type (Nyiredy, 1992). In the M (micro)-chamber, the plate

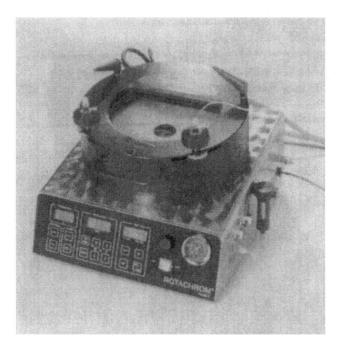

FIGURE 7.17   Rotachrom Model P. (Photograph courtesy of Petazon Ltd., Zug, Switzerland.)

FIGURE 7.18   Schematic drawing of RPC instruments. (a) Chromatotron Model 7924. 1 = Annular chromatographic chamber, 2 = pedestal, 3 = flat glass rotor, 4 = stationary phase, 5 = fixing screw, 6 = motor, 7 = circular channel for eluate collection, 8 = mobile-phase inlet, 9 = quartz glass lid, 10 = eluate outlet, 11 = inlet tube for inert gas. (b) ROTACHROM Model P. 1 = Upper part of the stationary chromatographic chamber, 2 = collector, 3 = tubes in the collector, 4 = flat glass rotor, 5 = stationary phase, 6 = vapor space, 7 = fixing screw, 8 = solvent delivery system, 9 = safety glass, 10 = UV lamp (254 nm), 11 = UV lamp (366 nm), 12 = motor shaft with tube, 13 = motor, 14 = lower part of the stationary chromatographic chamber, 15 = casing of the instrument, 16 = front panel for adjusting and controlling units with keyboard, 17 = eluate outlet. [Reproduced from Nyiredy (1991) with permission from Marcel Dekker, Inc.]

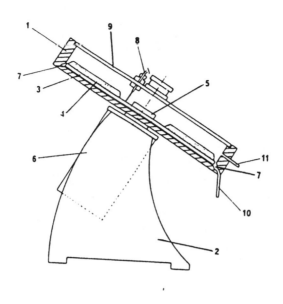

(a)

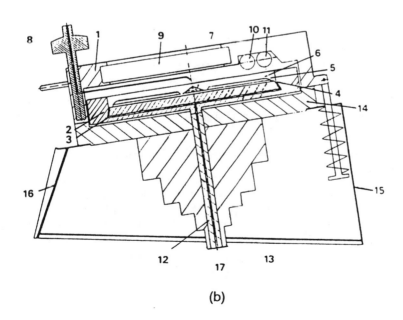

(b)

rotates together with a small chamber; the space between the layer and the chamber lid is less than 2 mm so that the vapor space is rapidly saturated. In the U (ultramicro)-chamber, the lid of the rotating chamber is directly on the plate so that essentially no vapor space exists (Szepesi and Nyiredy, 1996). Preparative and analytical RPC can be carried out in linear, radial, and antiradial development modes. Linear development is performed in an M- or U-chamber by scraping the layer to form lanes (Nyiredy et al., 1989).

Commercial instruments for RPC, the Chromatotron (Harrison Research Inc., Palo Alto, CA) and Rotachrom (Petazon Ltd., Zug, Switzerland), are illustrated in Figures 7.16 and 7.17, respectively, and schematic drawings of both instruments are given in Figure 7.18. The microprocessor-controlled Rotachrom allows variable rotation speed from 20 to 2000 rpm, with higher rotational speed accelerating mobile phase migration. With proper layer preparation, the Rotachrom can be used for linear, circular, or (by scraping appropriate channels in the layer) anticircular development and analytical or preparative separations. In the Chromatotron Model 7924, an annular chamber is fastened to a pedestal at an angle. A flat 24-cm-diameter glass rotor covered with stationary phase is mounted within the chamber and driven by a motor at a constant speed of 750 rpm. The Rotachrom Model P can be used for analytical or on-line preparative methods. The schematic (Figure 7.18b) shows the preparative U-RPC configuration. The plate is spun around a central axis at 80–1500 rpm to drive the solvent through the layer. Preparative RPC involves thicker layers and on-line collection of separated samples. Detailed descriptions of these instruments were provided by Nyiredy (1996, 1992) and Nyiredy et al. (1989).

A third RPC instrument, the CLC-5 (Hitachi), was described by Nyiredy (1996) for use in PLC. It consists of a rotating disk column comprising two removable disks of 30 cm diameter, UV detector, and automated fraction collector. Its speed can be continuously varied from 0 to 1000 rpm.

M. Robotics and Automation of the Overall TLC System

The application of automation and robotics in planar chromatography has been discussed in a number of recent publications (Delvorde and Postaire, 1993; Postaire et al., 1996; Buhlmann et al., 1994; Zieloff, 1995). Although various stages of the TLC process, such as multiple development, have been successfully automated, completely automated systems are not readily available.

SAFETY NOTE

When using mobile phases for development that contain carcinogenic or possibly carcinogenic solvents, such as carbon tetrachloride, chloroform, dioxane, or benzene, work in an efficient hood to avoid breathing vapors and avoid skin contact as much as possible.

REFERENCES

Bauer, K., Gros, L., and Sauer, W. (1991). *Thin Layer Chromatography—An Introduction*, E. Merck, Darmstadt, Germany.

Bhushan, R., and Martens, J. (1996). Peptides and proteins. In *Handbook of Thin Layer Chromatography*, 2nd ed., J. Sherma and B. Fried (Eds.). Marcel Dekker, New York, pp. 427–444.

Botz, L., Dallenbach-Toelke, K., Nyiredy, Sz., and Sticher, O. (1992). Characterization of band broadening in forced flow planar chromatography with circular development. *J. Planar Chromatogr.—Mod. TLC 5*: 80–86.

Brenner, M., and Niederwieser, A. (1961). Overrun thin layer chromatography. *Experientia 17*: 237–238.

Buhlmann, R., Carmona, J., Donzel, A., Donzel, N., and Gil, J. (1994). An open software environment to optimize the productivity of robotized laboratories. *J. Chromatogr. Sci. 32*: 243–248.

Buncak, P. (1984). Analytical experiences with a "MOBIL-R_F" chamber for sequence thin layer chromatography (STLC). *Fresenius' Z. Anal. Chem. 318*: 289–290.

Cserhati, T., and Forgacs, E. (1996a). Introduction to techniques and instrumentation. In *Practical Thin Layer Chromatography—A Multidisciplinary Approach*, B. Fried and J. Sherma (Eds.). CRC Press, Boca Raton, FL, pp. 1–18.

Cserhati, T., and Forgacs, E. (1996b). Thin layer chromatography. In *Chromatography Fundamentals, Applications, and Troubleshooting*, J.Q. Walker (Ed.). Preston Publications, Niles, IL, pp. 185–207.

Dallas, M. S. J. (1965). Reproducible R_F values in thin layer adsorption chromatography. *J. Chromatogr. 17*: 267–277.

De Leenheer, A. P., and Lambert, W. E. (1996). Lipophilic vitamins. In *Handbook of Thin Layer Chromatography*, 2nd ed. J. Sherma and B. Fried (Eds.). Marcel Dekker, New York, pp. 1055–1077.

Delvorde, P., and Postaire, E. (1993). Discussion: automation and robotics in planar chromatography. *J. Planar Chromatogr.—Mod. TLC 6*: 289–292.

Dethlefs, F., and Stan, H. J. (1996). Analysis of pulp mill EOP bleaching process effluent by thin layer chromatography with automated multiple development. *Acta Hydrochim. Hydrobiol. 24*: 77–84.

De Zeeuw, R. A. (1968a). Influence of solvent vapor in thin layer chromatography using sandwich chambers and unsaturated chambers. *Anal. Chem. 40*: 915–918.

De Zeeuw, R. A. (1968b). Vapor programmed thin layer chromatography, a new technique for improved separations. *Anal. Chem. 40*: 2134–2138.

De Zeeuw, R. A. (1970). The use of vapor-programmed thin layer chromatography in the separation and identification of sulfonamides. *J. Chromatogr. 48*: 27–34.

Dimov, N., and Filcheva, K. (1996). Two-dimensional, two phase TLC determination of parabens in pharmaceutical formulations. *J. Planar Chromatogr.—Mod. TLC 9*: 197–198.

Dongowski, G. (1997). Determination of saturated and unsaturated oligogalacuronic acids by means of thin layer chromatography. *J. Chromatogr., A 756*: 211–217.

Dzido, T. H. (1990). New modifications of the horizontal sandwich chamber for TLC: variants of the development technique. *J. Planar Chromatogr.—Mod. TLC. 3*: 199–200.

Dzido, T. H., Hawryl, M. A., Miroslaw, A., Golkiewicz, W., and Soczewinski, E. (1995). A device for high pressure thin layer chromatography. *J. Planar Chromatogr.—Mod. TLC 8*: 306–309.

Fenimore, D. C., and Davis, C. M. (1981). High performance TLC. *Anal Chem. 53*: 252A–266A.

Flodberg, G., and Roeraade, J. (1995). A device for high pressure thin layer chromatography. *J. Planar Chromatogr.—Mod. TLC 8*: 10–13.

Geiss, F. (1988). The role of the vapor phase in planar chromatography. *J. Planar Chromatogr.—Mod. TLC 1*: 102–115.

Geiss, F., and Schlitt, H. (1968). A new and versatile thin layer chromatography separation system (KS-Vario-Chamber). *Chromatographia 1*: 392–402.

Gocan, S., Cimpan, G., and Muresan, L. (1996). Automated multiple development thin layer chromatography of some plant extracts. *J. Pharm. Biomed. Anal. 14*: 1221–1227.

Golkiewicz, W. (1996). Gradient development in thin layer chromatography. In *Handbook of Thin Layer Chromatography*, 2nd ed., J. Sherma and B. Fried (Eds.). Marcel Dekker, New York, pp. 149–170.

Gonnord, M.-F., and Siouffi, A.-M. (1990). Ultrafast UV detection for bidimensional chromatography. *J. Planar Chromatogr.—Mod. TLC 3*: 206–209.

Grinberg, N. (1990). Gradients in thin layer chromatography. In *Modern Thin Layer Chromatography*, N. Grinberg (Ed.). Marcel Dekker, New York, pp. 368–398.

Jaenchen, D. E. (1996). Instrumental thin layer chromatography. In *Handbook of Thin Layer Chromatography*, 2nd ed., J. Sherma and B. Fried (Eds.). Marcel Dekker, New York, pp. 129–148.

Jork, H., Funk, W., Fischer, W., and Wimmer, H. (1990). *Thin Layer Chromatography, Reagents and Detection Methods*, Volume 1a. VCH Verlagsgesellschaft, Weinheim, Germany.

Kaiser, R. E. (1977). Simplified theory of TLC. In *High Performance TLC*. A. Zlatkis and R. E. Kaiser (Eds.). Elsevier, New York, pp. 15–38.

Kaiser, R. E. (1978). Anticircular high performance thin layer chromatography. *J. High Resolut. Chromatogr. Chromatogr. Commun. 1*: 164–168.

Kaiser, R. E. (1988). Scope and limitations of modern planar chromatography. Part 2: Separation modes. *J. Planar Chromatogr.—Mod. TLC 1*: 265–268.

Kaiser, R. E., and Rieder, R. I. (1986). High pressure TL circular chromatography HPTLCC—A new dimension in analytical chromatography ranging from GC to LC. In *Planar Chromatography*, Vol. 1, R. Kaiser (Ed.). Alfred Huethig Verlag, Heidelberg, pp. 165–191.

Katy, Gy., Mincsovics, E., Szokan, Gy., and Tyihak, E. (1997). Comparison of thin layer chromatography and overpressured thin layer chromatographic techniques for the separation of ascorbigen and 1'-methylascorbigen. *J. Chromatogr., A 764*: 103–109.

Kovar, K. -A. (1997). The efficiency of thin layer chromatographic systems: A comparison of separation numbers using addictive substances as an example. *J. Planar Chromatogr.—Mod. TLC 10*: 114–117.

Kralecik, P. (1996). A new TLC/AMD method for testing of textiles for carcinogenic amines arising from azo dyes. *GIT Spez. Chromatogr. 16*: 125–129.

Lautie, J.-P., and Stankovic, V. (1996). Automated multiple development of phenylurea herbicides in plants. *J. Planar Chromatogr.—Mod. TLC 9*: 113–115.

Lee, K. Y., Poole, C. F., and Zlatkis, A. (1980). Simultaneous multi-mycotoxin determination by HPTLC. In *Instrumental HPTLC*. W. Bertsch, S. Hara, R. E. Kaiser, and A. Zlatkis (Eds.). Alfred Huethig Verlag, Heidelberg, p. 245.

Liteanu, C., and Gocan, S. (1974). *Gradient Liquid Chromatography*. Ellis Horwood, Chichester.

Lodi, G., Betti, A., Menziani, E., Brandolini, V., and Tosi, B. (1991). Some aspects and examples of automated multiple development (AMD) gradient optimization. *J. Planar Chromatogr.—Mod. TLC 4*: 106–110.

Lodi, G., Bighi, C., Brandolini, V., Menziani, E., and Tosi, B. (1997). Automated multiple development HPTLC analysis of sugars on hydrophilic layers: II. Diol layers. *J. Planar Chromatogr.—Mod. TLC 10*: 31–37.

Marutoiu, C., and Fabian, F. (1995). Separation and identification of polycyclic aromatic hydrocarbons by high performance thin layer chromatography using vapor and temperature gradient. *Acta Chromatogr. 5*: 158–166.

Matysik, G. (1994). Gradient thin layer chromatography of extracts from digitalis lanata and digitalis purpurea. *Chromatographia 38*: 109–113.

Matysik, G., and Toczolowski, J. (1997). TLC-densitometric investigation of aucubin penetration from Herba Euphrasiae extract to the aqueous humor of the eye. *Chem. Anal. (Warsaw) 42*: 221–226.

Matysik, G., and Wojtasik, E. (1994). Stepwise gradient HPTLC analysis of frangula anthraquinones. *J. Planar Chromatogr.—Mod. TLC 7*: 34–37.

Matysik, G., Kowalski, J., Strzelecka, H., and Soczewinski, E. (1994). Investigation of the accumulation of lanatosides in Digitalis lanata by gradient elution TLC and densitometry. *J. Planar Chromatogr.—Mod. TLC 7*: 129–132.

Mincsovics, E., Ferenczi-Fodor, K., and Tyihak, E. (1996). Overpressured layer chromatography. In *Handbook of Thin Layer Chromatography*, 2nd ed., J. Sherma and B. Fried (Eds.). Marcel Dekker, New York, pp. 171–203.

Muething, J., and Ziehr, H. (1996). Enhanced thin layer chromatographic separation of G_{M1b}-type gangliosides by automated multiple development. *J. Chromatogr., B: Biomed. Appl. 687*: 357–362.

Newman, J. M. (1985). Overpressure layer chromatography: a review. *Am. Lab. 17*(4): 52–63.

Niederwieser, A. (1969). Recent advances in thin layer chromatography. IV. Gradient thin layer chromatography. Part II. Gradients of elution with special regard to vapor impregnation and flux of mobile phase. *Chromatographia 2*: 362–375.

Nurok, D., Becker, R. M., and Sassic, K. A. (1982). Time optimization in TLC. *Anal. Chem. 54*: 1955–1959.

Nurok, D., Tecklenburg, R. E., and Maidak, B. L. (1984). Separation of complex mixtures by parallel development TLC. *Anal. Chem. 56*: 293–297.

Nyiredy, Sz. (1992). Planar Chromatography. In *Chromatography*, 5th ed., E. Heftmann (Ed.). Elsevier, Amsterdam, pp. A109–A150.

Nyiredy, Sz. (1996). Preparative layer chromatography. In *Handbook of Thin Layer Chromatography*, 2nd ed., J. Sherma and B. Fried (Eds.). Marcel Dekker, New York, pp. 307–340.

Nyiredy, Sz. and Fater, Z. (1994). The elimination of typical problems associated with overpressured layer chromatography. *J. Planar Chromatogr.—Mod. TLC 7*: 329–333.

Nyiredy, Sz., Dallenbach-Toelke, K., and Sticher, O. (1988a). Analytical rotation planar chromatography. In *Recent Advances in Thin Layer Chromatography*, F. A. A. Dallas, H. Read, R. J. Ruane, and I. D. Wilson (Eds.). Plenum Press, New York, pp. 45–55.

Nyiredy, Sz., Meszaros, S. Y., Dallenbach-Toelke, K., Nyiredy-Mikita, K., and Sticher, O. (1988b). Ultra-microchamber rotation planar chromatography (U-RPC): a new analytical and preparative forced flow method. *J. Planar Chromatogr.—Mod. TLC 1*: 54–60.

Nyiredy, Sz., Botz, L., and Sticher, O. (1989). Rotachrom: a new instrument for rotation planar chromatography. *J. Planar Chromatogr.—Mod. TLC 2*: 53–61.

Nyiredy, Sz., Fater, Z., Botz, L., and Sticher, O. (1992). The role of chamber saturation in the optimization and transfer of the mobile phase. *J. Planar Chromatogr.—Mod. TLC 5*: 308–315.

Perry, J. A. (1979). A new look at solvent strength, selectivity, and continuous development. *J. Chromatogr. 165*: 117–140.

Perry, J. A., Jupille, T. H., and Glunz, L. H. (1975). Thin layer chromatography. Programmed multiple development. *Anal. Chem. 47*(1): 65A–74A.

Po, A., Li Wan, and Irwin, W. J. (1979). The identification of tricyclic neuroleptics and sulfonamides by HPTLC. *J. High Resolut. Chromatogr. Chromatogr. Commun. 2*: 623–627.

Poole, C. F., and Belay, M. T. (1991). Progress in automated development. *J. Planar Chromatogr.—Mod. TLC 4*: 345–359.

Poole, C. F., and Poole, S. K. (1991). *Chromatography Today*. Elsevier, Amsterdam, pp. 649–734.

Poole, S. K., and Poole, C. F. (1992). The influence of solvent entry position on resolution in unidimensional multiple development thin layer chromatography. *J. Planar Chromatography—Mod. TLC 5*: 221–228.

Poole, C. F., and Poole, S. K. (1995). Multidimensionality in planar chromatography. *J. Chromatogr., A 703*: 573–612.

Poole, C. F., Butler, M. E., Coddens, M. E., Khatib, S., and Vandervennt, R. (1984). Comparison of methods for separating polycyclic aromatic hydrocarbons by high performance thin layer chromatography. *J. Chromatogr. 302*: 149–158.

Poole, C. F., Poole, S. K., Fernando, W. P. N., Dean, T. A., Ahmed, H. D., and Berndt, J. A. (1989). Multidimensional and multimodal thin layer chromatography: pathway to the future. *J. Planar Chromatogr.—Mod. TLC 2*: 336–345.

Postaire, E. P. R., Delvorde, P., and Sarbach, C. (1996). Automation and robotics in planar chromatography. In *Handbook of Thin Layer Chromatography*, 2nd ed., J. Sherma and B. Fried (Eds.). Marcel Dekker, New York, pp. 373–385.

Robards, K., Haddad, P. R., and Jackson, P. E. (1994). *Principles and Practice of Modern Chromatographic Methods*. Academic Press, San Diego, CA, pp. 179–226.

Sandroni, S., and Schlitt, H. (1970). An arrangement for simultaneous elution of one thin layer chromatography plate with several solvents. *J. Chromatrogr. 52*: 169–171.

Schuette, S. A., and Poole, C. F. (1982). Unidimensional, sequential separation of PTH-amino acids by high performance thin layer chromatography. *J. Chromatogr. 239*: 251–257.

Snyder, L. R., and Saunders, D. L. (1969). Resolution in thin layer chromatography with solvent and adsorbent programming. Comparisons with column chromatography and normal thin layer chromatography. *J. Chromatogr. 44*: 1–13.

Soczewinski, E. (1986). Equilibrium sandwich TLC chamber for continuous development with a glass distributor. In *Planar Chromatography*, Vol. 1, R. E. Kaiser (Ed.). Alfred Huthig Verlag, Heidelberg, pp. 79–117.

Stahl, E. (1958). Thin layer chromatography: Standardization, detection, documentation, and application. *Chem. Z. 82*: 323–329.

Stahl, E. (1959). Thin layer chromatography. IV. Scheme, border effect, acidic and basic layers, step technique. *Arch. Pharm. 292*: 411–416.

Stahl, E. (1965). Effect of low temperatures and use of gradients in thin layer chromatography. *J. Pharm. Belg. 20*: 159–168.

Stahl, E. (1969). Apparatus and general techniques. In *Thin Layer Chromatography. A Laboratory Handbook*, 2nd ed. E. Stahl (Ed.). Springer-Verlag, New York, pp. 86–105.

Stahl, E., and Mangold, H. K. (1975). Techniques of thin layer chromatography. In *Chromatography*. E. Heftmann (Ed.). Van Nostrand, Reinhold, New York, pp. 164–189.

Stan, H. J., and Wippo, U. (1996). Pesticide residue analysis in plant food products by TLC with automated multiple development. *GIT Fachz. Lab. 40*: 855–858.

Stewart, G. H., and Wendel, C. T. (1975). Evaporative TLC, solvent flow, and zone migration. *J. Chromatogr. Sci. 13*: 105–109.

Szabady, B., Ruszinko, M., and Nyiredy, Sz. (1997). Prediction of retention data when using incremental multiple development techniques. *Chromatographia 45*: 369–372.

Szabady, B., Ruszinko, M., and Nyiredy, Sz. (1995). Prediction of retention data in multiple development. Part 1. Linearly increasing development distances. *J. Planar Chromatogr.—Mod. TLC 8*: 279–283.

Szepesi, G., and Nyiredy, Sz. (1996). Pharmaceuticals and drugs. In *Handbook of Thin Layer Chromatography*, 2nd ed., J. Sherma and B. Fried (Eds.). Marcel Dekker, New York, pp. 819–876.

Tecklenburg, R. E., and Nurok, D. (1984). Correlation in continuous development thin layer chromatography. *Chromatographia 18*: 249–252.

Tecklenburg, R. E., Becker, R. M., Johnson, E. K., and Nurok, D. (1983). Time optimized TLC in a chamber with fixed plate lengths. *Anal. Chem. 55*: 2196–2199.

Tecklenburg, R. E., Fricke, G. H., and Nurok, D. (1984). Overlapping resolution maps as an aid in parallel development TLC. *J. Chromatogr. 290*: 75–81.

Tyihak, E., and Mincsovics, E. (1988). Forced flow planar liquid chromatographic techniques. *J. Planar Chromatogr.—Mod. TLC 1*: 6–19.

Tyrpien, K. (1996). Analysis of hydroxy-PAH by thin layer chromatography and OPLC. *J. Planar Chromatogr.—Mod. TLC 9*: 203–207.

Vajda, J., Leisztner, L., Pick, J., and Anh-Tuna, N. (1986). Methodological developments in overpressured layer chromatography. *Chromatographia 21*: 152–156.

Waksmundzka-Hajnos, M., and Wawrzynowicz, T. (1994). Optimization of the separation of plant extracts with the aid of the horizontal DS chamber. *J. Planar Chromatogr.—Mod. TLC 7*: 58–62.

Wieland, T., and Determann, H. (1962). New applications of thin layer chromatography. II. Gradient chromatography on a mixture of diethylamino cellulose and Sephadex. *Experientia 18*: 431–432.

Witkiewicz, Z., Mazurek, M., and Bladek, J. (1993). A new instrument for overpressured layer chromatography. *J. Planar Chromatogr.—Mod. TLC 6*: 407–412.

Zellmer, S., and Lasch, J. (1997). Individual variation of human plantar stratum corneum lipids, determined by automated multiple development of high performance thin layer chromatography plates. *J. Chromatogr., B: Biomed. Appl. 691*: 321–329.

Zieloff, K. (1995). Automation in thin layer chromatography and good laboratory practice (GLP). *Oesterr. Chem. Z. 96*: 120–122.

8

Detection and Visualization

After development, the layer is removed from the chamber, and the mobile-phase solvents are evaporated in a well-ventilated area in ambient air, with warm or hot air from a hair dryer, or in an oven. The evaporation process must not cause loss of volatile solutes from the layer or be carried out at a temperature that is high enough to cause compound decomposition. In some cases, drying is carried out using nitrogen flow to preclude oxidation of air-sensitive solutes. Zones are then detected by various means. Colored substances may be viewed in daylight without any treatment. Detection of colorless substances is simplest if compounds show self-absorption in the short-wave ultraviolet (UV) region (254 nm) or if they can be excited to produce fluorescence by short-wave and/or long-wave (366 nm) UV radiation. Otherwise, detection can be achieved by means of chromogenic reagents (producing colored zones) or fluorogenic reagents (producing fluorescent zones), or by biological methods. In the ideal case, the detection reagent should produce high-contrast, high-sensitivity zones that are stable (unless the method is nondestructive or reversible) and proportional to the quantity present on the layer (for quantitative evaluation). A specialized approach for direct, nondestructive detection involves the use of autoradiography, fluorography, spark chamber, or scanning techniques to measure radioactively labeled solutes (see Chapter 13). Typical detection limits for TLC and HPTLC are shown in Table 2.1. The layer serves as a storage medium for the separated zones, which can be examined as long as necessary to obtain the maximum amount of information concerning the chromatograms. The freedom from time restraints and the ability to use a variety of techniques for detection (viewing under short and long UV

FIGURE 8.1 Cabinet with removable lamp for viewing TLC plates under 254-nm and 366-nm UV light. (Photography courtesy of Analtech.)

light, chromogenic and fluorogenic reagents, and biological methods) and identification (UV absorption, FTIR, Raman, and mass spectrometry, Chapter 9) in sequence are among the greatest advantages of TLC.

Books by Jork et al. (1990 and 1993) are excellent sources of general information on methods of detection for TLC and detailed procedures and results for use of numerous individual detection reagents. Strategies for detection were described in detail by Kovar and Morlock (1996).

I. INSPECTION UNDER ULTRAVIOLET LAMPS

Plates are best viewed in a darkened room or a small dark corner or area enclosed by dark curtains. Many suitable UV lamps are available commercially, as are viewing cabinets incorporating long- (366 nm) and short- (254 nm) wave lamps, as well as white light (Figure 8.1). A homemade viewing cabinet can be constructed from a corrugated cardboard box (about $30 \times 35 \times 40$ cm) painted black on the inside and set up with the flaps forward. The appropriate UV lamp is placed inside or set in a slot cut in the top of the box.

On normal sorbents not impregnated with phosphor, fluorescent substances absorb UV light (mostly long-wave) and emit longer-wave visible energy against the dark layer background. The specific excitation wavelength and the emitted color, which can be red, yellow, orange, green, blue, or violet, provide selectivity and aid compound identification. Fluorescent compounds, which can be derivatives formed by use of a fluorogenic reagent (e.g., zirconium salts, Kovar and Morlock, 1996) or naturally fluorescent (e.g., aflatoxins, polycyclic aromatic hydrocarbons, quinine, riboflavin), can often be detected at levels of 10 ng or less. Because relatively few compounds fluoresce, fluorescent solutes can be selectively detected in complicated chromatograms containing nonfluorescent zones.

An inorganic phosphor (e.g., zinc cadmium sulfide or Mn-activated zinc silicate) incorporated in the adsorbent allows many organic compounds that absorb at or near 254 nm (aromatic, unsaturated, or conjugated compounds) to be visualized as black or dark violet zones on a yellow-green background when the layer is irradiated with short-wave UV radiation. This process, termed fluorescence quenching, inhibition, or diminution, is rather universal, with limits of detection ranging from ~50 ng to several micrograms, depending on compound structure. It has been the basis of many densitometric quantitative analyses of compounds such as food additives (Smith and Sherma, 1995), fragrance product ingredients (Anderton and Sherma, 1996), and pharmaceuticals (Lippstone et al., 1996). Commercial layers with phosphor emitting blue, yellow-green, or red radiation when activated at 366 nm and those with phosphors activated both at 366 and 254 nm are also available, as are layers containing acid-stable fluorescent indicators (alkaline earth tungstates) that emit a pale blue light upon radiation. Phosphor-containing plates are generally designated with the symbols F or UV. The fluorescent indicator is usually inert and does not interfere with the chromatography, but there are examples of undesirable effects caused by the indicator, that is, the formation of zinc complexes during the separation of porphyrins on a layer with zinc-containing phosphor (Saitoh et al., 1992).

A detailed discussion of fluorescence quenching, fluorescence, and other non-destructive TLC detection methods such as the use of reagents giving reversible color reactions and various biological methods was published by Barrett (1974). Although visualization with UV light is nondestructive with most substances and, therefore, well suited for preparative purposes, rearrangement of structure can sometimes occur—for example, with some steroid and vitamin molecules.

Fluorescence has been intensified and stabilized by dipping chromatograms into a viscous solution of a hydrophobic (e.g., a 25% solution of liquid paraffin in hexane) or hydrophilic (e.g., a 10% solution of polyethylene glycol 4000 in methanol) reagent. Enhancements ranging from 2 to 100% have been reported in various studies.

II. EQUIPMENT AND TECHNIQUES FOR VISUALIZATION WITH DETECTION REAGENTS

A. Sprayers and Spraying

The usual way of applying visualizing reagents to a layer is by spraying. Some reactions occur at room temperature; others require the application of a certain elevated temperature from a regulated heating plate or temperature-controlled oven for a given time period. It is often necessary to view the plate under UV light after spraying in order to evaluate the results of the reaction.

Commercial glass sprayers that can be attached to a compressed air line or tank of inert gas such as nitrogen (Figure 8.2) permit the application of a fine, uniform mist to the plate. Some widely used detection reagents, such as ninhydrin, rhodamine B, and phosphomolybdic acid, are available commercially as premixed solutions. The spraying operation should be carried out inside a cardboard spraying box contained in a fume hood with an efficient draft. The hood should be kept as chemically clean and free from dust as possible to reduce the transfer of contamination to the layer. The first spray should be directed beside the TLC plate, checking for the production of a very fine, uniform aerosol mist. Only then should the spray be directed onto the layer from a distance of 20–30 cm while moving the sprayer carefully back and forth and up and down in a uniform pattern. Reagent is generally sprayed until the layer begins to become translucent. Application of excess solution may cause spots to streak or run. The optimum degree of spraying varies for different reagent/layer combinations and must be determined by trial and error. The sprayer should be cleaned after each use by spraying with solvent to prevent clogging. Safety glasses and laboratory gloves should be worn when spraying TLC plates.

Cordless, rechargeable electro-pneumatic reagent sprayers that can apply solutions that are viscous (e.g., aqueous sulfuric acid) or of normal viscosity (water-organic solvent mixtures) in an even, fine (0.3–10 μm), reproducible aerosol are commercially available (e.g., from Camag and Desaga) to facilitate application of detection reagents (Figure 8.3).

It is often claimed that manual spraying cannot produce reproducible chromatograms, but many successful densitometric analyses have been based on this method—e.g., the use of phosphomolybdic acid for detection in the quantification of neutral lipids and phospholipids (Frazer at al., 1997).

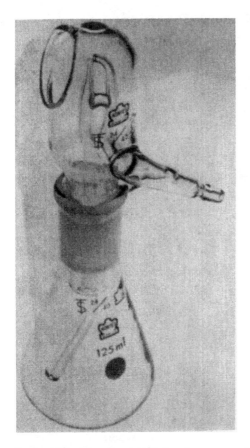

FIGURE 8.2 The Kontes detection-reagent sprayer. (Photograph courtesy of Kontes, Vineland, NJ.)

B. Dipping Vessels and Dipping

Glass dipping chambers are available from various sources (e.g., Desaga Dip Fix) in sizes that will accommodate 20 × 20-, 20 × 10-, and 10 × 10-cm plates. Glass, stainless steel, or enamel trays can also serve as crude dipping vessels. In use, detection reagent is poured into the chamber, and the developed and dried chromatogram is dipped into the reagent for a specified time, usually several seconds. The plate or sheet is removed slowly to allow excess reagent to drain back into the chamber, and the plate is air dried and often heated to visualize spots. The procedure of dipping in the detection reagent following development

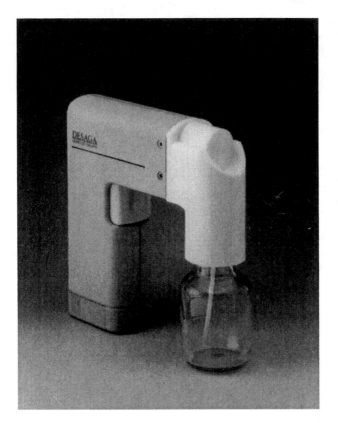

FIGURE 8.3 Desaga battery-operated Sprayer SG 1. (Photograph courtesy of Desaga GmbH.)

of the chromatogram is referred to as "postdipping." The dipping procedure is usually less dangerous than spraying with carcinogenic or corrosive reagents, and the distribution of reagent on the layer may be more uniform than that obtained through spraying. Because many dipping reagents can be reused, it is often more economical to dip than to spray. Commercial dip chambers are supplied with stainless steel lids, which are used to cover the reagent solution between uses. After several uses, it may be necessary to filter the dipping reagent if particles of sorbent are present. In quantitative TLC (Chapter 10), where the uniformity of the zones and background is very important, visualization by dipping is often preferred to spraying. Dipping chambers can be used for the application of impregnating agents that improve separations (Chapter 3) as well as for applying detection reagents. Layers bound with organic polymers are most resistant to damage during dipping procedures. The opportunity for zone diffusion or dissolving is greater when reagents are applied by dipping rather than spraying.

For the most reproducible reagent transfer, a battery-operated, automated dipping device is available for immersing and withdrawing a TLC plate with uniform speed (Figure 8.4). The vertical speed is selectable between 30 and 50 mm/sec and the immersion time between 1 and 8 sec. The device will accommodate 10-or 20-cm plates and can be used for plate prewashing, typically in dichloromethane–methanol (1 : 1) (Chapter 3). In general, application of detection

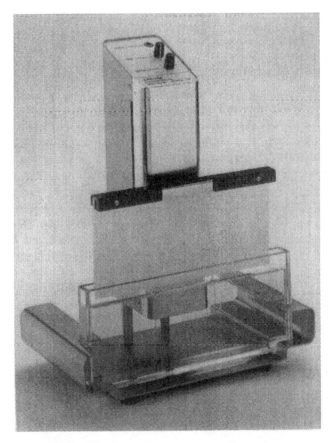

Figure 8.4 The Camag Chromatogram Immersion Device III. (Photograph courtesy of Camag.)

reagents by dipping is preferred to spraying, but it must be mechanically controlled to ensure reproducible results. Spraying must be used when two or more aqueous reagents are applied without intermediate drying, as in diazotization followed by coupling.

C. Overpressure Derivatization

Overpressure derivatization (OPD) is a relatively new method in which an absorbent polymeric pad prewetted with detection reagent is pressed onto the TLC plate. OPD was found to give more reproducible results when compared to spraying of ninhydrin reagent in the quantitative analysis of tryptophan on a cellulose layer using densitometric scanning. Use of OPD for a variety of other detection reagents and application to the analysis of a multivitamin preparation was described by Postaire et al. (1990).

D. Preimpregnation or Predipping

Detection reagents such as silver nitrate and phosphomolybdic acid may be incorporated into the layer prior to sample application and development. The actual detection reaction usually occurs upon heating or exposure to UV light after development and drying the mobile phase. If this procedure is used, it is imperative that the reagent not interfere with separation and that the mobile phase does not remove the reagent during development. When plates are prepared manually, a detection reagent may be incorporated into the slurry. With the aid of a dipping chamber, detection reagent may be applied to a coated plate (homemade or commercially precoated layers) prior to development and sample application. This procedure allows the incorporation of detecting reagent into a relatively large number of plates at one time. Other advantages of predipping are essentially as described for postdipping.

Charring reagents have been incorporated into layers by predipping to avoid spraying of corrosive chemicals. For example, silica gel plates have been impregnated with 4% sulfuric acid in methanol (Touchstone et al., 1972a) or ammonium bisulfate (Touchstone et al., 1972b).

E. Incorporation of the Detection Reagent in the Mobile Phase

In addition to application by exposure to vapor (see Section III.A.1), dipping, or spraying, detection reagents can be incorporated in the mobile phase if they spread uniformly over the layer and move with the solvent front during development. Reagents successfully added to the mobile phase include acids for detection of quinine alkaloids and fluorescamine for biogenic amines (Kovar and Morlock,

1996), as well as rhodamine B and 6G, fluorescein, and 2', 7'-dichlorofluorescein (see Section III.A).

F. Heating the Plate

Heating is often necessary to produce the required color or fluorescence with the detection reagent, usually at 110–115°C for 5–15 min. Certain reactions such as charring require higher temperatures for a longer period of time. Heating is usually carried out in a temperature-controlled electric oven, preferably exhausted into a hood, or a commercial hot-plate heating apparatus (Figure 8.5); IR sources, microwave apparatus, or drying cupboards are less often used. A plate heater has the advantage of providing a flat, evenly heated surface with precisely controlled temperature. In some cases, excessive temperature or prolonged heating can cause darkening of the background or decomposition of compounds, leading in either case to poorer detection sensitivity.

FIGURE 8.5 Camag TLC Plate Heater III. The heating area has a reagent-resistant, easily cleaned ceramic surface with a 20 × 20 cm grid to facilitate reproducible positioning of the plate. The temperature range is 25–200°C, and the programmed and actual temperatures are digitally displayed. (Photograph courtesy of Camag.)

With some compounds and layers, zone detection is possible by merely heating the plate, without application of a reagent. For example, glucose and fructose were detected at nanogram levels as fluorescent spots under 366-nm UV light on HPTLC amino-bonded plates heated for 3 min at 170°C (Klaus et al., 1989). In addition to sugars, catecholamines, steroid hormones, and fruit acids have been detected on amino layers by reagent-free thermochemical activation (see Hauck, 1995, for a review).

In place of heating, irradiation by UV light is necessary to complete the reaction of some chromogenic reagents. An example is the detection of chlorinated hydrocarbons (pesticides) by spraying with silver nitrate reagent and exposure to 254-nm light.

III. CHROMOGENIC AND FLUOROGENIC CHEMICAL DETECTION REAGENTS

Postchromatographic detection reagents are of two types: (1) general reagents that react with a wide variety of different compound types and can totally characterize an unknown sample and (2) specific reagents that indicate the presence of a particular compound or functional group. There are nondestructive (or reversible) and destructive reagents in each of the categories (1) and (2). When a series of detection reagents is used on the same plate to enhance detection and solute identification, it is imperative to use nondestructive reagents prior to the destructive test. Nondestructive reagents are also beneficial for use between runs in multidimensional or two-dimensional development procedures or to locate a substance that is to be recovered unchanged from the chromatogram, for example, in preparative TLC (Chapter 12). Reagents with a high proportion of water may give problems with layer stability or blotchy appearance after spraying. Use of plates with polymeric organic binder and reformulation of reagents, if possible, with ethanol or 1-propanol in place of some of the water can help in these situations.

A. General Reagents

1. Iodine

The universal detection reagent iodine can be used as 0.5–1% alcoholic spray or dip, but more frequently, the plate is simply placed in a closed jar or tank containing a few iodine crystals in the bottom and saturated with I_2 vapors. The iodine vapor dissolves in, or forms weak charge-transfer complexes with, most lipophilic organic compounds, which show up as dark brown spots on a pale yellow or tan background within a few minutes. After marking or scribing the zones for future reference, exposure of the plate to air causes the iodine to sublimate and the spots to fade in a few minutes, after which the plate can be sprayed with another reagent or the solute can be eluted from the plate for further analysis. Although the detection is usually nondestructive and reversible with a large variety of compounds such as amino acids, psychoactive drugs, indoles, and steroids, some compounds (e.g., polyunsaturated fatty acids and esters, aromatic hydrocarbons, isoquinoline alkaloids, phenolic steroids, and certain drugs and pharmaceuticals) may be altered through nonreversible derivatization or oxidation reactions with iodine. Sensitivities in the 0.1–0.5-µg range are often possible with iodine.

The fading of zones can be delayed and the detection sensitized by spraying the plate with a solution consisting of 0.3 g of 7,8-benzoflavone dissolved in 95 ml of ethanol and 5 ml of 30% sulfuric acid or dilute starch solution. When sprayed just after the color of the background has faded, these reagents produce an intense blue coloration with iodine. Some plates with organic binders give dark backgrounds when treated with iodine vapors. If the plate is first humidified in a water-vapor-saturated chamber for 15 min, the background will remain light, but organic spots will still be detected, although more slowly.

In addition to iodine, other detection reagents (e.g., bromine, chlorine, formaldehyde, ammonia, diethylamine, HCl, sulfuryl chloride, sulfur dioxide, oxides of nitrogen, and ammonium bicarbonate) (see Section 11) can be applied as vapors.

2. Charring Reagents

Many corrosive reagents have been devised for charring organic compounds upon spraying and heating. These reagents are suitable for glass-backed layers with inorganic adsorbents and inert binders (e.g., gypsum), but layers with organic sorbents and binders and plastic foil- or metal-backed layers must be tested for resistance to reaction with each individual reagent–temperature–time combination. Many charring reagents will produce colored or fluorescent zones when heating is carried out at a relatively low temperature for a short time. Heating at higher temperatures for longer periods will char the spots, that is, form black zones on a colorless background.

A classic charring reagent is prepared by dissolving 5 g of potassium dichromate in 100 ml of 40% sulfuric acid. Straight sulfuric acid, 50% aqueous sulfuric acid, nitric acid–concentrated sulfuric acid (1:1), and acetic anhydride–sulfuric acid (3:1) are other successful charring reagents. After spraying the solvent-free plate, charring is achieved by heating at 120–280°C for 15–40 min, depending on the reagent, layer, and solutes to be detected.

A newer reagent is prepared by dissolving 3 g of cupric acetate in 100 ml of an 8% aqueous phosphoric acid solution. Heating is carried out at 130–180°C for up to 30 min. This reagent is suitable for use with some commercial organic-bound layers, which will not react to give a dark background under the stated conditions.

Cupric sulfate reagent is prepared by dissolving 100 g of the salt plus 80 ml of phosphoric acid in enough water to make 1 L and magnetically stirring for 1 h. Plates are dipped and heated for 10–15 min at 160°C to produce charred spots on a white background. This reagent will detect both saturated and unsaturated phospholipids, whereas cupric acetate reagent will detect only unsaturated phospholipids (Sherma and Bennett, 1983).

Spraying of acid-charring reagents must be carried out in a hood using extreme caution. A less hazardous charring spray consists of 10–20% aqueous ammonium sulfate. Heating the sprayed chromatogram decomposes the salt to produce sulfuric acid in situ. Commercial plates are available with 10% ammonium sulfate incorporated into silica gel G or H (Analtech).

3. Phosphomolybdic Acid

Spraying of a chromatogram with a 5% solution of phosphomolybdic acid in ethanol followed by a brief (i.e., 5–20 min) heating at 110°C gives blue-gray spots against a yellow background with a large variety of organic compounds, including reducing substances, steroids, bile acids and conjugates, lipids and phospholipids, fatty acids and their methyl esters, substituted phenols, indole derivatives, prostaglandins, essential oil components, and drugs (Jork et al., 1990). The solution is sprayed until the layer is completely saturated, as indicated by a uniform yellow color visible on the back of the layer through the glass plate. The predipping (Touchstone et al., 1970) and postdipping application procedures are also widely used. Subsequent exposure of the plate to ammonia vapors decolorizes the background and increases the color contrast. The color is very stable, and chromatograms can be stored for long periods in the dark. The reagent is especially suitable for quantitative densitometry, for example, of lipids (Chapter 14).

4. Rhodamine B and 6G

A solution of 50 mg of rhodamine B dye dissolved in 100 ml of ethanol produces violet spots on a pink background for many organic compounds, including lipids. Contrast is improved by exposure of the layer to bromine vapor to decolorize the background. Orange spots are visible against a dark background under longwave UV light.

A solution of 100 mg of rhodamine 6G (50 mg for dipping) in 100 ml of 96% ethanol produces pink spots on a red-violet background for many lipophilic substances. Spots may show yellow-orange fluorescence under 366-nm UV light. Sensitivity may be increased by exposing the chromatogram to ammonia vapors. The reagent can be incorporated in the layer or added to the mobile phase (Jork et al., 1990).

5. Antimony Trichloride (SbCl₃) or Antimony Pentachloride (SbCl₅)

5. *Antimony Trichloride (SbCl₃) or Antimony Pentachloride (SbCl₅)*

Spraying with a 10–20% solution of the chlorides in chloroform or carbon tetrachloride followed by heating at 110°C for a brief period produces spots of different characteristic colors with many organic compounds on a white layer background. The spots often fluorescence in long-wave UV light. Compounds detected include vitamins, water- and fat-soluble pigments, sterols, terpenes, steroids, bile acids, sapogenins, glycosides, and phospholipids (Jost et al., 1990).

6. Anisaldehyde–Sulfuric Acid

The plate is dipped into a solution prepared by dissolving 1 ml of anisaldehyde and 2 ml of concentrated sulfuric acid in 100 ml of glacial acetic acid, or sprayed with a solution prepared by adding 8 ml of concentrated sulfuric acid and 0.5 ml of anisaldehyde to a mixture of 85 ml of methanol and 10 ml of glacial acetic acid while cooling in an ice bath. After heating the plate to 90–125°C for 1–15 min, variously colored spots appear on an almost colorless background, and some may fluoresce under 366–nm UV light. Compounds detected include antioxidants, steroids, prostaglandins, carbohydrates, phenols, glycosides, sapogenins, essential oil components or terpenes, antibiotics, and mycotoxins. This is a universal reagent for natural products, and different colors are produced that can aid compound identification. See Jork et al. (1990) for descriptions of procedures and results.

7. Vanillin–Phosphoric Acid and Vanillin–Potassium Hydroxide

The dried chromatogram is dipped into or sprayed with a solution containing 1 g of vanillin (4-hydroxy-3-methoxybenzaldehyde) in 25 ml ethanol + 25 ml water + 35 ml *o*-phosphoric acid, and then heated to 120–160°C for 5–15 min. Colored zones are produced on a pale background; some compounds fluoresce under 366–nm UV light. Steroids, sterols, triterpenes, cucurbitacins, digitalis glycosides, prostaglandins, and saponins are detected among other compounds.

Amines, amino acids, and aminoglycoside antibiotics form variously colored zones that are often fluorescent under 366-nm UV light when the plate is dipped into a 2% solution of vanillin in isopropanol, dried for 10 min at 110°C, dipped into 0.01*M* ethanolic KOH solution, and heated again. See Jork et al. (1990) for descriptions of procedures and results.

8. Fluorescein Sodium and 2′,7′-Dichlorofluorescein

Spraying with a 0.05–0.2% alcoholic solution of one of these reagents produces yellow-green fluorescent spots on a purple background in UV light for many saturated and unsaturated lipophilic organic compounds. The reagent has also been added to the mobile phase. The detection is nondestructive.

9. pH Indicators

Spraying with a 0.01–1% aqueous or aqueous alcohol solution of an acid–base indicator (e.g., bromocresol green, bromothymol blue, bromophenol blue, methyl red, malachite green) can detect acid or basic compounds on the layer based on a change in color at the location of the zones and knowledge of the pH transition range of the indicator.

10. Water

Spraying an alumina, kieselguhr, or silica gel plate with water until the layer appears translucent when held against light reveals lipophilic solutes such as steroids, hydrocarbons, and bile acids as white opaque spots on a semitransparent background because they take up less water than the layer. The spots may be the clearest if the plate is saturated by spraying and is then allowed to dry slowly while being viewed. After marking the spots and thorough redrying, the layer can be oversprayed with another reagent or the zones eluted for further analysis (Chavez, 1979).

An aqueous solution of a hydrophilic dye, such as methylene blue, can be used in place of water. The background turns blue, whereas pale blue spots occur at the locations that are "dry." Lipophilic dyes in aqueous alcohol solution have been used to reverse the process, producing darkly colored zones at the position of the lipophilic substances on a less-colored background (Jork et al., 1990).

11. Vapor-Phase Fluorescence

Heating a dry TLC plate in the presence of ammonium bicarbonate vapor produces fluorescent derivatives of most organic compounds, including alkaloids, sugars, steroids, and amino acids.

Sensitivity of detection is usually better than for charring, often in the nanogram or picogram range, the fluorescent zones are stable, and fluorodensitometry with 366-nm excitation and 450–470-nm emission wavelengths can be used for quantification. The exact color of the fluorescent emission is a useful aid for compound identification. Inorganic layers with inorganic binders and absence of organic impurities give the best spot contrast. Figure 8.6 shows a commercial chamber for this technique.

A developed TLC plate is sealed into the chamber permeated by ammonium bicarbonate vapor. Vapor is generated by heating solid reagent in a stainless steel tray positioned in the chamber. The TLC plate is also heated, allowing the reaction to occur. Heat is supplied by a regulated hot plate upon which the chamber rests. The hollow tube, seen at left, allows venting of fumes via an aspirator and also aids in removing hot TLC plates from the apparatus. The tube can also be

FIGURE 8.6 Vapor phase fluorescence visualization chamber. (Photograph courtesy of Analtech.)

used to purge the chamber with inert gas for applications where TLC plates must be heated in an inert atmosphere to prevent oxidation of sensitive compounds.

The design of a leak-proof heating-detection chamber from a 2-L resin kettle was described by Maxwell (1988), and this apparatus was used for the comparison of the induced vapor-phase fluorescence detection of four polycyclic ether antibiotics and lipids on C_{18} and silica gel layers (Maxwell and Unruh, 1992).

B. Selective Reagents

Stahl (1969), Kirchner (1978), Randerath (1963), Bobbitt (1963), Gasparic and Churacek (1978), Touchstone (1992), Jork ct al. (1990 and 1993), booklets by E. Merck (1976) and Macherey-Nagel (1997), and the CRC *Handbook of Chromatography* (Zweing and Sherma, 1972) list hundreds of spray reagents to which the reader of this book is referred for details. These reagents will react more or less specifically with different functional groups to reveal natural products and organic or biochemicals as colored or fluorescent zones. Selective reagents simplify the demands on the TLC separation because overlapped zones are not a problem if only the analyte is detected. Usually spray reagents are described in the cited references; the same solutions may be applicable for dipping TLC plates, or the preparation of the reagent may have to be modified. In some cases, spray reagents are made less concentrated, water is replaced by an alcohol or other lipophilic solvent, or the solution is made less polar for preparation of dipping solutions (Kovar and Morlock, 1996). Older references often cite benzene as a solvent, which in many cases can be replaced by less toxic toluene without changing results.

Fluorescamine is an example of a widely used reversible, selective detection reagent (Sherma and Touchstone, 1974; Sherma and Marzoni, 1974; Touchstone et al., 1976). When a developed and dried chromatogram is sprayed with fluorescamine and viewed under long-wave UV light, compounds with primary amine groups will be detected as brightly fluorescent, greenish spots; amino acids, drugs, and pesticides have been detected using this reagent with sensitivities in the nanogram range. An analogous reagent is *o*-phthalaldehyde, which, in the presence of a strong reducing agent such as 2-mercaptoethanol, produces strongly fluorescent compounds (Lindeberg, 1976). 2,4-Dinitrophenylhydrazine hydrochloride is another selective reagent, producing orange spots for carbonyl compounds. Selective reagents are not specific, and care must be exercised in interpreting results. For example, ninhydrin, a widely used selective chromogenic reagent for detecting amines, also produces colored derivatives with other reducing substances such as ascorbic acid (Robards et al., 1994).

Table 8.1 contains a list of selective detection reagents that have proven to be satisfactory over many years of use. Additional reagents are described in the detailed experiments contained in Chapters 15–23. Important new detection reagents for TLC are reviewed biennially in the *Analytical Chemistry* Fundamentals Review issue, the latest of which is by Sherma (1998).

IV. DERIVATIVE FORMATION PRIOR TO TLC

Detection may be facilitated by preparation of colored or, more often, fluorescent derivatives prior to spotting and development. Examples include dansylation (reaction with 1-dimethyl aminonaphthalene-5-sulfonyl chloride) of amino acids and formation of 2,5-dinitrophenylhydrazones of amino acids and ketosteroids. Other reactions that have been used include oxidation, reduction, hydrolysis, halogenation, enzymatic, nitration, diazotization, esterification, and etherification (Szepesi and Nyiredy, 1996; Kovar and Morlock, 1996). Prechromatographic derivatization sometimes improves separation, stability, and/or quantification of the analytes in addition to providing visualization.

Occasionally, the derivative can be formed in situ at the origin of the layer, after spotting but before development (Junker-Buchheit and Jork, 1989), rather than in solution prior to spotting. In practice, the reagent is applied first and the sample applied to the same initial zone, or the reagent can be applied over the initial zones of sample. Use of the automated Linomat spray-on applicator (Chapter 5) facilitates precise location of the solutions. If heat is required to accelerate the reaction, the origin area is covered with a glass strip before placing the layer in an oven or on a plate heater. If the reagent is applied as a vapor, the layer is covered with a glass plate, except for the origin area, before exposure.

Many derivatization reactions and techniques have been described in an extensive review by Lawrence and Frei (1974) and books by Jork et al. (1990 and 1993), and this topic is discussed in Chapter 4.

V. BIOCHEMICAL DETECTION METHODS

The following methods are based on the biological-physiological properties of the substances to be detected rather than their chemical or physical properties. These methods have been applied to the following compound types, among others: antibiotics, animal and vegetable growth substances, pesticides, mycotoxins, cytotoxic substances, hot and bitter substances, vitamins, saponins, and alkaloids with enzyme-inhibiting effects (Jork et al., 1990). The methods are highly specific and sensitive and can indicate information about the toxicology and degradation products of the detected substances (Kovar and Morlock, 1996).

A. Enzyme Inhibition

Enzymatic reactions can be monitored on a TLC plate, and the end products can be detected. For example, cholinesteraselike enzymes associated with nervous tissue will catalyze the substrate indoxyl acetate to indoxol and acetic acid. In air, indoxol will be converted to indigo blue, which can be visualized on a TLC plate. Various pesticides (i.e., organophosphates and carbamates) inactivate or inhibit enzymes associated with animal nervous tissue. The presence of minute amounts of such cidal substances may interfere with the substrate reaction described above, producing colorless spots on a blue background when the layer is treated with a mixture of indoxyl acetate and cholinesterase. When naphthylacetate/Fast blue salt B is used as the substrate solution, inhibition of the enzyme-substrate reaction by the pesticide causes formation of an azo dye, resulting in bright zones on a rose-red colored background (Kovar and Morlock, 1996). Amine compounds were detected on thin layers at 1-ng levels using enzymic reactions (Kucharska and Maslowska, 1996). For further details on enzyme inhibition TLC, see Mendoza (1972, 1974) and Touchstone (1992).

B. Bioautography

An example of bioautography involves the detection of hemolyzing components (such as saponins) by casting a blood–gelatin suspension on the layer and observing hemolytic zones, that is, transparent and nearly colorless zones on the turbid red gelatin-layer background. Bacteriostatic and bactericidal compound zones can be detected in a similar way by pouring an appropriate gelatin–bacterial suspension on the plate, incubating, and observing zones where there is no bacte-

TABLE 8.1 Detection Reagents for Different Functional Groups

Compound class	Reagent	Procedure	Results
Alcohols	Ceric ammonium sulfate	6% ceric ammonium sulfate in $2N$ HNO_3	Brown spots on yellow background
Aldehydes and ketones	Dinitrophenylhydrazine	Dissolve 0.4 g 2,4-dinitrophenylhydrazine in 100 ml $2N$ HCl	Yellow to red spots on pale orange background
	o-Dianaisidine	1% solution of reagent in acetic acid; spray	Yellow-brown spots
Alkaloids and choline-containing compounds	Dragendorff-Munier modification	(a) Dissolve 1.7 g bismuth subnitrate and 20 g tartaric acid in 80 ml H_2O (b) Dissolve 16 g KI in 40 ml H_2O Color reagent: mix one volume (a) and one volume (b), then mix 5 ml of this solution with 10 g tartaric acid in 50 ml H_2O	Orange spots that are intensified by spraying with $0.05N$ H_2SO_4
Alkaloids	Iodoplatinate	(a) 5% platinic chloride in H_2O (b) 10% aqueous KI Color reagent: mix 5 ml (a) with 45 ml (b), dilute to 100 ml with H_2O	Basic drugs yield blue or blue-violet spots that turn brownyellow; addition of conc. HCl, 1 part to 10 parts of spray solution, will often increase sensitivity of reagent
Amides	Hydroxylamine–ferric nitrate	(a) Dissolve 1 g hydroxylamine hydrochloride in 9 ml H_2O	Various colors on white background

Substance	Reagent	Preparation	Results
		(b) Dissolve 2 g sodium hydroxide in 8 ml H_2O (c) Dissolve 4 g ferric nitrate in 60 ml H_2O and 40 ml acetic acid. Color reagent: mix one volume (a) and one volume (b); spray and dry at 110°C for 10 min; spray with 45 ml (c) and 6 ml HCl	
Amines and amino acids	Indanetrione hydrate (ninhydrin)	Mix 95 parts 0.2% ninhydrin in n-butanol with 5 parts 10% acetic acid	Yellow, pink-red, or violet spots on white background
Amines, aromatic	Ferric chloride	1:1 mix of $FeCl_3$ and 0.1M $K_3[Fe(CN)_6]$, freshly prepared	Blue spots
Antioxidants	Dichloroquinone-4-chlorimine	Dissolve 1 g reagent in 100 ml ethanol; respray after 15 min with 2% borax in 4% ethanol	Various colors in white background
Barbiturates	s-Diphenylcarbazone	0.1% reagent in 95% ethanol	Purple spots
Carboxylic acids and bases	Bromocresol green	Dissolve 0.04 g BCG in 100 ml ethanol; add 0.1N NaOH until blue color just appears	Acids: yellow on blue background; bases: blue spots on green background
Chlorinated hydrocarbons and chlorine-containing pesticides	Silver nitrate	Dissolve 0.1 g reagent in 1 ml H_2O, 10 ml 2-phenoxyethanol, and 190 ml acetone; add one drop 30% hydrogen peroxide; spray and expose 15 min to long-wave UV light	Gray spots on colorless background

TABLE 8.1 Continued

Compound class	Reagent	Procedure	Results
Cholesterol and cholesteryl esters	Phosphotungstic acid	10% reagent in 90% ethanol; heat 15 min at 100°C	Red spots on white background
3,5-Dinitrobenzoate esters	α-Naphthylamine	Dissolve 1 g reagent in 100 ml ethanol	Orange spots on white background
Ethanolamines	Benzoquinone	Dissolve 1 g reagent in 20 ml pyridine and 80 ml *n*-butanol	Red spots on pale background
Flavonoids	Uranium acetate	1% solution in water; spray	Brown spots
Glycolipids	Trichloroacetic acid	Spray with 25% reagent in chloroform; heat at 100°C for 2 min; view under UV light	Yellow fluorescence
Heterocyclic nitrogen compounds	Malonic acid	Spray with 0.2 g reagent and 0.1 g salicylaldehyde in 100 ml absolute ethanol; heat 15 min at 120°C; view by UV light	Yellow spots
Heterocyclic oxygen compounds	Aluminum chloride	1% $AlCl_3$ in ethanol; spray and view under long-wave UV light	Flavonoids produce yellow fluorescent spots
Hydrocarbons	Tetracyanoethylene	Dissolve 10% reagent in benzene; spray immediately after development, heat at 100°C	Aromatic hydrocarbons yield various colors
Hydroxamic acids	Ferric chloride	Dissolve 2 g $FeCl_3$ in 100 ml 0.5 N HCl	Red spots on colored background

	Reagent	Method	Result
Indoles	Ehrlich reagent	Dissolve 10% *p*-dimethylamino-benzaldehyde in conc. HCl; mix this solution 1:4 with acetone; color develops within 20 min of spraying	Indoles: purple; hydroxyindoles: blue; aromatic amines and ureides: yellow; tyrosine: purple-red
	Prochazka reagent	(a) Mix 10 ml 35% formaldehyde, 10 ml 25% HCl, and 20 ml 96% ethanol; spray with (a). heat to 100°C, observe in daylight and UV light	Yellow spots
Inorganic ions	Ammonium sulfide	Saturated aqueous H_2S made alkaline with NH_3	Cations detected black: Ag, Hg, Co, Ni; brown: Au, Pd, Pt, Pb, Bi, Cu, V, Ti; yellow: Cd, As, Sn; yellow-orange: Sb
Ketoses	Urea	5 g reagent in 20 ml 2M HCl plus 100 ml ethanol	Blue spots
Lipids	2′,7′-Dichlorofluorescein	Spray with 0.2% solution in 96% ethanol. Observe in UV light	Saturated and unsaturated polar lipids give green spots on purple background
Nitrate esters	Diphenylamine	Dissolve 1 g reagent in 100 ml ethanol; spray and expose to short-wave UV light	Yellow-green spots on white background
Nitrosamines	Diphenylamine	A: 1.5% reagent in ethanol; B: 0.1% $PdCl_2$ in 0.2% saline solution. Spray with 5 parts A to 1 part B; expose to 240 nm UV light	Blue-violet spots
Oligosaccharides	Thymol	5 g reagent in 20 ml HCl plus 100 ml ethanol	Blue spots

TABLE 8.1 Continued

Compound class	Reagent	Procedure	Results
Organo-tin compounds	Pyrocatechol violet	100mg reagent in 100 ml ethanol; expose chromatogram to UV light for 20 min; spray with reagent	Dark blue spots on grey-brown background
Penicillins	Iodine azide	Dissolve 3.5 g of sodium azide in 100 ml of $0.1N\,I_2$ (explosive when dry); spray	Yellow spots
Peroxides	Ferrous thiocyanate	(a) Dissolve 4 g ferrous sulfate in 100 ml H_2O (b) Dissolve 1.3 g ammonium thicyanate in 100 ml acetone Color reagent: mix 10 ml (a) and 15 ml (b)	Red-brown spots on pale background
Pesticides, chlorinated	o-Toludine	0.5% reagent in ethanol; dry and expose to UV light	Green spots on white background
Phenols	4-Amino antipyrine	(a) Dissolve 3 g reagent in 100 ml ethanol (b) Dissolve 8 g potassium ferricyanide in 100 ml H_2O Spray with (a), then (b) and expose to ammonia vapor	Red, orange, or pink spots on pale background
Phenols and aromatic amines that will couple	Pauly's reagent	(a) Dissolve 0.5 g sulfanilic acid and 0.5 g KNO_2 in 100 ml $1N$ HCl; spray with (a) and then with $1N$ NaOH	Yellow-orange spots

Phosphorus-containing insecticides	4-(p-Nitrobenzyl)-pyridine	(a) Dissolve 2 g reagent in 100 ml acetone (b) Dissolve 10 g ammonium carbonate in 50 ml H_2O and 50 ml acetone. Spray heavily with (a), then lightly with (b)	Blue spots on white background
Plasmalogens	Fuchsin–sulfuric acid	Dissolve 1 g parafuchsin in 700 ml water and 50 ml 2 M HCl. Dilute to 1 L with water and allow to stand overnight with occasional shaking. Add 1% $HgCl_2$ and spray	Violet spots
Polynuclear aromatic hydrocarbons	Formaldehyde–sulfuric acid	Dissolve 2 ml 37% formaldehyde solution in 100 ml conc. H_2SO_4; spray	Various colors on white background
Sapogenins	Chlorosulfonic acid	Spray with reagent–acetic acid (1:2); heat 1 min at 130°C	Blue fluoresence
Streroids	p-Toluenesulfonic acid	Dissolve 20% reagent in $CHCl_3$; spray and heat at 100°C for a few min; observe under 360-nm UV	Steroids, flavonoids, and catechins fluoresce
Steroid glycosides	Trichloroacetic acid–chloramine T	(a) Dissolve 3% chloramine T in H_2O (b) Dissove 25% trichloroacetic acid in 95% ethanol. Color reagent: mix 10 ml (a) and 40 ml (b); spray and heat 5–10 min at 110°C and observe under UV light	Digitalis glycosides: blue spots

TABLE 8.1 Continued

Compound class	Reagent	Procedure	Results
Sugars	Aniline phthalate	Dissove 0.93 g aniline and 1.66 g phthalic acid in 100 ml of n-butanol saturated with water. Spray and heat to 105°C for 10 min	Reducing sugars give various colors
Sugars, reducing	3,5-Dinitrosalicylic acid	0.5% reagent in 4% NaOH; dry in air then 4–5 min at 100°C	Brown spots on yellow background
Sulfonamides	N-(1-Naphthyl)-ethylene diamine dihydrochloride (NED)	(a) 1NHCl (b) 5% NaNO₂ (c) 0.1% NED in ethanol; spray consecutively with (a), (b), (c)	Red-purple spots
Terpenes	Diphenylpicrylhydrazyl	15 mg reagent in 25 ml chloroform; spray, heat 10 min at 110°C	Yellow spots on purple background
Terpenoids	Vanillin–sulfuric acid	1% solution of vanillin in conc. H₂SO₄; spray and observe in daylight and UV light	Various colored spots

Thiobarbituric acids and thio-acids	Diethylamine	0.5 g $CuSO_4$ in 100 ml methanol, add 3 ml reagent, shake and spray	Green spots
Triterpenes and cardiac glycosides	Chlorosulfonic acid–acetic acid	Prepare a (1:2) solution of the acids; spray, heat at 130°C for 5 min, and observe in UV light	Violet-brown spots
Vitamins	α,α'-Dipyridyl-ferric chloride	(a) Dissolve 2% α,α'-dipyridyl in chloroform (b) Dissolve 0.5% ferric chloride in H_2O. Spray with (a), then with (b)	Detects vitamin E, phenols, and other compounds with reducing properties
	Iodine – starch	Dissolve 0.001–0.005% I_2 in 1% KI, with 0.4% starch solution; spray	Ascorbic acid: white area on blue background
Vitamin B_6	N, 2, 6-Trichloro-p-benzoquinoneimine	0.1% reagent in ethanol; expose to ammonia vapor	Blue spots

rial growth. A variant of bioautography is based on growth promotion. The layer is placed on a seeded agar medium that is complete except for the analyte. After incubation, the plate is examined for growth areas.

As a specific recent example (Saxena et al., 1995), an improved bioautography agar overlay method was developed for detection of antimicrobial compounds such as gram-positive and -negative bacteria and fungi. Phenol red and the test organisms were incorporated directly into nutrient agar that was overlayed onto a thin layer chromatogram of plant extracts. In the case of bacteria, after spraying with thiazolyl blue, dark red inhibition zones appeared against a blue background. The method was useful for activity-guided fractionation or target-directed isolation of plant active constituents. An agar overlay assay was also described for detection and activity-guided fractionation of antifungal compounds (Rahalison et al., 1991).

Working techniques and details are described in Stahl's TLC laboratory handbook (1969) and by Touchstone (1992), Betina (1973), and Durackova et al. (1976).

C. Immunostaining

TLC-immunostaining is a relatively new biochemical detection method for glycolipids having strict structural selectivity and nanogram or lower sensitivity (Saito and Yu, 1990). The separation is carried out by HPTLC [e.g., glycosphingolipids separated on silica gel 60 with chloroform–methanol–water (55:45:10) + 0.02% $CaCl_2$ mobile phase]; the plate is dried completely in vacuo, treated with 0.4% polyisobutylmethacrylate solution for 1 min, and incubated with anti-glycolipid antibody, usually for 1 h at room temperature, followed by washing with phosphate-buffered saline (PBS). For the immunoperoxidase technique, the plate is incubated with a second antibody that is peroxidase-conjugated, and then incubated again with staining solution (3,3'-diaminobenzidine tetrahydrochloride + imidazole + hydrogen peroxide) for 30–60 sec to produce colored zones. For the alternative autoradiography technique, the plate is incubated with a second antibody (if necessary), incubated with ^{125}I-staphylococcal protein A, washed with PBS containing 0.1% Triton X-100, and exposed to x-ray film at −70°. For additional information on the techniques of immunostaining, consult Towbin et al. (1984) and Bethke et al. (1986). TLC immunostaining has been reported for the assay of gangliosides (Ritter et al., 1991) and antiphospholipid antibodies (Sorice et al., 1994) as well as glycosphingolipids (Kushi et al., 1990; Muething and Cacic, 1997).

VI. FLAME IONIZATION DETECTION

A novel detection method for TLC involves use of a flame ionization detector (FID) in conjunction with quartz rods coated with a thin layer of sintered silica

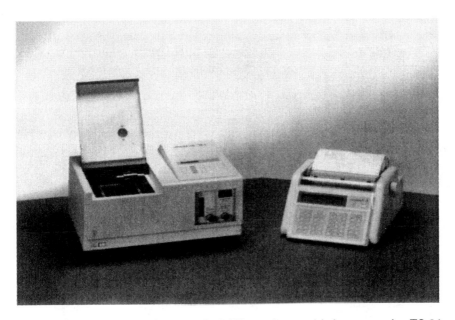

FIGURE 8.7 Iatroscan new MK-5 TLC/FID analyzer with Iatrorecorder TC-21. (Photograph courtesy of Bioscan, Inc.)

or alumina on which the sample is developed and separated (Ranny, 1987; Mukherjee, 1996). For some separations, the layer is impregnated by immersing the rods for a few minutes in solutions of boric acid, oxalic acid, or silver nitrate. This method has been designated Iatroscan TLC, after the name of the manufactuter of commercial instruments and rods (Chromarod-S III), Iatron Laboratories, Inc., Tokyo, Japan, or TLC/FID. In a typical analysis, the rods are cleaned and activated by blank scanning in the instrument, 3–30 μg of sample is applied in a volume of 0.1–1 μl, and up to 10 rods are simultaneously developed in a saturated N-chamber while held in a support frame. The rod holder is placed in an oven to remove the mobile-phase solvents, then in the instrument for measurement. The rods are advanced automatically at a constant speed through the hydrogen flame of the FID, the separated analytes on the rod are ionized by the flame, and the ions generate an electric current proportional to the amount of each organic substance entering the flame. The resulting peaks are recorded and processed like those obtained with similar detectors in gas chromatography.

A variation of this technique involves the use of quartz tubes coated on the inner surface with silica gel or silica gel + CuO instead of adsorbent-coated glass rods. In the tubular TLC (TTLC) method, fractionation on the chromato-tubes is carried out as in conventional TLC, followed by drying of solvent by heating in a stream of inert gas. In situ quantification of the resolved fractions with the FID is obtained by means of vaporization, both through pyrolysis/evaporation (on quartz tubes coated with silica gel) and by combustion (on glass tubes coated with silica gel and cupric oxide). The details of TTLC are described by Mukherjee (1996).

The Iatroscan TLC/FID MK-5 commercial apparatus is shown in Figure 8.7

Recent applications of rod TLC/FID include a quantitative method for compound class characterization of coal-tar pitch without previous fractionation (Cebolla et al., 1996); analysis of four oil and condensate samples (Kabir et al., 1997); determination of polar lipids in spinach, flagellate, and dinoflagellate samples (Parrish et al., 1996); and analysis of mixtures of triglycerides and oleic acid (Peyrou et al., 1996).

Information on TLC-FID is available by mail from Bioscan, Inc., the sole U.S. distributor of the Iatroscan, and a TLC-FID database with references to books, abstracts, and articles in journals and magazines is available by downloading JEPRS software available on the Bioscan website, <http://www.bioscan.com>.

REFERENCES

Anderton, S. M., and Sherma, J. (1996). Quantitative thin layer chromatographic determination of benzylideneacetone in fragrance products. *J. Planar Chromatogr.—Mod. TLC 9*: 136–137.

Barrett, G. C. (1974). Nondestructive detection methods in paper and thin layer chromatography. *Adv. Chromatogr. 11*: 146–179.

Bethke, U., Muething, J., Schauder, B., Conradt, P., and Muehlradt, P. F. (1986). An improved semi-quantitative enzyme immunostaining procedure for glycosphingolipid antigens on high performance thin layer chromatograms. *J. Immunol. Methods 89*: 111–116.

Betina, V. (1973). Bioautography in paper and thin layer chromatography and its scope in the antibiotic field. *J. Chromatogr. 78*: 41–51.

Bobbitt, J. M. (1963). *Thin Layer Chromatography*. Reinhold Publishing Corporation, New York (Chapman & Hall, Ltd., London).

Cebolla, V. L., Vela, J., Membrado, L., and Ferrando, A. C. (1996). Coal-tar pitch characterization by thin layer chromatography with flame ionization detection. *Chromatographia 42*: 295–299.

Chavez, M. N. (1979). Thin layer chromatographic separation of keto derivatives of free bile acids. *J. Chromatogr. 162*: 71–75.

Durackova, Z., Betina, V., and Nemec, P. (1976). Bioautographic detection of mycotoxins on thin layer chromatograms. *J. Chromatogr. 116*: 155–161.

Frazer, B. A., Reddy, A., Fried, B., and Sherma, J. (1997). HPTLC determination of neutral lipids and phospholipids in *Lymnaea elodes* (Gastropoda). *J. Planar Chromatogr.—Mod. TLC 10*: 128–130.

Gasparic, J., and Churacek, J. (1978). *Laboratory Handbook of Paper and Thin Layer Chromatography*. Halsted Press, a Division of John Wiley & Sons, New York.

Hauck, H. E. (1995). Aluminum-backed amino-modified layers for economical separations and easy detection in TLC. *J. Planar Chromatogr.—Mod. TLC 8*: 346–348.

Jork, H., Funk, W., Fischer, H., and Wimmer, H. (1990). *Thin Layer Chromatography— Reagents and Detection Methods*, Vol. 1a. VCH-Verlag, Weinheim, Germany.

Jork, H., Funk, W., Fischer, H., and Wimmer, H. (1993). *Thin Layer Chromatography— Reagents and Detection Methods*, Vol. 1b. VCH-Verlag, Weinheim, Germany.

Junker-Buchheit, A., and Jork, H. (1989). Monodansyl cadaverine as a fluorescent marker for carboxylic acids—in situ prechromatographic derivatization. *J. Planar Chromatogr.—Mod. TLC 2*: 65–70.

Kabir, A., Rafiqul, I., Mustafa, A. I., and Manzur, A. (1997). Quantitative separation of crude petroleum into main structural types: comparison of three analytical techniques. *Dhaka Univ. J. Sci. 45*: 1–5.

Kirchner, J. G. (1978). *Thin Layer Chromatography*. Wiley-Interscience, New York.

Klaus, R., Fischer, W., and Hauck, H. E. (1989). Use of a new adsorbent in the separation and detection of glucose and fructose by HPTLC. *Chromatographia 28*: 364–366.

Kovar, K. -A., and Morlock, G. E. (1996). Detection, identification, and documentation. In *Handbook of Thin Layer Chromatography, 2nd ed.*, J. Sherma and B. Fried (Eds.). Marcel Dekker, New York, pp. 205–239.

Kucharska, U., and Maslowska, J. (1996). Detection of trace quantities of amines with the use of enzymic reactions and chromatographic techniques. *Chem. Anal. (Warsaw) 41*: 975–982.

Kushi, Y., Ogura, K., Rokukawa, C., and Handa, S. (1990). Blood group A–active glycosphingolipids analysis by the combination of TLC immunostaining assay and TLC/ SIMS mass spectrometry. *J. Biochem. (Tokyo) 107*: 685–688.

Lawrence, J. F., and Frei, R. W. (1974). Fluorometric derivatization for pesticide residue analysis. *J. Chromatogr. 98*: 253–270.

Lindeberg, E. G. G. (1976). Use of *o*-phthalaldehyde for detection of amino acids and peptides on thin layer chromatograms. *J. Chromatogr. 117*: 439–441.

Lippstone, M. B., Grath, E. K., and Sherma, J. (1996). Analysis of decongestant pharmaceutical tablets containing pseudoephedrine hydrochloride and guaifenesin by HPTLC with UV absorption densitometry. *J. Planar Chromatogr.—Mod. TLC 9*: 456–458.

Macherey-Nagel (1997). *TLC Applications*. Macherey-Nagel GmbH & Co., Dueren, Germany.

Maxwell, R. J. (1988). An efficient heating-detection chamber for vapor phase fluorescence TLC. *J. Planar Chromatogr.—Mod. TLC 1*: 345–346.

Maxwell, R. J., and Unruh, J. (1992). Comparison of induced vapor phase fluorescent responses of four polycyclic ether antibiotics and several lipid classes on RP-18 and silica gel HPTLC plates. *J. Planar Chromatogr.—Mod. TLC 5*: 35–40.

Mendoza, C. E. (1972). Analysis of pesticides by thin layer chromatographic-enzyme inhibition technique. *Residue Rev. 43*: 105–142.

Mendoza, C. E. (1974). Analysis of pesticides by thin layer chromatographic-enzyme inhibition technique. II. *Residue Rev. 50*: 43–72.

Merck, E. (1976). *Dyeing Reagents for Thin Layer and Paper Chromatography*. E. Merck, Darmstadt, Germany.

Muething, J., and Cacic, M. (1997). Glycosphingolipid expression in human skeletal and heart muscle assessed by immunostaining thin layer chromatography. *Glycoconjugate J. 14*: 19–28.

Mukherjee, K. D. (1996). Applications of flame ionization detectors in thin layer chromatography. In *Handbook of Thin Layer Chromatography*, 2nd ed., J. Sherma and B. Fried (Eds.). Marcel Dekker, New York, pp. 361–372.

Parrish, C. C., Bodennec, G., and Gentien, P. (1996). Determination of glycoglycerolipids by Chromarod thin layer chromatography with Iatroscan flame ionization detection. *J. Chromatogr., A 741*: 91–97.

Peyrou, G., Rakotondrazafy, V., Mouloungui, Z., and Gaset, A. (1996). Separation and quantitation of mono-, di-, and triglycerides and free oleic acid using thin layer chromatography with flame ionization detection. *Lipids 31*: 27–32.

Postaire, E., Sarbach, C., Delvordre, P., and Regnault, C. (1990). A new method of derivatization in planar chromatography: overpressure derivatization. *J. Planar Chromatogr.—Mod. TLC 3*: 247–250.

Rahalison, L., Hamburger, M., Hostettmann, K., Monod, M., and Frank, E. (1991). A bioautographic agar overlay method for the detection of antifungal compounds from higher plants. *Phytochem. Anal. 2*: 199–203.

Randerath, K. (1963). *Thin Layer Chromatography*. Verlag Chemie GmbH, Weinheim, Germany (Academic Press, San Diego, CA)

Ranny, M. (1987). *Thin Layer Chromatography with Flame Ionization Detection*. D. Reidel Publishing Co., Dodrecht.

Ritter, K., Schaade, L., and Thomssen, R. (1991). Electroblotting from a silica gel thin layer chromatogram to a polyvinylidenedifluoride membrane: a new method demonstrated by the transfer and immunostaining of gangliosides. *J. Planar. Chromatogr.—Mod. TLC 4*: 150–153.

Robards, K., Haddad, P. R., and Jackson, P. E. (1994). *Principles and Practice of Modern Chromatographic Methods*. Academic Press, San Diego, CA, pp. 179–226.

Saito, M., and Yu, R. K. (1990). TLC-immunostaining of glycolipids. In *Planar Chromatography in the Life Sciences*, J. C. Touchstone (Ed.). Wiley-Interscience, New York, pp. 59–68.

Saitoh, K., Kiyohara, C., and Suzuki, K. (1992). Formation of zinc complexes during chromatography of porphyrins on fluorescent thin-layer plates. *J. Chromatogr. 603*: 231–234.

Saxena, G., Farmer, S., Towers, G. H. N., and Hancock, R. E. W. (1995). Use of specific dyes in the detection of antimicrobial compounds from crude plant extracts using a thin layer chromatography agar overlay technique. *Phytochem. Anal. 6*: 125–129.

Sherma, J. (1998). Planar chromatography. *Anal. Chem. 70*(12): 7R–26R.

Sherma, J., and Bennett, S. (1983). Comparison of reagents for lipid and phospholipid detection and densitometric quantification on silica gel and C_{18} reversed phase thin layers. *J. Liq. Chromatogr. 6*: 1193–1211.

Sherma, J., and Marzoni, G. (1974). Detection and quantitation of anilines by thin layer chromatography using fluorescamine reagent. *Am. Lab. 6*(10): 21–30.

Sherma, J., and Touchstone, J. C. (1974). Quantitative thin layer chromatography of amino acids using fluorescamine reagent. *Anal. Lett. 7*: 279–287.

Smith, M. C., and Sherma, J. (1995). Determination of benzoic and sorbic acid preservatives by solid phase extraction and quantitative TLC. *J. Planar Chromatogr.—Mod. TLC 8*: 103–106.

Sorice, M., Griggi, T., Circella, A., Garafalo, T., d'Agostino, F., Pittoni, V., Pontieri, G. M., Lenti, L., and Valesini, G. (1994). Detection of antiphospholipid antibodies by immunostaining on thin layer chromatography plates. *J. Immunol. Methods 173*: 49–54.

Stahl, E. (1969). *Thin Layer Chromatography. A Laboratory Handbook*, 2nd ed. Springer-Verlag, New York, pp. 568–577.

Szepesi, G., and Nyiredy, Sz. (1996). Pharmaceuticals and drugs. In *Handbook of Thin Layer Chromatography, 2nd ed.*, J. Sherma and B. Fried (Eds.). Marcel Dekker, New York, pp. 819–876.

Touchstone, J. C. (1992). *Practice of Thin Layer Chromatography*, 3rd ed. Wiley-Interscience, New York, pp. 311–317.

Touchstone, J. C., Balin, A. K., Murawec, T., and Kasparow, M. (1970). Quantitative densitometry on thin layer plates of silica gel impregnated with phosphomolybdic acid. *J. Chromatogr. Sci. 8*: 443–445.

Touchstone, J. C., Murawec, T., Kasparow, M., and Wortmann, W. (1972a). Quantitative spectrodensitometry on silica gel thin layers impregnated with sulfuric acid. *J. Chromatogr. Sci. 10*: 490–493.

Touchstone, J. C., Murawec, T., Kasparow, M., and Wortman, W. (1972b). The use of silica gel modified with ammonium bisulfate in thin layer chromatography. *J. Chromatogr. 66*: 172–174.

Touchstone, J. C., Sherma, J., Dobbins, M. F., and Hansen, G. R. (1976). Optimal use of fluorescamine for in situ thin layer chromatographic quantitation of amino acids. *J. Chromatogr. 124*: 111–114.

Towbin H., Schoenenberger, C., Ball, R., Braun, D. G., and Rosenfelder, G. (1984). Glycosphingolipid-blotting: an immunological detection procedure after separation by thin layer chromatography. *J. Immunol. Methods 72*: 1471–479.

Zweig, G., and Sherma, J. (1972). *Handbook of Chromatography, General*, Vol. II. CRC Press, Boca Raton, FL.

9

Qualitative Evaluation and Documentation

I. IDENTIFICATION OF COMPOUNDS

A. Traditional Methods

Positions of separated zones on thin-layer (and paper) chromatograms are described by the R_F value of each substance, where

$$R_{F_a} = \frac{\text{distance traveled by solute ``a''}}{\text{distance traveled by solvent front}}$$

The point of greatest density is used for measurement of the location of the zone; for a symmetrical zone, that point is the center of the zone (see Figure 2.2). R_F values range from 1.0 for zones migrating at the solvent front to 0.0 for a zone not leaving the point of application. Only two decimal places can be determined with validity. When R_F data are tabulated, it is more convenient to report $R_F \times$ 100 or hR_F values. In conventional thin-layer chromatography, or especially for the techniques of overrun (continuous) or multiple development, where the solvent front is not measurable, R_X values can be recorded. R_X is defined by the equation

$$R_{X_a} = \frac{\text{distance traveled by solute ``a''}}{\text{distance traveled by standard substance X}}$$

Unlike R_F, R_X values can be greater than 1.

Neither R_F nor R_X values are true constants, but R_X values are greatly more reproducible than absolute R_F values and should be preferred for purposes of identification when comparing sample mobilities to tabulated values. The reproducibility of R_F values depends on many factors, such as quality of the sorbent, humidity, layer thickness, development distance, and ambient temperature (see Chapter 11).

The optimum method for obtaining tentative identification of a substance is to spot the sample and a series of reference compounds on the same chromatogram. In this way, mobilities of all compounds are compared under the same experimental conditions, and a match in R_F values between a sample and standard is evidence for the identity of the sample. Experimental conditions should be chosen so that the compound to be identified moves to a point near the center of the layer (R_F ~0.5) and resolution between spotted standards is as good as possible. If R_F values on silica gel are higher than desired, the polarity of the mobile phase is reduced; for R_F values that are too low, the polar component of the mobile phase is increased. If the spots of interest from the sample do not line up with the standards, they are either not the same as any of these compounds or their mobilities have been affected by accompanying extraneous material from the sample. If the latter is the case, sample cleanup (see Chapter 4) or two-dimensional TLC will be necessary.

Identification of a compound based on mobility in one chromatographic system is only tentative and must be confirmed by other evidence. It is common practice to chromatograph the sample and standards in several thin-layer systems to compare their mobilities, but this provides useful corroborative evidence only if the systems are truly independent. Connors (1974) has defined TLC systems that differ sufficiently in properties to yield independent data; for example, an adsorption layer in combination with a reversed-phase layer. Two-dimensional TLC on a two-phase silica gel and C_{18}-bonded silica gel layer has been used to provide fingerprints of complex mixtures for confirmation of identity (Beesley and Heilweil, 1982).

Instead of running the sample and standards side by side on the same plate, the sample can be mixed with the standard of the expected compound (co-chromatography). Nonseparability of this mixed sample in several independent systems is evidence that the compounds are identical.

Thin-layer chromatography can provide further information critical in compound identification. Colors from selective chemical detection reactions, behavior in long- and short-wave ultraviolet light, absorbance and/or fluorescence spectra obtained directly on the chromatogram by use of a spectrodensitometer or in solution after elution, and R_F values of derivatives prepared by reaction before, during, or after development (Dallas, 1970; Hais, 1970) can be combined with the R_F values of the sample to increase the degree of probability of correct identification. The maximum amount of information is obtained by applying, if possible, two or more compatible detection reagents in sequence to a single chromatogram, for example, both fluorescamine and ninhydrin for amino acid identification, or ninhydrin for amino acids followed by *m*-phenylene diamine for sugars. Information concerning identity can also often be obtained by comparing color changes upon heating and subsequent storing of the plates.

Various early methods were described for systematically analyzing large groups of compounds using R_F values in a series of solvent systems (e.g., Macek, 1965). A modern approach (Romano et al., 1994) used principal components analysis of standardized R_F values of 443 drugs and metabolites in four mobile phases to produce a "scores" plot that enabled either identification of an unknown or restriction of the possibilities to a few compounds. A computerized database was created containing R_F values in a number of mobile phases and detection-reagent colors for commonly used pesticides to aid in their identification (Ionov et al., 1995). Recently reported computerized methods use R_F values in a single mobile phase and responses to nine detection methods for identifying drugs and metabolites in serum, urine, and other specimens (Siek et al., 1997) or data from two parallel TLC analyses for identifying seized street drugs (Ojanpera et al., 1997).

One of the advantages of substance identity by TLC is the ability to work with a very small amount of sample. It is usually necessary, however, to have an authentic sample available for comparison. If such a standard is not available, the structure of a substance can sometimes be deduced from chromatographic data using the principle of constant differences and R_M values (Gasparic et al., 1966) as described below.

The logarithmic equation

$$R_M = \log\left(\frac{1}{R_F} - 1\right) = \log k' \quad \text{(see chapter 2)}$$

is an alternative expression for mobility that has additive and linear properties useful in studies of correlation between structure and chromatographic properties. R_M values can be calculated from a basic constant plus the values of functional group constants. These latter values are calculated by subtracting experimental R_M values of two substances differing in structure by only the group of interest. Group constants are useful not only for identification purposes (Dallas, 1970) but also as an aid for predicting a suitable solvent for separation of certain compounds.

R_M values are often used in conjunction with chemical reactions for identification of functional groups. Reactions can be carried out on one or two identical samples spotted on the origin of a chromatogram, and the difference in mobilities, expressed as ΔR_M, will measure the group formed by the reaction, the number of such groups, and sometimes the position in the molecule. Reagents causing reactions can be impregnated into the layer (Macek, 1965). If the reaction is carried out after the first development of two-dimensional chromatography with the same solvent, reaction products will lie off of a 45° diagonal line and indicate the presence of a group in a compound in the original mixture capable of reacting with the reagent (Hais et al., 1970). The use of chemical, biological, or photochemical reactions between two-dimensional development is known as reaction chromatography or SRS (TRT) technique (Dallas, 1970). This "diagonal chromatography" approach (Hais, 1970) can be used to detect the decomposition of solutes during chromatography; any degradation products will lie outside of a diagonal line after two-dimensional development with the same mobile phase (Figures 14.2 and 14.3). Reaction methods for identification have been reviewed by Gasparic (1970).

Comparison of zone sizes gives information on the quantitative ratios of identified zones (see Chapter 10). Because zone spreading always takes place during development, it is valid only to compare zones with the same R_F values. The zone size also depends on the detection reaction used; an insensitive detection will yield small, apparently well-defined zones, whereas a sensitive (e.g., fluorescence) detection often produces large, diffuse spots.

It is important to understand that evidence from TLC alone is not sufficient for unambiguous identification of unknowns. Correspondence of R_F values between sample and standard zones provides only presumptive identification, and non-universal detection reagents are not specific for one compound but selective for a type or class of compounds. Combined methods (Section B) must be used for increased certainty.

B. TLC Combined with Other Analytical Methods

TLC has been combined with spectrometric and other chromatographic methods after scraping or for direct, in situ measurement. As for PLC (Chapter 12), a miniature suction device can be used to collect scrapings directly in an extraction thimble held in a vacuum flask for elution with an appropriate solvent, or a recently reported (Skett, 1997) automated device for scraping and transfer of sorbent into empty solid-phase extraction tubes can be used. In situ methods have been referred to as direct, coupled, hyphenated, interfaced, or on-line methods. For reviews of coupled methods see Sherma (1996), Somsen et al. (1995), Kovar and Morlock (1996), Lickl (1995), Gocan and Simpan (1997), and Cserhati and Forgacs (1997). The term multidimensional denotes the interfacing of TLC with other chromatographic methods to improve separation capacity (Poole and Poole, 1995).

Infrared (IR) and mass spectrometry (MS) combined with chromatography can provide unequivocal identification if sufficient amounts of sample are available. This is not always true in trace analysis where, for example, nanogram or picogram amounts of material may be separated and detected by TLC, whereas microgram quantities would be needed for conventional IR spectrometric confirmation of the trace amount. Fourier transform IR (CECON Group, 1990) and nuclear magnetic resonance (NMR) (Bonnet et al., 1990) spectrometry allow confirmation of submicrogram amounts of separated solutes, whereas GC/MS requires only 1–5 ng or 1 pg for selected ion monitoring. Identification can be based on obtaining matching spectra from sample and standards or by deducing the structure of the unknown compound from the IR or mass spectrum it yields. Visible, ultraviolet (UV), and Raman spectrometry can also provide valuable information on structural characteristics and identity of compounds separated by TLC. Fluorescence spectrometry provides more characteristic information than visible or UV absorption spectrometry because two different spectra can be scanned, i.e., excitation and emission. Quantitative selectivity is also higher for fluorescence because two different wavelengths are used for each measurement.

1. Visible and UV Absorption and Fluorescence

Many densitometers have a spectral mode that allows automatic recording of in situ reflectance visible and UV absorption spectra and fluorescence spectra of one or more spots, in addition to determination of peak areas in the scanning mode. Solid-phase spectra obtained with densitometers are low resolution and lack detail, but they have some value for identification purposes when used with other information. An in situ sample spectrum cannot be reliably compared to a standard spectrum taken in solution because solid- and liquid-phase spectra of a given compound are often dissimilar. This noncorrespondence is especially evident in the case of fluorescence spectra. Ebel and Kang (1990) described background correction and normalization techniques to facilitate spectral comparison for identification purposes. As an alternative to full-spectrum scanning, comparison of absorption ratios at selected wavelengths can be used to compare unknowns and standards for characterization purposes. This is similar to the practice in HPLC, where the absorption ratios at the wavelengths 254 nm and 280 nm are often used. Highly resolved fluorescence line narrowing (FLN) spectra of pyrene zones on silica plates were obtained using an Ar-ion laser (363.8 nm) for excitation, a closed cycle helium refrigerator to achieve cryogenic temperatures down to 10 K, and a high resolution monochromator (Somsen et al., 1995).

Figure 9.1 shows an AMD (Chapter 7) separation of a complex mixture of pesticides, with multiple wavelength scanning to provide confirmation of identity (Butz and Stan, 1994). On-line TLC-UV and TLC-FTIR were used to identify drugs and metabolites (Chambuso et al., 1994; Kovar and Pisternik, 1996). Computer software for identification of separated zones by spectrodensitometry was described (Aginsky, 1995).

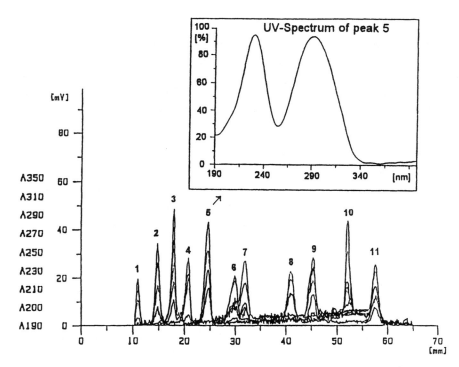

FIGURE 9.1 Multiwave chromatogram of a mixture of 11 pesticides (50 ng/compound) and in situ UV spectrum of peak 5. Compounds represented by peaks: 1, clopyralid acid; 2, triclopyr acid; 3, bitertanol; 4, atraton; 5, chlorida-zon; 6, sethoxydim; 7, atrazine; 8, iprodione; 9, desmedipham; 10, ethofumes-ate; 11, pendimethalin. [Reproduced from Butz and Stan (1994) with permission of the American Chemical Society.]

2. Infrared Spectrometry

Because of its highly discriminative capability, Fourier transform infrared (FTIR) spectrometry is valuable for identification of unknown compounds by comparison of sample spectra to reference spectra or by spectral interpretation (Somsen et al., 1995). IR spectra have been obtained on eluted samples or directly on TLC plates. About 5 µg is usually required for the elution method, which involves scraping of the zone and elution from the layer material onto an IR-transparent substance such as KBr (Issaq, 1983). Spectra can be measured directly on TLC plates by diffuse reflectance Fourier transform (DRIFT) IR spectrometry (Zuber et al., 1984). For in situ DRIFT-IR spectrometry, which requires ±1–10 µg of compound, solvents must be removed from the layer and spectra corrected for background absorption of the sorbent. FTIR spectrometry has been coupled with AMD-HPTLC for identification and quantification of complex mixtures (Kovar et al., 1991; Wolff and Kovar, 1994a). Shimadzu introduced an accessory for transfer of zones separated on a TLC sheet to a layer of KBr powder, without loss in resolution, for in situ measurement of spectra (Tajima et al., 1992).

An overview of TLC-FTIR evaluation was given by Frey et al. (1993), and qualitative and quantitative uses of near-IR reflectance densitometry were reviewed by Ciurczak et al. (1991), Mustillo and Ciurczak (1992), and Fong and Hieftje (1994). TLC on microchannels of zirconia was shown to provide ~500 times better sensitivity compared to IR detection on microscope slides (Bouffard et al., 1994). Recent applications of TLC coupled with FTIR include identification of adenosine in biological samples (Pfeifer et al., 1996), surfactants (Buschmann and Kruse, 1993), EDTA in water (Wolff and Kovar, 1994b), and drugs (Mink et al., 1995; Pfeifer and Kovar, 1995).

FTIR photoacoustic spectrometry has been used to analyze TLC spots after removal from the plate (White, 1985). It has also been applied to depth profiling of compound distribution on the sorbent (Vovk et al., 1997).

3. Raman Spectrometry

An advantage of Raman spectrometry for identification of TLC spots is the low background interference caused by the sorbents, which give weak Raman spectra (Somsen et al., 1995). FT-Raman spectrometry was shown to be a useful method for identifying a range of aliphatic and aromatic compounds separated on commercial metal-backed layers with or without a UV fluorescent indicator (Fredericks et al., 1992). Near-IR FT-Raman spectra were obtained for 200 µg of plastic additives at a loading of 3 µg/mm^2 (Everall et al., 1992). In situ surface-enhanced Raman spectrometry (SERS) and surface enhanced resonance Raman spectrometry (SERRS) have been used to acquire spectra from HPTLC spots down to 1 µm in size, containing as low as femtogram (10^{-15} g) levels of sample on layers prepared by depositing colloidal silver particles and exciting with Ar-ion or He-Ne laser (Koglin, 1989, 1990, 1994, 1996; Caudin et al., 1995). Sample and background fluorescence, which normally swamp the Raman spectra, are almost totally quenched by the silver colloidal particles (Somsen et al., 1995). Compared to SERS and SERRS, in situ regular Raman spectroscopy generates spectra that are identical to published Raman spectra measured on solids, and identification of unknowns can be made by reference to a Raman atlas. Detection limits are comparatively high, in the range of 0.5–5 µg per fraction, whereas SERS and SERRS are much more sensitive (Jaenchen, 1997). Aminotriphenylmethane dyes were characterized by TLC-SERRS (Somsen et al., 1997).

Use of near-IR (NIR) excitation was shown to eliminate the disturbing fluorescence of impurities characteristic of Raman spectroscopy in the visible range (Keller et al., 1993). NIR-SERS spectra were obtained for polycyclic aromatic hydrocarbons on Ag-activated silica plates (Matejka et al., 1996).

4. Mass Spectrometry

Mass spectra have been obtained for samples eluted from scraped stationary phase and deposited by evaporation onto an appropriate substrate, such as a Wick-Stick (Somogyi et al., 1990), for insertion into the mass spectrometer. Alternatively, the sample plus the stationary phase can be inserted into the ion source of the spectrometer. After volatilization from a heatable probe tip, ionization is carried out by electron impact or, more often, by use of secondary ion mass spectrometry (SIMS) or fast ion bombardment (FAB) (Busch, 1989). In this method, a high energy ion or atom beam is used to sputter molecules from the top layers of the sample spot into the gas phase for MS analysis. Liquid SIMS (LSIMS) can also be used after mixing the stationary phase with a low vapor-pressure liquid solvent, which acts to increase the intensity and duration of the sputtered secondary ion signal. In matrix-assisted laser desorption ionization (MALDI), sputtering of organic molecules from a surface occurs as a result of high thermal energy from a laser beam focused on the surface of a sample cocrystallized with an excess of a matrix that absorbs the photons rather than the sample molecules (Busch, 1996). The sensitivity of MALDI is in the picogram range and spatial resolution is between 250 and 500 μm (Gusev et al., 1995).

Successful imaging of samples directly from TLC plates with high spatial and mass resolution and sensitivity was demonstrated using FAB and a time-of-flight mass spectrometer (Busch et al., 1992). Sophisticated two-dimensional scanners have been based on LSIMS (Busch, 1996). The use of laser desorption (LD) in connection with TLC is of interest because very small areas of spots (a few micrometers) can be sampled, the majority of the sample is unchanged for further study, and a liquid matrix is not required (Somsen et al., 1995). The TLC/FAB-MS-MS technique has been applied to the analysis of nucleotides and bases in order to gain more spectral information and eliminate background matrix interferences found with TLC/FAB-MS (Morden and Wilson, 1995). Instrumentation, procedures, and applications of TLC/MS were reviewed in detail by Busch (1992, 1996).

The following are recent applications of TLC coupled with MS: polar lipids in bacteria (FAB-MS) (Yassin et al., 1993), glycosphingolipids directly (FAB-MS) (Lanne et al., 1996) and after detection by thin-layer chromatography blotting (SIMS) (Taki et al., 1995), unlawful food dyes (FAB-MS) (Oka et al., 1994a), antibiotics in milk (FAB-MS) (Oka et al., 1994b), heparin and chondroitin sulfate disaccharides (LSIMS) (Chai et al., 1995), ecdysteroids in plants and arthropods (MS-MS) (Lafont et al., 1993), rhamnolipids (FAB-MS-MS) (de Koster et al., 1994), peptides (LD-TOF-MS) (Krutchinsky et al., 1995), the drug naproxen (LD-TOF-MS) (Fanibanda et al., 1994), alkyl polyglucosides (SIMS) (Buschmann et al., 1996), and cocaine (MALDI) (Nicola et al., 1996).

5. Nuclear Magnetic Resonance (NMR) Spectrometry

High-resolution magic-angle-spinning (HR MAS) solid state NMR was shown to provide compound identification of separated TLC zones without elution (Wilson et al., 1997). Zones were removed by scraping from C_{18}-bonded plates, slurried with D_2O, and placed in a sealed 4-mm zirconia MAS rotor of a Brucker DRX-600 NMR spectrometer for ^1H NMR observation. Spectra were illustrated for 10 μg of salicylic acid and phenolphthalein glucuronide, which quantity can be obtained from spots on an analytical plate without the need for preparative layer chromatography.

6. Atomic Absorption Spectrometry (AAS) and Inductively Coupled Plasma Atomic Emission Spectrometry (ICP-AES)

TLC was combined off-line with AAS to study the stability of mixed zinc carboxylato complexes with two nicotinic acid ligands (Orinak et al., 1997). The zinc content, determined in scraped fraction from silica layers, was indicative of dissociation of the complexes during TLC. TLC separation, ion-exchange cleanup, and ICP-AES were combined for determination of rare earth elements in granites and greisens (safranova et al., 1995).

7. TLC Combined with other Chromatographic Methods

Column chromatographic methods can be applied to additional portions of the sample solution to confirm tentative identities based on R_F values, but again it must be known that these chromatographic systems are governed by different mechanisms before they will provide additional evidence. Elgar (1971) has examined the correlation or independence of different analytical methods for compound identification. Possibilities include retention times on nonpolar, slightly polar, or highly polar gas–liquid chromatography phases and on reversed-phase or normal-phase high-performance liquid chromatography columns.

Multidimensional separations have been achieved by combining reversed-phase HPLC with normal-phase AMD-HPTLC to achieve peak capacities of ~500. Samples can be applied to the plate by spraying using a modified Linomat or Automatic TLC Sampler (Chapter 5) connected to the column outlet. Because the maximum volume of liquid that can be successfully applied without disturbing the layer is 60 μl/min, a microbore HPLC column or a regular column with a splitter is most appropriate for this coupling (Banks, 1993; Jaenchen, 1997). As examples, RP-HPLC was coupled to HPTLC-AMD for the determination of pesticides in drinking water (Burger, 1996) and iprodione in vegetables

(Wippo and Stan, 1997). Other on-line couplings have included TLC with gas chromatography, supercritical fluid extraction, and the thermal separation technique (Kovar and Morlock, 1996). Somsen et al. (1995) have described couplings of column liquid chromatography with TLC and spectrometric methods (FTIR, SERS, and fluorescence spectrometry).

Poole and Poole (1995) provided detailed coverage of coupled chromatographic systems with TLC as one component. Among the techniques described were the use of TLC to confirm the identity of a GC peak or as a test of peak homogeneity, coupled TLC-GC using laser desorption to vaporize the sample zones separated by TLC, pyrolysis GC of scraped TLC zones, supercritical fluid extraction with deposition of the fluid extracts on a moving plate, packed column supercritical fluid chromatography with direct coupling to TLC, and various interfaces for coupling regular and narrow-bore HPLC columns with gravity and forced-flow TLC.

II. DOCUMENTATION AND STORAGE OF CHROMATOGRAMS

Individual needs often determine how chromatograms are documented and stored. For instance, the requirements of a biological research laboratory will be very different from those of a hospital clinical laboratory, a forensic or government regulatory agency laboratory where court evidence is generated, or an industrial laboratory operating under GLP or GMP standards. Laser-coded plates (Merck), with catalog, batch, and plate numbers stamped on top of each layer, significantly aid in documentation according to GLP standards by helping to eliminate mix-ups and misidentified plates. Documentation has been reviewed in detail by Kovar and Morlock (1996).

A. Listing of R_F Values

A list of R_F values does not give a true picture of the chromatogram because there is no description of the size and shape of the zones and the separation achieved. Neither do R_F values give any guide about the quantitative composition of the sample; that is, whether the solutes are present as major, minor, or trace components. To keep a record of these results, the actual chromatogram or a copy must be preserved as described below. Colors of zone detections in daylight and under long- and short-wave UV light can also be recorded along with R_F values.

B. Storage of the Chromatograms

Flexible layers on plastic or foil backing can be covered with Saran Wrap (or equivalent) and stored in a file or notebook. Plastic plates are unbreakable and lightweight, require little storage space, and can be cut with scissors to any desired size. Layers on glass plates are more difficult to store, but it is possible to do this, especially with the harder, polymer-bound layers. A darkened drawer is a convenient storage location so that colors or fluorescence of zones are less likely to fade. If the zones are visible only under UV light, their boundaries should be penciled in directly on "hard" layers or marked with a dissecting or disposable syringe needle on softer layers. The solvent front, and any irregularities in it, should be marked immediately as the plate is removed from the development chamber. Chromatographic parameters can also be written directly on most "hard" layers with a pencil or ballpoint pen. The detection reagent should produce stable colors; phosphomolybdic acid is a good general visualization reagent for this purpose because chromatograms can be preserved for years with little change if a glass cover plate is taped over the layer. If colors are expected to fade, they should be described as well as possible in the notebook, or the chromatogram should be copied by one of the procedures below.

A variation of chromatogram storage is the use of a liquid plastic dispersion that is sprayed or poured on the layer and allowed to dry. After hardening, the layer is peeled off and retained as a flexible film as a permanent record, or the separated spots can be cut from the layer and studied individually—e.g., by liquid scintillation counting (Chapter 13). The product is sold under the name Strip-Mix by Alltech-Applied Science.

C. Tracing

The results of a TLC separation can be sketched, or tracing paper can be placed on top of a glass-covered chromatogram, and the spots can be traced directly. Colors can be reproduced with a large set of colored pencils, or the color-key system devised by Nybom (1967) for recording the colors of fluorescent spots can be used and extended to colored spots. Notation can be made of the apparent relative amounts of substances based on spot size and intensity. This method of documentation, although of value for internal use, tends to be subjective and is generally unsuitable for critical analyses of data. For a further discussion of tracing, consult Berlet (1966). Direct copying on Ozalid, Ultrarapid blue print, or AGFA-Copyrapid CpP or CpN papers has also been employed for documentation (Jork et al., 1990).

D. Photography

Layers can be photographed in black and white or color under daylight or UV light with appropriate filters. Commercial TLC camera stands are available for use with Polaroid instant image or 35 mm cameras. For optimal photographic reproduction of a TLC plate or sheet, both reflected (45°) and transmitted light and a combination of these should be available and tested. Workers familiar with close-up photography can usually design an arrangement suitable for photographing chromatograms using 35 mm or Polaroid film. Because glass plates are fragile and cumbersome to store, photographic records of layers on glass provide a simple and convenient method for documentation and storage. For an excellent discussion on photographic documentation of TLC plates, see Vitek (1991). Data sheets have been described for recording the conditions and results of photographing chromatograms (Caldwell et al., 1968). Figure 9.2 shows a commercial TLC photodocumentation system. The optimal conditions for use and applications of this system for black and white, colored, fluorescent, and fluorescence quenched zones were described (Layman et al., 1995).

FIGURE 9.2 Photodocumentation system for TLC plates that provides instant black and white or color photographs of 20 × 20 cm (or smaller) plates with visible or UV lighting. The system shown includes a darkroom viewing cabinet equipped with a Polaroid DS-34 camera, but adapters are available for use with a variety of cameras. Colored spots as well as those that fluoresce upon exposure to 366-nm UV light or quench 254-nm UV light can be photographed. (Photograph courtesy of Analtech.)

E. Image Processing (Video Documentation)

Commercial video documentation systems (e.g., Figure 9.3) consist of a CCD camera with zoom objective lens, illumination module (daylight or UV at 254 or 366 nm), computer with software, and printer. A color or gray-scale digital image of a chromatogram containing colored or UV-active zones is focused under computer control and then stored and printed out as analog peaks with data such as R_F, height, and area. Video scanning can be used for zone quantification (Chapter 10) as well as documentation.

F. Computer Imaging

Documentation using a computer, digital image scanner, and printer was recently described as an alternative to commercial photographic or video systems (Rozylo et al., 1997). The layer with the detected zones is protected by covering with a

FIGURE 9.3 Camag VideoStore system including Reprostar 3 transilluminator with cabinet cover, camera support, and video camera with zoom objective. (Photograph courtesy of Camag.)

transparent foil and placed in the scanner, which transmits the image of the layer to the computer. After processing of the graphic image, it is transmitted to a black and white laser printer to produce a black and white image, to a color ink-jet printer to produce a paper or transparency color image, or to a color thermal printer to produce a high-resolution color photograph.

G. Densitometer Traces

A photodensitometer scan of a chromatogram can be used as a record of results as well as for quantification (Chapter 10). For this purpose, the origin of the layer should be marked with a needle prior to scanning. Also, the relative rate of travel of the plate and recorder should be known so that the distance of a peak from the origin can be related to actual spot distance on the plate.

H. Autoradiography

Autoradiography can be used to document chromatograms containing samples labeled with radioactive isotopes (see Chapter 13). Digital autoradiography combined with TLC-MS-MS was used for the structure determination of drug metabolites (Ludanyi, 1997).

I. Recording of Conditions

Examples of experimental parameters to be recorded for documentation of results are listed in Chapter 11. It is most usual today to collect and store this information by computer in order to allow the demonstration of the reliability of results for purposes such as laboratory accreditation, auditing, and certification.

REFERENCES

Aginsky, V. N. (1995). Increasing the reliability of the identification of separated substances by spectrodensitometric thin layer chromatography. *Adv. Forensic Sci., Proc. Meet. Int. Assoc. Forensic Sci., 13th 5*: 58–60 and 315–318.

Banks, C. T. (1993). The development and application of coupled HPLC-TLC for pharmaceutical analysis. *J. Pharm. Biomed. Anal. 11*: 705–710.

Beesley, T. E., and Heilweil, E. (1982). Two phase, two-dimensional TLC for fingerprinting and confirmation procedures. *J. Liq. Chromatogr. 5*: 1555–1566.

Berlet, H. H. (1966). Permanent records of thin layer chromatograms on transparent paper. *J. Chromatogr. 21*: 484–487.

Betina, V. (1961). pH chromatography. *Chem. Zvesti 15*: 661–667, 750–759.

Bonnett, R., Czechowski, F., and Latos-Grazynski, L. (1990). TLC–NMR of iron porphyrins from coal: the direct characterization of coal hemes using paramagnetically shifted proton NMR spectroscopy. *Energy Fuels 4*: 710–716.

Bouffard, S. P., Katon, J. E., Sommer, A. J., and Danielson, N. D. (1994). Development of microchannel thin layer chromatography with infrared microspectroscopic detection. *Anal. Chem. 66*: 1937–1940.

Burger, K. (1996). Automated multiple development (AMD). Applications and online coupling with reversed-phase HPLC. Part 1. Principles and application of AMD. Multiple methods for ultratrace determination: plant-protection agents in groundwater and drinking water analyzed by thin layer chromatography (DC) with AMD. In *Duennschicht-Chromatogr.*, R. E. Kaiser (Ed.). InCom-Bureau, Duesseldorf, Germany, pp. 31–52.

Busch, K. L. (1989). Planar chromatography coupled with secondary ion mass spectrometry. *J. Planar Chromatogr.—Mod. TLC 2*: 355–361.

Busch, K. L. (1992). Coupling thin layer chromatography with mass spectrometry. *J. Planar Chromatogr.—Mod. TLC 5*: 72–79.

Busch, K. L (1996). Thin layer chromatography coupled with mass spectrometry. In *Handbook of Thin Layer Chromatography*, 2nd ed., J. Sherma and B. Fried (Eds.). Marcel Dekker, New York, pp. 241–272.

Busch, K. L., Mullis, J. O., and Chakel, J. A. (1992). High-resolution imaging of samples in thin layer chromatograms using a time-of-flight secondary ion mass spectrometer. *J. Planar Chromatogr.—Mod. TLC 5*: 9–15.

Buschmann, N., and Kruse, A. (1993). In-situ TLC-IR and TLC-SIMS: powerful tools in the analysis of surfactants. *Comun. Jorn. Com. Esp. Deter. 24*: 457–468.

Buschmann, N., Merschel, L., and Wodarczak, S. (1996). Analytical methods for alkyl polyglucosides. Part II. Qualitative determination using thin layer chromatography and identification by means of in-situ secondary ion mass spectrometry. *Tenside, Surfactants, Deterg. 33*: 16–20.

Butz, S., and Stan, H. -J. (1994). Screening of 265 pesticides in water by thin layer chromatography with automated multiple development. *Anal. Chem. 67*: 620–630.

Caldwell, R. L., Shields, J. K., and Stith, L. S. (1968). Recording of thin layer chromatograms. *J. Chromatogr. 36*: 372–374.

Caudin, J. P., Beljebbar, A., Sockalingum, G. D., Angiboust, J. F., and Manfait, M. (1995). Coupling of FT-Raman and FT-SERS with TLC plates for in situ identification of chemical compounds. *Spectrochim. Acta Part A 51A*: 1977–1983; Caudin, J. P., Beljebbar, A., Sockalingum, G. D., Nabiev, I., Angiboust, J. F., and Manfait, M. (1995). NIR FT-Raman and FT-SERS microspectroscopy combined with thin layer chromatography for biological applications. *Spectrosc. Biol. Mol., Eur. Conf., 6th*: pp. 31–32.

CECON Group (1990). Thin layer chromatography/Fourier transform infrared spectrometry. *Pract. Spectrosc. 10*: 137–162.

Chai, W., Rosankiewicz, J. R., and Lawson, A. M. (1995). TLC-LSIM of neoglycolipids of glycosaminoglycan disaccharides and of oxymercuration cleavage products of heparin fragments that contain unsaturated uronic acid. *Carbohydr. Res. 269*: 111–124.

Chambuso, M., Kovar, K. -A., and Zimmermann, W. (1994). The systematic separation of unknown mixtures of drugs by multi-column solid-phase extractions analogous to the Stas-Otto extraction process. *Pharmazie 49*: 142–148.

Ciurczak, E. W., Murphy, W. R., and Mustillo, D. M. (1991). The use of near infrared spectroscopy in thin layer chromatography, part II: qualitative applications. *Spectroscopy (Eugene, OR) 6*: 34, 36–40.

Connors, K. A. (1974). Use of multiple R_F values for identification by paper and thin layer chromatography. *Anal. Chem. 46*: 53–58.

Cserhati, T., and Forgacs, E. (1997). Trends in thin layer chromatography: 1997. *J. Chromatogr. Sci. 35*: 383–391.

Dallas, M. S. J. (1970). In situ chemical reactions on thin layers in the identification of organic compounds. *J. Chromatogr. 48*: 193–199.

de Koster, C. G., Vos, B., Versluis, C., Heerma, W., and Haverkamp, J. (1994). High performance thin layer chromatography/fast atom bombardment (tandem) mass spectrometry of Pseudomonas rhamnolipids. *Biol. Mass Spectrom. 23*: 179–185.

Ebel, S., and Kang, J. S. (1990). UV/vis spectra and spectral libraries in TLC/HPTLC—background correction and normalization. *J. Planar Chromatogr.—Mod. TLC 3*: 42–46.

Elgar, K. E. (1971). The identification of pesticides at residue concentrations. In *Advances in Chemistry Series 104*. American Chemical Society, Washington, DC, pp. 151–161.

Everall, N. J., Chalmers, J. M., and Newton, I. D. (1992). In situ identification of thin layer chromatography fractions by Fourier transform Raman spectroscopy. *Appl. Spectrosc. 46*: 597–601.

Fanibanda, T., Milnes, J., and Gormally, J. (1994). Thin layer chromatography–mass spectrometry using infrared laser desorption. *Mass Spectrom. Ion Processes 140*: 127–132.

Felici, R., Franco, E., and Cristalli, M. (1974). Xerox copying, a simple method for recording chromatograms. *J. Chromatogr. 90*: 208–214.

Fong, A., and Hieftje, G. M. (1994). Near-IR spectroscopic examination of thin layer chromatography plates in the diffuse transmittance mode. *Appl. Spectrosc. 48*: 394–399.

Fredericks, P., DeBakker, C., Martinez, E. (1992). In situ characterization by FT–Raman spectroscopy of compounds separated on a thin layer chromatography plate. *Proc. SPIE–Int. Soc. Opt. Eng. 1575*: 468–469.

Frey, O. R., Kovar, K. -A., and Hoffmann, V. (1993). Possibilities and limitations by on-line coupling of thin layer chromatography and FTIR spectroscopy. *J. Planar Chromatogr.—Mod. TLC 6*: 93–99.

Gaenshirt, H. (1969). Documentation of thin layer chromatograms. In *Thin Layer Chromatography. A Laboratory Handbook*, 2nd ed., E. Stahl (Ed.). Springer-Verlag, New York, pp. 125–133.

Gasparic, J. (1970). Identification of organic compounds by chromatography after decomposition and degradation. *J. Chromatogr. 48*: 261–267.

Gasparic, J., Gemzova, I., and Snobl, D. (1966). Investigations of the relationship between structure and chromatographic retention of water insoluble azo dyes. *Collection Czech. Chem. Commun. 31*: 1712–1741.

Gocan, S., and Cimpan, G. (1997). Compound identification in thin layer chromatography using spectrometric methods. *Rev. Anal. Chem. 16*: 1–24.

Gusev, A. I., Vasseur, A., Proctor, A., Sharkey, A. G., and Hercules, D. M. (1995). Imaging of thin layer chromatograms using matrix-assisted laser desorption/ionization mass spectrometry. *Anal. Chem. 67*: 4565–4570; Gusev, A. I., Proctor, A., Rabinovich, Y. I., and Hercules, D. M. (1995). Thin layer chromatography combined with matrix-assisted laser desorption/ionization mass spectrometry. *Anal. Chem. 67*: 1805–1814.

Hais, I. M. (1970). Diagonal techniques. Introductory remarks. *J. Chromatogr. 48*: 200–206.

Hais, I. M., Lederer, M., and Macek, K. (Eds.). (1970). *Identification of Substances by Paper and Thin Layer Chromatography.* Elsevier, New York.

Heinz, D. E., and Vitek, R. K. (1975). Photographic documentation in the laboratory. *J. Chromatogr. Sci. 13*: 570–576.

Inoue, Y. M., Noda, M., and Hirayama, O. (1955). Paper chromatography of unsaturated fatty acid esters as their mercuric acetate addition compounds. *J. Am. Oil Chem. Soc. 32*: 132–135.

Ionov, I., Kapitanova, D., and Blajev, K. (1995). Computer-aided identification of TLC-separated pesticides. *Adv. Forensic Sci., Proc. Meet. Int. Assoc. Forensic Sci., 13th, 5*: 114–117.

Issaq, H. J. (1983). A combined TLC-micro IR method. *J. Liq. Chromatogr. 6*: 1213–1220.

Jaenchen, D. (1997). Thin layer (planar) chromatography. In *Handbook of Instrumental Techniques for Analytical Chemistry*, F. Settle (Ed.). Prentice Hall PTR, Upper Saddle River, NJ, pp. 221–239.

Jork, H., Funk, W., Fischer, W., and Wimmer, H. (1990), *Thin Layer Chromatography—Reagents and Detection Methods*, Vol. 1a. VCH-Verlag, Weinheim, Germany.

Keller, S., Loechte, T., Dippel, B., and Schrader, B. (1993). Quality control of food with near-infrared-excited Raman spectroscopy. *Fresenius' J. Anal. Chem. 346*: 863–867.

Koglin, E. (1989). Combining HPTLC and micro-surface-enhanced Raman spectroscopy (micro-SERS). *J. Planar Chromatgr.—Mod. TLC 2*: 194–197.

Koglin, E. (1990). In situ identification of highly fluorescent molecules on HPTLC plates by SERRS. *J. Planar Chromatogr.—Mod. TLC 3*: 117–120.

Koglin, E. (1994). Combining TLC (HPTLC) and SERS. *GIT Fachz. Lab. 38*: 627–628, 630–632.

Koglin, E. (1996). Combination of thin layer chromatography (TLC) and surface-enhanced Raman scattering (SERS). *CLB Chem. Labor Biotech. 47*: 257–261.

Kovar, K. -A., and Morlock, G. E. (1996). Detection, identification, and documentation. In *Handbook of Thin Layer Chromatography*, 2nd ed., J. Sherma and B. Fried (Eds.). Marcel Dekker, New York, pp. 205–239.

Kovar, K. -A., and Pisternik, W. (1996). Determination of the designer drug MDE and its most important metabolites in urine by direct HPTLC-UV/FTIR combination. *Pharm. Unserer Zeit 25*: 275.

Kovar, K. A., Ensslin, H. K., Frey, O. R., Rienas, S., and Wolff, S. C. (1991). Applications of online coupling of thin layer chromatography and FTIR spectroscopy. *J. Planar Chromatogr.—Mod. TLC 4*: 246–250.

Krutchinsky, A. N., Dolgin, A. I., Utsal, O. G., and Khodorkovski, A. M. (1995). Thin layer chromatography/laser desorption of peptides followed by multiphoton ionization time-of-flight mass spectrometry. *J. Mass Spectrom. 30*: 375–379.

Lafont, R., Porter, C. J., Williams, E., Read, H., Morgan, E. D., and Wilson, I. D. (1993).

The application of off-line HPTLC-MS-MS to the identification of ecdysteroids in plant and arthropod samples. *J. Planar Chromatogr.—Mod. TLC 6*: 421–424.

Lanne, B., Olsson, B. -M., Larson, T., Karlsson, K. -A. (1996). Fast atom bombardment mass spectrometric analysis of glycosphingolipids by direct desorption from thin layer chromatography plates: application of the method to a receptor-active 8-sugar glycolipid. *Eur. Mass Spectrom. 2*: 361–368.

Layman, L. R., Targan, D. A., and Sherma, J. (1995). Photodocumentation of thin layer chromatography plates. *J. Planar Chromatogr.—Mod. TLC 8*: 397–400.

Lickl, E. (1995). Coupling techniques for thin layer chromatography. *Oesterr. Chem. Z. 96*: 123–124.

Ludanyi, K., Gomory, A., Klebovich, I., Monostory, K., Vereczkey, L., Ujszaszy, K., and Vekey, K. (1997). Applications of TLC-FAB mass spectrometry in metabolism research. *J. Planar Chromatogr.—Mod. TLC 10*: 90–96.

Macek, K. (1965). Impregnation of reagents in paper and thin layer chromatography. In *Stationary Phase in Paper and Thin Layer Chromatography*, K. Macek and I. M. Hais (Eds.). Elsevier, New York, pp. 309–315.

Matejka, P., Stavek, J., Volka, K., and Schrader, B. (1996). Near-infrared surface-enhanced Raman scattering spectra of heterocyclic and aromatic species adsorbed on TLC plates activated with silver. *Appl. Spectrosc. 50*: 409–414.

Mink, J., Horvath, E., Kristof, J., Gal, T., and Veress, T. (1995). Direct analysis of thin layer chromatographic spots by means of diffuse reflectance Fourier transform infrared spectroscopy. *Mikrochim. Acta 119*: 129–135.

Morden, W., and Wilson, I. D. (1995). Separation of nucleotides and bases by high performance thin layer chromatography with identification by tandem mass spectrometry. *J. Planar Chromatogr.—Mod. TLC 8*: 98–102.

Mustillo, D. M., and Ciurczak, E. W. (1992). The development and role of near-infrared detection in thin layer chromatography. *Appl. Spectrosc. Rev. 27*: 125–141.

Nicola, A. J., Gusev, A. I., and Hercules, D. M. (1996). Direct quantitative analysis from thin layer chromatography plates using matrix-assisted laser desorption/ionization mass spectrometry. *Appl. Spectrosc. 50*: 1479–1482.

Nybom, N. (1967). Key for marking the color of fluorescent spots on chromatograms. *J. Chromatogr. 26*: 520–521.

Oka, H., Ikai, Y., Ohno, T., Kawamura, N., Hayakawa, J., Harada, K. -i., and Suzuki, M. (1994a). Identification of unlawful food dyes by thin layer chromatography–fast atom bombardment mass spectrometry. *J. Chromatogr., A 674*: 301–307.

Oka, H., Ikai, Y., Hayakawa, J., Masuda, K., Harada, K. -i., and Suzuki, M. (1994b). Improvement of chemical analysis of antibiotics. Part XIX: Determination of tetracycline antibiotics in milk by liquid chromatography and thin layer chromatography/fast atom bombardment mass spectrometry. *J. AOAC Int. 77*: 891–895.

Ojanpera, I., Nokua, J., Vuori, E., Sunila, P., and Sippola, E. (1997). Combined dual-system R_F and UV library search software: Application to forensic drug analysis by TLC and RPTLC. *J. Planar Chromatogr.—Mod. TLC 10*: 281–285.

Orinak, A., Orinakova, R., Slesarova, L., and Gazdova, V. (1997). Off-line hyphenation of thin layer chromatography with atomic absorption spectrometry in the study of stability of zinc complexes. *J. Planar Chromatogr.—Mod. TLC 10*: 44–47.

Pfeifer, A. M., and Kovar, K. -A. (1995). Identification of LSD, MBDB, and atropine in real samples with online HPTLC-FTIR coupling. *J. Planar Chromatogr.—Mod. TLC 8*: 388–392.

Pfeifer, A. M., Tolimann, G., Ammon, H. T. P., and Kovar, K. -A. (1996). Identification of adenosine in biological samples by HPTLC-FTIR on-line coupling. *J. Planar Chromatogr.—Mod. TLC 9*: 31–34.

Poole, C. F., and Poole, S. K. (1995). Multidimensionality in planar chromatography. *J. Chromatogr., A 703*: 573–612.

Romano, G., Caruso, G., Masumarra, G., Pavone, D., and Cruciani, G. (1994). Qualitative organic analysis. Part 3. Identification of drugs and their metabolites by PCA of standardized TLC data. *J. Planar Chromatogr.—Mod. TLC 7*: 233–241.

Rozylo, J. K., Siembida, R., and Jamrozek-Manko, A. (1997). A new simple method for documentation of TLC plates. *J. Planar Chromatogr.—Mod. TLC 10*: 225–228.

Safronova, N. S., Matveeva, S. S., Fabelinsky, Y. I., and Ryabukhin, V. A. (1995). Inductively coupled plasma atomic emission spectrometric determination of rare earth elements in granites and greisens using thin layer chromatography preconcentration. *Analyst (Cambridge, U.K.) 120*: 1427–1432.

Sherma, J. (1996). Basic techniques, materials, and apparatus. In *Handbook of Thin Layer Chromatography*, 2nd ed., J. Sherma and B. Fried (Eds.). Marcel Dekker, New York, pp. 3–47.

Siek, T. J., Stradling, C. W., McCain, M. W., and Mehary, T. (1997). Computer-aided identification of thin layer chromatography patterns in broad-spectrum drug screening. *Clin. Chem. (Washington, D.C.) 43*: 619–626.

Skett, P. (1997). The design and operation of a novel device for collecting fractions from TLC plates for biological and spectroscopic analysis. *J. Planar Chromatogr.—Mod. TLC 10*: 263–266.

Somogyi, G., Dinya, Z., Lacziko, A., and Prokai, L. (1990). Mass spectrometric identification of thin layer chromatographic spots: implementation of a modified Wick-Stick method. *J. Planar Chromatogr.—Mod. TLC 3*: 191–193.

Somsen, G. W., Morden, W., and Wilson, I. D. (1995). Planar chromatography coupled with spectroscopic techniques. *J. Chromatogr., A 703*: 613–665.

Somsen, G. W., ter Riet, P. G. J. H., Gooijer, C., Velthorst, N. H., and Brinkman, U. A. Th. (1997). Characterization of aminotriphenylmethane dyes by TLC coupled with surface-enhanced resonance Raman spectroscopy. *J. Planar Chromatogr.—Mod. TLC 10*: 10–17.

Tajima, T., Wada, K., and Ichimura, K. (1992). Zone transfer technique for diffuse reflectance Fourier transform infrared spectrometry of analytes separated by thin layer chromatography. *Vib. Spectrosc. 3*: 211–216.

Taki, T., Ishikawa, D., Handa, S., and Kasama, T. (1995). Direct mass spectrometric analysis of glycosphingolipid transferred to a polyvinylidene difluoride membrane by thin layer chromatography blotting. *Anal. Biochem. 225*: 24–27.

Vitek, R. K. (1991). Photographic documentation of thin layer chromatograms. In *Handbook of Thin Layer Chromatography*, J. Sherma and B. Fried (Eds.). Marcel Dekker, New York, pp. 211–248.

Vovk, I., Franko, M., Gibkes, J., Prosek, M., and Bicanic, D. (1997). Depth profiling of

TLC plates by photoacoustic spectroscopy. *J. Planar Chromatogr.—Mod. TLC 10*: 258–262.

White, R. L. (1985). Analysis of thin layer chromatographic adsorbates by Fourier transform infrared photoacoustic spectroscopy. *Anal. Chem. 57*: 1819–1822.

Wilson, I. D., and Morden, W. (1996). Advances and applications in the use of HPTLC-MS-MS. *J. Planar Chromatogr.—Mod. TLC 9*: 84–91.

Wilson, I. D., Spraul, M., and Humpfer, E. (1997). Thin layer chromatography combined with high-resolution solid-state NMR for compound identification without substance elution: Preliminary results. *J. Planar Chromatogr.—Mod. TLC 10*: 217–219.

Wippo, U., and Stan, H. J. (1997). Determination of iprodione residues in vegetable food samples using online coupling of RP-HPLC followed by AMD-TLC. *Dtsch. Lebensm.-Rundsch. 93*: 144–148.

Wolff, S. C., and Kovar, K. -A. (1994a). Direct HPTLC-FTIR measurement in combination with AMD. *J. Planar Chromatogr.—Mod. TLC 7*: 344–348.

Wolff, S. C., and Kovar, K. -A. (1994b). Determination of edetic acid (EDTA) in water by on-line coupling of HPTLC and FTIR. *J. Planar Chromatogr.—Mod. TLC 7*: 286–290.

Yassin, A. F., Haggeni, B., Budzikiewicz, H., and Schaal, K. P. (1993). Fatty acid and polar lipid composition of the genus Amycolatopsis: Application of fast atom bombardment–mass spectrometry of underivatized phospholipids. *Int. J. Syst. Bacteriol. 43*: 414–420.

Zuber, G. E., Warren, R. J., Begash, P. P., and O'Donnell, E. L. (1984). Direct analysis of TLC spots by diffuse reflectance Fourier transform infrared spectrometry. *Anal. Chem. 56*: 2935–2939.

10

Quantification

Methods for the quantitative evaluation of thin-layer chromatograms can be divided into two categories. In the first, solutes are assayed directly on the layer, either by visual comparison, area measurement, or densitometry. In the second, solutes are eluted from the sorbent before being examined further. Quantification of radioactive zones is not considered here but is discussed in Chapter 13.

Requirements for the various steps of TLC are more stringent when quantitative analysis is to be carried out. The accurate and precise application of samples is especially critical, and disposable Microcap micropipets (Emanuel, 1973) or commercial automatic spot or streak applicators (Chapter 5) are recommended for use. The development step should completely separate the compound of interest, and no loss of substance by decomposition, evaporation, or irreversible adsorption during application, chromatography, or drying can occur. Standards and samples are always chromatographed on the same plate. Unresolved zones can sometimes be quantified by sequential densitometric scanning using two wavelengths chosen so that neither over-lapping component absorbs or fluoresces at the wavelength used to measure the other.

In situ methods of quantification are based on visual, manual, or instrumental measurements of spots directly on the layer. In addition to the above requirements, in situ determination requires that a detection reagent, if needed, be chosen that produces a zone (colored or fluorescent) contrasting sharply with the background; the reagent must be applied uniformly, usually by dipping; the intensity of the zone must be reproducible and proportional to the amount of substance present; and the reagent must react completely with the highest concentration of substance spotted. The samples should be spotted within an optimum concentration range that provides maximum sensitivity consistent with the greatest linearity of response. It is best to apply the same small volumes of sample and standard solutions of various concentrations within this range, so that all initial zone areas are as compact and uniform as possible. This latter requirement may not be important if layers with a preadsorbent dispensing area (Chapter 3) are used because all initial zones, regardless of their analyte content and volume, are concentrated to a constant size prior to separation.

The following books and articles are recommended for information on quantitative TLC: Kirchner (1973), Hezel (1978), Coddens et al. (1983a), Poole and Schuette (1984), Treiber (1987), Poole and Poole (1989, 1991), and Szepesi (1990).

I. SCRAPING AND ELUTION

Quantification can be carried out after scraping the separated analyte zone, collection of the sorbent, and recovery of the compound by elution. If the compound of interest is colored or fluorescent, quenches the fluorescence of a phosphor-impregnated layer, or can be detected nondestructively (as with I_2 vapors), the location of the spot is obvious. Otherwise, spots applied to the sides of the layer are detected after masking the central part of the layer. An area at least 10% larger than the visual size of the spot should be scraped if possible to compensate for three-dimensional spreading of the spot below the layer surface. The scrapings are transferred to either a Soxhlet extractor or centrifuge tube and homogenized with a suitable eluting solvent, and the clear extract or supernatant is analyzed (Robards et al., 1994). Detection, scraping, and collection techniques are in general essentially the same as those described in Chapter 12 for preparative layer chromatography. The only difference is that collection and elution of samples and standards must be complete, or at least consistent, if reliable quantitative comparison of the samples and standards spotted on each plate is to be made.

The eluates from sample and standard spots can be analyzed by any convenient and appropriate procedure, for example, titration, electroanalysis, gas chromatography, or spectrophotometry. Ultraviolet, visible, or fluorescence spectrophotometry is most often the method of choice; however, densitometry can be carried out directly on the plate in these modes, and the compound will usually be more concentrated on the chromatogram than it will be in solution in a cuvette. In addition, the scraping and elution approach is time-consuming and labor-intensive and subject to losses of material—for example, by irreversible adsorption. The extraction method has been shown to have a reproducibility that is markedly poorer than with the use of a modern scanning densitometer (Matysik et al., 1994).

If the sorbent contains materials that can interfere in the subsequent analysis or if nonspecific background absorption of colloidal silica gel in the analytical solution is a problem, a blank area of the layer should be eluted and analyzed, and the blank value subtracted from the sample and standards results. This value can often be lowered by predevelopment of the layer with dichloromethane–methanol (1:1) for organic impurities or ethanol–concentrated hydrochloric acid (9:1) for inorganic ions, with subsequent drying.

An automatic TLC elution apparatus, the Eluchrom, was described in the previous edition of this book. This instrument is no longer sold by Camag.

Although direct densitometry is becoming increasingly important, the spot elution method is still very widely used throughout the world. Many selective and sensitive analyses can be performed in a simple and relatively inexpensive fashion by using TLC for separation and a basic absorption or fluorescence spectrometer for quantification. For example, assay of some drugs by TLC with elution is specified in the United States Pharmacopeia (USP).

II. VISUAL COMPARISON

The basic requirements for all in situ quantification methods are as follow: (a) The area of the starting spot must be the same for all samples and standards; (b) samples and standards are chromatographed side by side on the same plate with the same mobile phase; (c) the concentrations must be adjusted so that the standards and samples are closely bracketed, and so that the response is maximum and linear; (d) the R_F values for samples and standards of a given compound should be as identical as possible and within the range of 0.3 to 0.7 (Touchstone et al., 1971). The accuracy of the results can be improved by repeating the analysis with the sample(s) and a narrowed range of standards.

The simplest kind of in situ quantitative TLC is based on visual or "eyeball" comparison of size and/or intensity of colored, fluorescent, or quenched spots. This method is quick and requires no equipment, but results are only semiquantitative ($\pm$ 10–30%). Greatest accuracy is obtainable when the samples and standards are in the concentration range where the intensities and areas of the spots are changing most rapidly with changes in the weight of solute, and when standards closely bracketing the sample are spotted. Comparison is easier on TLC plates with wider particle size range sorbents than on HPTLC plates. With the latter, spot size will not change as much during migration, so intensity rather than size must be used for quantitative estimation. The human eye can judge size differences more accurately than intensities.

As an example of applications, semiquantitative estimation of impurities by visual comparison is generally used in different pharmacopeias for the purity testing of active raw materials and formulated products (Szepesi and Nyiredy, 1996).

III. AREA MEASUREMENT

Measurement of spot areas by methods such as planimetry; photocopying or photographing, cutting, and weighing; or copying onto millimeter-squared graph paper and counting the number of squares are described in earlier editions of this book. These methods, which can reduce the error of quantification to a range of 5–15%, are tedious and time-consuming and seldom used today.

IV. IN SITU DENSITOMETRY

Instruments for quantitative TLC were reviewed by Jaenchen (1996, 1997) and the theory of optical quantification by Pollak (1991), Ebel (1996), and Prosek and Pukl (1996).

A. Introduction

Densitometry is the instrumental measurement of visible or UV absorbance, fluorescence, or fluorescence quenching directly on the thin layer. Measurements are made either through the plate (transmission), by reflection from the plate, or by reflection and transmission simultaneously, and either single-beam, double-beam, or single-beam–dual wavelength operation of scanning instruments has been used.

Compounds that are not naturally colored or fluorescent can be derivatized prior to or after chromatography (Duez et al., 1991) by treatment with a chromogenic or fluorogenic reagent. The purpose of a scanner is to convert the spots on a layer into a chromatogram consisting of a series of peaks similar in appearance to a gas or high-performance liquid chromatogram. The positions of the scanner peaks on the recorder chart are related to the migration distance of the spots on the layer, and the peak heights or areas are related to the concentrations of the substances in the spots. The signal that is measured represents the absorption of transmitted or reflected light that passes through the spot compared to a blank portion of the layer. In addition to absorption (or fluorescence) by the solute, light is diffusely scattered by the layer particles, leading to differences in the theory of densitometry and solution spectrometry. Modern densitometric scanners are connected to computers and are able to perform such data handling tasks as automatic peak location, multiple wavelength scanning, and spectral comparison of spots, as well as automated scanning in a variety of operating modes (reflectance, transmission, absorption, fluorescence, zigzag, etc.) (Poole and Poole, 1989).

B. Slit-Scanning Densitometer Design and Performance

The design of all densitometers is similar in principle but differs in detail. Virtually all instruments have a light source, a monochromator or filters or both, an optical system to form the light beam into a fixed slit, one or more photosensing detectors, a readout system, and a stage controlled by a stepping motor that moves the mounted plate through the fixed beam of monochromatic light past the detector(s). Common accessories include a baseline corrector, linearizer, integrator, or computer.

Figure 10.1 shows schematically the common features and differences in the design of several types of scanners that have been produced commercially. Halogen or tungsten lamps are used as sources for the visible range, a deuterium lamp for the UV, and high-intensity mercury or xenon lamps for exciting fluorescence. Filters (glass, plastic, or interference) or monochromators (prism or grating) are used for wavelength selection, and a cutoff filter is used between the plate and detector to reject the exciting UV light for fluorescence measurement. Photo-multiplier or photodiode detectors are employed; the former has the advantages of a wider wavelength range and linear output as a function of the exciting energy. Fiber-optic based instruments have recently been described for in situ scanning (Aponte et al., 1996). Monochromators, light sources, and detectors used in various commercial densitometers are described by Jork et al. (1990).

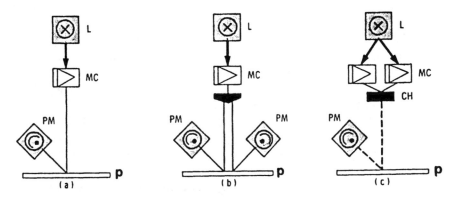

FIGURE 10.1 Schematic diagrams showing the optical arrangements of different types of scanning densitometers: (a) single beam, (b) double beam in space with beam splitter, (c) single beam, dual wavelength. L = source; MC = monochromator; P = TLC plate; PM = photomultiplier; CH = chopper. [Reproduced with permission of the American Chemical Society from Fenimore and Davis (1981).]

The single-beam, single-wavelength scanning mode (Figure 10.1a) can give excellent quantitative results with homogeneous high-performance plates but is subject to baseline drift due to background noise from surface irregularities in non-HPTLC plates, extraneous adsorbed impurities, or fluctuations in source output. These problems can be minimized by good analytical techniques and by electronic correction (Poole and Poole, 1991). The single-wavelength, double-beam arrangement can be "double beam in space," with a split light beam passing over the spot and an adjacent blank area and matched detectors incorporated in the densitometer design (Figure 10.1b), or "double beam in time" (Figure 10.2b with an additional beam collector and a single detector). Densitometers with the double beam in space design are apparently no longer manufactured. Double-beam operation can at least partly compensate for background and source fluctuations.

A third arrangement is the single-beam, dual-wavelength mode (Figure 10.1c) (Cheng and Poole, 1983). Two monochromators and a chopper alternatively furnish the sample lane with a reference wavelength (minimal absorption by the spot) and analytical wavelength (maximum absorption), separated in time. The beams are recombined into a single beam to provide the difference signal beam at the detector. This mode corrects for scattering (Figure 10.2), spurious absorption by plate impurities, and plate irregularities. For optimal scattering correction, the two chosen wavelengths should be as close together as possible. In general, the best baselines are obtained in this mode for irregular, impure layers.

Some densitometers may also provide for scanning in a zigzag (flying spot) rather than linear fashion using a small rectangle of light in place of a slit (point scanning) and computer integration. This method provides smooth, reproducible scans for irregularly shaped zones, but measurement time is long compared to linear scanning, spatial resolution and signal-to-noise ratio are lower, and error propagation in quantification is unfavorable (Jaenchen, 1996). Scanning of circular and anticircular HP chromatograms requires a special densitometer with a stage that can turn the plate with a circular motion for radial scanning (in the direction of development) or peripheral scanning (at right angles to the direction of development) (Jaenchen and Schmutz, 1978). The Camag TLC Scanner II could be used to scan circular and anticircular chromatograms, but the Scanner 3 (Figure 10.3) cannot. Simultaneous measurement of reflection and transmission, or transmission alone, can be carried out by means of a second detector positioned on the opposite side of the plate. It has been claimed (Robards et al., 1994) that combined transmission-reflectance scanning provides a better signal-to-noise ratio for spots with an absorption maximum greater than 380 nm.

Many modern scanners have a computer-controlled motor-driven monochromator that allows automatic recording of in situ absorption and fluorescence excitation spectra. These spectra can aid compound identification by comparison

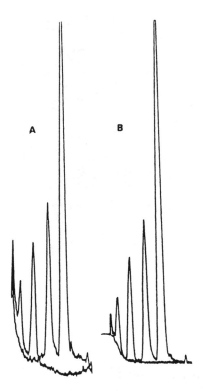

FIGURE 10.2 Use of background correction to improve baseline stability in the analysis of a mixture containing metroprolol and some potential contaminants. A was measured using the single wavelength mode, $\lambda = 280$ nm and B by the single-beam, dual-wavelength mode, $\lambda_1 = 280$ nm, $\lambda_2 = 300$ nm. [Reproduced with permission of the Elsevier Scientific Publishing Co. from Cheng and Poole (1983).]

of unknown spectra with stored standard spectra obtained under identical conditions or spectra of standards measured on the same plate. They can test for identity by superimposition of spectra from different zones on a plate, and check zone purity by superimposition of spectra from different areas of a single zone. The fluorescence mode is more selective for identification because different excitation and emission wavelengths are used for each measurement.

The Camag TLC Scanner 3 with CATS software, a representative modern TLC scanning system, is pictured in Figure 10.3. A schematic diagram of the light path is shown in Figure 10.4. This scanner provides for reflectance or transmission scanning of absorbance or fluorescence on plates up to 20×20 cm. The spectral range is 190–800 nm with use of deuterium and tungsten-halogen lamps;

FIGURE 10.3 TLC Scanner 3 with personal computer and CATS software. (Photograph courtesy of Camag.)

a mercury vapor lamp is also available for fluorescence. Scanning speed is selectable up to 100 mm/sec and spatial resolution between 25 and 200 µm. The monochromator (concave holographic grating) has selectable bandwidth of 5 nm (used for spectral recording and multiwavelength scanning) or 20 nm (for high intensity when spectral selectivity is not required); it can be flushed with nitrogen to eliminate oxygen and avoid formation of ozone, which absorbs in the UV region). A maximum of 36 tracks with up to 100 peak windows can be evaluated. Integration with either automatic baseline correction and/or video integration and single- or multi-level calibration with linear or nonlinear regression and internal or external standards can be performed. The CATS spectral library program for qualitative identification is run under MS Windows. The Desaga CD 60 is another PC-controlled single-beam densitometer that operates in absorbance, fluorescence, transmission, and reflection modes between 200 and 700 nm and offers features such as spectrum and multiwavelength scanning and three-dimensional color graphics. Shimadzu manufactures a dual-wavelength flying spot densitometer (CS-930 1 PC) utilizing the company's original photometric system with advanced Windows software. It operates in transmission and reflection modes for absorption or fluorimetry and offers image analysis software for densitometry of 2D chromatograms and an attachment for measuring microplates.

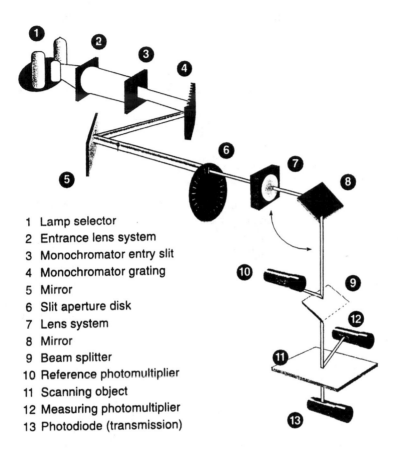

1 Lamp selector
2 Entrance lens system
3 Monochromator entry slit
4 Monochromator grating
5 Mirror
6 Slit aperture disk
7 Lens system
8 Mirror
9 Beam splitter
10 Reference photomultiplier
11 Scanning object
12 Measuring photomultiplier
13 Photodiode (transmission)

FIGURE **10.4** Light path diagram of TLC Scanner 3. (Photograph courtesy of Camag.)

C. Kubelka–Munk Equation

Detection limits with commercial scanners and HP plates are typically low nanogram levels for compounds that absorb visible or UV light strongly and low picogram amounts for compounds that fluoresce strongly. Linear response ranges are typically 10^2–10^3 for fluorescence and at least one order of magnitude less for absorption. Nonlinearity is the result of the scattering plus absorption that occurs on a thin-layer plate. These combined processes are not adequately described by the familiar Beer–Lambert law for dilute solutions, but the following

Kubelka–Munk equation (Jork et al., 1990) predicts the nonlinear relation between reflection and concentration:

$$\frac{(1 - R)^2}{2R} = 2.303\ \varepsilon\ \frac{C}{S}$$

where

R = light reflected from the plate surface
ε = molar extinction (absorption) coefficient of the sample
C = spot concentration
S = plate scatter coefficient

With transmitted light, response may follow the Beer–Lambert law more closely and be more linear if conditions are idealized. Although the Kubelka–Munk equation is considered to give a fair approximation for measurements on a particulate surface by absorbance (Jaenchen, 1996), Prosek and Pukl (1996) state that this equation is not very appropriate in quantitative thin-layer chromatography but that hyperbolic equations developed by Kubelka in 1948 are suitable for calculation of concentrations in TLC.

For fluorescence densitometry, the following relationship is applicable when the amount of substance is small (Jork et al., 1990):

$$I = kI_0\ \varepsilon\ ad$$

where

I = fluorescence intensity emitted
k = proportionality factor
I_0 = intensity of irradiationg light
ε = molar absorption coefficient
a = amount of substance applied
d = thickness of adsorbent layer

According to this equation, I is directly proportional to the amount of substance applied (a). This is a much simpler relationship than the Kubelka–Munk equation for absorption (above).

The exact determination of solute amount requires a reproducible and accurate calibration curve. Both peak height and peak area can be used for construction of the calibration curve; it is generally accepted that either method gives reliable results for narrow and symmetrical peaks, but in all other cases the use of peak area is preferred (Cserhati and Forgacs, 1996). In practice, a plot of peak area (Y-axis) versus weight (pg, ng, μg) or concentration spotted (X-axis) for a series of standards is usually linear at low weights and curves toward the X-axis higher weights. Some older densitometers had an electronic linearizer circuit to extend the linear range of calibration curves. Otherwise, linearity has been enhanced by plotting (peak area)^2 versus concentration or log area versus log concentration rather than area versus concentration, or using an internal standard with properties similar to the analyte and plotting the peak area ratio versus concentration. Reciprocal transformation of densitometric data has also been used successfully. Today there is no need for linearizing the relationship between optical response and standard amount because integrator and computer programs can handle nonlinear functions. Nonlinear response curves are valid for quantification if samples are bracketed closely with standards. The use of a second-order polynomial calibration curve in densitometric absorption/reflection measurements is not unusual (Windhorst and de Kleijn, 1992; Poole and Poole, 1994). Automated integration processes sometimes give unsatisfactory results and must be used with care. Visual inspection of the integrated peak areas and modification of the integration parameters in accordance with these observations is recommended (Cserhati and Forgacs, 1996).

D. Variables in Densitometric Measurements

The fluorescence mode of measurement is generally the most advantageous because of its selectivity, achieved through the choice of both excitation and emission wavelengths, and its sensitivity. Minimum detection levels for fluorescent spots are typically 10–1,000 times lower than for colored or quenched spots, with determinations of low nanograms or picogram amounts being common. Nonfluorescent compounds can often be derivatized to a fluorescent product. For example, bile acids are rendered fluorescent by spraying with 5% sulfuric acid in methanol followed by oven heating (Taylor et al., 1979). Prior removal or chromatographic resolution of background fluorescence from biological samples or from impurities in the layer or solvents is a critical requirement if the excitation and emission wavelengths of these interferences are similar to those of the fluorogenic compound being measured. This requirement does not hold if the available densitometer has monochromators that allow overlapping fluorescent zones not to be measured. Peak areas of fluorescent spots are usually linear with concentration over two or three orders of magnitude starting with the lowest detectable level, in the absence of interferences. At high concentrations, fluorescence is quenched through self-absorption, and the calibration curve curves downward toward the X-axis. The fluorescence intensity emitted is independent of spot shape if the spot is completely contained within the measuring light beam (Poole and Poole, 1991). Fluorescence may deteriorate with time, so area–time characteristics as well as a calibration curve must be determined for each substance being analyzed. For a particular compound, the mode of scanning (reflectance of transmittance), the nature of the sorbent (Pataki and Wang, 1968), and the presence of small amounts of moisture can drastically affect the level of fluorescence measured. Dipping the chromatogram in a 20–50% solution of mineral oil or liquid paraffin in ethyl ether or hexane prior to scanning has been reported by many workers to improve the sensitivity of fluorescence densitometry in some analyses (Uchiyama and Uchiyama, 1978). Triton X-100, glycerol, triethanolamine, and Fomblin H-Vac are other reagents that have been reported to enhance emission by 10- to 200-fold by reducing fluorescence quenching (Poole et al., 1985). An antioxidant (e.g., BHT, butylated hydroxytoluene) can be added to the layer and mobile phase to stabilize the fluorescence of compounds that are susceptible to quenching by oxygen (Poole et al., 1989; Poole and Poole, 1989; Dean and Poole, 1988). Matrix interferences may cause either enhancement or quenching of fluorescence; for example, coelution of free fatty acids with aflatoxins increased the fluorescence response of the aflatoxins by 14–36% (Robards et al., 1994). A two-point calibration method involving only one standard for each analyte has been recommended for the quantification of fluorescent compounds with high sample throughput (Butler et al., 1985).

Ultraviolet-absorbing spots can be measured directly at 254 nm (or some other absorption maximum) on unmodified layers, or as dark spots against a bright background on layers impregnated with a phosphor if the UV-absorption spectrum of the compound and the excitation spectrum of the phosphor overlap. A mercury or xenon lamp is usually used to excite the fluorescence of the phosphor, and a cutoff filter is placed before the detector to eliminate any stray exciting UV light. The direct determination of absorption at the wavelength of maximum absorption is more sensitive and specific than indirect measurement of absorption by fluorescence quenching, and it has been claimed that the latter technique suffers from severe background fluctuations resulting from inhomogeneous distribution of the phosphor in the sorbent layer (Robards et al., 1994). However, many excellent quantitative analyses have been reported based on fluorescence quenching (e.g., Greshock and Sherma, 1997), and both of these possibilities should be evaluated for optimum quantitative results because reports in the literature are contradictory in regard to their relative convenience, accuracy, and detectability. The presence of an F-254 nm indicator in a layer has no negative influence on scanning by visible absorbance nor by fluorescence.

Reflection must be used for measuring UV absorbance at wavelengths below 320 nm due to absorption by the glass backing plate and the silica gel itself at short UV wavelengths. Fluorescence quenching can be measured by transmission (exciting UV wavelength blocked by the glass plate) or reflectance (UV cutoff filter needed to discriminate exciting UV and emitted visible fluorescence). For visible absorbance, transmission gives a greater signal (peak height or area), up to a factor of 2, and also usually greater noise compared to reflection (Coddens et al., 1983b). Therefore, densitometric sensitivity, defined as millimeter peak height/nanogram spot, is greater for transmittance. However, detectability (2 $\times$

noise/sensitivity) can be higher or lower, depending on the instrument being used and layer homogeneity (Coddens and Poole, 1984).

Given all of the available scanning variables, the best practice is to evaluate and optimize each determination under local conditions with the densitometer available by testing all possible measurement modes (absorption, fluorescence, transmittance, reflection); different filter or monochromator settings, slit sizes and positions, and scan directions; and the electronic time constant to obtain the best signal-to-noise ratio for the peaks, the best linearity and slope of the calibration curve, and the highest degree of reproducibility in scanning. Spatial resolution is inversely proportional to the slit width, the optimal resolution being obtained with a width of 0.2 nm (Jaenchen, 1996). Scanning should always be done in or against the direction of development with newer PC-controlled densitometers. Scanning perpendicular to the direction of development was sometimes done with advantage using older instruments that were manually adjusted for each track. Application of sample bands is advantageous for quantitative TLC because the band can be made longer than the slit length (height) of the scanner, which minimizes errors due to positioning of the sample within the light beam (Poole and Poole, 1991). For spots, it is usually best to choose a slit length equivalent to the diameter of the largest spot. For fluorescent spots, peak area increases linearly with increasing slit width, but the maximum allowable slit width may be determined by the distance of separation between zones. Scan rate is inversely related to both resolution and scan area for both absorption and fluorescence. Reflection scanning is superior for layers of uneven thickness (Hezel, 1978) and has a wider measurement range ($\sim$185–2500 nm). Procedures for comparing densitometer sensitivities under standard conditions have been described by Coddens and Poole (1983) and Allwohn and Ebel (1989).

Figure 10.5 shows the variation in peak heights obtained by scanning an amino acid derivative at different wavelengths. The peaks are very narrow because an HPTLC layer was used. A line drawn through the peak apexes represents the in situ absorption spectrum of the compound. As stated earlier, some densitometers provide a spectral mode for scanning of UV and visible absorption spectra of nonfluorescent compounds directly on the TLC plate by comparison with a substance-free portion of the layer to facilitate determination of the wavelength of maximum absorption for quantitative analysis. The absorption maximum is usually, but not always, the same for in situ and solution spectra of a given compound. Determination at wavelengths other than the maximum is sometimes advantageous despite lower sensitivity, e.g., to eliminate interferences or improve the baseline. Densitometers must be equipped with excitation and emission monochromators to generate in situ absorption or fluorescence spectra of fluorescent compounds.

In addition to those cited above, studies of optimization of scan variables

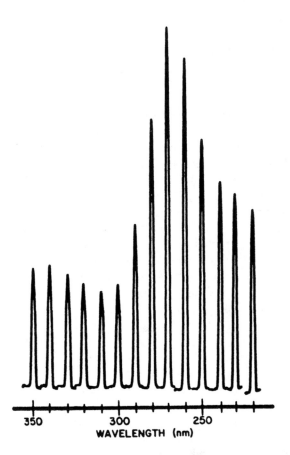

Figure 10.5 Repetitive single beam, reflectance scans of absorption of PTH-L-asparagine on a silica gel HPTLC plate at 10-nm intervals. [Reprinted with permission of Elsevier from Poole and Schuette (1984).]

have been published by Butler et al. (1982, 1983, 1984) and Butler and Poole (1983a, 1983b).

E. Performing a Densitometric Analysis

To carry out a TLC or HPTLC densitometric analysis, three or four standards and adequately purified samples are applied to the same plate, development and detection are carried out, and the spots are scanned. A calibration curve consisting of the scan areas of the standards versus the amounts of analyte spotted is constructed, and the amount (weight) of analyte in the sample, represented by its scan area, is interpolated from the curve. The concentration of analyte in the sample is calculated by consideration of the original amount of sample taken, any dilution steps during sample workup, and the portion (aliquot) of the final concentrated solution that is spotted. The sample spots must have areas that are bracketed by the areas of the standards for accurate quantification, i.e., areas that are lower than the highest standard and higher than the lowest standard.

A radioactive internal standard can be added to the sample to monitor analyte loss during the workup steps, and a nonradioactive internal reference standard can be added to all samples and standards to correct for variables such as spotting errors and layer variations and improve accuracy and precision (Aginsky, 1994). Relative standard deviations can reportedly be improved by a factor of 5–10 by use of an internal reference standard (Windhorst and de Kleijn, 1992), although other sources claim that internal standards are normally not required in quantitative HPTLC for good precision because samples and standards can be processed by external standardization on the same plate under essentially identical conditions. As an example, *p*-hydroxybenzaldehyde was used as the internal standard in the quantitative HPTLC determination of purine alkaloids in aqueous solutions such as daily foods and health drinks (Kunugi and Tabei, 1997).

The quality of the plates used is an important factor in obtaining good quantitative results. For example, Liang et al. (1996a) found that fluorescent impurities present in the layer were the major limitation to the sensitivity and dynamic range of quantitative HPTLC with fluorescence densitometry. They recommended that for best results, plates should be cleaned by development with methanol or chloroform–methanol (1 : 1) before use and stored in a clean environment to minimize contamination. In addition, use of pure solvents and a clean laboratory environment was found to be essential to obtain good quantitative results, especially at low sample concentration.

Figure 10.6 shows calibration curves for three phospholipids detected by charring with cupric acetate–phosphoric acid reagent. The differences in slopes of these lines clearly indicate the importance of generating a separate calibration curve for each compound that is to be determined. When the highest possible precision is not required, quantification can be based on the peak area ratio of the sample spot to an adjacent single standard spot of known concentration, if the peak areas agree within ±10% and are known to lie within the linear calibration range.

Modern scanners allow automatic, microprocessor-controlled scanning of an entire plate in any pattern chosen by the analyst. With a fully automated system, the computer can perform the following tasks: data acquisition, automated peak searching and optimization of scanning for each fraction located, multiple wavelength scanning, baseline location and correction, computation of peak areas and/or heights of samples and codeveloped standards, calculation of calibration curves by linear or polynomial regression, interpolation of sample concentrations,

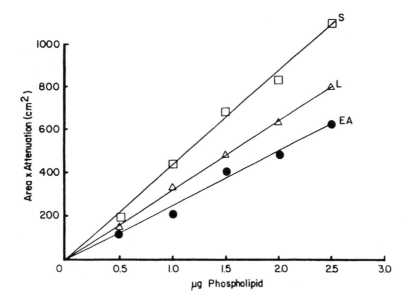

FIGURE 10.6 Calibration curves for 50–2500 ng of phosphatidyl ethanolamine (EA), lecithin (L), and sphingomyelin (S). Standards were developed on preadsorbent silica gel plates with ethyl acetate–n-propanol–chloroform–methanol–0.25% aqueous KCl (25:25 :25:13:9) and scanned in the single-beam transmission mode with a Kontes densitometer after detection by charring. [Reproduced with permission of Wiley-Interscience from Zelop et al. (1985).]

statistical analysis of reproducibility, presentation of a complete analysis report, and storage of data on disk. Multiwavelength scanning allows each zone to be quantified at its optimum wavelength and optical resolution of fractions insufficiently separated by TLC (Jaenchen, 1996; Treiber, 1987; Kaiser, 1989).

Initial zones of samples and standards should be homogeneously distributed, their sizes should be small, positions accurately referenced for subsequent lane positioning in scanning densitometry, and volumes accurately known (Poole and Poole, 1994). These criteria are best accomplished by spotting equal volumes of samples and standards to a single plate by means of a mechanical or automated commercial spotting device (Chapter 5). Manual spotting with Drummond microcaps (Emanuel, 1973) or a digital microdispenser has also been widely used, but the starting positions are not as precisely known in this case. Calibration curves resulting from the application of variable volumes of a single standard solution are more likely to have poor linearity and intercepts that do not go through the origin (x = 0, y = 0 point), but this standardization procedure has often been used for successful analyses, especially with manual application on preadsorbent plates or automated application with the Linomat (Chapter 5). When quantifying by densitometry, it is best to apply the smallest amount of material that can be detected reliably.

A more uniform distribution of the detection reagent on the plate will usually result if it can be applied by dipping rather than by spraying. Organic polymer-bound layers are especially sturdy with regard to dipping into detection reagents, including even some water-based reagents. The time-intensity (area) characteristics of the detection reaction must be determined and all work done on a definite time schedule so that scanning is performed when the spot areas and intensities have become constant (Pataki, 1968). If heating is part of the detection procedure, the temperature and time of heating should be carefully controlled and reproduced.

Examples of densitometric analyses on high performance layers are given in Figures 10.7 through 10.9. Figure 10.7 illustrates the almost baseline separation of six phenothiazine drugs. Figure 10.8 is a chromatogram of an extract of human blood serum containing 45 ng/ml of chlorpromazine. The extract was cleaned up by solvent partitioning at controlled pH, and butaperazine was added as an internal standard. Figure 10.9 shows scans of aspartame (NutraSweet) standards and a soda sample applied directly to a preadsorbent high-performance silica gel layer with fluorescent indicator. The analysis indicated an aspartame concentration of approximately 10 mg/100 ml of beverage.

Relative standard deviations of determinations by quantitative HPTLC are usually less than 5% and often fall below 1%. The ability to chromatograph samples and standards simultaneously leads to statistical improvements in data handling for TLC and is one of the major reasons for the excellent reproducibilities reported for densitometric TLC. Ebel (1996) found the accuracy and precision of UV/visible spectrophotometry, HPLC, and HPTLC to be comparable if samples were spotted in duplicate and calibration and data evaluation performed correctly; relative standard deviations of 0.6–1.5% were reported in this study. The data pair technique (Bethke et al., 1974) was designed to reduce systematic errors due to chromatographic parameters that tend to influence reproducibility as a function of varying R_F values. In this method, pairs of samples and standards are spotted so they are about one-half plate width apart. Each spot is scanned four times (two scans up the track and two down), and the average area value is used. Relative standard values of 1.1–1.9% were reported for TLC plates. On HPTLC plates, 13 mycotoxins were determined with RSD values between 0.7 and 2.2% using the data pair technique (Lee et al., 1980). The advent of powerful TLC evaluation software for densitometry and improved commercial layers have reduced the need for the data pair technique in order to obtain a high degree of reproducibility, and use of the method is seldom reported today.

One advantage often cited for quantitative HPTLC compared to HPLC is higher sample throughput. To investigate this, a content uniformity test for diclo-

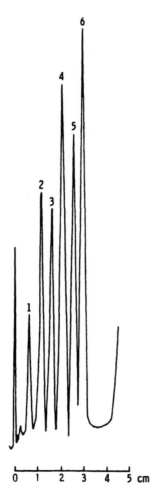

Figure 10.7 HPTLC separation of 1, acetophenazine; 2, perphenazine; 3, trifluoroperazine; 4, promazine; 5, thioridazine; and 6, chlorpromazine on a high-performance silica gel plate using multiple development in the mobile phase benzene–acetone–ammonium hydroxide (80:20:0.2). [Reproduced with permission of The American Association for Clinical Chemistry from Fenimore et al. (1978).]

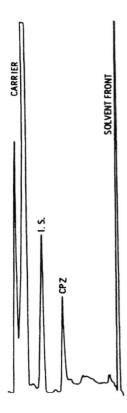

FIGURE 10.8 HPTL chromatogram of a serum sample from a patient receiving chlorpromazine (CPZ). The concentration of chlorpromazine was 45 ng/ml. [Reproduced with permission of The American Association for Clinical Chemistry from Fenimore et al. (1978).]

fenac sodium was performed by both methods (Morlock et al., 1997). HPLC resulted in 180 determinations in 24 h, whereas the same number could be performed in 2 h by HPTLC. The use of multilevel calibration gave more reliable results than single-level calibration.

The calibration and calculation operations in a quantitative TLC analysis are handled automatically by the software in modern computer-controlled densitometers. Chromatographic and measurement errors in scanning densitometry were discussed by Allwohn and Ebel (1989), Ebel and Glaser (1979), and Pollak (1989).

Schemes for validation of quantitative TLC methods under guidelines for

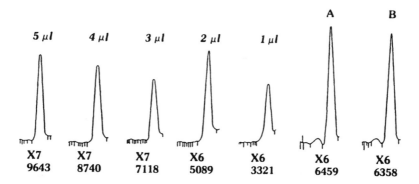

FIGURE **10.9** Scans of zones resulting from application of 1.00 to 5.00 µl of aspartame standard solution (0.976 µg/µl) and duplicate 25.0 µl cola samples (A, B) using a Kontes Model 800 scanner and HP 3390A integrator. The relative integrator areas and attenuation settings are given below the peaks. The integrator peak width setting was 0.01, chart speed 5, and integration mode 2. The high-performance preadsorbent silica gel F layer was developed with *n*-butanol–acetic acid–water (3:1:1). Scanning of the quenched zones was carried out at 254 nm. [Reprinted with permission of International Scientific Communications, Inc. from Sherma et al. (1985).]

good laboratory practice (GLP) have been developed (Prosek et al., 1993; Prosek and Pukl, 1996; Rischer et al., 1997). In the latter study, precision of 0.05–0.30% was obtained despite the use of copper sulfate detection reagent before densitometric scanning. An ANOVA method was used to validate linear regression in quantitative TLC, and it was proved that the ordinary least squares regression method may be successfully applied (Sarbu and Cobzac, 1997). See Section II in Chapter 11 for additional information on testing and validation of results.

F. Image Analysis (Video Densitometry)

Most quantitative TLC analyses reported in the literature make use of slit-scanning densitometers that perform electromechanical scanning of a moving plate with a rectangular light beam, as described above. In electronic scanning densitometry (Pollak, 1989; Poole and Poole, 1989; Cosgrove and Bilhorn, 1989), a stationary plate is point-scanned with an instrument consisting of a computer with video digitizer; light source to illuminate the complete layer; an optical system including components such as a shutter, lens, filter, and monochromator; and an imaging detector such as a vidicon tube or charge-coupled video camera, which functions as a two-dimensional array of unit detectors that are periodically discharged and the signal digitized for analysis by the computer (Nyiredy, 1992; Poole and Poole, 1995; Robards et al., 1994).

Commercial video scanners are available from a number of companies, including Camag; Analtech (Uniscan); Alpha Innotech, San Leanardo, CA (IS-1000-M system); and Photometrics Ltd., Tucson, AZ (Scientific CCD Camera System) (Liang and Denton, 1996). The Camag VideoScan is a software that is combined with the VideoStore (Figure 9.3) and appropriate hardware for quantitative evaluation of chromatograms recorded with a video camera and stored in digitized form. The following comparisons are made in Camag literature: slit-scanning densitometric evaluation is better than video quantification in terms of superior accuracy and precision, the ability to use the entire UV range to 190 nm with excellent monochromaticity, and spectral recording capability, but video densitometry provides superior speed of evaluation. Video technology functions only in the visible range, with spectral selectivity being influenced only marginally by means of electronic filters. UV-absorbing substances are measured only indirectly by fluorescence quenching on layers containing fluorescent indicator, which shifts detection into the visible region. Fluorescent substances excited by 366 nm UV light and emitting visible radiation are detected, but video technology lacks the variable-excitation-based selectivity of conventional densitometry.

Poole and Poole (1995) stated that the main attractions of image analysis for detection in TLC are fast data acquisition, simple instrument design, and the absence of moving parts, but that the available measuring range, sample detectability, and dynamic signal range are significantly restricted for comparable system cost compared to slit-scanning densitometers. Poole and Poole (1995) and Prosek and Pukl (1996) have predicted that image analysis will become the preferred method of densitometry as the equipment improves and more research is carried out over time. Pollak and Schulze-Clewing (1990) and Vovk and Prosek (1997) have also compared classical electromechanical densitometry and electronic scanning with a CCD camera.

The following qualitative and quantitative determinations are examples of recent applications of TLC with videodensitometry: lower halogenated subsidiary colors or the ethyl ester in multiple dye samples (Wright et al., 1997); two-dimensional TLC coupled to fluorescence analysis for the study of complexes that contain at least one fluorescent component (Guilleux et al., 1994); photochemically derivatized phenothiazines (Garcia Sanchez et al., 1993); methanol and ethanol in biological fluids as xanthate derivatives, tryptophan derivatives, and gold and platinum (Huynh and Leipzig-Pagani, 1996); aflatoxins in peanut butter (Liang et al., 1996b); neutral lipids and phospholipids after separation by multiple development on an EDTA-impregnated layer (Ruiz and Ochoa, 1997); and a neurotoxic steroidal alkaloid in death camas (Makeiff et al., 1997).

G. Thin Layer Chromatography–Flame Ionization Detection (TLC-FID)

Many publications have reported the use of the Iatroscan TLC-FID apparatus for quantitative analysis (e.g., Bharati et al., 1994). See Chapter 8, Section VI for discussion of this technique.

REFERENCES

Aginsky, V. N. (1994). A new version of the internal standard method in quantitative thin layer chromatography. *J. Planar Chromatogr.—Mod. TLC 7*: 309–314.

Allwohn, J., and Ebel, S. (1989). Testing and validation of TLC scanners. *J. Planar Chromatography—Mod. TLC 2*: 71–75.

Aponte, R. L., Diaz, J. A., Perera, A. A., and Diaz, V. G. (1996). Simple thin layer chromatography method with fiber optic remote sensor for fluorometric quantification of tryptophan and related metabolites. *J. Lig. Chromatogr. Relat. Technol. 19*: 687–698.

Bethke, H., Santi, W., and Frei, R. W. (1974). Data pair technique, a new approach to quantitative TLC. *J. Chromatogr. Sci. 12*: 392–397.

Bharati, S., Roestrum, G. A., and Loeberg, R. (1994). Calibration and standardization of the Iatroscan (TLC-FID) using standards derived from crude oils. *Org. Geochem. 22*: 835–862.

Butler, H. T., and Poole, C. F. (1983a). Optimization of a scanning densitometer for fluorescence detection in HPTLC. *J. High Resolut. Chromatogr. Chromatogr. Commun. 6*: 77–81.

Butler, H. T., and Poole, C. F. (1983b). Two point calibration method applied to fluorescence scanning densitometry and HPTLC. *J. Chromatogr. Sci. 21*: 385–388.

Butler, H. T., Pacholec, F., and Poole, C. F. (1982). Reference plate calibration scheme for sensitivity and wavelength selection in scanning densitometry. *J. High Resolut. Chromatogr. Chromatogr. Commun. 5*: 580–581.

Butler, H. T., Schuette, S. A., Pacholec, F., and Poole, C. F. (1983). Characterization of a scanning densitometer for HPTLC. *J. Chromatogr. 261*: 55–63.

Butler, H. T., Coddens, M. E., and Poole, C. F. (1984). Qualitative identification of polycyclic aromatic hydrocarbons by HPTLC and fluorescence scanning densitometry. *J. Chromatogr. 290*: 113–126.

Butler, H. T., Coddens, M. E., Khatib, S., and Poole, C. F. (1985). Determination of polycyclic aromatic hydrocarbons in environmental samples by high performance thin layer chromatography and fluorescence scanning densitometry. *J. Chromatogr. Sci. 23*: 200–207.

Cheng, M.-L., and Poole, C. F. (1983). Minor component tablet analysis by HPTLC. *J. Chromatogr. 257*: 140–145.

Coddens, M. E., and Poole, C. F. (1983). Sensitivity of scanning densitometers for TLC. *Anal. Chem. 55*: 2429–2431.

Coddens, M. E., and Poole, C. F. (1984). A protocol for measuring the sensitivity of slit-scanning densitometers. *LC, Liq. Chromatogr. HPLC Mag. 1*: 34–36.

Coddens, M. E., Butler, H. T., Schuette, S. A., and Poole, C. F. (1983a). Quantitation in HPTLC. *LC, Liq. Chromatogr. HPLC Mag. 1*: 282–289.

Coddens, M. E., Khatib, S., Butler, H. T., and Poole, C. F. (1983b). Mode selection and optimization of parameters for recording high performance thin layer chromatograms by transmission measurements. *J. Chromatogr. 280*: 15–22.

Cosgrove, J. A., and Bilhorn, R. B. (1989). Spectrometric analysis of planar separations using charged-coupled device detection. *J. Planar Chromatogr.—Mod. TLC 2*: 362–367.

Cserhati, T., and Forgacs, E (1996). Thin layer chromatography. In *Chromatography Fundamentals, Applications, and Troubleshooting*, J. Q. Walker (Ed.). Preston Publications, Niles, IL, pp. 185–207.

Dean, T. A., and Poole, C. F. (1988). Robust derivative for the determination of 1-nitropyrene by fluorescence scanning densitometry. *J. Planar Chromatogr.—Mod. TLC 1*: 70–72.

Duez, P., Chamart, S., and Hanocq, M. (1991). Postchromatographic derivatization in quantitative thin layer chromatography: Pharmaceutical applications. *J. Planar Chromatogr.—Mod. TLC 4*: 69–76.

Ebel, S. (1996). Quantitative analysis in TLC and HPTLC. *J. Planar Chromatogr.—Mod. TLC 9*: 4–15.

Ebel, S., and Glaser, E. (1979). Systematic and statistical errors in quantitative evaluation in TLC and high performance thin layer chromatography. III. Determination of the primary statistical errors. *J. High Resolut. Chromatogr. Chromatogr. Commun. 2*: 36–38.

Emanuel, C. F. (1973). Delivery precision of micropipets. *Anal. Chem. 45*: 1568–1569.

Fenimore, D. C., and Davis, C. M. (1981). High performance TLC. *Anal. Chem. 53*: 252A–266A.

Fenimore, D. C., Davis, C. M., and Meyer, C. J. (1978). Determination of drugs in plasma by HPTLC. *Clin. Chem. 24*: 1386–1392.

Garcia Sanchez, F., Navas Diaz, A., and Fernandez Correa, M. R. (1993). Image analysis of photochemically derivatized and charge-couple-device-detected phenothiazines separated by thin layer chromatography. *J. Chromatogr. 655*: 31–38.

Greshock, T., and Sherma, J. (1997). Analysis of decongestant and antihistamine pharmaceutical tablets and capsules by HPTLC with ultraviolet absorption densitometry. *J. Planar Chromatogr.—Mod. TLC 10*: 460–463.

Guilleux, J. C., Barnouin, K. N., Ricchiero, F. A., and Lerner, D. A. (1994). Two-dimensional TLC and fluorescence analysis with CCD video camera used to determine the dissociation of diphenylhexatriene included in beta-cyclodextrin. *J. Liq. Chromatogr. 17*: 2821–2831.

Hezel, U. (1978). Quantitative photometry of thin layer chromatograms for research and routine analysis. *Am. Lab. 10(5)*: 91–107.

Huynh, T. K. X., and Leipzig-Pagani, E. (1996). Quantitative thin layer chromatography on cellulose: II. Selected applications: lower alcohols, tryptophan enantiomers, gold and platinum. *J. Chromatogr., A 746*: 261–268.

Jaenchen, D. E. (1996). Instrumental thin layer chromatography. In *Handbook of Thin Layer Chromatography*, 2nd ed. J. Sherma and B. Fried (Eds.). Marcel Dekker, New York, pp. 129–148.

Jaenchen, D. E. (1997). Thin layer (planar) chromatography. In *Handbook of Instrumental Techniques for Analytical Chemistry.* F. Settle (Ed.). Prentice Hall PTR, Upper Saddle River, NJ, pp. 221–239.

Jaenchen, D. E., and Schmutz, H. (1978). Considerations of mechanical precision required for optical in situ scanning of circular HPTL chromatograms. *J. High Resolut. Chromatogr. Chromatogr. Commun. 1:* 315–316.

Jork, H., Funk, W., Fischer, W., and Wimmer, H. (1990). *Thin Layer Chromatography, Reagents and Detection Methods,* Volume 1a. VCH Verlagsgesellschaft, Weinheim, Germany.

Kaiser, R. E. (1989). Modern quantitation in thin layer chromatography, part 1. *J. Planar Chromatogr.—Mod. TLC 2:* 323–326.

Kirchner, J. G. (1973). Thin layer chromatographic quantitative analysis. *J. Chromatogr. 82:* 101–115.

Kunugi, A., and Tabei, K. (1997). Simultaneous determination of purine alkaloids in daily foods by high performance thin layer chromatography. *J. High Resolut. Chromatogr. 20:* 456–458.

Lee, K. Y., Poole, C. F., and Zlatkis, A. (1980). Simultaneous multi-mycotoxin determination by HPTLC. In *Instrumental HPTLC,* W. Bertsch, S. Hara, R. E. Kaiser, and A. Zlatkis (Eds.). Huthig, Heidelberg, p. 245.

Liang, Y., and Denton, M. B. (1996). Application of a CCD to planar chromatographic analysis. *Spec. Publ.—R. Soc. Chem. 194:* 161–169.

Liang, Y., Baker, M. E., Gilmore, D. A., and Denton, M. B. (1996a). Evaluation of commercial silica gel HPTLC plates for quantitative fluorescence analysis. *J. Planar Chromatogr.—Mod. TLC 9:* 247–253.

Liang, Y., Baker, M. E., Yeager, B. T., and Denton, M. B. (1996b). Quantitative analysis of aflatoxins by high performance thin layer chromatography utilizing a scientifically operated charge-coupled device detector. *Anal. Chem. 68:* 3885–3891.

Makeiff, D., Majak, W., McDiarmid, R. E., Reaney, B., and Benn, M. H. (1997). Determination of zygacine in Zigadenus venenosus (death camas) by image analysis on thin layer chromatography. *J. Agric. Food Chem. 45:* 1209–1211.

Matysik, G., Glowniak, K., Soczewinski, E., and Garbacka, M. (1994). Chromatography of esculin from stems and bark of Aesculus hippocastanum L. for consecutive vegetative periods. *Chromatographia 38:* 766–770.

Morlock, G. E., Handloser, D. M., and Altorfer, H. R. (1997). New applications in thin layer chromatography: Content uniformity test of diclofenac sodium. *GIT Spez. Chromatogr. 17:* 14–16.

Nyiredy, Sz. (1992). Planar chromatography. In *Chromatography, 5th Edition,* E. Heftmann (Ed.). Elsevier, Amsterdam, pp. A109–A150.

Pataki, G. (1968). Some recent advances in thin layer chromatography. II. Application of direct spectrophotometry and direct fluorometry in amino acid, peptide, and nucleic acid chemistry. *Chromatographia 1:* 492–503.

Pataki, G., and Wang, K.-T. (1968). Quantitative thin layer chromatography. VII. Further investigations of direct fluorometric scanning of amino acid derivatives. *J. Chromatogr. 37:* 499–507.

Pollak, V. A. (1989). Sources of error in the densitometric evaluation of thin layer separations with special regard to nonlinear problems. *Adv. Chromatogr. 30:* 201–246.

Pollak, V. A. (1991). Theoretical foundations of optical quantitation. In *Handbook of Thin Layer Chromatography*, J. Sherma and B. Fried (Eds.), Marcel Dekker, New York, pp. 113–134.

Pollak, V. A., and Schulze-Clewing, J. (1990). Electronic scanning for the densitometric evaluation of flat bed separations. *J. Planar Chromatogr.—Mod. TLC 3*: 104–110.

Poole, C. F., and Poole, S. K.(1989). Progress in densitometry for quantitation in planar chromatography. *J. Chromatogr. 492*: 539–584.

Poole, C. F., and Poole, S. K. (1991). *Chromatography Today*. Elsevier, Amsterdam, pp. 649–734.

Poole, C. F. and Poole, S. K. (1994). Instrumental thin layer chromatography. *Anal. Chem. 66*: 27A–37A.

Poole, C. F. and Poole, S. K. (1995). Multidimensionality in planar chromatography. *J. Chromatogr., A 703*: 573–612.

Poole, C. F., and Schuette, S. A. (1984). *Contemporary Practice of Chromatography*. Elsevier, New York, pp. 656–665.

Poole, C. F., Coddens, M. E., Butler, H. T., Schuette, S. A., Ho, S. S. J., Khatib, S., Piet, L., and Brown, K. K. (1985). Some quantitative aspects of scanning densitometry in high performance thin-layer chromatography. *J. Liq. Chromatogr. 8*: 2875–2926.

Poole, C. F., Poole, S. K., Dean, T. A., and Chirco, N. M. (1989). Sample requirements for quantitation in thin layer chromatography. *J. Planar Chromatogr.— Mod. TLC 2*: 180–189.

Prosek, M., and Pukl, M. (1996). Basic principles of optical quantitation in TLC. In *Handbook of Thin Layer Chromatography*, 2nd ed., J. Sherma and B. Fried (Eds.). Marcel Dekker, New York, pp. 273–306.

Prosek, M., Pukl, M., Ljubica, M., and Golc-Wondra, A. (1993). Quantitative thin layer chromatography. Part 12: Quality assessment in QTLC. *J. Planar Chromatogr.— Mod. TLC 6*: 62–65.

Rischer, M., Schnell, H., Greguletz, R., Wolf-Heuss, E., and Engel, J. (1997). Validation of an HPTLC method for impurities testing and determination of log P value of a new phospholipid. *J. Planar Chromatogr.—Mod. TLC 10*: 290–297.

Robards, K., Haddad, P. R., and Jackson, P. E. (1994). *Principles and Practice of Modern Chromatographic Methods*, Academic Press, San Diego, CA, pp. 179–226.

Ruiz, J. I., and Ochoa, B. (1997). Quantification in the subnanomolar range of phospholipids and neutral lipids by monodimensional thin layer chromatography and image analysis. *J. Lipid Res. 38*: 1482–1489.

Sarbu, C., and Cobzac, S. (1997). Validation of the linear calibration function in thin layer chromatography by analysis of variance. *Rev. Chim. (Bucharest) 48*: 239–245.

Sherma, J., Chapin, S., and Follweiler, J. M. (1985). Quantitative TLC determination of aspartame in beverages. *Am. Lab. 17(7)*: 131–133.

Szepesi, G. (1990). Quantitation in thin layer chromatography. In *Modern Thin Layer Chromatography*, N. Grinberg (Ed.). Marcel Dekker, New York, pp. 249–283.

Szepesi, G., and Nyiredy, Sz. (1996). Pharmaceuticals and drugs. In *Handbook of Thin Layer Chromatography*, 2nd ed., J. Sherma and B. Fried (Eds.). Marcel Dekker, New York, pp. 819–876.

Taylor, W. A., Blass, K. G., and Ho, C. S. (1979). Novel thin layer chromatography separation and spectrofluorometric quantitation of lithocholic acid. *J. Chromatogr. 168*: 501–507.

Touchstone, J. C., Levin, S. S., and Murawec, T. (1971). Quantitative aspects of spectrodensitometry of thin layer chromatograms. *Anal. Chem. 43*: 858–863.

Treiber, L. R. (1987). *Quantitative Thin Layer Chromatography and Its Industrial Applications*. Marcel Dekker, New York.

Uchiyama, S., and Uchiyama, M. (1978). Fluorescent enhancement in thin layer chromatography by spraying viscous organic solvents. *J. Chromatogr. 153*: 135–142.

Vovk, I, and Prosek, M. (1997). Quantitative evaluation of chromatograms from totally illuminated thin layer chromatography plates. *J. Chromatogr., A 768*: 329–333.

Windhorst, G., and de Kleijn, J. P. (1992). Some technical recommendations for improving the performance of quantitative thin layer chromatography. *J. Planar Chromatogr.— Mod. TLC 5*: 229–233.

Wright, P.R., Richfield-Fratz, N., Rasooly, A., and Weisz, A. (1997). Quantitative analysis of components of the color additives D&C Red nos 27 and 28 (Phloxine B) by thin layer chromatography and videodensitometry. *J. Planar Chromatogr.—Mod. TLC 10*: 157–162.

Zelop, C. M., Koska, L. H., Sherma, J., and Touchstone, J. C. (1985). Quantification of phospholipids and lipids in saliva and blood by TLC with densitometry. In *Techniques and Applications of TLC*, J. C. Touchstone and J. Sherma (Eds.). Wiley-Interscience, New York, Chap. 21, pp. 291–304.

11

Reproducibility and Validation of Results

I. REPRODUCIBILITY OF RESULTS

The major sources of poor reproducibility of results (i.e., R_F values, development time, zone shape, detection characteristics, etc.) in TLC can be broadly characterized as contamination, poor technique, and variations in chromatographic materials and/or conditions. Geiss (1987), Smith et al. (1973), and Jork et al. (1990) listed the following parameters that can influence a TLC separation. These parameters should be recorded as completely as possible as part of the documentation procedure for TLC results (Chapter 9).

1. The sorbent or stationary phase (type of sorbent; brand; batch number; layer thickness; particle size; binder in layer; precoated or homemade; type of backing; size of channels, if any; activation method and temperature; impregnation, if any)
2. The solvent or mobile phase (the individual solvents, including any impurities; pH; method of preparing mobile-phase mixture; volume and height of solvent pool in the chamber; migration distance and time; gradients, if any)
3. Sample and standard solutions (weights; solvent and volumes; extraction and cleanup procedure for sample; derivative preparation, if any; location of samples and standards across the layer)
4. Sample and standard application (spotting apparatus and technique,

sample volume and size and shape of initial zones, distance between
the initial zone location and the feeding point of the solvent)

5. Development technique and conditions (type of chamber, method of
   placing and supporting the layer in the chamber, i.e., upright versus
   angled; preadsorption of a solvent mixture or conditioning; saturation
   of chamber with solvent vapors; temperature; humidity; convection
   currents in the gas phase inside the chamber; mode of development)

6. Detection method (method of drying plate after development, description of reagent preparation, method of reagent application, temperature
   and time of heating, if any)

7. R_F range

This extensive list of variables illustrates the importance of controlling and
reporting as many experimental conditions as possible when describing the results
of TLC so that others wanting to duplicate results will know how to proceed.
All the stated conditions under which the initial results were obtained should be
followed as carefully as possible for best reproducibility.

The following discussion examines some of the specific causes and reme-
dies for the nonreproducibility of TLC.

A. The Stationary Phase (Sorbent)

It is very important to recognize that layers of any particular sorbent purchased
from different manufacturers can vary drastically in characteristics such as particle size (average and range), pore size (average and distribution), layer thickness,
activity, pH, and binder (nature and amount), leading to differences in selectivity
and efficiency. Single manufacturers have available different formulations of the
same sorbent, each of which may give diverse results (Geiss, 1968). It has even
been observed that the same product from a single manufacturer can vary in
performance from batch to batch, but this is not as likely as discrepancies among
different products and manufacturers. As an example of plate variation, Gumprecht (1992) compared silica gel plates from five different manufacturers for
the separation of dyes, analgesics, and phenols and found significant differences
in R_F values as well as in relative migration order.

Commercial precoated layers are activated by manufacturers before being
shipped, but the method of packing, length of storage, method of handling, and
laboratory temperature and humidity will determine the activity at the time of
use (Reichel, 1967). It is possible to attempt to bring the layer to an exact degree
of activation by oven heating or storage in a desiccator prior to use, but this
operation is time-consuming and inconvenient and of little practical value because a dried plate equilibrates with water in the atmosphere within a few minutes
(Dallas, 1965; Robards et al., 1994). Therefore, during the time it takes to apply
the initial zones, a dried plate will pick up an amount of water dependent on the

ambient relative humidity. Simply breathing on the plate during spotting can change layer activity and markedly affect R_F values. The most reproducible results are obtained with special chambers (Chapter 7) that prevent interaction of a layer with ambient air and provide activity control, such as the Vario-KS or Vario-HPTLC-chamber (Geiss, 1987). In practice, it is most successful and convenient to use plates as received from the manufacturer with the type of chamber available and to apply a standard reference compound for calculation of relative R_F values, which will be constant regardless of sorbent activity (Dallas, 1968a). If possible, TLC should be performed in a room with controlled temperature and relative humidity to maintain constant layer activity.

Inorganic and organic binders of many different types (e.g., gypsum, starch, glass, polyvinyl alcohol, organic polymers) have been utilized by plate manufacturers to produce "hard" and "soft" layers. Different binders can alter R_F values, selectivity, and zone detection to a considerable degree, even with a constant type of sorbent. When the binder is changed, however, the specifications of the sorbent are often also changed, making the likelihood of plate reproducibility very small (Turano and Turner, 1974). The presence of binders also makes it difficult to transfer results from TLC to a column containing the same sorbent without a binder.

A number of phosphors have been added to make layers fluorescent for detection of compounds by fluorescence quenching. It is important for good reproducibility that these UV or fluorescent indicators do not modify the chromatographic properties of the layer.

Manufacturers are currently supplying highly purified layers with carefully controlled sorbent particle and pore size, layer thickness, pH, fluorescent indicator, and binder. Once a plate is found that provides the desired results, reproducibility should be obtained if this same product is repeatedly ordered from the same source. These parameters are more carefully controlled in high-performance plates compared to conventional layers, and HPTLC plates should provide the most reproducible results. The quality and reproducibility of homemade plates will not be as good as with commercial precoated plates.

When plates are homemade rather than purchased precoated from the manufacturer, nonreproducibility caused by layer thickness variation will be much more noticeable. R_F values are greatly affected by layer thickness (Pataki and Keller, 1963; Dallas, 1968b) and particle size distribution (Halpaap, 1973), as is the speed of solvent flow. Many workers have found no dependence of R_F values on layers with thicknesses between 0.2 and 0.5 mm, and Perry et al. (1972) claimed that layers of constant activity developed in an S-chamber (Chapter 7) show no variation in R_F values with layer thickness in the range 0.1–3 mm. However, Geiss (1987) reported that all systems he tested showed a decrease of R_F values with increasing layer thickness. The decrease was relatively strongest between 0.2 and 0.5 mm (-20%), but was still noticeable up to 2 mm.

B. Contamination of Sorbents and Solvents

It is essential to obtain commercial sorbents or precoated plates that are highly purified and not to introduce contaminants during preparation, use, or storage steps. In addition to adsorption of water, the large, exposed surface area of thin layers can cause rapid pickup of organic vapors in the atmosphere, e.g., from solvents, cigarette smoke, floor wax, or other extraneous chemicals. Storage by wrapping in plastic does not solve the problem because many contaminants can penetrate from air through the plastic covering, and compounds may migrate from the plastic itself into the layer. Some of these organic impurities may be detected on the layer along with the analytes, especially at the solvent front as a dark or fluorescent band that can interfere with the detection of quickly moving zones. It is a good precaution to preclean plates by development to the top with methylene chloride–methanol (1:1) just before use, followed by drying. If storage of a prewashed plate is necessary, it should be wrapped in aluminum foil or placed in an airtight container. Storage over a desiccant is generally not helpful, as described above.

Careless handling can introduce oil from the hands to the edges and back of the plate. The oil can be dissolved by the solvent in the chamber and subsequently deposited throughout the layer as the solvent rises during the development step.

Attempting to blow off the thin film of adsorbent dust that often covers the surface of the plate (especially on "soft" sorbent layers) can deposit impurities from the mouth or moisten the surface, causing dust particles to adhere. These impurities may show up as anomalous spots upon detection by sulfuric acid charring or with other reagents.

Care should be taken to use only the purest grade of solvent available because impurities can alter R_F values and spot definition and interfere with detection steps. Some solvents contain stabilizers that may vary in amount, as can the water content of various solvents on standing. Geiss (1987) stated that polar impurities in even very low concentrations can drastically change chromatograms in an N-tank where layer preloading is involved. A good precaution is to redistill "reagent-grade" solvents in an all-glass apparatus, or to select "pesticide-grade," "HPLC-grade," or other specially purified commercial solvents if their additional cost allows. All solvents should be tested for suitability in the TLC procedure before use. Solvents used for extracting samples and in extract cleanup steps are often employed in large volumes and then concentrated to prepare the final spotting solution. Impurities in these solvents will show up as artifacts and generally cause greater problems than impurities in mobile-phase solvents used for development. The quality and uniformity of the laboratory water supply is a factor in reproducibility and should not be overlooked; water purified in a Millipore Milli-Q system (or equivalent) is recommended.

C. Water Content

The unsuspected, unwanted, or variable presence of water in samples, solvents, and sorbents is one of the most important causes of nonreproducibility in TLC. Increased water content in an adsorbent layer lowers activity and raises R_F values. It has been shown that the R_F of a substance can vary by as much as 300% when developed in a system containing 1% relative humidity compared to 80% humidity (Geiss et al., 1963, 1965a, 1965b). The magnitude of the change in R_F with layer activity decreases with increasing polarity of the mobile phase, and is not constant for different substances; therefore, not only zone spacing but elution order may change in extreme cases (Robards et al., 1994).

The presence of moisture leading to lowered activity is not necessarily detrimental and, in fact, may be responsible for the favorable separations obtained in some "deactivated" chromatographic systems. The important point to recognize is that the water level should be as consistent as possible from plate to plate over the entire layer surface if reproducible resolution is to be achieved. As stated above, heat activation of a precoated plate before use (110°C is often prescribed) is essentially worthless because the layer will equilibrate with the ambient atmosphere within a few minutes. Homemade layers prepared from sorbent slurries are usually heated to accelerate drying. Heating silica gel at 150°C removes physically and reversibly bound water and provides maximum activity without irreversible chemical change, which can occur at temperatures above 200°C. Attainment of constant water content is aided by working in a humidity-controlled room and using plates stored and equilibrated with the air in the same room.

D. Sample Application and Sample Size

Initial zones should be thoroughly dried in a consistent manner to achieve reproducible TLC results. Removal of organic solvents and moisture from initial zones can be accomplished by simple air drying, drying with warm or hot air from a hair dryer, or oven heating, depending on the nature of the sample. The heat stability of the analyte is a prime concern when choosing the drying method. Water entering the sample during preliminary extraction and purification steps can be removed before spotting by the addition of an anhydrous drying agent (e.g., Na_2SO_4) that has been preextracted with organic solvents to remove impurities. This operation will reduce the amount of water deposited at the origin and lower the amount of initial zone drying required. The last traces of moisture in a dried initial zone can be removed by applying an equal volume of ethanol over the initial zone, followed by reheating to allow the alcohol to azeotrope off any remaining water. For example, apply 20 μl of ethanol if 20 μl of sample has been spotted.

In addition to residual water in the initial zone and decomposition of the sample during initial zone drying, as already mentioned, other possible problems associated with sample application include irreversible precipitation, low solubility, and sample degradation on highly active sorbents. Grease deposited at the origin with the sample can cause streaked zones and completely ruin separations. Grease can enter the sample from contact of lipophilic solvents with greased stopcocks or standard-taper stoppers in apparatus used for preliminary sample preparation. Grease should not be used to seal desiccators in which plates may be stored before use, or to seal the tops of development chambers.

Application of too much sample will overload the chromatographic system, resulting in trailing spots with R_F values that are increased or decreased compared to the value when the spotted concentration is low (Geiss, 1968). The R_F value can also be affected by ionic bonding between different solutes in the sample. To detect this effect on R_F due to close proximity of other solutes in the chromatogram, single pure standards should always be chromatographed on the same plate.

Proper spotting technique is undoubtedly the most important factor in achieving good precision and accuracy in quantitative TLC. All initial zones of samples and standards must be uniform and small for proper area comparison by eye, manual measurement, or densitometry (see Chapter 10). Preadsorbent plates (Chapter 3) and sample application instruments (Chapter 5) facilitate reproducibility in quantification and R_F values.

Samples applied as bands generally broaden to a smaller degree during development than those applied as spots, leading to improved resolution. Overloading is also less likely and evaluation by scanning more successful with band application.

Samples and standards should be dissolved in the same solvent because equal volumes of solutions of the same substance at a constant concentration spotted in different solvents will produce initial zones with greatly varying sizes. Sample sizes are generally 1–10 μl for TLC and 100–1000 nl for HPTLC. When larger volumes are spotted in increments on the same point, consistent and complete drying between applications is necessary. Samples and standards may not chromatograph uniformly if the sample contains ingredients that can alter R_F values or zone shape. Preparation of standards in solutions containing these additional constituents rather than pure solvent will improve reproducibility.

It is clear from the number of variables discussed above that attention to proper solution preparation and a consistent application method for sample and standard solutions is required to obtain reproducible TLC results.

E. Chromatographic Development

In order to obtain reproducible results, the same type and size of chamber (e.g., N-tank or sandwich) with the same degree of vapor saturation (with or without paper liner or saturation layer or pad) must always be used (Jaenchen, 1968). The plate should always be placed in the chamber at the same angle and development carried out across the direction of layer coating, if this can be determined.

Each time a chromatogram is developed, the saturation changes, and the composition of the mobile phase changes due to preferential absorption by the sorbent of certain constituents of the solvent mixture, differential evaporation rates of the constituents, chemical interaction between constituents, or introduction of moisture into the solvent and/or layer (De Zeeuw, 1968; Geiss, 1987). Reproducibility will be compromised by reuse of a mobile phase; therefore, fresh mobile phase with carefully controlled composition should be used for each run.

Good activity control is obtained by plate development in a chamber equilibrated with the vapors of a humidity-controlling solution such as aqueous sulfuric acid or a saturated salt solution. A solution of 50% sulfuric acid will provide a relative humidity of ~40%, which has proven optimum for reproducibility. In addition, Hahn-Deinstrop (1993) has tubulated saturated salt solutions that will establish constant relative humidities ranging from 15 to 95% in closed chambers. These solutions can be used by placement in either section of a twin-trough chamber or in the inner portion of a horizontal chamber (Chapter 7).

Because solvent demixing will cause a gradient up the plate that will determine R_F values, the mobile-phase migration distance should always be standardized, usually 10–15 cm for TLC and some shorter distance (3–7 cm) for HPTLC.

Temperature is almost never measured or even considered when performing TLC, and virtually all work is done at ambient temperature. Temperature changes of $\pm 5°C$ cause changes in R_F that do not exceed ± 0.02 unit (Geiss, 1987), which is within the range of normal experimental error, but large temperature variations can affect R_F values by up to one-half unit. Temperature gradients in the chamber resulting from solvent evaporation, thermal nonequilibrium, and proximity to cold windows or heating vents can cause serious distortions in separations. When performing partition separations on impregnated layers (Chapter 3), the mobile phase is saturated by shaking with the stationary liquid phase, and the chamber atmosphere is saturated with the vapors of both phases. Temperature control is especially important with such partition systems because mutual solubilities of the two phases are affected by any variations. Partition TLC with impregnated layers has been superseded in many cases by bonded-phase TLC, for which temperature control is not critical.

It is important to standardize the depth of the mobile-phase pool in the developing chamber (usually 0.5 cm) and the distance of the origins from the lower edge of the plate (usually 2.0 cm for 20 × 20-cm plates), especially when using mobile-phase mixtures whose components have different polarities. If this is not done, R_F values will change because of the variable solvent demixing on the layer during each development (Shellard, 1964).

Dallas (1965) has suggested that more constant and reproducible (although slightly higher) R_F values are obtained if sorbent is scraped away above the point of maximum desired solvent flow and the plate is allowed to stand in the tank for 15 min after the solvent reaches this level. This allows the ratio of the liquid to the solid phase to become more constant over the entire length of the chromatogram, as indicated by uniform transparency of the layer.

pH variation affects the chromatography of substances that can dissociate in water, whose R_F values depend on both their distribution coefficient and ionization constant (Geiss, 1987). For these compounds, control of pH in the TLC system is necessary for reproducible results.

II. VALIDATION OF QUANTITATIVE RESULTS

The requirements of programs such as GLP (good laboratory practice) that are aimed at evaluating and documenting analytical results have led to great interest in validation (quality assurance) protocols for instrumental quantitative HPTLC, especially for analyses in the pharmaceutical industry. The validation of the purity analysis of drug substances in pharmaceutical analysis by quantitative TLC has been described in a number of papers (Ferenczi-Fodor, 1995 and 1993; Szepesi, 1993a and 1993b; Sun et al., 1994).

Laboratory validation procedures are often performed according to recommendations of external agencies, such as CPMP (Committee for Proprietary Medicinal Products, EEC) or ICH (International Conference on Harmonization) and with consideration of the special features of the TLC procedure (Szepesi and Nyiredi, 1992). The following validation parameters are typically tested: accuracy (correctness; indication of systematic errors), precision (repeatability; indication of random errors), specificity or selectivity (absence of interferences), stability of the analyte (before, during, and after TLC development), limit of detection (lowest analyte concentration that can be detected but not quantified in a sample), limit of quantification (lowest analyte concentration that can be quantified with acceptable precision and accuracy), sensitivity (ability to measure small variations in concentration), linearity (ability of the method to give results that are directly proportional to analyte concentration for specified analyte levels, as well as linearity of the calibration graph), range (concentration levels within which the analyte can be quantified satisfactorily), ruggedness (degree of reproducibility obtained under different conditions, such as in different laboratories on different days), and robustness (capacity of a method to remain unaffected by small variations in method parameters) (Krull and Swartz, 1997; Sherma, 1997) in order to meet the standards necessary for a particular application.

Each step of the analysis must be validated through error analysis and a suitability test, including sample preparation and the TLC procedures. TLC plates, which must meet high standards of performance and uniformity, are available from at least one source (Merck) with laser coding, catalog number, batch number, and plate number, to aid in documentation for GLP work. Modern densitometers can provide automatic instrument validation. For example, the software controlling the densitometer shown in Figure 10.3 compares the operation of the monochromator, plate stage, lamps, optical system, and electronics to target conditions for each of these instrument components, and a validation report of the check results is printed out for proof of compliance with the rules of GLP/GMP.

Validated modern instrumental HPTLC methods, for which relative standard deviations may be routinely as low as 1.5–3.0% (Renger, 1993) are especially important in the pharmaceutical industry for product control. The basic working rules for validation, as stated by Szepesi and Nyiredy (1992), include scanning every spot in triplicate to establish the instrumental error, spotting the same volume of test solution in triplicate, and spotting three bracketing standards in triplicate that contain a known relationship to the expected test solution value (e.g., 80, 100, and 120%). A computer program, ROUTINE-QTLC, was described (Prosek et al., 1993) for reporting quantitative TLC results and documenting the quality of the data, which is appropriate for validation of TLC methods under the guidelines of GLP (Poole and Poole, 1994). Definitions, general principles, and practical approaches for validation are covered in detail in the previously cited sources and an additional book chapter (Szepesi and Nyiredy, 1996).

REFERENCES

Dallas, M. S. J. (1965). Reproducible R_F values in thin layer adsorption chromatography. *J. Chromatogr. 17*: 267–277.

Dallas, M. S. J. (1968a). Reproducibility of R_K values on silica gel in thin layer chromatography. *J. Chromatogr. 33*: 58–61.

Dallas, M. S. J. (1968b). The effect of layer thickness on R_F values in thin layer chromatography. *J. Chromatogr. 33*: 193–194.

De Zeeuw; R. A. (1968). The role of solvent vapor in TLC. *J. Chromatogr. 32*: 43–52.

Ferenczi-Fodor, K., Vegh, Z, and Pap-Sziklay, Z. (1993). Validation of the quantitative planar chromatographic analysis of drug substances. 1. Definitions and Practice in TLC. *J. Planar Chromatogr.—Mod. TLC 6*: 198–203.

Ferenczi-Fodor, K., Nagy-Turak, A., and Vegh, Z. (1995). Validation and monitoring of quantitative thin layer chromatographic purity tests for bulk drug substances. *J. Planar Chromatogr.—Mod. TLC 8*: 349–356.

Geiss, F. (1968). Reproducibility in TLC. Introduction. *J. Chromatogr. 33*: 9–24.

Geiss, F. (1987). *Fundamentals of Thin Layer Chromatography*. Alfred Huthig Verlag, New York.

Geiss, F., Schlitt, H., and Klose, A. (1965a). Reproducibility in thin layer chromatography: Influence of atmospheric humidity, chamber shape, and atmosphere. *Z. Anal. Chem. 213*: 331–346.

Geiss, F., Schlitt, H., and Klose, A. (1965b). Pretreatment of thin layer adsorbents and their chromatographic properties. *Z. Anal. Chem. 213*: 321–330.

Geiss, F., Schlitt, H., Ritter, R. J., and Weimar, W. M. (1963). Analysis of polyphenyls by TLC. *J. Chromatogr. 12*: 469–487.

Gumprecht, D. L. (1992). Comparison of commercially available thin layer chromatography plates with mixtures of dyes, analgesics, and phenols. *J. Chromatogr. 595*: 368–374.

Hahn-Deinstrop, E. (1993). TLC plates: initial treatment, prewashing, activation, conditioning. *J. Planar Chromatogr.—Mod. TLC 6*: 313–318.

Halpaap, H. (1973). Achievement of reproducible separations by application of standardized sorbents in specified systems. *J. Chromatogr. 78*: 77–87.

Jaenchen, D. (1968). The apparent influence of layer thickness on R_F values of thin layer chromatograms. *J. Chromatogr. 33*: 195–198.

Jork, H., Funk, W., Fischer, W., and Wimmer, H. (1990). *Thin Layer Chromatography— Reagents and Detection Methods*, Vol. 1a. VCH-Verlag, Weinheim Germany.

Krull, I., and Swartz, M. (1997). Validation viewpoint—introduction: national and international guidelines. *LC-GC 15*: 534–538.

Pataki, G., and Keller, M. (1963). Influence of the thickness of the layer on the R_F value in thin layer chromatography. *Helv. Chim. Acta 46*: 1054–1056.

Perry, S.G., Amos, R., and Brewer, P.I. (1972). *Practical Liquid Chromatography*. Plenum Press, New York, p. 135.

Poole, C.F., and Poole, S.K. (1994). Instrumental thin layer chromatography. *Anal. Chem. 66*: 27A–36A.

Prosek, M., Pukl, M., Miksa, L., and Golc-Wondra, A. (1993). Quantitative thin layer chromatography. Part 12: Quality assessment in QTLC. *J. Planar Chromatogr.— Mod. TLC 6*: 62–65.

Reichel, W.L. (1967). Effects of humidity in the laboratory on TLC of insecticides. *J. Chromatogr. 26*: 304–306.

Renger, B. (1993). Quantitative planar chromatography as a tool in pharmaceutical analysis. *J. AOAC Int. 76*: 7–13.

Robards, K., Haddad, P. R., and Jackson, P. E. (1994). *Principles and Practice of Modern Chromatographic Methods*. Academic Press, San Diego, CA, pp. 179–226.

Shellard, E. J. (1964). R_F values in thin layer chromatography. *Lab. Pract. 13*: 290–294.

Sherma, J. (1997). Thin layer chromatography. In *Encyclopedia of Pharmaceutical Technology*, vol. 15, J. Swarbrick and J.C. Boylan (Eds.). Marcel Dekker, New York, pp. 81–106.

Smith, I., Baitsholts, A. D., Boulton, A. A., and Randerath, K. (1973). Recommendations concerning the publication of chromatographic data. *J. Chromatogr. 82*: 159–163.

Sun, S. W., Fabre, H., and Maillols, H. (1994). Test procedure validation for the TLC assay of a degradation product in a pharmaceutical formulation. *J. Liq. Chromatogr. 17*: 2495–2509.

Szepesi, G. (1993a). Some aspects of the validation of planar chromatographic methods used in pharmaceutical analysis. I. General principles and practical approaches. *J. Planar Chromatogr.—Mod. TLC 6*: 187–197.

Szepesi, G. (1993b). Some aspects of the validation of planar chromatographic methods used in pharmaceutical analysis. II. Ruggedness testing. *J. Planar Chromatogr.— Mod. TLC 6*: 259–268.

Szepesi, G., and Nyiredy, Sz. (1992). Planar chromatography: Current status and future perspectives in pharmaceutical analysis. I. Applicability, quantitation, and validation. *J. Pharm. and Biomed. Anal. 10*: 1007–1015.

Szepesi, G., and Nyiredy, S. (1996). Pharmaceuticals and drugs. In *Handbook of Thin Layer Chromatography*, 2nd ed. J. Sherma and B. Fried (Eds.). Marcel Dekker, New York, pp. 819–876.

Turano, P., and Turner, W. J. (1974). Variations in the manufacture of commercial silica gel plates. *J. Chromatogr. 90*: 388–389.

12

Preparative Layer Chromatography

I. INTRODUCTION

Preparative layer chromatography (PLC) is used to separate and isolate larger amounts of material than are normally separated by analytical thin-layer chromatography. The amounts handled are usually 10–500 mg or occasionally 1 g or more. The material is isolated for the purpose of further chromatography; analytical studies such as melting point determination or infrared (IR), nuclear magnetic resonance (NMR), ultraviolet (UV) or mass spectrometry for identification; use as purified material for biological activity or chemical studies (e.g., synthesis); or obtaining pure standards for comparison with unknown samples.

The procedures of PLC are generally similar to analytical TLC, the major differences being the use of thicker layers. Layer thicknesses ranging from 500 μm (0.5 mm) to 10 mm have been reported, but layers measuring between 0.5 and 2 mm give the best results. PLC is faster and more convenient than classical column chromatography, and it is much less expensive and less complicated than high-performance column liquid chromatography. The equipment and operational skills are very simple to master and easy to apply. The use of small amounts of solvent, ease of estimating parameters, direct detection on the layer, easy removal of separated zones from the plate, and the ability to run reference compounds simultaneously under identical conditions to assist in locating the desired material are distinct advantages for PLC. Resolution on layers and in columns should be comparable if equivalent sorbents and loading (sample-to-sorbent ratio) are used. A disadvantage of PLC is the possibility of decomposition of labile compounds when in contact with atmospheric oxygen on the exposed layer surface (Halpaap, 1969). PLC has been reviewed by Felton (1982), Sherma and Fried (1987), and Nyiredy (1996).

II. PREPARATIVE LAYERS AND PRECOATED PLATES

Most PLC is performed on layers thicker than the normal 100–250 μm used for analytical TLC so that more load can be applied. Precoated commercial silica gel layers (designated P) are available in 500, 1000, 1500, and 2000 μm thicknesses, the 1000 and 2000 μm being by far the most widely used. Preparative layers are usually formed using particles averaging 25 μm, with a range of 5–40 μm. These coarse, nonuniform particles and the increased thickness of the layer lead to lower-resolution separations compared to analytical TLC and HPTLC. PLC plates can be commercially obtained in 5 × 20-, 10 × 20-, 20 × 20-, 20 × 40-, and 20 × 100-cm sizes. The latter three are most commonly employed because of their higher loading capacity. Manufactured layers of alumina, cellulose (MN-300 and Avicel), and C_2- and C_{18}-bonded silica gel (for reversed-phase PLC) are also available. Because silica gel is most widely used for preparative TLC, the rest of the chapter will refer mainly to this sorbent.

Preparative layers can also be made in the laboratory with commercial "thick-layer" spreading equipment or by a pouring technique (Schmid et al., 1967) using tape to build up the edges of the plate or a special plastic frame sealed to the plate with foam rubber (Halpaap, 1963). Special silica gel and alumina powders containing different size particles to improve adhesion and 254-nm or 254- and 366-nm phosphors, but no $CaSO_4$ binder, are commercially available for making preparative layers of 500–2000 μm (P series). Sorbents designated P + $CaSO_4$ are suitable for making layers 2–10 mm thick, but resolution is limited on these very-high-capacity layers. The manufacturer's instructions for slurrying the powders and drying and activating the layers must be followed carefully to avoid pitting, cracking, or flaking of the layers. Advantages of preparing plates in the laboratory include the ability to incorporate salts, buffers, or other additives to improve separations and to produce layers of any desired thickness (Nyiredy, 1996). Commercial precoated preparative layers are more convenient and reproducible.

Preparative layers with preadsorbent spotting area are described in Section III.

III. SAMPLE APPLICATION

Up to 15 mg of sample has been applied to a 20 × 20-cm analytical layer, so that if this quantity is sufficient for the purpose at hand, thicker preparative layers are not needed. Overloading of the analytical plate will be evident from the appearance of tailed or otherwise distorted zones. Analytical plates with an inert preadsorbent or dispensing area (Chapter 3) can be loaded with more sample than standard adsorbent plates.

As a rule of thumb, the capacity of a PLC plate increases as the square root of thickness with little or no degradation of separation; a 1000-μm layer will, therefore, have twice the loading capacity of a 250-μm layer. The loading can be increased beyond this amount if deterioration of the separation can be tolerated. Another general guideline is that the maximum load for a 1-mm, 20 ×20-cm silica gel layer is 100 mg (5 mg/cm); cellulose and reversed-phase layers have a lower sample capacity. The actual weight that can be spotted will depend on the specific separation and chromatographic system.

The sample should be dissolved in a volatile solvent that is as nonpolar as possible, at a concentration that allows the sample components to be adsorbed on the coating and not deposited as crystals. Crystal overload can give badly distorted bands or spots because the components' rate of dissolving in the moving carrier solvent becomes a limiting factor (Felton, 1978). A concentration of 2–10% is typical.

Samples are usually applied as a narrow streak across the plate. Streak application is not only convenient for loading larger amounts onto the layer, but resolution of streaks is usually superior to that obtained from initial spots. Manual application of up to 10 ml of solution can be achieved using a syringe or ordinary pipet and a ruler as a guide or a commercial chromatography streaking pipet (Figure 12.1). The Camag Linomat IV (Chapter 5) is an automated spray device that can apply uniformly as much as 500 μl of sample solution in a streak with selectable lengths up to 190 mm. The streak should be as straight and narrow as possible; most available streakers can give an initial zone only 1–3 mm high. The streak should terminate at least 1–3 cm in from each side of the plate to avoid "edge effects" that often make the solvent move faster or slower (usually the former) at the edges than in the center 15–17 cm of a 20 × 20-cm plate.

Repeated sample applications are usually made to deposit the desired amount of solution. It is important to remove the solvent completely between streaks and before development. If the streak becomes excessively broad (high) during sample application, it can be sharpened by a predevelopment for a distance of 1–2 cm with a very polar solvent, after which the solvent is thoroughly evaporated and development is carried out with an appropriate chromatographic mobile phase (Fessler and Galley, 1964).

Botz et al. (1990) and Nyiredy (1996) described a solid-phase sample application method and device that is useful for both capillary-and forced-flow PLC. The initial zone produced is homogeneous within the entire cross section of the preparative layer and has an extremely sharp edge leading to the chromatographic layer, with the advantage of in situ sample concentration and cleanup.

Several manufacturers provide plates with 400- or 1000-μm-thick silica gel

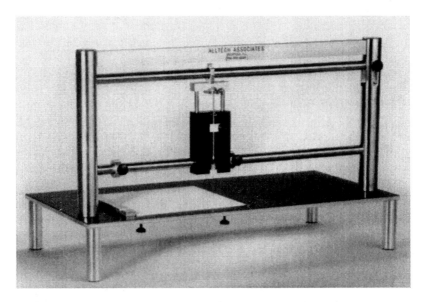

FIGURE 12.1 TLC Sample Streaker. A manually operated device used to apply large volumes of solution as a continuous, narrow band (1 mm width) onto a preparative plate up to 40 cm wide. Solution is applied through the mechanical action of forcing the plunger of a 250 or 500 μl syringe downward as it is pushed across a sloping stainless steel bar. (Photograph courtesy of Alltech—Applied Science Labs.)

layers and a preadsorbent spotting area (Chapter 3). Larger volumes of samples can be applied to these plates and less care and time in streaking can be tolerated because the plate will automatically produce a narrow sample zone at the layer junction. However, the concentration applied across the preadsorbent should be as uniform as possible to prevent local overloading and optimize the separation. Resolution of zones is improved on precoated plates with preadsorbent zone compared to those without, as illustrated by Nyiredy (1996).

A unique plate for PLC combining a 700-μm-thick preadsorbent area and a wedge-shaped silica gel G layer is manufactured by Analtech (Uniplate-T Taper Plate). The silica gel layer is thin (300 μm) adjacent to the preadsorbent and varies uniformly to a thickness of 1700 μm at the top. Sample concentration occurs in the preadsorbent, and low-R_F zones are better separated on the thin, bottom area of the silica gel than they would be on a 1000-μm constant-thickness preparative layer. The solvent flow pattern in the tapered layer causes the lower portions of the zones to travel at a faster rate than the top portions, reducing vertical band spreading and leading to tighter, better resolved zones. Nyiredy

(1996) presented schematic diagrams of the construction of the taper plate and comparison of a separation on the plate versus a plate without a layer-thickness gradient.

IV. DEVELOPMENT

A. Capillary-Flow Controlled PLC

Development is carried out in the usual ascending manner as for analytical-scale plates, using a solvent system established as being optimum by preliminary evaluations on analytical plates or microscope slides developed in a paper-lined, solvent-saturated chamber. For development of 20 × 20 cm PLC plates, rectangular glass N-chambers lined on four sides with thick, mobile-phase soaked filter paper and equilibrated with solvent vapors by standing for 1–2 h are used. Two plates leaning against opposite side walls of the chamber can be developed simultaneously if sufficient mobile phase, ~50–100 ml, is added. Special large-volume N-chambers can be used for larger plates, and racks are available for simultaneous development of more than two plates. Because of the thicker layer, PLC plates will more quickly deplete the mobile phase in the chamber, and additions, if necessary, should be made carefully to maintain an adequate liquid level and chamber equilibration. Thicker layers generate more heat during development, leading to increased analyte solubility and higher R_F values. This effect may require the use of weaker mobile phases compared to analytical layers. Ascending development times of 1–2 h or more for distances up to 18 cm on a plate 20 cm in length are common in PLC; mobile-phase speed is increased if the chamber is fully saturated and the optimum angle at which the plate leans against the chamber wall (75°) is used (Nyiredy, 1996). A typical preparative separation is shown in Figure 12.2.

It is important to use optimized mobile phases in PLC because of inevitable poorer resolution caused by the larger particle size of the sorbent and the large amount of sample applied. Mobile phase optimization procedures, such as PRISMA, are discussed in Chapter 6. Resolution of complex mixtures can be improved by multiple development with less polar solvents that initially give low R_F values (Halpaap, 1963). Care must be taken to evaporate the solvent carefully after each run. After detection and recovery by elution (below), fractions incompletely resolved in the initial multiple development are rechromatographed on a more efficient, thinner layer. Devices for performing circular and anticircular PLC have been described by Nyiredy (1996), but these methods have not been used nearly as much as linear development despite advantages in resolution for compounds that have low R_F or high R_F values, respectively. Gradient elution has been shown to improve preparative separations carried out in sandwich chambers with zonal application of large sample volumes (Matysik et al., 1994).

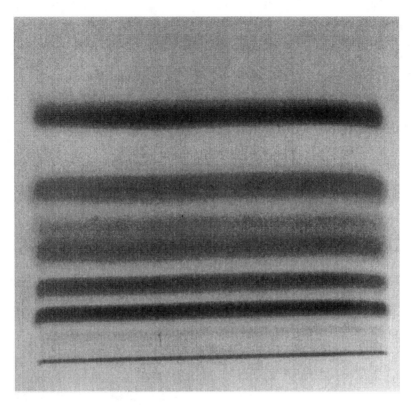

FIGURE 12.2   A typical preparative layer separation. (Photograph courtesy of Analtech.)

A sequential development technique (Nyiredy, 1996) can be used in PLC to improve resolution and, additionally, to reduce separation time because separations always occur under the influence of the highest initial mobile-phase velocity. After the first development, the same or a different mobile phase is applied to different plate regions at different times, depending on where zones requiring additional separation are located. A special Mobile-R_F Chamber has been described for performing the sequential developments.

B.   Forced-Flow Preparative Planar Chromatography

Forced-flow planar chromatography (FFPC) with the use of external pressure (overpressured layer chromatography, OPLC) or centrifugal force (rotation planar chromatography, RPC) can be used for off-line or on-line PLC (Mincsovics et al., 1988; Nyiredy, 1990). In off-line FFPC, the procedures after development are similar to conventional TLC—i.e., drying, scraping of the sorbent layer, elution, and crystallization (Mincsovics et al., 1996). In on-line FFPC, separated compounds are eluted from the layer and isolated from the instrumentation; this enables connection of a flow detector, recording of chromatograms, and collection of separated compounds with a fraction collector (Nyiredy, 1996). Fine-particle layers can be used to improve separations. Instruments for FFPC are described in Chapter 7.

Preparative OPLC (Zogg et al., 1988) can be performed with the Chrompres 10 or Chrompres 25 (Figure 7.14) instruments. The Chrompres 10 allows on-line linear separations on 20 × 20-cm or 20 × 40-cm plates and off-line circular and anticircular separations. The Chrompres 25 allows a higher pressure and use of more viscous solvents, which are suitable for C_2- and C_{18}-bonded RP layers. Regular commercial precoated preparative plates are mostly used for preparative OPLC.

OPLC may be used for separation of two to seven compounds, and resolution is improved compared with capillary-flow PLC because of longer separation distances. Specially prepared commercial precoated plates with an average particle size of 25 μm and a 5–40 μm range are mostly used for preparative OPLC. Analytical plates with 0.25 μm layer thickness have been used for 2–10 mg sample amounts and thicker layers for amounts up to 300 mg. Preparative plates with concentrating zone are normally developed for distances of 18 or 36 cm, depending on the length of the plate, for linear, on-line separations. Precoated preparative plates have been reused up to eight times without loss of resolution if proper intermediate washing and reconditioning steps are used. The usual mobile phase velocity is 3–6 ml/min for 2-mm thick layers; lower rates increase separation time and decrease efficiency. The generally accepted method for a preparative OPLC separation, in order to eliminate the negative effect of adsorbed air and gas, is to start the separation with a hexane-equilibrated layer. Isocratic and step-gradient mobile phases have been used for preparative OPLC separations, and several specially prepared chromatoplates have been developed simultaneously for separating a large sample size during a single chromatographic run. The solid-phase sample application method (Section III) has been used successfully for linear OPLC, and on-line sample application has also been employed. Readers should consult the book chapter by Nyiredy (1996) for details of the techniques and applications of preparative OPLC.

Preparative RPC is carried out in five different on-line modes: normal chamber, microchamber, ultramicrochamber (Nyiredy et al., 1988), column, and sequential. The first four methods differ in the size of the vapor space in the chamber. Circular development is used in all cases. The sample is applied to the layer near the center, and the solvent is driven by centrifugal force through the stationary phase from the center to the outside of the round plate or planar column, and separated compounds are eluted from the plate or column as a result of centrifugal force and recovered in a fraction collector. Sequential RPC involves the combination of circular and anticircular development. The selection of the RPC mode to be used is based on an analytical preassay and depends on the nature of the mobile phase (one or several components), chamber (saturated or unsaturated), and layer (particle size, binder), and whether a single mobile phase can provide the needed separation. The normal chamber (N-RPC) has been used in the majority of published applications: in this method, the layer rotates slowly (100–300 rpm) in a large-vapor space, stationary chamber; the sample is applied near the outside edge of the round layer; the mobile phase migrates through the sorbent by capillary action, against the centrifugal force, from the outside to the center layer; and the separated compounds are eluted from the center of the layer in a fraction collector. Amounts between 50 mg and 1 g of substance have been applied to a single plate for preparative RPC. Preparation of layers and planar columns, application of samples for preparative RPC, commercial centrifugally accelerated instruments (Chromatotron, CLC-5, and Rotachrom), and other aspects of preparative RPC were described in detail by Nyiredy (1996). Precast silica rotors with 1000–8000 μm thickness for the Chromatotron PLC apparatus (Chapter 7) are available from Analtech.

Forced-flow methods require expensive instruments and increased preparation time but have significantly improved resolution compared to capillary-flow PLC. Capillary-flow PLC is the simplest and still the most widely used PLC method (Nyiredy, 1992, and 1996).

V. LOCATION OR DETECTION OF THE DESIRED COMPONENT

If the desired components are naturally colored or fluorescent, location can be made simply by eye in daylight or under ultraviolet (UV) light, respectively. A PLC plate containing a fluorescent phosphor will indicate separated, UV-absorbing components as dark zones on a bright background when examined under 254- and/or 366-nm UV light. Most commercial fluorescent impregnating agents are insoluble in the solvents used for compound elution (below) and will add no significant contamination.

The best chemical detection method is exposure to iodine vapor in a closed chamber, which will visualize compounds of many chemical classes as light or dark brownish zones on a tan background. In most cases, the vapor can be subsequently evaporated, leaving compounds unchanged. The zones must, of course, be marked, usually by outlining with a sharply pointed object, before the iodine evaporates, making the materials invisible.

If other, nonreversible or destructive reagents are required for detection, additional small spots of sample are applied to the outer side margins of the layer. These are then sprayed carefully with the required reagent, after masking or covering the major streaked portion of the plate with glass or cardboard and scribing vertical channels between the streak and the spots. The desired components are then visible on each side to serve as a guide for scribing horizontal lines across the plate to outline the area in which they are contained (Figure 12.3). A similar alternative procedure is to streak the sample almost entirely across the layer and to use the outer edges of the streaked sample, outside the mask, as a guide area to be sprayed. Analtech supplies preparative plates that are prescored 1 in. in from each side. The sample is streaked into the scored areas, and after development, the side portions are snapped free, detections made, and the edge strips returned next to the center section to mark the areas for scraping (Figure 12.4). These plates eliminate the possibility of sprayed reagent wicking into the desired component in the center area and any undesirable effect that heating, if required as part of the visualization procedure, might cause. The same thing can be done with any plate if a glass cutter is available.

As a general rule, heating of preparative plates at any stage in the chromatographic process should be avoided so that substances can be recovered in an unaltered state.

VI. REMOVAL OF THE DESIRED MATERIAL BY ELUTION

The material detected in the scribed, central area of the plate is recovered by physical removal of the adsorbent zone, elution of the substance from the adsorbent with a solvent, separation of the residual adsorbent, and concentration of the solution.

The outlined layer areas containing the materials are scraped off cleanly down to the glass or other backing with a spatula, razor blade, or commercial scraper (Figure 12.4), and then the loosened adsorbent is transferred to a sheet of glassine weighing paper and poured into a small glass vial or tube having a solvent-resistant cap. If a centrifuge is available, the container can be a centrifuge tube. One or two drops of water can be added to displace the component from the active adsorbent sites. Wetting the chromatogram with a spray of ethanol can reduce loss of fine silica particles by flaking and blowing during the scraping and transfer steps.

The organic solvent used to elute the compound must be one that is adequately polar and a good solvent for the component. Methanol is not recommended because silica gel itself and some of its common impurities (Fe, Na, SO_4) are soluble in this solvent and will contaminate the isolated material after removal of the methanol. Ethanol, acetone, or chloroform, or the mobile phase originally used for the TLC, are frequently used choices for solute recovery.

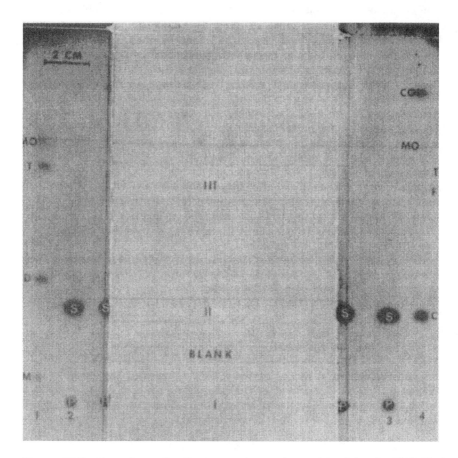

FIGURE 12.3 Sample application procedure using guide strips for PLC. Up to 1 mg of lipid was streaked across the center 10 cm of a 20 × 20-cm analytical plate, which was developed in the dual solvent system of Skipski et al. (Chapter 15). Major bands from the organism contained sterol (II) and phospholipid (I). The band labeled "blank" was lipid negative. Lanes 1 and 4 of the edge strips (5 cm wide) were spotted with neutral lipid standards, lanes 2 and 3 with sample. The edge strips were cut and visualized with 10% phosphomolybdic acid in ethanol and were then matched with the center to determine the position of the bands.

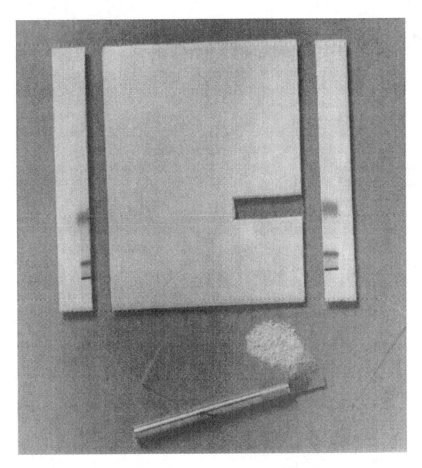

FIGURE 12.4 Preparative scored plate with adsorbent scraper and scraped zone. (Photograph courtesy of Analtech.)

Water is not desirable because it is so difficult to remove by evaporation during the concentration step. A formula that has been used to calculate the volume of solvent to employ when the TLC mobile phase is chosen for elution is as follows:

Solvent volume $= (1.0 - R_F)$ (10)(volume of the scrapings)

The selected solvent is added to the tube with the scrapings, and the tube is shaken mechanically or by hand or is Vortex mixed for about 5 min. The tube is centrifuged and the supernatant liquid is carefully removed by pipet, decantation, or filtration. A second portion of solvent can be added to the scraping to repeat the extraction, and the two solutions combined. If it is definitely necessary to exclude all silica gel particles from the solution, the filtering medium should be capable of retaining 2-μm particles (most TLC silica gel particles range from 5 to 20 μm). An apparatus for isolation of compounds from thin layers by elution and direct Millipore filtration has been described (Dekker, 1979).

An alternative method for removal of scraped adsorbent zones is the use of a vacuum suction collector. These devices can be used like a miniature vacuum cleaner to collect the scrapings in a Soxhlet extraction thimble (Analtech) or a recovery tube containing a sintered glass disk (Kontes). In the latter case, the compound is eluted from the tube with solvent with the help of vacuum after collection of the adsorbent.

A similar homemade combination scraping–collection device can be constructed. A Pasteur pipet (225 mm × 0.7 mm o.d.) is cut with a file 60 mm from the top and 65 mm from the tip, to produce a 100-mm-long pipet that is plugged with glass wool. The tip is attached to an aspirator or vacuum pump, and the top is used to scrape the chromatographic zone containing the compound of interest. Scraping and vacuum collection can be done simultaneously. The pipet is removed from the vacuum line and is then used as a column. The column (pipet) is placed in a vial (60 mm × 10 mm i.d., 8 ml capacity) through an aluminum foil cover, and elution solvent is then percolated through the column. If the glass wool plug is adequately tight, virtually all of the silica gel is retained in the column.

Once a solution of the material is obtained free from adsorbent, the solvent is evaporated to dryness. This should be done at low temperature in an inert-gas stream, such as nitrogen, so that the collected sample is not decomposed or otherwise changed.

VII. Applications of PLC

The various PLC techniques were compared and trends in PLC summarized by Nyiredy (1996). The following recent applications for compound isolations by classic, capillary-flow PLC have been published: photodegradation products of primaquine formed in an aqueous medium (Kristensen et al., 1993); synthesized isoprenoid diphosphates (20 mg of pure compound recovered from a single plate) (Kennedy Keller and Thompson, 1993); brain lipids (Deleva et al., 1992); components from aloe sap (0.75 mm layers of silanized silica gel and 0.5 mm of silica gel were developed in horizontal sandwich chambers) (Wawrzynowicz et al., 1994); taxol and cephalomannine from *Taxus cuspidata* (silica gel preparative plates developed with heptane–dichloromethane–ethyl acetate, 11:8:1) (Glowniak et al., 1996); fenvalerate in a pesticide formulation (1-mm silica gel layers developed with hexane–acetone, 9:1, marker spots detected with *o*-dinitrobenzene–*p*-nitrobenzaldehyde reagent) (Gupta et al., 1996); and L-ornithine–L-aspartate in pharmaceutical dosage forms (silica gel 60F layer developed with 1-butanol–acetic acid–water, 50:25:25, and derivatization with ninhydrin) (Gaba et al., 1997).

REFERENCES

Botz, L., Nyiredy, Sz., and Sticher, O. (1990). A new solid phase sample application method and device for preparative planar chromatography. *J. Planar Chromatogr.—Mod. TLC 3*: 10–14.

Dekker, D. (1979). Apparatus for the isolation of microgram amounts of compounds from thin layers by elution and direct Millipore filtration.*J. Chromatogr. 168*: 508–511.

Deleva, D. D., Ilinov, P. P., and Zaprianova, E. T. (1992). Modified procedure for determination of lipid-bound sialic acid in rat brain. *Bulg. Chem. Commun. 25*: 514–519.

Felton, H. R. (1978). Preparative thin layer chromatography. *Analtech. Technical Report No. 7903*, Newark, DE.

Felton, H. R. (1982). Preparative TLC. In *Advances in TLC—Clinical and Environmental Applications*, J. C. Touchstone (Ed.). Wiley-Interscience, New York, pp. 13–20.

Fessler, J. H., and Galley, H. (1964). Thin layer chromatography of relatively voluminous samples. *Nature 201*: 1056–1057.

Gaba, S., Agarwal, S., Omray, A., Eapan, D., and Sethi, P. (1997). Estimation of L-ornithine-L-aspartate by post-derivatization preparative TLC in combined dosage forms. *East. Pharm. 40*: 137–138.

Glowniak, K., Furmanowa, M., Zgorka, G., Jozefczyk, A., and Oledzka, H. (1996). Isolation and determination of taxol and cephalomannine in twigs, needles, and tissue cultures from Taxus cuspidata Sieb et Zucc. *Herba Pol. 42*: 309–316.

Gupta, S., Sharma, K. K., and Handa, S. K. (1996). Fourier transform infrared spectroscopic determination of fenvalerate in emulsifiable concentrate formulation. *J. AOAC Int. 79*: 1260–1262.

Halpaap, H. (1963). Preparative thin film chromatography. *Chem.-Ing.-Tech. 35*: 488–493.

Halpaap, H. (1969). Preparative layer chromatography. In *Chromatographic and Electrophoretic Techniques*, I. Smith (Ed.), Vol. 1. Interscience, New York, pp. 834–886.

Kennedy Keller, R., and Thompson, R. (1993). Rapid synthesis of isoprenoid diphosphates and their isolation in one step using either thin layer or flash chromatography. *J. Chromatogr. 645*: 161–167.

Kristensen, S., Grislingaas, A. L., Greenhill, J. V., Skjetne, T., Karlsen, J., and Toennesen, H. H. (1993). Photochemical degradation of primaquine in an aqueous medium. *Int. J. Pharm. 100*: 15–23.

Matysik, G., Soczewinski, E., and Polak, B. (1994). Improvement of separation in zonal preparative thin layer chromatography by gradient elution. *Chromatographia 39*: 497–504.

Mincsovics, E., Ferenczi-Fodor, K., and Tyihak, E. (1996). Overpressured layer chromatography. In *Handbook of Thin Layer Chromatography*, 2nd ed., J. Sherma and B. Fried (Eds.). Marcel Dekker, New York, pp. 171–203.

Mincsovics, E., Tyihak, E., and Siouffi, A. M. (1988). Comparison of off-line and on-line overpressured layer chromatography. *J. Planar Chromatography–Mod. TLC 1*: 141–145.

Nyiredy, Sz. (1990). Sample preparation and isolation using planar chromatography. *Anal. Chim. Acta 236*: 83–97.

Nyiredy, Sz. (1992). Planar chromatography. In *Chromatography, 5th Edition*, E. Heftmann (Ed.). Elsevier, Amsterdam, pp. A109–A150.

Nyiredy, Sz. (1996). Preparative layer chromatography. In *Handbook of Thin Layer Chromatography*, 2nd ed., J. Sherma and B. Fried (Eds.). Marcel Dekker, New York, pp. 307–340.

Nyiredy, Sz., Meszaros, S. Y., Dallenbach-Tolke, K., Nyiredy-Mikita, K., and Sticher, O. (1988). Ultra-microchamber rotation planar chromatography (U-RPC): a new analytical and preparative forced flow method. *J. Planar Chromatogr.—Mod. TLC 1*: 54–60.

Schmid, H. H. O., Jones, L. L., and Mangold, H. K. (1967). Detection and isolation of minor lipid constituents. *J. Lipid Res. 8*: 692–693.

Sherma, J., and Fried, B. (1987). Preparative thin layer chromatography. In *Preparative Liquid Chromatography*, B. A. Bidlingmeyer (Ed.). Elsevier, Amsterdam, pp. 105–127.

Wawrzynowicz, T., Waksmundzka-Hajnos, M., and Mulak-Banaszek, K. (1994). Isolation of aloine and aloeemodine from Aloe (Lillaceae) by micropreparative TLC. *J. Planar Chromatogr.—Mod. TLC 7*: 315–317.

Zogg, G. C., Nyiredy, Sz., and Sticher, O. (1988). Influence of the operating parameters in preparative overpressured layer chromatography (OPLC). *J. Planar Chromatogr.—Mod. TLC 1*: 261–264.

13

Radiochemical Techniques

I. INTRODUCTION

Radioactive isotopes are widely used as tracers or labels for substances separated by thin-layer chromatography for following the course of chemical and biochemical reactions and determining the qualitative and quantitative distribution of substances in a reaction mixture; for elucidating metabolic pathways of drugs (Inoue et al., 1997), pesticides (Tal and Rubin, 1993; Matsunaga et al., 1997), pollutants, and natural substances in human, animal, and plant tissues or soil; assessing the purity of isotopes (Gattavecchia et al., 1994; Mallol and Bonino, 1997); and for biochemical studies such as elucidation of biochemical pathways, radiotracer binding (Zamora et al., 1996), and assessment of enzyme activity. Because of the low detection limits and variety of methods for detecting and quantifying tagged compounds, thin-layer radiochromatography (TLRC) is often the only reasonable means of carrying out such studies. TLRC is advantageous compared to gas and column liquid chromatography with radiodetectors because all of the radioactivity can be accounted for on the layer (except for volatile components); therefore, recovery experiments to prove complete elution of radioactivity from the column are not required (Clark and Klein, 1996). Early general reviews of isotope techniques in TLC were given by Mangold (1969), Roberts (1978), Snyder (1969), and Prydz (1973). More recent information on procedures, instruments, and applications of radio-TLC were published by Shulman and Weaner (1991), Filthuth (1990), Rapkin (1990), Dallas et al. (1988), Touchstone (1992), and Clark and Klein (1996).

Prior to working with radioactive materials, the scientist should attend a pertinent training course or consult a standard source of information on radioisotope methodology to become familiar with the precautions used in the safe handling of radiochemicals in the laboratory (e.g., DuPont, 1988a; Shapiro, 1972; Stewart, 1981). As much as possible, experiments should be carried out, and plates with radioactive zones stored, in a fume hood.

The most widely used labeled substances in TLRC contain the weak β emitter ^{3}H (tritium) and the stronger ^{14}C, both with long half-lives. Other labeled substances often used contain ^{35}S, ^{32}P, ^{131}I, all with relatively short half-lives but stronger radioactive emissions. The sensitivity of detection of these isotopes is strongly related to the energies of the particles from the nuclides considered. The basic unit of radioactivity is the curie (Ci), usually measured in disintegrations per minute (dpm). Because of the small amounts separated by TLRC, activities are usually in the millicurie (mCi), microcurie (μCi), or nanocurie (nCi) range. A millicurie represents 2.22×10^9 dpm. The specific activity of a substance is the activity per unit of mass, most frequently, millicuries per millimole (mCi/mmole). Each isotope has a specified "half-life," which is the length of time needed for one-half of the radioactivity to decay. The half-life of ^{14}C is ~5600 years and of ^{3}H is ~12 years. For a more detailed discussion of the terminology of TLRC, consult Mangold (1969).

It is critical that labeled substances used for TLRC be assessed for purity immediately before use (Snyder and Piantadosi, 1966). The high sensitivity of the detection methods employed, especially scintillation counting, can reveal very low levels of extraneous matter, so that, for example, a radioactive contaminant or degradation product of the original labeled compound might be mistaken for a metabolite in a metabolism study. If the radiochemical is diluted with nonactive material to the required specific activity, the purity check should be carried out after dilution, and it is advisable to perform a chemical purity check as well (Clark and Klein, 1996). Chemical changes can be caused by exposure of compounds, both labeled or nonlabeled, to air and/or light on the surface of a chromatogram exposed to the atmosphere (Snyder, 1968). For this reason, it is best to assess the results of TLRC as soon as the chromatographic development step is complete.

Commercial glass-backed TLC and HPTLC plates, which contain flat, uniform layers, are best to use for instrumental TLRC. Plates with plastic or aluminum backing can be cut into sections, thereby facilitating zonal analysis by scintillation counting or preparative isolation of compounds. Layers must be clean and free from dust and other particles in order to measure radioactivity accurately. Samples should be spotted manually or automatically with an instrument (see Chapter 5) as small spots or narrow bands sufficiently far apart (minimum of 1.5 cm) to avoid measurement of radioactivity from adjacent lanes. Complete resolution of sample components is required for accurate quantification; therefore, the use of high-performance layers, optimized solvent systems, and small-volume samples is recommended. For maximum resolution, automated multiple development and one of the newer detectors (Section II.C) can be used. Artifacts that can form at the origin during the spotting and solvent-drying steps prior to development are minimized by placing the plate into the mobile phase before the sample solvent has fully dried.

Compound degradation on the plate can be checked by two-dimensional development of a mixture with the same mobile phase, as described in Chapter 14 (Figures 14.2–14.4). All of the zones will lie on a diagonal straight line if no decomposition occurs during development. The layer should be dried after development under conditions that completely remove the mobile phase constituents but do not volatilize sample components.

In metabolism studies by TLRC, one- and two-dimensional development has been used for qualitative identification and quantitative estimation of metabolites. Cochromatography with reference standards can help to confirm structures of unknown components. Isolation of larger quantities of metabolites for identification by spectrometry can be carried out by streaking a band of sample across an analytical or preparative layer (Chapter 11).

II. DETECTION OF RADIOACTIVE SUBSTANCES ON TLC PLATES

The three main ways of detecting radioactive isotopes on TLC plates are (1) film registration or autoradiography, (2) zonal analysis (plate scraping and liquid scintillation counting), and (3) direct, in situ methods using radiation detectors.

A. Autoradiography

A piece of x-ray or photographic film sensitive to α, β, γ, or x-rays is placed on top of the developed, thoroughly dried chromatogram. Radioactive spots appear as dark areas on the developed film, which can then be evaluated densitometrically. When properly exposed, the resolution of the spots on the film can be comparable to that on the original chromatogram and as high as any radioactivity detection method. The major disadvantage of the method is that exposure times can vary from a few hours to several weeks, depending on the level of radioactivity. Three different exposure methods are used—direct exposure (autoradiography), direct exposure with an intensifying screen, and fluorographic exposure (fluorography)—depending on the isotope involved and the amount of radioactivity (Shulman and Weaner, 1991).

Direct exposure is the simplest method and is suitable for all β emitters, except for low levels of tritium. A variety of films are available for use, including medical and industrial x-ray films (Eastman Kodak, 1986, 1988). The sandwich of the layer and similarly sized film is pressed tightly together (the edges can be taped) and wrapped in black cloth, or is placed in the black envelope in which films are purchased or in a commercially available autoradiography container (Mangold, 1969). Especially in the case of soft radiation, the chromatogram must be pressed tightly against the film (i.e., clamped or weighted) because even a thin layer of air can absorb this radiation (Prydz, 1973). Low energy isotopes such as tritium require the use of film that does not have the normal protective emulsion coating, which would prevent penetration and detection of the emitted radiation (Amersham Corp., 1988). The optimum length of exposure of the film to the plate depends on the level of radioactivity, the type of isotope, and the film emulsion and is determined by trial and error or by exposing the film to layers containing calibrated amounts of radioactivity. In general, weak β emitters are left exposed for relatively long periods (e.g., 1–4 weeks or longer), whereas strong β emitters are left in contact less than 1 week. For further details, consult Mangold (1969). The x-ray film is developed according to usual photographic procedures. For locating radioactive compounds on the original chromatogram, radioactive ink is applied as a marker to a few points on the layer after chromatography to facilitate lining up the film and the chromatogram after exposure. The compounds are marked by perforating the superimposed film around the dark spots with a needle.

Because artifacts, contaminants, radioactive background, and other factors may also darken the photographic emulsion, considerable care must be exercised in interpreting the results of autoradiography. It is especially important to dry the chromatogram well to prevent formation of spurious blackening from traces of organic solvents left in the layer. Photodensitometric scanning of the film is generally not considered a strict quantitative procedure but rather a semiquantitative estimation of amounts causing the darkening (Mangold, 1969).

Darkening of the film by nonradioactive substances can be tested for by going through an entire analysis with a chemically identical, but nonactive sample. No dark spots should result on the film from this "blank" run. Alternatively, a thin plastic sheet can be placed between the layer and the film to prevent darkening of the photographic emulsion through direct contact with chemical reducing agents. The sheet must be thin enough to allow weak emissions (e.g., from ^{14}C) through to the film, and no sheet can possibly be used with the very weakly emitting tritium isotope. The film should not be touched directly, to prevent contamination.

Within a limited range of exposure times, film blackening is proportional to the concentration of radioactive isotope on the chromatogram. Beyond a certain exposure time, blackening will not increase. If several isotopes with different emission intensities and exposure limits are present on the same chromatogram, a number of autoradiograms with different exposure times will have to be prepared for accurate relative quantitative estimation. Overexposure of high-activity spots not only can make quantification difficult but can cause merging of resolved spots and encourage appearance of dark background zones. Quantification requires comparison of the variations in optical densities, measured with a densitometer, to a radiation response curve (characteristic curve) generated with radioactive standards of the analyte on the same piece of film (Clark and Klein, 1996). For successful results, close control of all chromatographic factors must be exercised. Densitometry of autoradiographic film can be carried out with the same type of scanners used for measurement of colored spots on nonradioactive chromatograms (Chapter 10). However, it is most convenient to use a scanner with software that converts measured absorbance to data that corresponds directly with exposure intensities based on appropriate internal standards (e.g., the Molecular Dynamics Personal Scanner SI).

To improve detection efficiency for γ-emitting (e.g., ^{125}I) and high-energy β-emitting (e.g., ^{32}P) isotopes, the plates are exposed with commercially available, inorganic phosphor-coated intensifying screens placed behind the x-ray film. Radiation passing through the film strikes the phosphor and causes light to be emitted, increasing the exposure of the film. Enhancements in sensitivity of 7–16-fold have been obtained compared to direct exposure without the screen (Shulman and Weaner, 1991).

Enhancement of autoradiographic sensitivity can also be obtained by using Kodak NTB or Nuclear-Track emulsion films rather than by x-ray film, or a scintillator can be added to the layer before or after chromatography. In the latter case, the energy of the disintegrating atoms is converted to light, which is detected by the film to produce a "fluorogram." Increased sensitivity is especially critical for work with the important biochemical indicator tritium, which is difficult to detect with a high degree of efficiency because of the low energy of its soft β-ray emission. Scintillators employed have included anthracene added directly to the layer before chromatography (Luthi and Waser, 1965) or an ether solution of 2,5-diphenyloxazole (PPO) sprayed or poured onto the developed chromatogram (Wilson, 1985). The optimum amount of scintillator to prevent self-absorption of radiation must be determined by trial and error. Visualization of amounts as low as 2 to 3 nCi/cm^2/day of ^{3}H and 0.05 to 0.06 nCi/cm^2/day of ^{14}C have been reported for chromatograms with added scintillator (Randerath, 1969, 1970; Prydz, 1973). Solutions and spray reagents for fluorography are commercially available and allow simple and even application of the reagent (DuPont, 1988b).

As an example of a recent autoradiographic analysis (Koskinen et al., 1997), DNA isolated from livers of rats receiving tamoxifen was analyzed by the ^{32}P-postlabeling method using both reversed-phase HPLC with on-line detection and TLC on polyethyleneimine plates followed by autoradiography. The TLC method proved to be more sensitive than HPLC.

In a study of fluorography (Bartelsman et al., 1982), Kodak X-OMAT R film treated with EN[3]HANCE spray (New England Nuclear) and exposed at dry ice temperature for 24 h produced a 100- to 200-fold enhancement of the sensitivity of ^3H and ^{14}C detection compared to autoradiography. Methyl anthranilate scintillant enhanced detection of ^3H 1000-fold compared to autoradiography in a 24-h experiment (Bochner and Ames, 1983). The advantage of film exposure at low temperature due to increased light sensitivity of the photographic emulsion has been documented (Prydz, 1973). The sensitivity of fluorography (as well as the linearity of response at low exposure levels) can also be improved by partially exposing the film to a controlled preflashing of light before exposure to the radioactive spots (Shulman and Weaner, 1991).

A considerable amount of valuable technical information on autoradiography is available from the Technical Information Service of the Eastman Kodak Company.

B. Liquid Scintillation Counting (Zonal Analysis)

In liquid scintillation counting, a technique that is especially sensitive for weak β emitters such as ^3H and ^{14}C, a portion of the adsorbent is scraped off into a vial and mixed with a scintillation solution. The vial is then placed in a photomultiplier scintillation counter, which detects the radiation indirectly based on the light energy produced. The resolution depends on the width of adsorbent taken for the measurement, and the disadvantage of the method is the large number of samples that must be removed if unknown zones are present and the whole length of the chromatogram is to be examined (20–40 segments). Scintillation counting is applicable to naturally radioactive substances or inactive substances that are labeled with a radioactive reagent either before or after chromatography. The location of the areas of radiation to be removed by scraping or cutting (plastic- or aluminum-backed layers) can be determined by matching the dark zones on a predetermined autoradiograph to the corresponding layer areas.

In the past, automatic scrapers could be purchased that allowed sections down to 1 mm in width to be removed and transferred directly into counting vials, but apparently no commercial automatic scraper is presently available. Manual scraping and recovery techniques are the same as those outlined for PLC in Chapter 12. Manual adsorbent scrapers of various sizes are available commercially. Scraping should be done in a fume hood for safety purposes. Care is necessary to eliminate losses of adsorbent during transfer from the plate to the vial and from chipping during scraping. The latter can be minimized by lightly spraying the plate with water before scraping (Crosby and Dale, 1985).

The scrapings are shaken with a ''cocktail'' formulation as recommended by Snyder (1964). The recipe of a cocktail for measuring nonpolar compounds is as follows: mix 5.0 g of PPO and 0.3 g of dimethyl-POPOP with 40.0 g of Carb-O-Sil finely dispersed silicic acid suspending agent and add enough toluene to make exactly 1 liter. For measuring polar radioactive compounds, a recommended mixture is 6.5 g of PPO, 0.13 g of POPOP, 104.0 g of naphthalene, 5 ml of toluene, 500 ml of dioxane, 300 ml of methanol, and 32.0 g of Carb-O-Sil. The presence of certain detection or impregnating agents in the layer sorbent can quench the formation of light and reduce sensitivity (Snyder, 1968; Mangold, 1969). Fluorescent indicators present in silica gel F plates were found to interfere with accurate radioactivity quantification because of light-induced photoluminescence when scraped spots were counted (Stocklinski, 1980; Tobias and Strickler, 1981). Compounds that bind strongly to the adsorbent will be counted with lower efficiency than those that are readily dissolved. Commercial scintillation cocktails are available for a wide range of sample types (e.g., Peng, 1981).

As an alternative to counting scraped-off fractions in vials containing the adsorbent, the compound can be eluted from the adsorbent with an appropriate solvent into a vial and measured in solution ("homogeneous counting"). To facilitate elution, the adsorbent should be crushed to a fine powder. Three methods for elution are described by Clark and Klein (1996): removal of the adsorbent followed by elution with solvent, washing the zone with solvent, or immersing the zone in solvent. Recovery of sample is usually above 90%. Another alternative is to cut the radioactive spots from chromatograms developed on plastic- or aluminum-backed foils and to place the sections directly into counting vials without scraping.

As shown in Table 13.2, zonal analysis is among the most sensitive and quantitative TLRC methods, but it is very labor intensive, resolution is limited, and great care is necessary to achieve accurate and reproducible results. It is useful for analysis of samples containing low levels of radioactivity. As with other quantitative methods, standards and blank sections of the layer must be carried through the analytical procedure so that regression lines and correction factors can be determined. If only one or a few sample constituents are to be assessed, the process is much less tedious because these zones can be located on the chromatogram (e.g., by ultraviolet absorption or iodine vapor, which reagent does not appreciably quench scintillation), scraped and collected, and counted rather quickly. Low-level samples can be counted for longer periods of time to collect statistically valid quantitative data. Scintillation counting can be combined with pulse-height analysis for the simultaneous determination of different radionuclides in a sample.

Although liquid scintillation detection is rather common today, its expense may make it unavailable to some scientists. Interested readers are referred to standard textbooks on the subject, for example, Horrochs and Peng (1971).

C. Direct Measurement of Radioactivity Using Radiation Detectors

1. Radioscanners

Radioscanners were introduced in the 1960s for detection and subsequent quantification of radioactive zones. Compared to more recent detectors, speed, sensitivity, and resolution are not high. The chromatogram is scanned by a mechanically driven radiation-sensitive device such as a windowless gas-flow proportional counter or a thin end window Geiger–Muller counter to produce an output that is generally a plot of counts versus position on the lane. The size of the collimator slit is adjustable to control the size of the plate area being measured. The gas flow counter (Karlson et al., 1963) is more sensitive for low-energy β-emitting isotopes, especially tritium. For optimum results with radioscanners, Woods (1968) listed the following requirements: (1) a low scan speed, to enable the maximum number of counts to be recorded; (2) a narrow slit, for maximum peak separation; and (3) the minimum distance between the chromatogram and the detector, to improve resolution and detector efficiency. Mangold (1969) states that direct scanning is an insensitive technique because of the high self-absorption of the radiation and the softness of the glass background and that it is, therefore, an unreliable quantitative procedure. An advantage is that the chromatogram is available for spraying with chemical detection reagents after scanning. Some radioscanners are still available commercially, but they have been mostly replaced by the newer detection systems described below. The miniGITA TLC scanner (Raytest) is equipped with a proportional gas flow counter for beta measurement or a NaI(Tl) crystal scintillation probe with thin window entrance for measurement of gamma-emitting isotopes.

2. Linear Analyzers and Imaging-Proportional Counter Scanners

Linear analyzers have improved sensitivity, resolution, and automation compared to radioscanners. The instruments are computer controlled with software allowing measurement of multiple plates and tracks overnight, storage of data, comparison of chromatograms, and repeated quantification under different conditions. All zones on one chromatogram (track) are measured simultaneously without movement of the detector, and then the computer program can move the head automatically to another position on the layer for another measurement. At each position, the head is lowered to rest on the plate and form a gas-tight counting chamber, air is purged, and argon/methane counting gas is introduced. Two types of imaging counting systems are employed in linear analyzers, one using the resistive anode technique and the second the delay wire technique. The principles of these techniques are described in detail by Clark and Klein (1996).

The Berthold Linear Analyzer LB 285 (Figure 13.1) is a position-sensitive

FIGURE **13.1** Berthold Linear Analyzer LB 285 with IBM-AT computer. [Reproduced with permission of Wiley-Interscience from Filthuth (1990).]

proportional or wire chamber that measures radioactive zones using a detector with or without a thin entrance window, the latter for measurement of low energy ^3H and ^{125}I. The computer-controlled instrument includes a 1024-channel pulse height analyzer; choice of standard, high-sensitivity, or high-resolution detectors; and a 400-mm-wide table for holding TLC plates. The software package provides the following functions, among others: peak search, fit, and integration; background and chromatogram subtraction; and measurement of one- or two-dimensional chromatograms. The ability to measure radioactivity distribution in two dimensions allows the user to view a "digital autoradiograph" of the plate. Reports are generated in the form of counts, peak location, chromatogram plot, bar graph of relative masses, or plot of Gaussian fit of the peaks.

The Raytest RITA is another position-sensitive, proportional-counting thin layer scanner (Figure 13.2). The active width scanned can be limited from 20 mm down to 15, 10.3, and 1 mm by exchange of the diaphragm. Bioscan also manufactures imaging scanners (Rock et al., 1998; Shulman, 1982 and 1983), including the completely redesigned AR-2000.

The Digital Autoradiograph (DAR) LB 287, a two-dimensional, position-sensitive, multiwire proportional counter detector, was discontinued by Berthold, but the instrument is still being used in many analytical laboratories, and papers are published regularly citing its use. The DAR is described in Chapter 13 of the previous edition of this book.

The Packard InstantImager (Figure 13.3) directly images and quantifies radioactivity from TLC plates using a microchannel array detector (MICAD). Activity from ^{14}C, ^{35}S, ^{32}P, ^{33}P, ^{135}I, and other beta- and gamma-emitting radioiso-

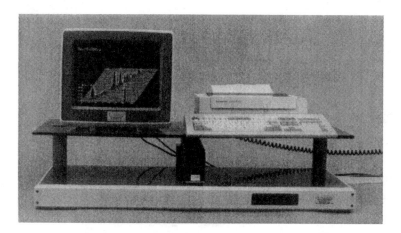

FIGURE 13.2 Raytest Radioactivity Intelligent Thin Layer Analyzer (RITA). (Photograph courtesy of Raytest.)

FIGURE 13.3 InstantImager Electronic Autoradiography System. (Photograph courtesy of Packard Instrument Company.)

topes is imaged with excellent resolution (0.5- or 0.25-mm pixels) by clicking a computer mouse. Speed of detection is 100 times faster than with film and 10 times faster than phosphor screens. The MICAD consists of 210,420 gas-filled cylindrical microchannels of 0.4 mm diameter honeycombed in a 20 × 24 cm laminated multilayer electron avalanche amplification plate that detects, collimates, and preamplifies ionizing radiation for readout with an interpolating multiwire proportional chamber. Beta-, gamma-, and positron-emitting radioisotopes produce ionizations in the microchannels that are electronically amplified and detected by high-speed electronics. The counts in each microchannel are displayed as a radioisotope image in real time, and can be reported during or after image acquisition. Quantitative results are obtained from many samples in less than 30 min. Because the background is low (0.015 CPM/mm^2), acquisition can be performed overnight for low activity samples. Reproducibility of counting is good, linearity is $>10^5$, and results correlate well with liquid scintillation counting. The above information and several applications notes were supplied by Packard Instrument Co.; the principle of operation of the InstantImager is described with a schematic diagram of the microchannels by Clark and Klein (1996).

Digital autoradiography was applied to the study of testosterone metabolites (Vingler et al., 1993), deramciclane metabolism (Hazal et al., 1995), and sebaceous neutral ^{14}C-lipids (Christelle et al., 1997).

3. Spark Chambers

A spark chamber gives a picture of 20 × 20-cm two-dimensional chromatograms containing compounds labeled with radioactive isotopes. The chamber, consisting of two planes of wires (electrodes) inside a chamber filled with ionizable counter gas, is placed directly above the TLC plate. Radioactive emissions from the plate (beta radiation) ionize the gas (90% argon + 10% methane), forming a spark between the two planes of wires. The spatial distribution of radioactivity on the plate is displayed as visible sparks, which are recorded on film with a camera. Only qualitative information can be obtained. The locations of radioactivity on the photograph can be related to positions on the layer using a built-in episcope print projector, thereby facilitating removal of sections for liquid scintillation counting (Clark and Klein, 1996). Spark chambers have been sold by Birchover Instruments (Letchworth, UK) and Raytest (BERTA, Dietzel et al., 1988), but current availability could not be confirmed.

4. Multidetector Radioanalytic Imaging System

The Ambis 4000 radioanalytic imaging system (Scanalytics-CSPI) (Figure 13.4) detects beta particles from 20 × 20 cm plates as much as 100 times faster than conventional autoradiography. The detector consists of 3696 individual elements (66 anode wires × 56 drift chamber tubes), arranged in a honeycomb-type, hexagonal array, each giving a data point. The array is filled with an ionizing gas

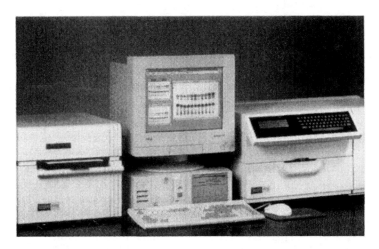

FIGURE 13.4 Ambis 4000 Radioisotopic Imager. (Photograph courtesy of Scanalytics-CSPI.)

mixture (90% argon + 10% methane), which is ionized when a beta particle enters the sensitive volume of the detector. The resulting free electrons are accelerated toward an anode wire in the detector element. As the electrons migrate, they gain sufficient kinetic energy to cause additional ionization in the gas. Within a few microseconds, a cloud of free electrons forms within the detector volume, causing pulses to occur on the anode and cathode of the detector. The ionizing events are amplified and coded for storage in the computer's memory. The plate can be moved through multiple positions that can yield a total of more than one million data points that are displayed as an image on a monitor. A separate detector corrects for background radiation, and zones can be quantified. Resolution and efficiency are controlled by the choice of aperture plate, according to the zone resolution and activity level on the chromatogram. The instrument is computer controlled and includes software for multiple data handling, calculation, and display functions. According to the manufacturer, the dynamic range is greater than five orders of magnitude, which is 1000 times greater than x-ray film, and speed of detection is 100 times greater than x-ray film.

5. Phosphor Imaging or Bioimaging Analyzers

Phosphor imaging systems [e.g., Fuji (available from Raytest), Molecular Dynamics, Packard] employ a sensitive, two-dimensional phosphor storage or imaging plate (IP) coated with a microcrystalline phosphor ($BaFBr:Eu^{+2}$), which is exposed to the layer as in conventional autoradiography. The IP accumulates and stores radioactive energy from beta and gamma emissions from radiolabeled zones on the layer as a latent image. The IP is inserted into the image-reading unit and scanned with a fine infrared laser beam, which stimulates the stored electrons to return to ground state through emission of blue fluorescent light at the locations of the radioactive spots. The fluorescence, the intensity of which is proportional to the recorded radiation intensity, is collected by a fiber-optic bundle or optical system of another type, measured with a photomultiplier tube, and digitized to produce a two-dimensional picture of the chromatogram. Plates for detection of weak (tritium) and stronger beta-emitters are available; they are reused after being erased by exposure to visible light. Exposure is carried out in a holder designed to shield the plate from background radiation. Phosphor imaging analyzers require a significantly shorter exposure time ($\sim$one-tenth) for equivalent response compared with photographic emulsions and have a wide linear range (from 10 to 10,000 dpm for an individual component), sensitivity that is 50–100 times higher than that of the linear analyzer or autoradiography, resolution (50–200 μm) that is better than that of the linear analyzer and MWPC and comparable to autoradiography, quantitative analysis capability is equivalent to zonal analysis, and the capability to count multiple samples simultaneously provides the highest sample throughput.

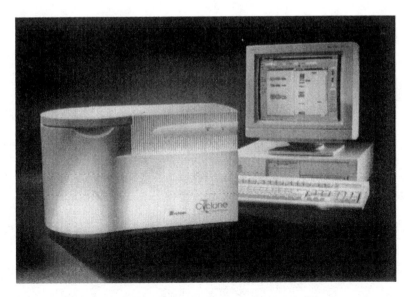

FIGURE 13.5 Cyclone Storage Phosphor System. (Photograph courtesy of Packard Instrument Company.)

TABLE 13.1 Comparison of Radiochromatogram Analysis Instrumentation

	Sensitivity	Resolution	Quantification	Turnaround time	Cost/sample	Health hazard	Sample preserved
1. Autoradiography	Poor	Excellent	Poor	Poor	Low	Low	Yes
2. Spark camera	Good	Moderate	Poor	Excellent	Low	Low	Yes
3. Scintill. counting	Excellent	Poor	Excellent	Moderate	High	High	No
4. Chrom. scanner	Moderate	Moderate	Moderate	Moderate	Low	Low	Yes
5. Chrom. imaging	Good	Moderate	Excellent	Excellent	Low	Low	Yes

For descriptive information on the principles of the phosphor imaging process, see Klein and Clark (1993) and Clark and Klein (1996) and literature available from the companies named above. An example of a commercial instrument is shown in Figure 13.5. The method was applied to studies of drug metabolism (Nagatsuka et al., 1993; Ueda et al., 1993), iodothyronine deiodinase activity (Koopdonk-Kool et al., 1993), and citrulline synthesis by nitric oxide synthase activity (Vyas et al., 1996). The synthesis of cholesterol in rat liver was studied using a combination of bioimaging and liquid scintillation counting (Okuyama et al., 1995).

III. COMPARISON OF METHODS

Comparisons of the various TLRC methods are summarized in Tables 13.1 (Shulman, 1983; Shulman and Weaner, 1991) and 13.2 (Clark and Klein, 1996). Linear analyzers are still widely used in many laboratories because they offer reasonable speed, resolution, sensitivity and quantitative accuracy that is adequate for many applications and are less expensive than the newer detection instruments. When the highest level of sensitivity, quantification, and resolution are required, use of a multiwire proportional counter or phosphor imager is in order despite their high price. The field appears to be moving rather quickly toward much wider use of phosphor imagers, and it is likely this trend will continue at the expense of proportional counters.

TABLE 13.2 Comparison of Thin-Layer Radiochromatographic Analysis Techniques

Parameters	Autoradiography	Zonal analysis	Linear analyzer	MWPC detector	Phosphor imager
Sensitivity	+	+++++	++	+++++	+++++
Resolution	+++++	+	++	++++	+++++
Quantitation	+	+++++	++++	+++++	+++++
Dynamic range	+++	+++	+++	+++++	+++++
Speed	+	++	+++++	+++++	++++
Sample throughput	+	+	+++	++++	+++++
Preserves sample	Yes	No	Yes	Yes	Yes

+++++ = Excellent
++++ = Very good
+++ = Good
++ = Satisfactory
+ = Poor.
Source: Clark and Klein (1996), reprinted with permission of Marcel Dekker, Inc.

REFERENCES

Amersham Corp., (1988). *Research Products Catalog, 1988/9*, Arlington Heights, IL, pp. 188–191.

Bartelsman, B. W., Ohringer, S. L., and Frost, D. L. (1982). Improved methods for detection of low energy beta emitters in TLC. In *Advances in TLC—Clinical and Environmental Applications*, J. C. Touchstone (Ed.). Wiley-Interscience, New York, Chap. 9, pp. 139–148.

Bochner, B. R., and Ames, B. N. (1983). Sensitive fluorographic detection of ^{3}H and ^{14}C on chromatograms using methyl anthranilate as a scintillant. *Anal. Biochem. 131:* 510–515.

Christelle, C., Vingler, P., Boyera, N., Galey, I., and Bernard, B. A. (1997). Direct quantitative digital autoradiography TLC coupling for the analysis of neutral ^{14}C-lipids neosynthesized by the human sebaceous gland. *J. Planar Chromatogr.—Mod. TLC 10:* 243–250.

Clark, T., and Klein, O. (1996). Thin layer radiochromatography. In *Handbook of Thin Layer Chromatography*, 2nd ed., J. Sherma and B. Fried (Eds.). Marcel Dekker, New York, pp. 341–359.

Crosby, S. D., and Dale, G. L. (1985). Simplified procedure for the quantitation of radioactive phosphoinositides by thin layer chromatography. *J. Chromatogr. 323:* 462–464.

Dallas, F. A. A., Read, H., Ruane, R. J., and Wilson, I. D. (Eds.) (1988). *Recent Advances in Thin Layer Chromatography*, Plenum Press, New York, pp. 77–135.

Dietzel, G. (1992). Personal communication dated September 3, 1992, Raytest Inc.

Dietzel, G., Kubisiak, H., and Seltzer, H. (1988). Development of 2-dimensional detectors for radio-thin layer chromatography. In *Recent Advances in Thin Layer Chromatography*, F. A. A. Dallas, H. Read, R. J. Ruane, and I. D. Wilson (Eds.). Plenum, New York, pp. 117–123.

DuPont. (1988a). *DuPont Guide to Radiochemical Storage and Handling*, Publication H-02631, DuPont/NEN Research Products, Boston, MA.

DuPont (1988b). *A Guide to Fluorography*, Publication E-92332, DuPont/NEN Research Products, Boston, MA.

Eastman Kodak Co. (1986). *Autoradiography of Macroscopic Specimens*, Publication M3-508, Eastman Kodak Co., Rochester, NY.

Eastman Kodak Co. (1988). *Kodak Products for Medical Diagnostic Imaging*, Publication M5-15, Eastman Kodak Co., Rochester, NY.

Filthuth, H. (1989). A new detector for radiochromatography and radio-labeled multi-sample distributions—the digital autoradiograph. *J. Planar Chromatogr.—Mod. TLC 2:*198–202.

Filthuth, H. (1990). Detection of radioactivity distribution with position-sensitive detectors, linear analyzer, and digital autoradiograph. *Chem. Anal. (NY) 108:*167–183.

Gattavecchia, E., Tonelli, D., Breccia, A., Fini, A., and Ferri, E. (1994). Radio TLC in quality control and development of radiopharmaceuticals. *J. Radioanal. Nucl. Chem. 181:* 77–84.

Hazal, I., Urmos, I., and Klebovich, I. (1995). Application of TLC–digital autography as a rapid method in a pilot study of deramciclane metabolism. *J. Planar Chromatogr.—Mod. TLC 8*: 92–97.

Horrochs, D. L., and Peng, C.-T. (Eds.) (1971). *Organic Scintillators and Liquid Scintillation Counting*. Academic Press, New York.

Inoue, C., Aijima, K., Nakamura, Y., Kaburagi, T., and Shigematsu, A. (1997). In vitro method for assessing hepatic drug metabolism. *Biol. Pharm. Bull. 20*: 1–5.

Karlson, P., Mauer, R., and Wenzel, M. (1963). A micromethod for labeling steroids and ecdysone with tritium. *Z. Naturforsch. 18*: 219–224.

Klein, O., and Clark, T. (1993). The advantages of a new bioimaging analyzer for investigation of the metabolism of ^{14}C-radiolabeled pesticides. *J. Planar Chromatogr.—Mod. TLC 6*: 368–371.

Koopdonk-Kool, J. M., van Lopik-Peterse, M. C., Veenboer, G. J. M., Visser, T. J., Schoenmakers, C. H. H., and de Vijlder, J. J. M. (1993). Quantification of type III iodothyronine deiodinase activity using thin layer chromatography and phosphor screen autoradiography. *Anal. Biochem. 214*: 329–331.

Koskinen, M., Rajaniemi, H., and Hemminki, K. (1997). Analysis of tamoxifen-induced DNA adducts by ^{32}P-postlabelling assay using different chromatographic techniques. *J. Chromatogr., B: Biomed. Sci. Appl. 691*: 155–160.

Luthi, U., and Waser, P. G. (1965). Low temperature fluorography induced by tritium labeled compounds on thin layer chromatograms. *Nature 205*: 1190–1191.

Mallol, J., and Bonino, C. (1997). Comparison of radiochemical purity control methods for technetium-99m radiopharmaceuticals used in hospital radiopharmacies. *Nucl. Med. Commun. 18*: 419–422.

Mangold, H. K. (1969). Isotope techniques. In *Thin Layer Chromatography. A Laboratory Handbook*, 2nd ed., E. Stahl (Ed.). Springer-Verlag, New York, pp. 155–180.

Matsunaga, H., Isobe, N., Kaneko, H., Nakasuka, I., and Yamane, S. (1997). Metabolism of Flumiclorac pentyl in rats. *J. Agric. Food Chem. 45*: 501–506.

Nagatsuka, S.-i., Ueda, K., Ninomiya, S.-i., and Esumi, Y. (1993). Application of bioimage analyzer on drug metabolism studies using thin layer chromatography. (I). Performance of the system coupled to an image analyzer. *Yakubutsu Dotai 8*: 1261–1271.

Okuyama, M., Tsunogai, M., Watanabe, N., Asakura, Y., and Shigematsu, A. (1995). Study of the de novo synthesis of cholesterol in rat bile: a newly developed radiotracer technique, "TLC-autoradioluminography." *Biol. Pharm. Bull. 18*: 1467–1471.

Peng, C. T. (1981). *Sample Preparation in Liquid Scintillation Counting*, Amersham Corp., Arlington Heights, IL.

Poole, C. F., and Poole, S. K. (1991). *Chromatography Today*, Elsevier, Amsterdam, pp. 649–734.

Prydz, S. (1973). Summary of the state of the art in radiochromatography. *Anal. Chem. 45*: 2317–2326.

Randerath, K. (1969). Improved procedure for solid scintillation fluorography of tritium labeled compounds. *Anal. Chem. 41*: 991–992.

Randerath, K. (1970). Analysis of nucleic acid derivatives at the sub-nanomole level. VI. Evaluation of film detection methods for weak beta emitters, particularly tritium. *Anal. Biochem. 34*: 188–205.

Rapkin, E. (1990). One- and two-dimensional scanning for phosphorus-32 and other uncommon tags. *Chem. Anal. (NY) 108*: 127–155.

Roberts, T. R. (1978). *Radiochromatography: The Chromatography and Electrophoresis of Radiolabeled Compounds*. Elsevier, New York.

Rock, C. O., Jackowski, S., and Shulman, S. D. (1988). Imaging scanners for radiolabeled thin layer chromatography. *BioChromatography 3*: 127–129.

Shapiro, J. (1972). *Radiation Protection, A Guide for Scientists and Physicians*. Harvard University Press, Cambridge, MA.

Shulman, S. D. (1982). Quantitative analysis by imaging radiation detection. In *Advances in TLC—Clinical and Environmental Applications*. J. C. Touchstone (Ed.). Wiley-Interscience, New York, Chap. 8, pp. 125–137.

Shulman, S. D. (1983). A review of radiochromatogram analysis instrumentation. *J. Liq. Chromatogr. 6*: 35–53.

Shulman, S. D., and Weaner, L. E. (1991). Thin layer radiochromatography. In *Handbook of Thin Layer Chromatography*, J. Sherma and B. Fried (Eds.). Marcel Dekker, New York, pp. 317–338.

Smith, I. (1988). The Ambis two dimensional beta scanner. In *Recent Advances in Thin Layer Chromatography*, F. A. A. Dallas, H. Read, R. J. Ruane, and I. D. Wilson (Eds.). Plenum Press, New York, pp. 77–86.

Snyder, F. (1964). Radioassay of thin layer chromatograms: A high resolution zonal scraper for quantitative ^{14}C and ^{3}H scanning of thin layer chromatograms. *Anal. Biochem. 9*: 183–196.

Snyder, F. (1968). Thin layer chromatography radioassay: A review. *Adv. Tracer Methodol. 4*: 81–104.

Snyder, F. (1969). Review of instrumentation and procedures for ^{14}C and ^{3}H radioassay by thin layer chromatography and gas liquid chromatography. *Isotop. Radiat. Technol. 6*: 381–400.

Snyder, F., and Piantadosi, C. (1966). Labeling and radiopurity of lipids. *Lipid. Res. 4*: 257–283.

Stewart, D. C. (1981). *Handling Radioactivity, A Practical Approach for Scientists and Engineers*, John Wiley and Sons, New York.

Stocklinski, A. W. (1980). Photoluminescence of silica gel GF as a potential problem in liquid scintillation counting. *Anal. Chem. 52*: 1005–1006.

Tal, A., and Rubin, B. (1993). Metabolism of EPTC by a pure bacterial culture isolated from thiocarbamate-treated soil. *Pestic. Sci. 39*: 207–212.

Tobias, B., and Strickler, R. C. (1981). Photoactivated zinc silicate in TLC plates: a potential cause for error in liquid scintillation counting. *Steroids 37*: 213–221.

Touchstone, J. C. (1992). *Practice of Thin Layer Chromatography, 3rd ed*. Wiley-Interscience, New York, pp. 230–255.

Ueda, K., Ninomiya, S.-i., Esumi, Y., Shimada, N., and Nagatsuka, S.-i. (1993). Application of bioimage analyzer on drug metabolism studies using thin layer chromatography. (I). Direct development of plasma samples containing [^{14}C] indomethacin and its metabolites. *Yakubutsu Dotai 8*: 1273–1282.

Vingler, P., Filthuth, H., Bague, A., Pruche, F., and Kermici, M. (1993). Direct quantitative digital autoradiography of testosterone metabolites in the pilosebaceous unit: an environmentally advantageous trace radioactive technology. *Steroids 58*: 429–438.

Vyas, P., Attur, M., Ou, G.-M., Haines, K. A., Abramson, S. B., and Amin, A. R. (1996). Thin layer chromatography: an effective method to monitor citrulline synthesis by nitric oxide synthase activity. *Portland Press Proc. 10* (Biology of Nitric Oxide, Part 5): 44.

Wilson, A. T. (1958). Tritium and paper chromatography. *Nature 182*: 524.

Wood, B. A. (1968). Quantitative scanning of radioactive thin layer chromatograms. In *Quantitative Paper and Thin Layer Chromatography*, E. J. Shellard (Ed.). Academic Press, New York, pp. 107–122.

Zamora, P. O., Stratesteffan, M., Guhlke, S., Sass, K. S., Cardillo, A., Bender, H., and Biersack, H. J. (1996). Radiotracer binding to brain microsomes determined by thin layer chromatography. *Nucl. Med. Biol. 23*: 61–67.

14

Basic TLC Design and TLC of Organic Dyes

I. BASIC TLC DESIGN

Basic TLC design involves spotting a plate, developing the plate in a suitable solvent, and detecting the spots after development. When substances such as dyes are used in the TLC process, no detection reagent is needed because the dyes are chromogenic, that is, have intrinsic color.

Basic TLC design is illustrated here with the aid of a three-component lipophilic dye mixture (Mallinkrodt dye #3082, containing 1 μg/μl of each dye), containing butter yellow, Sudan red G, and indophenol blue. Various other lipophilic dyes are available commercially, including four- to six-component mixtures. For further information, consult the Analtech catalog. One of Stahl's (1969) major contributions to TLC was the introduction of standard dyes to test basic TLC procedures.

In the recommended experiment, either 1, 5, or 10 μl of dye is applied as a spot or a streak to the origin of a TLC plate or sheet. The origin is set at 2 cm from the bottom of the plate. Each spot or streak is applied 2–3 cm apart. Application of spots or streaks is made with 1-, 5-, or 10-μl Microcap micropipets or with a 10-μl digital microdispenser (Drummond Scientific Co.). For applications of 5 or 10 μl, solvent is dried during sample application with the aid of a hair dryer. The solvent front line is drawn lightly with pencil on each plate or sheet, 10 cm from the origin. Two different commercially precoated plates may be used as follows: a glass-backed Merck silica gel 60, 20 × 20-cm plate and a plastic-backed silica gel IB 2 Baker-flex 20 × 20-cm plate.

The tank used for development is the Kontes Chromaflex N-Tank, and the mobile phase is "Baker Analyzed" reagent-grade toluene, 100 ml per tank. Two separate tanks are prepared as follows: One tank employs the saturated mode and is lined with filter paper (Chapter 7). The toluene is poured over the filter-paper, the lid is placed on the tank, and the tank is allowed to saturate (Chapter 7) before plates are placed inside. For the unsaturated mode, plates and toluene are placed in an unlined tank, and then the lid is placed on the tank. Each tank contains two plates, one plate being the Merck glass plate and the other the Baker-flex plastic sheet. The plates are developed for a distance of 10 cm from the origin and are then removed; the spots are examined and R_F values determined.

The results of all separations are shown in Table 14.1. Separation of all spots was achieved in both the saturated and unsaturated mode. Relatively small differences in separations were seen on the two different commercial plates. The shape after development of the streak assumed a "band effect," as opposed to the "circular effect" from the spot application. The distance between bands is usually greater than the distance between spots, indicating that for some separations the application of streaks may be advantageous for resolving substances that may migrate close to each other. In this experiment, the unsaturated mode produced better overall resolution of spots than development in the saturated mode.

This experiment is presented as a basic exercise that can be useful for introducing workers with little or no experience in TLC to techniques such as spotting,

TABLE 14.1 Separation of a Three-Dye Mixture on Merck Silica Gel 60 and Baker-flex IB 2 Plates Using Saturated and Unsaturated Modes of TLC

| | Average R_F values | | | |
| | Saturated mode | | Unsaturated mode | |
Component	IB 2	Merck	IB 2	Merck
Indophenol blue	0.14	0.15	0.08	0.08
Sudan red G	0.22	0.23	0.20	0.21
Butter yellow	0.45	0.48	0.57	0.58

Note: Development was at 28°C at a relative humidity of about 60%. The solvent consisted of toluene and the solvent front distance was 10 cm. Development time in the unsaturated mode was 38 min for the IB 2 plate and 34 min for the Merck plate. Development time in the saturated mode was 21 min for the IB 2 plate and 20 min for the Merck plate.

development, and determination of R_F values. Similar results should be obtained on other commercial silica gel layers—for example, Whatman or Analtech. Bauer et al. (1991) described a similar experiment and showed a chromatogram of the TLC separation of a mixture of lipophilic dyes (indophenol blue, Sudan red G, and 4-dimethylaminoazobenzene) on silica gel 60 sheets. Note that the common name of dimethylaminoazobenzene is butter yellow.

Of interest in regard to basic TLC design is an experiment that allows organic chemistry students to either determine or verify (depending on how much information they are given prior to the lab) the relative R_F characteristics of various functional groups. Such an experiment has been provided by Beauvais and Holman (1991) to introduce students to TLC and also familiarize them with the basis of organic functional group chemistry. The proposed experiment illustrates the concepts of relative dipole moments, polarity, and intermolecular forces for a series of organic functional group classes. The six functional group classes considered in this experiment are alkanes, alcohols, esters, carboxylic acids, ketones, and alkenes. Materials needed by the students to perform the experiments are well-described in the paper. Moreover, this experiment requires the use of UV light and/or iodine for visualization (detection) of the compounds. Since most organic compounds are detected this way (most organics are not highly colored as is the case with dyes), it will be good experience for students to make use of the above-mentioned visualization techniques.

Reynolds and Comber (1994) described how organic chemists can use TLC for separation, characterization, and isolation of organic compounds. They used a colorful demonstration that involved the separation of combinations of seven highly colored dyes by TLC. The dyes used were methylene blue, new methylene blue, basic fuchsin, methyl red, rhodamine B, bromocresol green, and fluorescein. The dyes were dissolved in ethanol and the solvent for TLC was isopropanol–acetic acid (15:2). Samples were applied with 1 μl micropipets on 20 × 20 cm silica gel G precoated glass plates and the plates were developed in a 27.5 × 27.5 × 7.5 cm TLC developing tank. Figures and tables were given in the paper that describe the templates needed to perform this rather unusual exercise. As a result of this experiment, students can spell out anything they desire. Reynolds and Comber (1994) concluded that dye writing is recommended as a medium to captivate an audience, and that this is a colorful and memorable way to show the meaning of separation to an organic chemist. Moreover, this technique is helpful in making a rather mundane coverage of fundamentals of TLC more exciting to a student audience.

II. TLC OF ORGANIC DYES

Organic dyes can be used to illustrate simple thin-layer chromatographic separations on silica gel. Two examples of such separations are given below. The first involves readily available dyes and is suitable for an elementary chemistry lab. The second experiment is with the organic dye Sudan black B, used in lipid histochemistry. This dye is usually available in most biology departments.

Scism (1985) used the following simple organic dye experiment to introduce TLC to undergraduates in organic and biochemistry courses. The dye mixture consisted of a 0.17% (w/v) solution of rhodamine B, methylene blue, and fluorescein in 95% ethanol. Sample was spotted onto 5×20-cm 250-μm-thick silica gel plates, Anasel Type O (Analabs, Inc.), and the solvent system consisted of acetone–n-propanol–water $(3:2:1)$. Presumably, the spotted plates were placed immediately into chambers or tanks and development time was about 25 min. The spots were resolved as follows: methylene blue (blue), $R_F = 0.01$; rhodamine B (pink red), $R_F = 0.51$; fluorescein (yellow), $R_F = 0.97$. For good descriptions of TLC separations of various other organic dyes, see the applications section in Bauer et al. (1991).

Biological stains often contain organic dyes. The Sudan dyes are organic dyes used to detect lipids by histochemical techniques. According to Humason (1979), Sudan black B is considered the most sensitive of lipid dyes. TLC on silica gel has been used to separate the components of Sudan black B (Lansink, 1968). Long standing or aging of the dye solution may result in extra components associated with certain staining characteristics (Frederiks, 1977).

To demonstrate the effects of aging on a Sudan black B solution in 70% ethanol, the following experiment can be done. Prepare 200 mg of Sudan black B (Merck) in 100 ml of 70% ethanol. Stir the mixture and then filter it. Use some of the mixture fresh and let some age for 2 months at room temperature. Use silica gel G (Merck) and a solvent system of chloroform–benzene $(1:1)$. Spot fresh and aged solution on a plate and use ascending one-dimensional TLC. The fresh solution should resolve into five spots (Figure 14.1) with components 5 and 7 associated with SBBI (stains neutral lipids) and SBBII (stains phospholipids), respectively. The aged solution should contain additional components (3 and 4, Figure 14.1). According to Frederiks (1977), the extra components are associated with certain staining characteristics. Therefore, aged stain may react differently on tissues than freshly prepared material.

Two-dimensional TLC can be used with organic dyes to show the effects of thermal decomposition on the dyes. A six-component test dye was made up in toluene at an approximate concentration of 0.2% per dye. Five microliters of the dye was spotted at the origin (Figure 14.2) of a Baker-flex IB 2 plate. The layer was placed in a rectangular tank containing benzene as the mobile phase and developed for a distance of 10 cm in the first direction (45 min development time). The plate was air dried between runs and developed in the same solvent in the second direction, also for 10 cm (45 min development time). The six spots fell along a diagonal and were resolved as follows: P, purple; Y, yellow, B, blue; BP, blue-purple; GY, gold-yellow; R, red (Figure 14.2). To show the effects of thermal decomposition on the dye, a second layer was spotted as described above

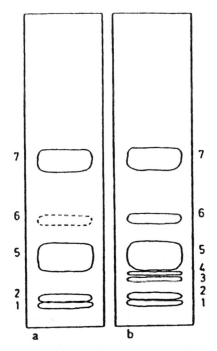

FIGURE 14.1 Drawings of chromatograms showing TLC separations of a fresh (left) and a 2-month-old (right) 0.1% Sudan black B solution in 70% ethanol. The sorbent was silica gel G and the mobile phase was chloroform–benzene (1:1). Note the extra components, 3 and 4, in the aged solution on the right. [Reproduced with permission of Springer-Verlag from Frederiks (1977).]

and heated at 140°C for 60 min between runs. The original spots have been circled completely and the decomposition spots have been incompletely circled (Figure 14.3). Note the presence of 10 additional spots as a result of thermal decomposition.

The improved resolving power of two-dimensional TLC with different mobile phases also can be illustrated using a multicomponent dye mixture. A Baker-flex IB 2 sheet was cut to a 10 × 10-cm size. Five microliters of dye mixture was applied in the lower left corner of the layer (O_1), and lines were drawn lightly with a pencil 7 cm above and to the right of the origin (F_1 and F_2) (Figure 14.4). Another 5 μl mixture spot was applied on the right edge of the sheet (O_2). The layer was developed for a distance of 7 cm with methylene chloride in a paper-lined, equilibrated chamber, which required about 7 min. The separation into seven zones, shown on the right side of the plate in Figure 14.4, was obtained. The layer was air dried thoroughly, a 5-μl spot of mixture was applied above

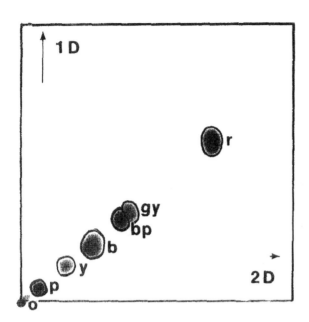

FIGURE 14.2 Photocopy of a two-dimensional chromatogram developed in both directions with benzene and air dried between runs. Note the components of the six-dye mixture on a diagonal line. O, origin; P, purple; Y, yellow; B, blue; BP, blue-purple; GY, gold-yellow; R, red; 1-D, first dimension; 2-D, second dimension.

the first solvent front (O_3), and the layer was rotated 90° and developed for a distance of 7 cm in chloroform (15 min). The separation in the second dimension, shown at the top of Figure 14.4, is in a different sequence than the first direction, because the selectivity of the two solvents differs. This is expected because these solvents are from different groups of the Snyder selectivity classification (Chapter 6, Section I.D), methylene chloride group V (dipole interactions) and chloroform in group VIII (proton donors). The two-dimensional separation resulting from development in both solvents resolved all major dye components completely, and also approximately six other trace components. The two-dimensional separation is clearly more complete than either of the one-dimensional runs. Because the layer was air dried in the dark between developments, the extra zones are undoubtedly natural components of dye mixture rather than decomposition products as shown in Figure 14.3.

Gupta (1996) has provided an extensive chapter on the separation of synthetic dyes by TLC. Included in this chapter is mention of TLC separations of organic dyes as follow: use of silica gel G plates for the separation of Martius

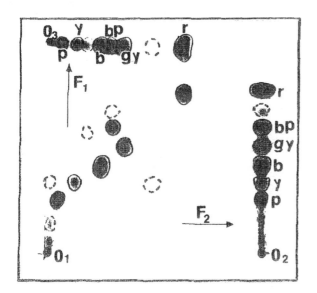

FIGURE 14.3 Photocopy of a two-dimensional chromatogram developed as described in Figure 14.2, but heated at 140°C for 60 min between runs. Note the incompletely circled spots, which are thermal decomposition products. Abbreviations as described in Figure 14.2.

yellow, dimethylaminoazobenzene, Ceres yellow, Ceres orange, Sudan red G, Ceres red BB, and indophenol; the separation of 15 fat-soluble dyes on silica gel layers in 12 different solvent systems (e.g., Sudan G, azobenzene, *p*-aminobenzene, butter yellow, *p*-methoxyazobenzene and *p*-hydroxyazobenzene resolved in 1,2 dichloroethane); xylene for the separation of oil yellow AB and oil yellow OB and chloroform to separate quinolene yellow SS from the remainder of the dyes.

Gupta (1996) considered the use of TLC to examine impurities in histological stains. He mentioned studies on impurities in xanthene stains for histological work using silica gel G layers and a solvent system of *n*-propanol–formic acid (8:2); he noted a study that investigated impurities in 11 Alcian blue and Alcian green samples by spotting samples on cellulose and developing the plates in *n*-butanol–water–acetic acid (3:3:1). Moreover, Gupta (1996) reported that in studies using various chromatographic techniques to resolve histological dye components, TLC on silica gel HF_{254} proved to be the best technique for these separations compared to studies using paper chromatography and thin-layer electrophoresis on agar.

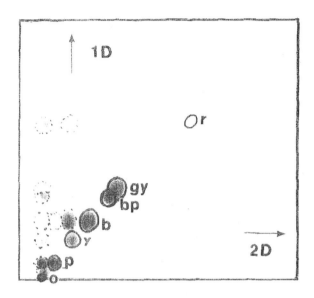

Figure 14.4 Photocopy of a two-dimensional separation of a multicomponent dye mixture applied at O_1 and developed with methylene chloride for 7 cm (to F_1) and then with chloroform for 7 cm (to F_2). One-dimensional separations of the mixture, applied at O_2 and O_3, are shown for comparison.

REFERENCES

Bauer, K., Gros, L., and Sauer, W. (1991). *Thin Layer Chromatography—An Introduction.* Huthing Buch Verlag GmbH, Heidelberg.

Beauvais, R., and Holman, R. W. (1991). An internal comparison of the intermolecular forces of common organic functional groups—a thin-layer chromatography experiment. *J. Chem. Educ. 68*: 428–429.

Frederiks, W. M. (1977). Some aspects of the value of Sudan black B in lipid histochemistry. *Histochemistry 54*: 27–37.

Gupta, V. K. (1996). Synthetic dyes. In *Handbook of Thin Layer Chromatography*, 2nd ed., J. Sherma and B. Fried (Eds.). Marcel Dekker, New York, pp. 1001–1032.

Humason, G. L. (1979). *Animal Tissue Techniques*, 4th ed. W. H. Freeman and Co., San Francisco.

Lansink, A. G. W. (1968). Thin layer chromatography and histochemistry of Sudan black B. *Histochemie 16*: 68–84.

Reynolds, R. C., and Comber, R. N. (1994). The ABC's of chromatography. *J. Chem. Educ. 71*: 1075–1077.

Scism, A. J. (1985). TLC of organic dyes in undergraduate labs. *J. Chem. Educ. 62*: 361.

Stahl, E. (Ed.). (1969). *Thin Layer Chromatography*, 2nd ed. Springer-Verlag, New York.

15

Lipids

I. DEFINITION

There is no universal definition of the term "lipids," but various lipidologists have provided useful definitions for workers interested in the chromatographic analysis of these compounds. Thus, Kates (1986) considered lipids as compounds generally insoluble in water, but soluble in a variety of organic solvents, e.g., ether, hexane, chloroform. He recognized various classes of lipids including hydrocarbons, alcohols, aldehydes, fatty acids, and derivatives such as glycerides, wax esters, phospholipids, glycolipids, and sulfolipids. His consideration of lipids also included the fat-soluble vitamins and their derivatives, carotenoids, sterols, and their fatty acids.

Christie (1987) noted that a variety of diverse compounds generally soluble in organic solvents are usually classified as lipids, i.e., fatty acids and their derivatives, steroids, terpenes, carotenoids, and bile acids. He suggested that many of these diverse compounds have little in the way of structure or function to make them related and that many substances regarded as lipids may be more soluble in water, e.g., glycolipids and gangliosides, than in organic solvents. Fried (1991, 1996) provided a list of numerous diverse lipophilic substances that have been examined by TLC and included typical sorbents, solvent systems, and references for these substances.

This chapter is concerned mainly with the more restrictive definition of lipids following Christie (1982). A convenient system of classification of lipids based on his scheme considers the *simple* lipids (compounds that on hydrolysis yield no more than two types or primary products/mole), also referred to as *neutral* or *apolar* lipids, and *polar* or *complex* lipids (compounds that on hydrolysis yield three or more primary products/mole). Complex lipids are the glycerophospho-lipids (or simply phospholipids) and the glycolipids (also termed glyceroglyco-lipids and glycosphingolipids), including gangliosides.

From the standpoint of practical TLC, this chapter considers the following simple lipids: free sterols; acyl-, diacyl-, and triacylglycerols; free fatty acids; methyl esters; sterol esters; and the following polar lipids: phosphatidyl choline, lysophosphatidyl choline, phosphatidyl ethanolamine, phosphatidyl serine, phosphatidyl inositol, sphingomyelin, cerebrosides, and sulfatides.

Gunstone and Henslof (1992) wrote an excellent glossary of lipid terminology with simple definitions of 900 terms. They considered lipids as ". . . compounds based on fatty acids or closely related compounds such as the corresponding alcohols or the sphingosine bases." They included in their glossary the names of fatty acids and lipids, the major oils and fats, terms associated with their analysis, refining, and modification, and the major journals and societies concerned with lipid chemistry.

II. FUNCTION

Lipids play an important role in many biological functions associated with plants, animals, and microbial organisms. The numerous functions of lipids have been studied at least in part with analytical and preparative liquid chromatographic methods. In the discussion that follows, references are given whenever possible to works that use TLC as a method of analysis.

Lipids are important as storage materials for energy reserve (Allen, 1976). In mammals and birds, the storage depot is mainly in the form of adipose tissue (Vague and Fenasse, 1965) and contains triacylglycerols along with lesser amounts of free fatty acids and mixed glycerides. In sharks, skates, and rays (the elasmobranchs or cartilaginous fishes), the fat deposits consist mainly of squalenes and glyceryl ethers, which are of a lower density than triacylglycerols and contribute to the buoyancy of these fishes (Malins and Wekell, 1970). In the teleosts (bony fishes), lipids are deposited in the liver, bone marrow, and muscles (Phleger et al., 1976). The presence of lipids in the bone of the teleost *Helicolenus dactylopterus lahiller* (blackbelly rosefish) was studied by Mendez et al. (1993). They determined the bone lipid composition of this fish by TLC combined with scanning densitometry. The study also used histological sections of bone to clarify the sites of lipid deposition in bone. Findings from the study suggest that lipid functions both as a hydrostatic agent and energy reserve in the blackbelly rosefish. Less information is available on the storage sites of lipids in invertebrates (Lawrence, 1976). During starvation, both vertebrates and invertebrates utilize lipids as an energy reserve. Duncan et al. (1987) used TLC to show that the medically important planorbid snail, *Biomphalaria glabrata*, depleted mainly triacylglycerols and free fatty acids during starvation. Fried et al. (1991) used TLC and transmission electron microscopy to show that lipid storage and accumulation occurred in the digestive gland cells of this snail.

Lipids are important in the structural integrity of cells and comprise the major components of all membranes (Vandenheuvel, 1971). Lipids of importance in these roles are sterols, phosphoglycerides, glycolipids, and sphingolipids.

Complex lipids in the membranes of neuronal tissues are involved in the transmission of electrical signals. Phosphoinositides and their metabolic products play a role in cellular chemical communication, and TLC has been used to study phosphoinositide metabolism in resting and stimulated cells (Mahadevappa and Holub, 1987).

Lipids are important as pheromones, precursors of pheromones, or carriers of pheromones in plants and animals. The topic has been reviewed for vertebrates and insects by Shorey (1976) and for invertebrates (mainly helminths) other than insects by Haseeb and Fried (1988). Lipophilic pheromones or their carriers are mainly glycerides, free fatty acids, and sterols. Insects excrete long-chain alcohols, alkyl acetates, aldehydes, and ketones that serve as intraspecific pheromones (Mahadevan and Ackman, 1984).

Lipids also function as antimicrobial agents in the skin of mammals (Nicolaides, 1974), antidesiccants in the cuticle of insects (Holloway, 1984), and bioacoustic lenses in dolphins (Ackman et al., 1975). A special structure known as the elaisome located in the seeds of numerous species of plants release lipids, particularly diacylglycerols, that attract ants (Marshall et al., 1979). A study by Skidmore and Heithaus (1988) used TLC to show that ants responded rapidly to the elaisomes or the diacylglycerol fraction of the elaisomes of seeds of the perennial herb *Hepatica americana*.

Christie (1987) discussed various functions of lipids as follow: their involvement in disturbances of lipid metabolism associated with specific lipid disorders; accumulation of various neutral, complex, and conjugated lipids in coronary artery and heart disease; the role of lipids in nutrition, disease, and human welfare; the importance of fats and oils as agricultural products and as major items in international trade; the role of fats as a major dietary component and supplier of calories for humans in developed countries; and the contribution that fats make to the taste and structure of foods. Ando and Saito (1987) contributed a review on TLC and HPTLC lipid analysis of normal and pathological tissues associated with specific lipidoses and gangliosidoses.

Glycolipids play a vital role in cellular metabolism, and TLC has been used in part to elucidate their functions (Breimer et al., 1981). Located mainly at the external surfaces of cell membranes, these lipids help to regulate cell growth and serve as receptors for toxins, hormones, viruses, and other substances. They also serve as differentiation markers and co-factors for ion transport (Hakomori,

1981), to modulate immune responses (Mehra et al., 1984), or possibly as antigens (Hunter and Brennan, 1981).

Gurr and Harwood (1991) summarized the functions of lipids as structural, storage, and metabolic. The structural lipids are important in membranes and at surfaces where they serve as barriers between one environment and another. The storage lipids, with their high energy density of triacylglycerols, are convenient for long-term storage of fuel. Relative to lipids in metabolic control, their barrier and storage functions take advantage of the bulk properties of lipids. Moreover, as individual molecules, lipids play a role as chemical messengers and participate in the control of metabolism.

Brouwers et al. (1996) have discussed some interesting functions of lipids in parasitic worms. For instance, the surface of adults of the human blood fluke *Schistosoma mansoni* consists of two closely apposed lipid bilayers (double bilayers), an apparent morphological and functional adaptation to parasitism. This membrane complex provides an effective tool for defeating the host's immune system in various ways. Brouwers et al. (1996) also presented a schematic overview of lipid metabolism in adult parasitic worms of interest because these parasites cannot synthesize fatty acids or cholesterol de novo and must obtain these lipids from the host.

III. APPLICATIONS OF TLC TO LIPIDS

The literature on the applications of TLC to lipids is extensive, and salient studies from 1986 to mid-1993 were reviewed by Fried and Sherma (1994). A summary of representative articles on applications of TLC to lipids from late 1993 to 1997 is as follows.

Dunstan et al. (1993) used the silica gel–coated quartz rod TLC/FID technique to examine the effect of lypholization on the solvent extraction of lipid classes, fatty acids, and sterols from Pacific oysters *Crassostrea gigas*. St. Angelo and James (1993) used TLC/FID to quantify lipids from cooked beef. Nickolova-Damyanova et al. (1993) used argentation TLC to separate isomeric triacylglycerols on silica gel impregnated with low concentrations of silver ion; development was with chloroform–methanol in open cylindrical tanks. Gerin and Goutx (1993) used TLC/FID to separate and quantify phospholipids from marine bacteria. Linard et al. (1993) used overpressured layer chromatography (OPLC) to separate the most common classes of phospholipids in animal lipid extracts. Cartwright (1993) reviewed methods for the extraction, purification, separation, and quantification of membrane phospholipids by TLC. Hamid et al. (1993) combined TLC with immunostaining to analyze glycolipids and mycolic acids from *Mycobacterium* species and strains. Ogawa-Goto et al. (1993) used TLC and TLC–immunostaining to analyze neutral and sulfated glucuronyl glycosphingolipids purified from human motor and sensory nerves. Wardas and Pyka (1993) described three visualizing agents (thymol blue, bromothymol blue, and bromophenol blue) useful for detecting certain glycerides and fatty acid derivatives after their separation by TLC.

Sajbidor et al. (1994) applied TLC to analyze baker's yeast for lipids using silica gel with hexane–diethyl ether–acetic acid (70:30:1) for class separation and by a two-step process using silica gel with acetone and chloroform–methanol–acetic acid–water (25:15:4:2) for phospholipids. Nikolova-Damyanova et al. (1994) used argentation–TLC to effect separation of positional isomers of monounsaturated fatty acids as their phenacyl derivatives. Martinez-Lorenzo et al. (1994) used argentation–TLC to separate polyunsaturated fatty acids with detection by self-staining after development with toluene–acetone mixtures. Tvrzicka and Votruba (1994) reviewed lipid analysis using the silica gel–coated quartz rod TLC–flame ionization detection (FID) technique. Marquez-Ruiz et al. (1994) used TLC-FID to evaluate susceptibility to oxidation of linoleyl derivatives. Przybylski et al. (1994) used TLC-FID for the quantitative analysis of complex cereal lipids. Perez et al. (1994) reported on a comparison of mobile phases and HPTLC qualitative and quantitative analysis, on preadsorbent silica gel plates, for the separation of phospholipids in *Biomphalaria glabrata* snails infected with larval stages of *Echinostoma caproni* (Trematoda). Brailoiu et al. (1994) used TLC to determine phosphatidylcholine for liposome characterization using silica gel 60 layers and *n*-butanol–*n*-propanol–water mobile phase. Lin et al. (1994) used silica gel TLC to identify four phospholipids associated with fetal lung maturity. They used one-dimensional, single development with chloroform–methanol–ethanol–acetic acid–water (24:4:2:6:0.5) mobile phase, detection by exposure to iodine vapor and scanning densitometry at 460 nm. Schnaar and Needham (1994) reviewed TLC analysis of glycosphingolipids. Taki et al. (1994) described a procedure for the purification of glycosphingolipids and phospholipids by TLC blotting. Muething (1994) described an improved TLC separation of gangliosides using automated multiple development (AMD). Wardas and Pyka (1994) reported on the use of a number of acid–base indicator dyes as visualizing agents for the detection of cholesterol following TLC.

Bonte et al. (1995) separated stratum corneum lipids by automated multiple development (AMD) on HPTLC silica gel plates using an initial isocratic step followed by a 25-step gradient from methanol–water to hexane. Tarandjiska et al. (1995) separated the molecular species of triacylglycerols from highly unsaturated plant oils by successive argentation and reversed-phase TLC. Watanabe and Mizuta (1995) detected glycosphingolipids from biological samples by TLC at the 5 pmol level using 5-hydroxy-1-tetralone as fluorescent labeling reagent. Arnsmeier and Paller (1995) detected gangliosides by TLC and noted that the process was simplified and made more sensitive by use of chemoilluminescence. Alvarez et al. (1995) used HPTLC–densitometry to determine dipalmitoylphosphatidylcholine in amniotic fluid as free dipalmitoylglycerol on silver nitrate–

modified silica gel HPTLC plates after enzymic hydrolysis. Schuerer et al. (1995) used densitometric quantification for the separation and analysis of human stratum corneum lipids by sequential one-dimensional TLC with detection by charring. Wiesner and Sweeley (1995) characterized complex mixtures of gangliosides from human plasma using two-dimensional TLC, resorcinol detection reagent, and computer-assisted image analysis densitometry. Davani and Olsson (1995) developed an HPTLC method for the detection of natural galactolipids of oat and wheat origin with 8-anolino-1-naphthalenesulfonate (ANS) as fluorogenic visualization reagent and scanning at an excitation wavelength of 375 nm. Smith et al. (1995) used HPTLC–densitometry to determine neutral lipids in regular and low-fat eggs using chloroform–methanol (2:1) extraction, separation of lipids in the extracts by development with the Mangold solvent system or a modified Mangold system for cholesteryl esters; lipids were detected by spraying with phosphomolybdic acid (PMA), and quantification was by reflectance scanning at 700 nm. Vecchini et al. (1995) described a method for quantifying phospholipids separated by TLC and detected with molybdate detection reagent; a personal computer–assisted desk-top scanner that was much less expensive than a conventional densitometer was used for quantification. Baldoni et al. (1995) developed a method to eliminate neutral lipids that interfere with the usual polar lipid chloroform–methanol extraction and subsequent TLC separation. A solvent more hydrophobic than chloroform (e.g., hexane) was used to remove neutral lipids from the rabbit parotid gland; the *n*-hexane phase when analyzed by TLC showed abundant amounts of cholesteryl esters, triacylglycerols, diacyl- and monoacylglycerols. The methanol phase, now devoid of neutral lipids, was analyzed by TLC to separate and visualize the polar lipid fractions. This method is particularly useful if sulfatides are being visualized in the polar lipid fraction. Brunn-Jensen et al. (1995) used HPTLC and scanning densitometry to detect and quantify phospholipid hyperoxides as primary oxidation products in cooked turkey meat. The meat was treated with antioxidants, and subsequently total lipids were extracted in chloroform–methanol (2:1) and the phospholipids were separated into classes on silica gel HPTLC plates. A phospholipid-hydroperoxide–specific reagent, N,N-dimethyl-p-phenylenediamine, and a charring reagent, cupric sulfate–phosphoric acid, were used separately to visualize and identify phospholipid hyperoxides and their parent lipid classes.

Peyrou et al. (1996) described separation and quantification of mono-, di-, and triacylglycerols, and oleic acid using TLC with flame-ionization detection (FID) (see Chapter 8) and a hexane–diethyl ether–formic acid (65:35:0.04) mobile phase. Saha and Das (1996) reported on a simple reflectance densitometric method to quantify polar and nonpolar lipids by HPTLC. They used an *n*-hexane–diethyl ether–glacial acetic acid (80:20:1) mobile phase for the nonpolar lipids and a chloroform–methanol–water (65:25:4) mobile phase for the polar lipids. Detection was by use of iodine vapor and the plates were scanned at 365 nm.

Rupcic et al. (1996) used silica gel TLC to analyze cell lipids of *Candida lipolytica* yeast grown on methanol. The dry cell mass contained 5% lipid, of which 52% were polar lipids, mainly phospholipids and sphingolipids. An interesting finding of the study was the high content of the sphingolipid, 19-phytosphingosine (about 91% of the total long-chain bases). Rabinowitz (1996) used silica gel TLC to analyze lipids in the saliva of the medicinal leech, *Hirudo medicinalis*. The total lipid content of the saliva was about 3 mg of lipids/100 ml of saliva. Neutral lipids consisted of about 67% of the lipids, with polar lipids making up the remainder. TLC was used to determine the profiles of the polar and nonpolar lipids. The largest percentages of the identified lipids were phosphatidic acids and free fatty acids. This leech contains a unique lipid distribution in its saliva, and probably some of these components play a role-in the makeup of the anticoagulants present in the saliva. Lu and Lai (1996) used silica gel–TLC to determine phosphatidylethanolamine in yolk phospholipid. The mobile phase was chloroform–methanol–ethanol–acetic acid–water (50:16:4:12:1); the compound was detected with iodine vapor and quantified by densitometry at 385 nm. Gennaro et al. (1996) used HPTLC to determine phospholipids in snail-conditioned water (SCW) from *Helisoma trivolvis* and *Biomphalaria glabrata* gastropods. Lipids were extracted from the water in chloroform–methanol (2:1), and extracts and standards were applied to silica gel plates developed in a chloroform–methanol–water (65:25:4) mobile phase. Lipids were detected by spraying the plates with 10% cupric sulfate in 8% phosphoric acid and heating, and the black zones were quantified by scanning densitometry at 370 nm. The major phospholipids in SCW were phosphatidylethanolamine and phosphatidylcholine at concentrations ranging from 0.18 to 0.4 µg/ml/snail. Hegewald (1996) used one-dimensional TLC for the separation of all known D-3 and D-4 isomers of phosphoinositides. The chromatography was done on two different types of HPTLC plates under identical conditions and depended on the ability of the D-4, but not D-3, isomers to form complexes with boric acid. Vuorela et al. (1996) described a quantitative TLC assay for the analysis of phospholipids in pharmaceutical products obtained from mammalian brains; samples were extracted in tetrahydrofuran (THF)–water (75:25), applied to HPTLC silica gel plates, and mobile phases were optimized using the PRISMA system (see Chapter 6). Optimal phospholipid separation was achieved in a methyl acetate–chloroform–1-propanol–methanol–0.25% aqueous KCl solution in an unsaturated horizontal sandwich chamber. Xu et al. (1996) described two efficient systems for the separation of phospholipids and lysophospholipids using one-dimensional systems for the analysis of phospholipids from the hippocampus region of the rat brain. They found that a chloroform–methanol–acetic acid–acetone–water (35:25:4:14:2) mobile phase was suitable for the separation of 10 phospholipids on silica gel G plates and that a chloroform–methanol–28% aqueous ammonia (65:35:8) mobile phase also gave clear separation of major phospholipids and their lyso forms. Iwamori et al.

(1996) described a sensitive method to determine the pulmonary surfactant phospholipid/sphingomyelin ratio in human amniotic fluid for the diagnosis of respiratory distress syndrome (RDS) by TLC–immunostaining. The method clearly distinguished the surfactant phospholipid/sphingomyelin ratios in normal amniotic fluid versus fluid from women who delivered babies with RDS syndrome. Muething (1996) provided an updated review with 234 references on the uses of TLC for the analyses of gangliosides. The review covers basic techniques for the successful separation of glycosphingolipids, new approaches concerning continuous and multiple development, and includes several preparative TLC methods. The review also includes TLC immunostaining and related techniques—i.e., practical applications of carbohydrate-specific antibodies, toxins and bacteria, viruses, lectins, and eukaryotic cells. Parrish et al. (1996) used a Chromarod TLC procedure to separate polar lipids in spinach and various phytoflagellates. Monogalactosyl diacylglycerol, digalactosyl diacylglycerol, and sulfoquinovosyl diacylglycerol were separated from each other, as well as from chlorophyll *a*, carotenoids, monoacylglycerol, phosphatidylethanolamine, and other phospholipids. Quantification of 0.5–10 μg amounts of individual glycoglycerolipids and pigments was done by scanning the rods through the flame-ionization detector (FID) of an Iatroscan (see Chapter 8).

Fried (1996) reviewed TLC and HPTLC of lipids, including studies on neutral lipids, phospholipids, glycolipids, and gangliosides. Coverage included detailed information on sample preparation prior to TLC; selected chromatographic systems for the separation and determination of all major classes of lipids; and consideration of quantitative TLC of lipids mainly by use of scanning densitometry. The review has nine figures, 10 tables, and 143 references. Hammond (1996) has contributed an interesting review on the chromatographic analysis of lipids, with coverage including TLC, GC, and HPLC. The chapter has 184 references, of which about one-third are related to TLC. There are three line drawings showing lipid separations in various mobile phases on plain silica gel plates and also plates treated with silver nitrate. Maloney (1996) has reviewed uses of TLC in bacteriology and has provided useful methods for sample preparation of lipids and TLC protocols for working with bacteria. TLC is used to determine microbial lipid composition and to study microbial lipases, of interest in identifying bacteria and bacterial strains. TLC is also useful to determine host lipids that function as receptors for microbial pathogens. Because it is mainly the glycosphingolipids and phospholipids that are of interest to the bacteriologist, TLC of these classes are well covered in this chapter, which contains 64 references and two figures. Fried and Haseeb (1996) described studies on the TLC analysis of neutral lipids, phospholipids and glycolipids in protozoan and helminth parasites. They also gave selected methods and protocols for the TLC analysis of lipids in animal parasites. Fell (1996) has provided extensive information on the uses of TLC and HPTLC in studies on entomology (insects and related groups). The technique has been used in this discipline not only for the separation and identification of insect lipids, but also for preparative isolation of lipids for use in other analytical procedures. Lipid analysis of insects usually includes studies on hemolymph, body tissue and cuticular lipids. Fell's chapter will be of interest to invertebratists other than entomologists who are interested in the TLC of lipids. Coverage includes considerable information on solvent systems, methods of visualization and protocols based on Fell's personal experience. Weldon (1996) has contributed a chapter on TLC of skin secretions of vertebrates that contains considerable information on obtaining lipid samples from the skin and exocrine glands of vertebrates, of lipid sample preparation and of chromatographic systems and detection compounds useful for the analysis of both nonpolar and polar lipids. A discussion of TLC analysis of lipids as used in vertebrate systematics is included along with selected TLC protocols using numerous vertebrate examples mainly based on Weldon's personal experience with the subject. Jain (1996) has described uses of TLC in studies on lipids in clinical chemistry. There is considerable interest in TLC analysis of human serum, cutaneous lipids (mainly sebum) and lipids from patients with alcoholic diseases. A very important use of TLC continues to be the analysis of amniotic fluid to determine the lecithin (phosphatidylcholine)/sphingomyelin ratio in children who may possibly suffer form respiratory distress syndrome (RDS). Protocols and methods useful for the above lipid analyses are provided by Jain.

Frazer et al. (1997a) used silica gel–HPTLC to determine neutral lipids and phospholipids in the medically important freshwater snail *Lymnaea elodes*. The mobile phase for neutral lipids was petroleum ether–diethyl ether–acetic acid (40:20:1), with compounds detected by spraying with 5% ethanolic phosphomolybdic acid with quantification by densitometry at 700 nm. The major neutral lipids and their mean percentage wet weights were triacylglycerols (0.11%), free sterols (0.50%) and free fatty acids (0.18%). The phospholipids were separated with a mobile phase of chloroform–methanol–water (65:25:4) and detected by spraying with 10% cupric sulfate in 8% phosphoric acid with quantification by densitometry at 370 nm. The major phospholipids were phosphatidylcholine (0.45%) and phosphatidylethanolamine (0.34%). Frazer et al. (1997b) used high-performance thin-layer chromatography (HPTLC) to determine neutral lipids and phospholipids in the intestinal trematode *Echinostoma caproni* from experimentally infected ICR mice fed a high-fat diet (hen's egg yolk) as compared with worms from mice fed a standard laboratory diet. Analysis by TLC–densitometry showed significantly greater amounts of triacylglycerols and free sterols at 2 weeks postinfection in worms from mice on the high-fat diet as compared with worms from mice on the standard laboratory diet. Significantly greater amounts of phosphatidylcholine and phosphatidylethanolamine were found in worms from mice on the high-fat diet as compared with worms from those on the standard diet at 2 weeks postinfection. The results of the study suggest that the host diet influences the lipid content of *E. caproni* adults. Innocente et al. (1997) used silica gel–TLC to determine triacylglycerols and phospholipids in milk fats; the mobile phase was petroleum ether–diethyl ether–formic acid (120:80:3) with detection of compounds under UV light. Zellmer and Lasch (1997) used automated multiple development (AMD) of HPTLC plates in combination with a 25-step gradient, based on methanol–diethyl ether and *n*-hexane to separate the six major human plantar stratum corneum lipids. Lipids were detected with either $MnCl_2–H_2SO_4$ or $CuSO_4–H_3PO_4$ and quantified by scanning densitometry. Asmis et al. (1997) described TLC procedures for the concurrent quantification of cellular cholesterol, cholesteryl esters and triacylglycerols from small biological samples. Their sensitive assay allowed for the detection of nanogram quantities of lipids, and quantification was done with laser densitometry. Pchelkin (1997) provided a review with 49 references on the evaluation of purity of fractions after separating unsaturated polar lipids by adsorption TLC in the presence of silver ions. Wardas and Pyka (1997) evaluated 11 new visualizing agents for the detection of unsaturated higher fatty acids by TLC. The agents were aniline blue, alkaline blue, brilliant green, neutral red, bromothymol blue, thymol blue, bromophenol blue, phenol red, helasol green, bromocresol green, and brilliant cresyl blue. Aniline blue provided the most advantageous visualizing effects in adsorption and partition TLC.

Rementzis et al. (1997) used silica gel–TLC to study gangliosides from the muscle of the common Atlantic mackerel, *Scomber scombrus*. They identified monosialogangliosides and disialogangliosides with a mobile phase of propanol–water (7:3), and the compounds were detected by spraying with resorcinal reagent. Gunstone and Padley (1997) edited a book on lipid technologies and applications that contains 31 chapters in six parts as follows: Part I, Introduction; Part II, Processing; Part III, Food Emulsions; Part IV, Non-Aqueous Foods; Part V, Special Food Applications; Part VI, Nonfood Uses. Although not directly related to TLC, the book provides valuable information on lipid structure and sample preparation and will be of use to chromatographers interested in the analysis of lipids.

IV. PRACTICAL LIPID TLC STUDIES

Practical TLC of lipophilic material usually involves separating neutral from complex lipids of plant or animal tissues following extraction of the material in chloroform–methanol (2:1) (Folch et al., 1957). Glycolipids [a contraction of glycosyllipids, see Lundberg (1984)] are removed in the aqueous wash of the Folch et al. (1957) procedure (Christie, 1982, 1984). From the standpoint of practical TLC, glycolipids have been studied less frequently than neutral and phospholipids. For coverage of practical TLC of glycolipids, see Ando and Saito

(1987) and Fried (1996). This chapter emphasizes TLC separations of neutral lipids and phospholipids.

A commonly used system for neutral lipids is that of Mangold (1969), consisting of petroleum ether–diethyl ether–acetic acid (80:20:1) as the solvent and silica gel as the sorbent. Neutral lipids migrate and are resolved upon development with the mobile phase, whereas phospholipids remain at the origin. Experiment 1 describes use of the Mangold (1969) system to separate neutral lipids.

Because the Mangold (1969) system does not separate all commonly occurring neutral lipids in biological materials, the dual-solvent system of Skipski et al. (1965) is often used with silica gel as the sorbent. In this system, glycerides and free fatty acids are clearly separated from free sterols by double development in the same direction with two different solvents. Solvent 1 consists of isopropyl ether–acetic acid (65:4). Following development in solvent 1, the plate is dried and then developed in the same direction in solvent 2, which consists of petroleum ether–diethyl ether–acetic acid (90:10:1). The Skipski et al. (1965) procedure is useful for both analytical and preparative TLC. Experiment 2 describes use of this technique for analytical TLC, and Experiment 9 describes its use for preparative TLC.

Phospholipids are more polar than neutral lipids and remain at the origin when nonpolar solvents are used. To separate phospholipids, a polar solvent such as chloroform–methanol–water (65:25:4) (Wagner et al., 1961) is frequently employed. This system separates the more common phospholipids of animal and plant tissues and moves the neutral lipids mainly as a single band at or near the solvent front. Experiment 3 describes use of the Wagner et al. (1961) system to separate phospholipids.

Numerous extraction procedures are available for lipids, and this subject has been reviewed by Phillips and Privett (1979), Christie (1984), and Fried (1993). Of all extraction procedures, that of Folch et al. (1957) is used most frequently. This procedure involves extracting fresh animal or plant material with chloroform–methanol (2:1), usually in a volume ratio of 20 parts of the solvent to 1 part tissue or fluid. There are many variations of the Folch et al. (1957) procedure, and for descriptions of these variations, see Christie (1982, 1984). Experiment 4 involves the use of the Folch et al. (1957) procedure on 100 mg of animal tissue and 100 µl of serum. The Folch et al. (1957) procedure extracts nonlipid material along with lipids. Nonlipids are usually removed by an aqueous wash with either water or a dilute salt solution. Phillips and Privett (1979) devised an extraction procedure for brain tissue in which nonlipid material is first removed with dilute acetic acid. The brain tissue is then treated with chloroform–methanol essentially as described in Folch et al. (1957) to get a relatively pure lipid fraction. Because so many different lipid extraction procedures have been used prior to TLC analysis, different results for a given separation may reflect technique differences rather than actual differences in lipid constituents of a sample.

General methods for the detection of lipids involve use of iodine, cupric acetate, dichlorofluorescein, H_2SO_4, and phosphomolybdic acid [see review by Sherma and Bennett (1983)]. These reagents are used in a variety of ways, including spray and dip procedures. Dip procedures involve both predipping and post-dipping techniques. Experiment 5 explores various general detection reagents for lipids and examines different ways of applying the detection reagents to the TLC plate.

Specific chemical detection tests play an important role in identifying lipid components. From a practical standpoint, Kates (1986) has provided excellent guidelines for specific chemical identification tests. Specific chemical tests are very helpful in the determination of cholesterol, free fatty acids, cholesteryl esters, and several phospholipids. Experiment 6 employs specific chemical detection tests in TLC analysis.

TLC is often used as a preparative technique to isolate individual lipid fractions for subsequent studies by a variety of analytical procedures. Preparative isolation of the sterol (Chitwood et al., 1985; Shetty et al., 1990) and triacylglycerol (Horutz et al., 1993) fractions have been reported. Preparative isolation of sterols may allow these lipids to be analyzed further by argentation TLC (Ditullio et al., 1965; Morris, 1966) or by GLC (Shetty et al., 1990). Moreover, preparative isolation of triacylglycerols will allow for TLC analysis of glyceryl ethers (Snyder, 1971) and HPLC analysis of specific triacylglycerols (Horutz et al., 1993). Experiment 7 is concerned with a preparative procedure for the analysis of sterols by argentation TLC.

The newest area of TLC involves quantitative in situ densitometry of compounds (Chapter 10). The area has been well explored in lipid TLC, and reference is made to the works of Privett et al. (1973), Downing (1979), and Fried (1991). For in situ densitometric quantification, the chromatographer will need a suitable densitometer and recorder, integrator, or computer. We have described one quantitative experiment (Experiment 8) using cholesterol as the lipid fraction quantified. The experiment described utilizes a Kontes densitometer.

A number of experiments have been described in detail on lipids in the Machery–Nagel (MN) "TLC Applications" catalog. Because of the availability of this catalog from MN (see Chapter 23 in this book for details), the experiments are not described herein. The following applications of TLC experiments are recommended (the application number of the experiment as listed in the MN catalog is given in parentheses): analysis of complex natural phospholipids and neutral lipid mixtures in tissues (66); deacylation of cardiolipin in vitro by various avian and mammalian tissues (67); investigation of amniotic fluid extracts for assessment of the neonatal respiratory syndrome (68); separation of skin lipids (69); and analysis of neutral lipids in tissues and hair by methylamine O-deacylation (70).

V. DETAILED EXPERIMENTS

A. Experiment 1. Separation of Neutral Lipids According to Mangold (1969)

1. Introduction

This is a simple experiment that serves as a useful guide to introduce lipid TLC. The first part of the experiment shows the separation of neutral lipid standards and the second part describes the TLC of natural products.

2. Preparation of Standards

Obtain the neutral lipid standard 18-4A from Nu-Chek Prep, Inc. This standard contains equal weights of cholesterol, oleic acid, triolein, methyl oleate, and cholesteryl oleate. The standard is shipped in an ampoule containing approximately 25 mg of the neutral lipids. (Note: Nu-Chek standards do not contain exact weights of lipids and are not useful for quantitative studies as received unless the weight is determined by dissolving and quantitative transfer to a tared vial followed by evaporation of the solvent.) Score the ampoule with a glass file and add analytical-grade chloroform to the ampoule with a disposable pipet. Carefully transfer the chloroform to a 25-ml volumetric flask. Repeat this procedure three to five times to assure removal of all standard from the ampoule. Adjust the volume in the flask to 25 ml by the addition of chloroform. The standard now contains approximately 1 µg/µl of the five neutral lipids. Each neutral lipid is present at a concentration of about 0.2 µg/µl. Frequent mention of the approximate 1-µg/µl 18-4A standard will be made in the detailed experiments on lipids. Transfer the standard to smaller glass vials, preferably having Teflon-lined caps. Store standards in a freezer at −20°C when not in use.

    Also, obtain neutral lipid standard 18-1A from Nu-Chek Prep., Inc. This standard contains equal weights of monolein, diolein, triolein, and methyl oleate. Prepare an approximate 1-µg/µl standard of 18-1A as described for 18-4A. The concentration of each neutral lipid in this standard will be about 0.25 µg/µl.

3. Layer

Silica gel layers, 20 × 20 cm, such as Baker-flex IB-2 (J. T. Baker Chemical Co.) should be used as received with no pretreatment.

4. Application of Initial Zones

Apply 1, 5, and 10 µl of the 18-4A standard to the origin of the plate using 1-, 5-, and 10-µl Microcap micropipets or a Drummond digital microdispenser. Locate the origin 2.5 cm from the bottom of the plate. Draw the solvent front line

10 cm from the origin. Place spots at least 3 cm from each edge of the plate and about 1.5–2 cm from each other. Repeat the above steps for the 18-1A standard.

5. Mobile Phase

Petroleum ether (60–70°C)–diethyl ether–acetic acid (80:20:1). Prepare 100 ml of this solvent in a 250-ml Erlenmeyer flask. Use graduated cylinders to measure large volumes of solvent. Use a 1-ml pipet to measure the acetic acid. (*Caution*: Do not pipet organic solvents or corrosive acids by mouth; use a pipet bulb.)

6. Development

Develop the plate in a saturated developing tank (e.g., Kontes Chromaflex, K-416100). To saturate the tank, line it with Whatman No. 1 filter paper. Pour the mobile phase over the filter paper and gently rock the tank. Place the lid on the tank. Allow the tank to saturate for at least 15 min prior to inserting the plate rapidly and closing the lid immediately. Allow the solvent to ascend to the pre-drawn solvent front line and then remove the plate. Evaporate the solvent with the aid of nitrogen gas or cool air from a hair dryer.

7. Detection

Spray the plate lightly in a well-ventilated fume hood with either 50% aqueous H_2SO_4 or 5% ethanolic phosphomolybdic acid (PMA) (Kritchevsky and Kirk, 1952). Allow the plate to dry, and then heat it for 5–10 min at 100–120°C in an oven or on a hot plate.

8. Results and Discussion

Charring with H_2SO_4 will produce brown to black spots. However, within 1–2 min of spraying, cholesterol and cholesteryl oleate will appear rose-violet and can be distinguished from the other neutral lipids. The rose-violet color changes to brown-black with continued heating. Visualization with phosphomolybdic acid (PMA) will result in purple-blue spots on a yellow-green background.

The order of migration of spots in the 18-4A mixture with approximate R_F values is as follows: cholesterol, 0.08; oleic acid, 0.24; triolein, 0.37; methyl oleate, 0.49; cholesteryl oleate, 0.62. The R_F values will be variable depending on factors such as temperature, humidity, saturation conditions, and so on. The order of migration of the spots, however, will remain the same. Occasionally, triolein and methyl oleate will appear very close to each other and, in some instances, may not be resolved. The 18-1A mixture will be resolved. Monolein will stay at the origin and diolein will migrate with a mobility similar to cholesterol.

9. TLC of Lipids from Hen's Eggs

The following experiments from the work of Butler and Fried (1977) is suggested to show use of the Mangold (1969) solvent for practical TLC with a natural product. Obtain 0.1 ml of yolk with a pipet and emulsify the yolk in 1 ml of sterile saline (0.85% NaCl) (Paul, 1975). Extract the yolk–saline mixture with 2 ml of chloroform–methanol (2 : 1). Allow the mixture to settle. Remove the supernatant and pass it through a glass wool filter. The supernatant, upon standing, will separate into two phases. Remove and discard the upper phase with a disposable pipet. Retain the bottom phase (chloroform layer) and dry it under nitrogen. Dissolve the lipid residue with 100 μl of chloroform. Spot 5-, 10-, or 20-μl aliquots of this sample on a TLC plate along with 1, 5, or 10 μl of neutral lipid standard 18-4A. Develop the plate in the mobile phase and detect lipids as described above. The most abundant neutral lipids in the yolk–saline mixture will be triacylglycerols and free sterols (Figure 15.1). If saline solution alone (devoid of yolk) is extracted and spotted for TLC, it should be lipid negative (Figure 15.1).

Work in our laboratory [see Masterson et al. (1993)] has been concerned with the neutral lipid profiles of various natural products including hen's egg

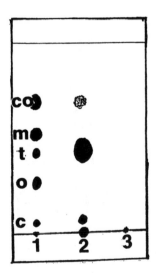

FIGURE 15.1 Separation of yolk-saline medium (see text) following extraction in chloroform-methanol (2:1). Lipids were developed in petroleum ether-diethyl ether–acetic acid (80:20:1) and detected by spraying with PMA. Lane 1 contains neutral lipid standard mixture 18-4A, which consists of cholesterol (c), oleic acid (o), triolein (t), methyl oleate (m), and cholesteryl oleate (co). Lane 2 shows presence of triacylglycerols and free sterols that are the predominant neutral lipids in the yolk-saline medium. Lane 3 contains saline alone that is neutral lipid negative.

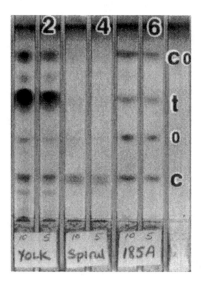

FIGURE 15.2 Separation of hen's egg yolk and *Spirulina* (see text) following extraction in chloroform–methanol. Development and detection was as described in Figure 15.1. Lanes 1 and 2 show the predominant lipids in the yolk. Lanes 3 and 4 show mainly the presence of free sterols in the *Spirulina*. Lanes 5 and 6 show the neutral lipid standards. Abbreviations as in Figure 15.1.

yolk. Extracts of the natural products have been separated on channeled preadsorbent HPTLC plates (Whatman LHP-KDF, 10×20 cm) using the Mangold solvent system and detection with phosphomolybdic acid. Results of a typical separation of egg yolk (lanes 1 and 2) and *Spirulina*, an algae with minimal lipids except for structural sterols (lanes 3 and 4), and neutral lipid standards (lanes 5 and 6) are shown in Figure 15.2. Note the presence of abundant amounts of triacylglycerols and free sterols in the yolk fractions. An abundant sterol ester fraction can also be seen in the yolk samples. Other trace lipids are also apparent. Note the absence of most lipids except for free sterols in the *Spirulina* samples. Neutral lipid standards are in lanes 5 and 6.

B. Experiment 2. Separation of Neutral Lipids in a Dual-Solvent System

1. Introduction

This experiment from the work of Skipski et al. (1965) shows the separation of neutral lipids on a TLC plate using dual development in the same direction. This procedure provides unequivocal separation of acylglycerols from free sterols, which was not obtained in the last experiment.

2. Preparation of Standards

Use standards 18-4A and 18-1A as described in Experiment 1, Section 2.

3. Layer

Silica gel layer, as described in Experiment 1, Section 3.

4. Application of Initial Zones

Draw an origin line 2.5 cm from the bottom of the plate. Draw (very lightly) the first solvent front line 14 cm from the bottom of the plate and the second solvent front line 19 cm from the bottom of the plate. Using Microcap micropipets or a Drummond digital microdispenser spot 1-, 5-, and 10-µl aliquots of the 18-4A standard at the origins. Repeat this procedure for the 18-1A standard. (See Experiment 1, Section 4 for additional hints on spotting.)

5. Mobile Phase

Solvent 1: isopropyl ether–acetic acid (96:4). Solvent 2: petroleum either (60–70°C)–diethyl ether–acetic acid (90:10:1).

6. Development

Saturate a tank with 100 ml of solvent 1 as described in Experiment 1, Section 6. Saturate a second tank with 100 ml of solvent 2. Develop the plate in solvent 1 to the first solvent front line. Remove the plate and dry it under nitrogen or cool air from a hair dryer. Develop the plate in solvent 2 to the second solvent front line. Remove the plate and dry it as described above.

7. Detection

Detect neutral lipids by spraying the plate with 50% H_2SO_4 or 5% ethanolic PMA (see Experiment 1, Section 7).

8. Results and Discussion

This procedure should provide unequivocal separation of the neutral lipids. In Figure 15.3, lanes 1 and 3 show the separation of the standards. The order of migration and approximate R_F values are as follows: monolein, 0.05; cholesterol, 0.26; diolein, 0.34; oleic acid, 0.55; triolein, 0.65; methyl oleate, 0.69; cholesteryl oleate, 0.84. These R_F values are approximate and will vary based on factors such as temperature, relative humidity, and saturation time.

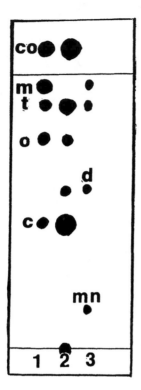

Figure 15.3 Separation of snail liver tissue and neutral lipid standards 18-4A and 18-1A using the dual solvent system of Skipski et al. (1965) (see text). Lane 1 shows separation of the 18-4A standard. Abbreviations for this standard are as described in Figure 15.1. Lane 3 contains equal amounts of the 18-1A standard, which consists of monolein (mn), diolein (d), triolein (t), and methyl oleate (m). Lane 2 shows separation of snail neutral lipids. Note the abundance of free sterols in this tissue.

9. TLC of Neutral Lipids of Snail Liver Tissue

The liver (hepatopancreas or digestive gland) of freshwater snails contains mainly free sterols, triacylglycerols, and sterol esters (Figure 15.3, lane 2). Additionally, diacylglycerols may also be detected. The study described here is based on un-published observations in our laboratory on the TLC analysis of the common viviparid snail, *Campeloma decisum*, found in freshwater ponds in the United States. Similar experiments can be done with the hepatopancreas of other available marine or freshwater snails (see Chapter 4). For this experiment, obtain 100 mg of snail liver and extract it in chloroform–methanol (2:1) as described in Experiment 4. Application of about 20% of the total snail lipids to a plate should present a lipid profile similar to that shown in Figure 15.3.

Duncan et al. (1987) used the dual development system of Skipski et al. (1965) to determine the presence of monoacylglycerols and diacylglycerols in whole body extracts of the medically important snail *Biomphalaria glabrata*. They failed to detect the presence of these minor neutral lipids in the whole-body extract. Thompson et al. (1991) used two-dimensional TLC on extracts of the digestive gland–gonad complex of *B. glabrata* snails and showed the presence of monoacylglycerols and diacylglycerols as minor lipid components in the extracts.

C. Experiment 3. Separation of Phospholipids

1. Introduction

The first part of this experiment describes separation of phospholipid standards on a TLC plate using the solvent system of Wagner et al. (1961). The second part describes use of this technique for the separation of phospholipids from natural products.

2. Preparation of Standards

Obtain the following phospholipid standards from Matreya: a polar lipid standard containing equal amounts of lysophosphatidylcholine, phosphatidylcholine, phosphatidyl ethanolamine, and cholesterol; a sphingomyelin standard containing equal amounts of sphingomyelin, sulfatides, and three cerebrosides; individual phosphatidyl inositol and phosphatidyl serine standards. Each standard is supplied in 25-mg amounts dissolved in 1 ml of chloroform–methanol (2:1). Add 4 ml of chloroform–methanol (2:1) to each standard to achieve a final concentration of 5 μg/μl.

3. Layer

Precoated 20 × 20-cm silica glass plate (silica gel 60, layer thickness 0.25 mm, E. Merck). As discussed in Fried and Shapiro (1979), other layers may not be as satisfactory for phospholipid separations.

4. Application of Initial Zones

Draw an origin line 2 cm from the bottom plate. Draw a solvent front line 10 cm from the origin. With Microcap micropipets or a Drummond digital micro-dispenser, apply 5 and 10 μl of each standard to separate origins as a 1-cm streak (Chapter 5).

5. Mobile Phase

Chloroform–methanol–water (65:25:4).

6. Development

Saturate a tank lined with filter paper (see Experiment 1, Section 6) with 100 ml of the mobile phase for 20–40 min prior to development. Develop the plate for a distance of 10 cm past the origin. Remove the plate and dry it thoroughly under nitrogen or with cool air from a hair dryer.

7. Detection

Spray the plate in a ventilated fume hood with 50% H_2SO_4. Place the plate in an oven or on a hot plate at a temperature of 120–140°C for 5–10 min.

8. Results and Discussion

This procedure should detect narrow bands of lipid that char with H_2SO_4. The use of the Merck plate and a saturation time of 20–40 min is important for achieving separations (Fried and Shapiro, 1979). Expected R_F values and the order of migration of compounds are as follow: lysophosphatidyl choline, 0.06; sphingomyelin, 0.12; phosphatidyl serine, 0.13; phosphatidyl inositol, 0.13; phosphatidyl choline, 0.20; sulfatides, 0.25; phosphatidyl ethanolamine, 0.40; cerebroside 1, 0.48; cerebroside 2, 0.51; cerebroside 3, 0.58; cholesterol, 0.80. Specific phospholipid spray tests (Experiment 6) should also be used.

9. TLC of Phospholipids of Parasite Tissue and in Snail-Conditioned Water (SCW)

The technique just described for standards along with specific phospholipid spray tests have been used by Fried and Shapiro (1979) to analyze phospholipids in a parasitic flatworm. Sample preparation employed the Folch et al. (1957) extraction procedure (Experiment 4), and aliquots of sample were spotted essentially as described for standards. The results of that study showed that phosphatidyl choline, phosphatidyl serine, and phosphatidyl ethanolamine were the major phospholipids in the flatworm. Some variation should be expected in phospholipid analysis of animal tissues, although phosphatidyl choline, phosphatidyl serine, and phosphatidyl ethanolamine are usually major phospholipids in animal tissues. The TLC procedure described here following Folch et al. (1957) extraction of tissue should serve as a guide for the TLC analysis of phospholipids in biological materials.

Gennaro et al. (1996) used HPTLC to determine phospholipids in snail-conditioned water (SCW) from planorbid snails (see Section III). The SCW was extracted in chloroform methanol, spotted on silica gel plates and separated in a mobile phase of chloroform–methanol–water (65:25:4). Lipids were de-

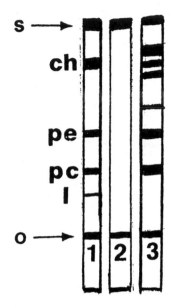

s →
ch
pe
pc
l
o →
1 2 3

FIGURE 15.4 Separation of phospholipids from snail-conditioned water (SCW) on a silica gel plate (see text for other details). Lane 1 contains a phospholipid standard; lane 2 is a blank of extracted water; lane 3 contains SCW. Abbreviations: ch = cholesterol, 1 = lysophosphatidylcholine, o = origin, pc = phosphatidylcholine, pe = phosphatidylethanolamine, s = solvent front.

tected by spraying with 10% cupric sulfate in 8% phosphoric acid. Figure 15.4 shows a line drawing of a typical phospholipid separation from the planorbid SCW.

D. Experiment 4. Extraction of Lipids from Small Samples of Biological Tissues or Fluids

1. Introduction

This experiment uses the classical Folch et al. (1957) procedure as described in Christie (1982, 1984) and Fried (1993) to extract lipids from biological tissues of 100 mg or less or from biological fluids of 100 μl or less. The procedure described is simple, requires no special equipment, and is basic to many of the described TLC experiments on lipids.

2. Extraction Procedures

Obtain about 100 mg of animal tissue or 100 μl of animal fluid (suggested materials are mouse, frog, or snail liver; human or chicken serum; or snail hemolymph).

Place the tissue or fluid in a glass test tube, and immediately add 1 ml of chloroform–methanol (2:1). Homogenize the tissue or fluid sample with a glass stirring rod. A protein precipitate will form. Allow the precipitate to settle (1–5 min). Remove the supernatant liquid with a disposable pipet, pass the liquid through a glass wool filter, and collect it in another test tube. Repeat the extraction procedure twice, each time adding 1 ml of chloroform–methanol to the residue. Collect this supernatant by passing it through a glass wool filter into the tube containing the original supernatant. At the end of this procedure, about 3 ml of a clear supernatant will be obtained. Although this procedure is not strictly quantitative, the supernatant should contain most of the lipids of the original sample, along with nonlipid impurities.

To separate the nonlipid impurities, add approximately 0.7 ml of 0.88% KCl (Christie, 1982) to the tube. Agitate the mixture by hand or with a Vortex mixer. This procedure will produce a biphasic mixture consisting of a top hydrophilic layer and a bottom lipophilic layer. If an emulsion forms in the tube, allow to stand at room temperature or in a refrigerator at 4°C for up to 4 h.

With small amounts of tissue or fluid, the emulsion should break after this time period, leaving a clear top and bottom layer. A slight flocculent layer may appear at the interface of the biphasic mixture. Remove the top layer (mainly amino acids, sugars, and organic acids) with a disposable pipet and either discard or save for TLC analyses of hydrophilic substances (Chapters 16–18). If flocculent material is present at the interface of the biphasic layers, carefully remove it with a disposable pipet and discard. The bottom layer contains mainly lipids. Remove this layer to a small vial with the aid of a disposable pipet. To determine the approximate amount of lipid in this layer, weigh an empty shell vial on an analytical balance before the layer is added. Dry the material under nitrogen and then weigh the vial again. The weight of the evaporated residue in the vial will represent the weight of the lipid as a percentage of wet weight of a tissue sample.

General values from animal tissue experiments indicate that the percentage of lipid based on wet weight of tissue is in the range 2–10%. The sample should then be reconstituted with 100 µl of chloroform–methanol (2:1) and placed in a freezer at −20°C until TLC procedures are employed. The amount of lipid in the sample will be in the range 2–10 µg/µl.

E. Experiment 5. General Detection Procedures for Lipids

1. Introduction

The purpose of this experiment is to show the efficacy of different general detection procedures for lipids. The experiment first examines spraying procedures and then considers predipping versus postdipping techniques.

2. Preparation of Standards

Use 1, 5, 10, and 20 µl of the 18-4A standard (see Experiment 1).

3. Layer

Baker-flex silica gel sheets, 20 × 20 cm (see Experiment 1).

4. Application of Initial Zones

a. For Spraying Procedures

With a pencil, locate an origin line 2 cm from the bottom of the sheet and a solvent front line 10 cm from the origin. Divide the sheet in half by drawing a vertical line 10 cm from the edges. Spot 1, 5, and 10 µl of the 18-4A standard on the origins on one half of the sheet. Repeat this procedure on the other half of the sheet.

b. For Dipping Procedures

1. Postdipping procedure. Locate an origin line 2 cm from the bottom of a silica gel sheet and a solvent front line 10 cm from the origin. Spot 1, 5, 10, and 20 µl of the 18-4A standard at the origins.
2. Predipping procedure. Locate an origin line 1.5 cm from the bottom of the sheet. Locate another line 2 cm from the bottom of the sheet. This line will serve as a mark to which the PMA is predipped on the sheet. This sheet will be spotted after it is predipped with PMA as described in Section 7.b. After the sheet has dried following predipping, spot 1, 5, 10, and 20 µl of the 18-4A standard on the origin line.

5. Mobile Phase

Petroleum ether–diethyl ether–acetic acid (80 : 20 : 1) for neutral lipid separations (see Experiment 1).

6. Development

Develop sheets, two per saturated tank, in 100 ml of the solvent as described in Experiment 1.

7. General Detection Procedures

a. Gaseous or Spray Application

1. Iodine. Place iodine crystals in a tank and close the lid. Allow vapors of iodine to saturate the tank. Place the sheet that had been penciled with a vertical line down the middle in the iodine chamber. Brown spots will soon appear on the sheet. Remove the sheet and quickly

mark the spots with a pencil (closed circle). The spots will fade as the iodine sublimates upon exposure to air.

2. 2′,7′-Dichlorofluorescein. Prepare a 0.1% solution of 2′,7′-dichlorofluorescein in 95% methanol. With a glass sprayer attached to a compressed-air line, spray the sheet that had first been treated with iodine with the dichlorofluorescein solution and then examine this sheet under short- and long-wave ultraviolet light. Lipids should appear as dark spots against a fluorescent background under both lamps. If the sheet is sprayed with dichlorofluorescein without the iodine pretreatment and then examined under short UV light, lipids will appear as yellow spots. Circle spots (open circles) with a pencil and compare the TLC results with previous observations from the iodine vapor detection study. The iodine and dichlorofluorescein techniques are both nondestructive.

3. PMA versus H_2SO_4 spraying procedure. These two techniques are destructive. Cut the sheet in half along the vertical line with either a scissors or a razor blade. Spray one half with 50% aqueous H_2SO_4 and the other half with 5% ethanolic PMA (see Experiment 1). Heat these sheets in an oven as described in Experiment 1, Section 7. Compare the detection of spots using the H_2SO_4 spray technique versus the PMA spray technique.

b. Dipping Procedures

Prepare enough 5% phosphomolybdic acid in ethanol to fill a dipping or developing tank or a large glass tray.

1. Postdipping. Dip the developed chromatogram in the PMA for about 5 sec. Remove the sheet, allow the excess PMA to drain on a paper towel, and then air dry. Heat the sheet in an oven at 110°C for 2–5 min to detect the lipids.

2. Predipping. Dip the sheet in PMA to the 2-cm predip line. To accomplish this, the top of the sheet must be first inserted into the tank or tray while the sheet is held at the bottom. (*Caution*—Use disposable gloves when handling the sheet at the bottom.) The predipping procedure will require some trial and error to achieve sheets that are properly dipped. Allow the dipped sheets to air dry before spotting with standards as described in Section 4. Following development, heat the predipped sheets at 110°C for 2–4 min to detect the lipids.

8. Results and Discussion

Neither iodine nor 2′,7′-dichlorofluorescein is as sensitive as the PMA and H_2SO_4 spray procedures. Material not detected with these two nondestructive reagents will be detected using the destructive procedures. The sensitivity of lipid detection obtained with the predip procedure is superior to that of the postdip procedure, making the extra effort involved in predipping worthwhile.

F. Experiment 6. Specific Chemical Detection Procedures for Lipids

1. Introduction

The purpose of this experiment is to use specific chemical detection procedures for the identification of neutral lipids and phospholipids on a TLC plate.

2. Preparation of Standards

Use neutral lipid standard 18-4A described in Experiment 1. Use the polar lipid standard and the individual phosphatidyl serine standard described in Experiment 3.

3. Layer

Use a 20 × 20-cm plastic silica gel (Baker-flex) sheet for neutral lipid separations (Experiment 1) and the glass Merck silica gel plate (Experiment 3) for the phospholipid experiments.

4. Application of Initial Zones

With a pencil, locate an origin line 2.5 cm from the bottom of each plate, and a solvent front line 10 cm from the origin. Divide each plate in half by drawing a vertical pencil line down the center, 10 cm from the side edges of the plate. Place 5 μl of the 18-4A standard on the origin of the Baker-flex plate to the left of the vertical line. Repeat this procedure on the other half of the plate.

On the Merck plate, apply 20 μl of the polar lipid standard as a 1-cm streak on the origin to the left of the vertical line. Allow this material to dry and then streak 5 μl of the phosphatidyl serine standard on top of it. Repeat the above procedure on the other side of the plate.

5. Mobile Phase

Petroleum ether–diethyl ether–acetic acid (80:20:1) for the neutral lipid separation (see Experiment 1).

Chloroform–methanol–water (65:25:4) for the phospholipid separation (see Experiment 3).

6. Development

Develop plates for neutral lipid separations as described in Experiment 1. Develop plates for phospholipid separations as described in Experiment 3.

7. Specific Chemical Detection Procedures

 a. Cholesterol and Cholesteryl Ester Detection with Acidic Ferric Chloride Solution

1. Preparation. Dissolve 50 mg of ferric chloride ($FeCl_3 \cdot 6H_2O$) in 90 ml of water along with 5 ml of acetic acid and 5 ml of sulfuric acid as described by Lowry (1968).
2. Use. Cut the developed Baker-flex plate in half and spray one half with the acidic ferric chloride solution. Heat the plate for 2–3 min at about 100°C at which time the presence of cholesterol and cholesteryl esters will be indicated by the appearance of a red-violet color. The color of cholesterol appears before that of the cholesteryl esters.

 b. Free Fatty Acid Detection

1. Preparation. Prepare three separate solutions according to Dudzinski (1967) as follows: solution, 1, 0.1% solution of 2′,7′-dichlorofluorescein in 95% methanol; solution 2, 1% aluminum chloride in ethanol; solution 3, 1% aqueous ferric chloride.
2. Use. Spray the second half of the Baker-flex plate in turn with solution 1, solution 2, and then solution 3. Warm the plate briefly at about 45°C after each spray. Free fatty acids will give a rose color.

 c. Phosphate Detection Stain

1. Preparation. Solution 1: dissolve 16 g of ammonium molybdate in 120 ml of water according to Vaskovsky and Kostetsky, as described in Kates (1986). Solution 2: add 10 ml of mercury and 80 ml of solution 1 to 40 ml of concentrated HCl. Shake for 30 min and filter. To the remainder of solution 1, add 200 ml of concentrated H_2SO_4 and all of solution 2. Cool and dilute to 1 liter with H_2O.
2. Use. Cover one half of the Merck plate with aluminum foil. Spray the plate with the phosphate reagent. Blue spots should appear in the presence of phosphatides within a few minutes of heating at 100°C.

 d. Ninhydrin Stain

1. Preparation. Dissolve 0.2 g of ninhydrin in 100 ml of n-butanol saturated with water according to Marinetti, as described in Christie (1982).
2. Use. Remove the aluminum foil from the second half of the Merck plate, and spray the plate with this reagent. Lipids having free amino groups will appear as red-violet spots when the plate is heated to 100°C.

8. Results and Discussion

The neutral lipid spray tests should be conducted along with Experiment 1. The spray test for cholesterol and cholesteryl oleate will show a distinct rose-violet color for these neutral lipids. Other neutral lipids will not become rose-violet. The free fatty acid test will detect only the oleic acid in the standard as a distinct rose-colored zone.

The phospholipid spray tests should be done along with Experiment 3. The phosphate test will produce very distinct blue spots with the following phospholipids: lysophosphatidyl choline, phosphatidyl serine, phosphatidyl choline, and phosphatidyl ethanolamine. The ninhydrin test will produce purple spots only with phosphatidyl serine and phosphatidyl ethanolamine.

Various other tests can be done on the phospholipid and sphingolipid mixtures as discussed in Fried and Shapiro (1979). These tests should be tried on natural products run along with standards. For an excellent discussion of additional specific chemical detection tests for lipids, see Kates (1986). Other suggested detection procedures that can be tried with phospholipids are the choline (Dragendorff), sphingolipid (Bischel and Austin), and α-naphthol (Siakatos and Rouser) tests, all as described in Kates (1986).

G. Experiment 7. Separation of Sterols by Argentation (Silver Ion) Chromatography

1. Introduction

This experiment from the work of Morris et al. (1987) describes the use of silver ion chromatography to separate closely related sterols and application of the technique to the examination of sterols from natural products.

2. Preparation of Standards

Obtain the following sterol standards from Serdary Research Labs: cholestanol, cholesterol, ergosterol. Prepare each standard in chloroform at a concentration of 1 μg/μl. Prepare a mixed standard to contain a total of 3 μg/μl of total sterols or 1 μg/μl of each (referred to here as the mixed sterol standard).

3. Layer

Baker-flex silica gel sheet (see Experiment 1). Dip the sheet into a dip tank or a glass tray containing 2.5% ethanolic $AgNO_3$. Allow the sheet to air dry at room temperature in subdued light.

4. Application of Initial Zones

Locate an origin line 2 cm from the bottom of the plate and a solvent front line 15 cm from the origin. Use plates that have been treated with silver nitrate and other plates without $AgNO_3$ treatment as controls. Use Microcap micropipets or a Drummond 10 μl microdispenser to apply 1 μl each of the mixed sterol standard and the individual ergosterol, cholesterol, and cholestanol standards to the plate.

5. Mobile Phase

Chloroform–acetone (95:5)

6. Development

Saturate a tank with 100 ml of the solvent as described in Experiment 1. Place two Baker-flex plates (one treated with $AgNO_3$ and the other untreated) in the tank. Allow the solvent to ascend to the solvent front line. Remove and air dry the plates.

7. Detection

Detect the sterols by spraying with 50% H_2SO_4 as described in Experiments 1 and 3.

8. Results and Discussion

The sterol mixture will not be resolved on untreated plates. On plates treated with silver ion, the mixed sterol standard should show three separate spots with the following order of migration: ergosterol (least migration), cholestanol (intermediate migration), cholesterol (greatest migration).

The spots can also be recognized by different color reactions following treatment with the detection reagent. Colors should be as follow: purple for ergosterol; blue for cholestanol; gray for cholesterol. Consult Morris et al. (1987) for information on the separation of other sterols using argentation TLC.

9. Suggested Experiments with Natural Products

Separate sterols from other lipids on a silica gel sheet using the dual-solvent system of Skipski et al. (1965). Locate the sterol band by placing the plate in an iodine chamber or by using markers at the edge of the plate (see Chapter 12, Section V). Remove the sterol band by scraping it from the silica gel plate as described in Chapter 12. Dissolve the sterol fraction in chloroform and apply various aliquots of this fraction to silica gel sheets. Perform similar experiments as described in Sections 1–8. A variety of other sterol standards may be needed (see Supelco, Serdary, or Matreya catalogs for listings of more available standards) to confirm sterols from natural products by argentation TLC. The results of argentation TLC should be verified independently with the aid of other analytical techniques for the identification of sterols as discussed by Chitwood et al. (1985), Morris et al. (1987), and Shetty et al. (1990).

The use of argentation–TLC may be of limited value if one component of the sterol fraction is predominant. Attempts to obtain free sterols other than cholesterol in the parasitic flatworm *Echinostoma trivolvis* (previously referred to as *E. revolutum*) using argentation–TLC were not successful (Fried et al., 1980). Chitwood et al. (1985), using the powerful technique of capillary GLC as well as capillary GLC/mass spectrometry, showed that the free sterol fraction of *Echinostoma trivolvis* contained about 97% cholesterol; the remainder consisted of nine other free sterols.

H. Experiment 8. Quantitative TLC of Cholesterol

1. Introduction

The purpose of this experiment is to describe a simple TLC procedure for the quantification of cholesterol. This procedure is based on the densitometric determination of steroids detected with phosphomolybdic acid (Wortmann and Touchstone, 1973; Wortmann et al., 1974).

2. Preparation of Standard

Use the Non-Polar Lipid Mix-B, 25 mg/ml-1 ml standard from Matreya (catalog number 1130). This standard contains equal amounts of cholesteryl oleate, methyl oleate, triolein, oleic acid, and cholesterol. The standard is shipped in an ampoule containing a total of 25 mg of the neutral lipids in 1 ml of chloroform. Quantitatively transfer the 1 ml of chloroform to a 50-ml volumetric flask by rinsing with chloroform. Adjust the volume in the flask to 50 ml by the addition of chloroform to achieve a final total lipid concentration of exactly 0.5 µg/µl. The weight of cholesterol in this standard will be 0.1 µg/µl.

3. Layer

Baker-flex silica gel plate (see Experiment 1). Preclean the plate by development in chloroform–methanol (1:1) prior to use.

4. Application of Initial Zones

Draw an origin line 2 cm from the bottom of the plate. Draw a "predip" line (see Experiment 7) 2.3 cm from the bottom of the plate. Draw a solvent front line 10 cm from the origin. Predip the plate in 5% ethanolic PMA to the 2.3-cm mark as described in Experiment 5. After the plate has dried, apply 1-, 2-, 3-, 4-, and 5-µl spots of standard to the origin line with Microcap micropipets or a Drummond 10 µl microdispenser.

5. Mobile Phase

Petroleum ether–diethyl ether–acetic acid (80:20:1) in a saturated tank, as described in Experiment 1.

6. Development

Develop the plate to a position 10 cm above the origin in the mobile phase.

7. Detection

PMA has already been incorporated into the layer. Therefore, to detect neutral lipids, thoroughly air dry the plate after development and heat at 110°C for 5 min.

8. Quantification

This experiment is based on the use of a Kontes fiber optics densitometer (K-49500) with baseline corrector and strip-chart recorder. The description given in the second paragraph of Experiment 8, Section 8 is based on the use of that equipment. Readers will have to modify their approaches to in situ densitometry based on the available instrumentation in their laboratories. In our laboratory, we no longer use the K-49500 model and do most of our lipid in situ densitometry with either a Kontes Model 800 densitometer equipped with a Hewlett-Packard Model 3992A integrator/recorder (Morris et al., 1987) or a Shimadzu CS-930 computerized TLC densitometer operated in the single-beam, reflectance mode. [See Morris et al. (1987), Park et al. (1991), and Masterson et al. (1993) for further information on the use of these instruments in lipid quantification by TLC-densitometry.]

Operate the densitometer using the visible light source without a filter and the reflectance, single-beam mode, scanning with the small head (reference head). Operate the densitometer at a scanning speed of 10 cm/min and an attenuation of 10. Operate the baseline corrector with an input of 2 mV and an output of 10 mV. Set both the positive drift knob and noise suppression knob at 2. Set the recorder input at 10 mV and the chart speed at 30 cm/min.

The order of migration of spots will be as described in Experiment 1, Section 8. The predipping procedure should produce uniform, compact purple-blue spots against a yellow-green background. Cholesterol will appear as the most compact spot, located near the origin. The R_F value of cholesterol will range from 0.08 to 0.10.

The five cholesterol spots should be scanned perpendicularly to the direction of mobile-phase development. A recorder tracing of standard spots ranging from 0.1 to 0.5 µg should appear as shown in Figure 15.5. To calculate the peak area in square centimeters, measure the height of each peak and the width at half-height, and multiply the peak height times the width at half-height. To obtain a calibration curve for the cholesterol standards in the 0.1–0.5-µg range, plot weight of cholesterol spotted in micrograms versus peak area (square centimeters) on arithmetic graph paper and draw the line that best fits the points. The proper

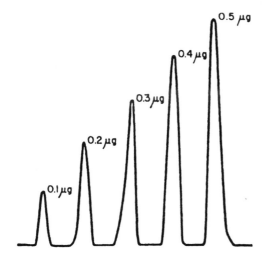

FIGURE 15.5   Densitometric scans of cholesterol spots. Spots were measured at right angles to the direction of plate development.

graph is best obtained by statistical regression analysis, although simple visual placement averaging out the scatter of the points is often adequate for the accuracy and precision required in the quantitative TLC analysis. Figure 15.6 shows a calibration curve for cholesterol using the aforementioned procedures.

9.   Results and Discussion

Published work (Fried et al., 1983; Bennett and Fried, 1983) suggests that this quantitative technique is satisfactory for the analysis of samples with cholesterol in the 100–500-ng range. Worm excretions including cholesterol have been analyzed. By judicious use of sample material and comparison with cholesterol standards in the 100–500-ng range, the amount of cholesterol excreted per worm per unit of time was determined based on this simple in situ densitometric technique. For example, 1–2-week-old *Echinostoma trivolvis* worms excreted 100–300 ng of cholesterol per worm when incubated for 1 h in a nonnutrient medium (Bennett and Fried, 1983). Although the predip procedure has been described in this experiment, excellent sensitivity can also be obtained using postdipping or spraying [see chromatograms reproduced in Park et al. (1991) or Masterson et al. (1993)] PMA procedures with silica gel preadsorbent plates and the Mangold (1969) solvent system for densitometric TLC analysis of neutral lipids.

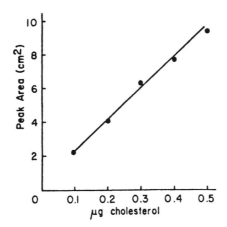

FIGURE 15.6   Linear calibration curve for cholesterol.

REFERENCES

Ackman, R. G., Eaton, C. A., Kinneman, J., and Litchfield, C. (1975). Lipids of freshwater dolphin *Sotalia fluviatilis*: Comparison of odontocete bioacoustic lipids and habitat. *Lipids 10*: 44–49.

Allen, W. V. (1976). Biochemical aspects of lipid storage and utilization in animals. *Am. Zool. 16*: 631–647.

Alvarez, J. G., Slomovic, B., and Ludmir, J. (1995). Analysis of dipalmitoylphosphatidyl-choline in amniotic fluid by enzymic hydrolysis and high-performance thin-layer chromatography reflectance spectrodensitometry. *J. Chromatogr., B: Biomed. Appl. 665*: 79–87.

Ando, S., and Saito, M. (1987). TLC and HPTLC of phospholipids and glycolipids in health and disease. *J. Chromatogr. Libr. 37*: 266–310.

Arnsmeier, S. L., and Paller, A. S. (1995). Chemiluminescence detection of gangliosides by thin-layer chromatography. *J. Lipid Res. 36*: 911–915.

Asmis, R., Buehler, E., Jelk, J., and Gey, K. F. (1997). Concurrent quantification of cellular cholesterol, cholesteryl esters and triglycerides in small biological samples. Reevaluation of thin layer chromatography using laser densitometry. *J. Chromatogr., B: Biomed. Sci. Appl. 691*: 59–66.

Baldoni, E., Bolognani, L. and Vitaioli, L. (1995). A rapid procedure for elimination of non-polar lipids hampering the usual polar lipid extraction and TLC separation. *Eur. J. Histochem. 39*: 253–257.

Bennett, S., and Fried, B. (1983). Densitometric thin-layer chromatographic analyses of free sterols in *Echinostoma revolutum* (Trematoda) adults and their excretory-secretory products. *J. Parasitol. 69*: 789–790.

Bonte, F., Pinguet, P., Chevalier, J. M., and Meybeck, A. (1995). Analysis of all stratum corneum lipids by automated multiple development high-performance thin-layer chromatography. *J. Chromatogr., B: Biomed. Appl. 664*: 311–316.

Brailoiu, E., Saila, L., Huhurez, G., Costuleanu, M., Filipeanu, C. M., Slatineanu, S., Cotrutz, C., and Branisteanu, D. D. (1994). TLC—a rapid method for liposome characterization. *Biomed. Chromatogr. 8*: 193–195.

Breimer, M., Hansson, G. C., Karlsson, K. A., and Leffler, H. (1981). Blood group type glycosphingolipids from the small intestine of different animals analyzed by mass spectrometry and thin-layer chromatography. A note on species diversity. *J. Biochem. 90*: 589–609.

Brouwers, J. F., Van Hellemond, J. J., and Tielens, A. G. M. (1996). Adaptations in the lipid and energy metabolism of parasitic helminths. *Netherlands J. Zool. 46*: 206–215.

Brunn-Jensen, L., Colarow, L., and Skibsted, L. H. (1995). Detection and quantification of phospholipid hydroperoxides in turkey meat extracts by planar chromatography. *J. Planar Chromatogr.—Mod. TLC 8*: 475–479.

Butler, M. S., and Fried, B. (1977). Histochemical and thin layer chromatographic analyses of neutral lipids in *Echinostoma revolutum* metacercariae cultured *in vitro*. *J. Parasitol. 63*: 1041–1045.

Cartwright, I. J. (1993). Separation and analysis of phospholipids by thin layer chromatography. *Methods Mol. Biol. 19*: 153–167.

Chitwood, D. J., Lusby, W. R., and Fried, B. (1985). Sterols of *Echinostoma revolutum* (Trematoda) adults. J. Parasitol. *71*: 846–847.

Christie, W. W. (1982). *Lipid Analysis*, 2nd ed. Pergamon Press, Oxford.

Christie, W. W. (1984). Extraction and hydrolysis of lipids and some reactions of their fatty acid components. In *Handbook of Chromatography—Lipids*, Vol. I, H. K. Mangold (Ed.). CRC Press, Boca Raton, FL, pp. 33–46.

Christie, W. W. (1987). *High Performance Liquid Chromatography and Lipids*. Pergamon Press, Oxford.

Davani, B., and Olsson, N. U. (1995). Detection of natural galactolipids by thin layer chromatography with 8-anilino-1-naphthalenesulfonate (ANS) as fluorogenic visualization reagent. *J. Planar Chromatogr—Mod. TLC. 8*: 33–35.

Ditullio, N. W., Jacobs, C. S. Jr., and Holmes, W. L. (1965). Thin layer chromatography and identification of free sterols. *J. Chromatogr. 20*: 354–357.

Downing, D. T. (1979). Lipids. In *Densitometry in Thin Layer Chromatography—Practice and Applications*, J. C. Touchstone and J. Sherma (Eds.). Wiley, New York, pp. 367–391.

Dudzinski, A. E. (1967). A spray sequence specific for the detection of free fatty acids. *J. Chromatogr. 31*: 560.

Duncan, M., Fried, B., and Sherma, J. (1987). Lipids in fed and starved *Biomphalaria glabrata* (Gastropoda). *Comp. Biochem. Physiol. 86A*: 663–665.

Dunstan, G. A., Volkman, J. K., and Barrett, S. M. (1993). The effect of lyophilization on the solvent extraction of lipid classes, fatty acids and sterols from the oyster *Crassostrea gigas*. *Lipids. 28*: 937–944.

Fell, R. D. (1996). Thin-layer chromatography in the study of entomology. In *Practical Thin-Layer Chromatography—A Multidisciplinary Approach*, B. Fried and J. Sherma (Eds.). CRC Press, Boca Raton, FL, pp. 71–104.

Folch, J., Lees, M., and Sloane-Stanley, G. H. (1957). A simple method for the isolation and purification of total lipids from animal tissue. *J. Biol. Chem. 226*: 497–509.

Frazer, B. A., Reddy, A., Fried, B., and Sherma, J. (1997a). HPTLC determination of neutral lipids and phospholipids in *Lymnaea elodes* (Gastropoda). *J. Planar Chromatogr.—Mod. TLC 10*: 128–130.

Frazer, B. A., Reddy, A., Fried, B., and Sherma, J. (1997b). Effects of diet on the lipid composition of *Echinostoma caproni* (Trematoda) in ICR mice. *Parasitol. Res. 83*: 642–645.

Fried, B. (1991). Lipids. In *Handbook of Thin-Layer Chromatography*, J. Sherma and B. Fried (Eds.). Marcel Dekker, Inc., New York, pp. 353–387.

Fried, B. (1993). Obtaining and handling biological materials and prefractioning extracts for lipid analyses. In *Handbook of Chromatography—Analysis of Lipids*, K. D. Mukherjee and N. Weber (Eds.). CRC Press, Boca Raton, FL, pp. 1–10.

Fried, B. (1996). Lipids. In *Handbook of Thin-Layer Chromatography*, 2 ed. J. Sherma and B. Fried (Eds.). Marcel Dekker, New York, pp. 683–714.

Fried, B., Cahn-Hidalgo, D., Fujino, T., and Sherma, J. (1991). Diet-induced differences in the distribution of neutral lipids in selected organs of *Biomphalaria glabrata* (Gastropoda: Planorbidae) as determined by thin-layer chromatography and light and electron microscopy. *Trans. Am. Microsc. Soc. 110*: 163–171.

Fried, B, and Haseeb, M. A. (1996). Thin-layer chromatography in parasitology. In *Practical Thin-Layer Chromatography—A Multidisciplinary Approach*, B. Fried and J. Sherma (Eds.). CRC Press, Boca Raton, FL, pp. 51–70.

Fried, B., and Shapiro, I. L. (1979). Thin-layer chromatographic analysis of phospholipids in *Echinostoma revolutum* (Trematoda) adults. *J. Parasitol. 65*: 243–245.

Fried, B., Shenko, F. J., and Eveland, L. K. (1983). Densitometric thin-layer chromatographic analysis of cholesterol in *Schistosoma mansoni* (Trematoda) adults and their excretory–secretory products. *J. Chem. Ecol. 9*: 1483–1489.

Fried, B., and Sherma, J. (1994). *Thin-Layer Chromatography, Techniques and Applications*, 3rd ed., Marcel Dekker, New York, pp. 245–285.

Fried, B., Tancer, R. B., and Fleming, S. J. (1980). In vitro pairing of *Echinostoma revolutum* (Trematoda) metacercariae and adults, and characterization of worm products involved in chemoattraction. *J. Parasitol. 66*: 1014–1018.

Gennaro, L., Fried, B., and Sherma, J. (1996). HPTLC determination of phospholipids in snail-conditioned water from *Helisoma trivolvis* and *Biomphalaria glabrata*. *J. Planar Chromatogr.—Mod. TLC 9*: 379–381.

Gerin, C., and Goutx, M. (1993). Separation and quantification of phospholipids from marine bacteria with the Iatroscan Mark IV TLC-FID. *J. Planar Chromatogr.—Mod. TLC 6*: 307–312.

Gunstone, F. D., and Herslof, B. G. (1992). *A Lipid Glossary*, The Oily Press, Ltd., Ayr, Scotland, 117 pages.

Gunstone, F. E., and Padley, F. B. (Eds.) (1997). *Lipid Technologies and Applications*. Marcel Dekker, New York, 834 pages.

Gurr, M. I., and Harwood, J. L. (1991). *Lipid Biochemistry—An Introduction*, 4th ed. Chapman S. Hall, London, pp. 1–9.

Hakomori, S. (1981). Glycosphingolipids in cellular interaction, differentiation, and oncogenesis. *Ann. Rev. Biochem. 50*: 733–764.

Hamid, M. E., Minnikin, D. E., Goodfellow, M., and Ridell, M. (1993). Thin-layer chromatographic analysis of glycolipids and mycolic acids from *Mycobacterium farcinogenes, Mycobacterium senegalense* and related taxa. *Zentralbl. Bakteriol. 279*: 354–367.

Hammond, E. W. (1996). The analysis of lipids. In *Advances in Applied Lipid Research: A Research Annual*, Vol. 2, F. B. Padley (Ed.). Greenwich, UK, pp. 35–94.

Haseeb, M. A., and Fried, B. (1988). Chemical communication in helminths. *Adv. Parasitol. 27*: 169–207.

Hegewald, H. (1996). One-dimensional thin-layer chromatography of all known D-3 and D-4 isomers of phosphoinositides. *Anal. Biochem. 242*: 152–155.

Holloway, P. J. (1984). Surface lipids of plants and animals. *Handbook of Chromatography—Lipids*, Vol. I, H. K. Mangold (Ed.). CRC Press, Boca Raton, FL, pp. 347–380.

Horutz, K., Layman, L. R., Fried, B., and Sherma, J. (1993). HPLC determination of triacylglycerols in the digestive gland–gonad complex of *Biomphalaria glabrata* snails fed hen's egg yolk versus leaf lettuce. *J. Liq. Chromatogr. 16*: 4009–4017.

Hunter, S. W., and Brennan, P. J. (1981). A novel phenolic glycolipid from *Mycobacterium leprae* possibly involved in immunogenicity and pathogenicity. *J. Bacteriol. 147*: 728–735.

Innocente, N., Blecker, C., Deroanne, C., and Paquot, M. (1997). Langmuir film balance study of the surface properties of a soluble fraction of milk fat–globule membrane. *J. Agric Food Chem. 45*: 1559–1562.

Iwamori, M., Hirota, K., Utsuki, T., Momoeda, K., Ono, K., Tsuchida, Y., Okumura, K., and Hanaoka, K. (1996). Sensitive method for the determination of pulmonary surfactant phospholipid/sphingomyelin ratio in human amniotic fluids for the diagnosis of respiratory distress syndrome by thin-layer chromatography–immunostaining. *Anal. Biochem. 238*: 29–33.

Jain, R. (1996). Thin-layer chromatography in clinical chemistry. In *Practical Thin-Layer Chromatography—A Multidisciplinary Approach*, B. Fried and J. Sherma (Eds.). CRC Press, Boca Raton, FL, pp. 131–152.

Kates, M. (1986). *Techniques of Lipidology, Isolation, Analysis, and Identification of Lipids*, 2nd ed. Elsevier, Amsterdam.

Kritchevsky, D., and Kirk, M. R. (1952). Detection of steroids in paper chromatography. *Arch. Biochem. Biophys. 35*: 346–351.

Lawrence, J. M. (1976). Patterns of lipid storage in postmetamorphic marine invertebrates. *Am. Zool. 16*: 747–762.

Lin, L., Zhang, J., Huang, S., He, C., and Tang, J. (1994). Identification of fetal lung maturity by one-dimensional thin layer chromatographic determination of several phospholipids in amniotic fluid. *J. Planar Chromatogr.—Mod. TLC 7*: 25–28.

Linard, A., Guesnet, P., and Durand, G. (1993). Separation of the classes of phospholipids by overpressured layer chromatography (OPLC). *J. Planar Chromatogr.—Mod. TLC 6*: 322–323.

Lowry, R. R. (1968). Ferric chloride spray detector for cholesterol and cholesteryl esters on thin layer chromatograms. *J. Lipid Res. 9*: 397.

Lu, P., and Lai, B. (1996). Determination of phosphatidylethanolamine in yolk phospholipid by thin-layer chromatography. *Chinese J. Pharm. Ind. 27*: 222–224.

Lundberg, W. O. (1984). Lipidology. In *Handbook of Chromatography—Lipids*, Vol. I, H. K. Mangold (Ed.). CRC Press, Boca Raton, FL, pp. 1–32.

Mahadevan, V., and Ackman, R. G. (1984). Alcohols, aldehydes, and ketones. In *Handbook of Chromatography—Lipids*, Vol. I. H. K. Mangold (Ed.). CRC Press, Boca Raton, FL, pp. 73–93.

Mahadevappa V. G., and Holub, B. J. (1987). Chromatographic analysis of phosphoinositides and their breakdown products in activated blood platelets/neutrophils. *J. Chromatogr. Libr. 37*: 225–265.

Malins, D. C., and Wekell, J. C. (1970). The lipid biochemistry of marine organisms. In *The Chemistry of Fats and Other Lipids*, Vol. 10, R. T. Holman (Ed.). Pergamon Press, London, pp. 339–369.

Maloney, M. (1996). Thin-layer chromatography in bacteriology. In *Practical Thin-Layer Chromatography—A Multidisciplinary Approach*, B. Fried and J. Sherma (Eds.). CRC Press, Boca Raton, FL, pp. 19–31.

Mangold, H. K. (1969). Aliphatic lipids. In *Thin Layer Chromatography*, 2nd ed., E. Stahl (Ed.). Springer-Verlag, New York, pp. 363–421.

Marquez-Ruiz, G., Perez-Camino, M. C., and Dobarganes, M. C. (1994). Evaluation of susceptibility to oxidation of linoleyl derivatives by thin-layer chromatography with flame ionization detection. *J. Chromatogr. 662*: 363–368.

Marshall, D. L., Beattie, A. J., and Bollenbacher, W. E. (1979). Evidence for diglycerides as attractants in an ant–seed interaction. *J. Chem. Ecol. 5*: 335–344.

Martinez-Lorenzo, M. J., Marzo, I., Naval, J., and Pineiro, A. (1994). Self-staining of polyunsaturated fatty acids in argentation thin-layer chromatography. *Anal. Biochem. 220*: 210–212.

Masterson, C., Fried, B., and Sherma, J. (1993). High performance thin-layer chromatographic analysis of neutral lipids in *Biomphalaria glabrata* (Gastropoda) fed yolk-lettuce versus lettuce-Tetramin diets. *Microchem. J. 47*: 134–139.

Mehra, V., Brennan, P. J., Rada, E., Convit, J., and Bloom, B. R. (1984). Lymphocyte suppression in leprosy induced by unique *Mycobacterium leprae* glycolipid. *Nature 308*: 194–196.

Morris, G. D. (1966). Separations of lipid by silver ion chromatography. *J. Lipid Res. 7*: 717–732.

Morris, K., Sherma, J., and Fried, B. (1987). Determination of cholesterol in egg yolk by high performance TLC with densitometry. *J. Liq. Chromatogr. 10*: 1277–1290.

Muething, J. (1994). Improved thin-layer chromatographic separation of gangliosides by automated multiple development. *J. Chromatogr., B: Biomed. Appl. 657*: 75–81.

Muething, J. (1996). High-resolution thin-layer chromatography of gangliosides. *J. Chromatogr., A 720*: 3–25.

Nicolaides, N. (1974). Skin lipids: Their biochemical uniqueness. *Science 186*: 19–26.

Nikolova-Damyanova, B., Chobanov, D., and Dimov, S. (1993). Separation of isomeric triacylglycerols by silver ion thin-layer chromatography. *J. Liq. Chromatogr. 16*: 3997–4008.

Nikolova-Damyanova, B., Christie, W. W., and Herslof, B. (1994). Improved separation of some positional isomers of monounsaturated fatty acids, as their phenacyl derivatives, by silver-ion thin layer chromatography. *J. Planar Chromatogr.—Mod. TLC 7*: 382–385.

Ogawa-Goto, K., Ohta, Y., Kubota, K., Funamoto, N., Abe, T., Taki, T., and Nagashima, K. (1993). Glycosphingolipids of human peripheral nervous system myelins isolated from cauda equina. *J. Neurochem. 61*: 1398–1403.

Park, Y. Y., Fried, B., and Sherma, J. (1991). Densitometric thin-layer chromatographic studies on neutral lipids in two strains of *Helisoma trivolis* (Gastropoda). *Comp. Biochem. Physiol. 100B*: 127–130.

Parrish, C. C., Bodennec, G., and Gentien, P. (1996). Determination of glycoglycerolipids by Chromarod thin-layer chromatography with Iatroscan flame ionization detection. *J. Chromatogr., A 741*: 91–97.

Paul, J. (1975). *Cell and Tissue Culture*, 5th ed. Churchill Livingstone, Edinburgh.

Pchelkin, V. P. (1997). Evaluation of purity of fractions after separating unsaturated polar lipids by adsorption thin-layer chromatography in the presence of silver ions. *J. Anal. Chem. 52*: 302–307.

Perez, M. K., Fried, B., and Sherma, J. (1994). Comparison of mobile phases and HPTLC qualitative and quantitative analysis, on preadsorbent silica gel plates, of phospholipids in *Biomphalaria glabrata* (Gastropoda) infected with *Echinostoma caproni* (Trematoda). *J. Planar Chromatogr.—Mod. TLC 7*: 340–343.

Peyrou, G., Rakotondrazafy, V., Mouloungui, Z., and Gaset, A. (1996). Separation and quantitation of mono-, di-, and triglycerides and free oleic acid using thin-layer chromatography with flame-ionization detection. *Lipids 31*: 27–32.

Phillips, F., and Privett, O. S. (1979). A simplified procedure for the quantitative extraction of lipids from brain tissue. *Lipids 14*: 590–595.

Phleger, C. F., Patton, J., Grimes, P., and Lee, R. F. (1976). Fish bone oil: Percent total body lipid and carbon-14 uptake following feeding of 1-^{14}C-palmitic acid. *Mar. Biol. 35*: 85–90.

Privett, O. S., Dougherty, K. A., and Erdahl, W. L. (1973). Quantitative analysis of the lipid classes by thin layer chromatography via charring and densitometry. In *Quantitative Thin Layer Chromatography*, J. C. Touchstone (Ed.). Wiley, New York, pp. 57–78.

Przybylski, R., and Eskin, N. A. M. (1994). Two simplified approaches to the analysis of cereal lipids. *Food Chem. 51*: 231–235.

Rabinowitz, J. (1996). Salivary lipid profiles of the leech (*Hirudo medicinalis*). *Lipids 31*: 887–888.

Rementzis, S., Antonopolou, C. A., and Demopoulos, C. A. (1997). Identification and study of gangliosides from *Scomber scombrus* muscle. *J. Agric Food. Chem. 45*: 611–615.

Rupcic, J., Blagovic, B. and Maric, V. (1996). Cell lipids of the *Candida lipolytica* yeast grown on methanol. *J. Chromatogr A, 755*: 75–80.

Saha, S. K., and Das, S. K. (1996). A simple densitometric method for estimation of polar and non-polar lipids by thin layer chromatography with iodine vapor visualization. *J. Liq. Chromatogr. Relat. Technol. 19*: 3125–3134.

Sajbidor, J., Certik, M., and Grego, J. (1994). Lipid analysis of bakers' yeast. *J. Chromatogr., A 665*: 191–195.

Schnaar, R. L., and Needham, L. K. (1994). Thin-layer chromatography of glycosphingolipids. *Methods Enzymol. 230*: 371–389.

Schuerer, N. Y., Schliep, V., and Barlag, K. (1995). Quantitative screening of human stratum corneum lipids by sequential one-dimensional thin layer chromatography. *Exp. Dermatol. 4*: 46–51.

Sherma, J. (1996). Planar chromatography. *Anal. Chem. 68*: 1R–19R.

Sherma, J., and Bennett, S. (1983). Comparison of reagents for lipid and phospholipid detection and densitometric quantitation on silica gel and C_{18} reversed phase thin layers. *J. Liq. Chromatogr. 6*: 1193–1211.

Shetty, P. H., Fried, B., and Sherma, J. (1990). Sterols in the plasma and digestive gland-gonad complex of *Biomphalaria glabrata* snails, fed lettuce versus hen's egg yolk, as determined by GLC. *Comp. Biochem. Physiol. 96B*: 791–794.

Shorey, H. H. (1976). *Animal Communication by Pheromones*. Academic Press, New York.

Skidmore, B. A., and Heithaus, E. R. (1988). Lipid cues for seed-carrying by ants in *Hepatica americana. J. Chem. Ecol. 14*: 2185–2196.

Skipski, V. P., Smolowe, A. F., Sullivan, R. C., and Barclay, M. (1965). Separation of lipid classes by thin-layer chromatography. *Biochim. Biophys. Acta 106*: 386–396.

Smith, M. C., Webster, C. L., Sherma, J., and Fried, B. (1995). Determination of neutral lipids in regular and lowfat eggs by high-performance TLC with densitometry. *J. Liq. Chromatogr. 18*: 527–535.

Snyder, F. (1971). The chemistry, physical properties and chromatography of lipids containing ether bonds. In *Progress in Thin-Layer Chromatography and Related Methods*, Vol. II. A. Niederwieser and G. Pataki (Eds.). Ann Arbor Science Publishers, Ann Arbor, MI, pp. 105–141.

St. Angelo, A. J., and James, C. Jr. (1993). Analysis of lipids from cooked beef by thin-layer chromatography with flame-ionization detection. *J. Am. Oil Chem. Soc. 70*: 1245–1250.

Taki, T., Kasama, T., Handa, S., and Ishikawa, D. (1994). A simple and quantitative purification of glycosphingolipids and phospholipids by thin-layer chromatography blotting. *Anal. Biochem. 223*: 232–238.

Tarandjiiska, R., Marekov, I., Nikolova-Damyanova, B., and Amidzhin, B. (1995). Determination of molecular species of triacylglycerols from highly unsaturated plant oils by successive application of silver ion and reversed phase TLC. *J. Liq. Chromatogr. 18*: 859–872.

Thompson, S. N., Mejia-Scales, V., and Borchardt, D. B. (1991). Physiologic studies of snail–schistosome interactions and potential for improvement of *in vitro* culture of schistosomes. *In Vitro Cell Dev. Biol. 27A*: 497–504.

Tvrzicka, E., and Votruba, M. (1994). Thin-layer chromatography with flame-ionization detection. *Chromatogr. Sci. Ser. 65*: 51–73.

Vague, J., and Fenasse, R. (1965). Comparative anatomy of adipose tissue. In *Handbook of Physiology*, Section 5: Adipose Tissue, A. E. Renold and G. F. Cahill, Jr. (Eds.). Williams & Wilkens, Baltimore, pp. 25–36.

Vandenheuvel, F. A. (1971). Structure of membranes and role of lipids therein. *Adv. Lipid Res. 9*: 161–248.

Vecchini, A., Chiaradia, E., Covalovo, S., and Binaglia, L. (1995). Quantitation of phospholipids on thin layer chromatographic plates using a desk-top scanner. *Mol. Cell. Biochem. 145*: 25–28.

Vuorela, P., Vuorela, H., Suppula, H., and Hiltunen, R. (1996). Development of quantitative TLC assay for phospholipid products. *J. Planar Chromatogr.—Mod. TLC 9*: 254–259.

Wagner, H., Horhammer, L., and Wolff, P. (1961). Thin layer chromatography of phosphatides and glycolipids. *Biochem. Z. 334*: 175–184.

Wardas, W., and Pyka, A. (1993). New visualizing agents for selected fatty derivatives in thin-layer chromatography. *J. Planar Chromatogr.–Mod. TLC 6*: 320–322.

Wardas, W., and Pyka, A. (1994). Visualizing agents for cholesterol in TLC. *J. Planar Chromatogr.—Mod. TLC 7*: 440–443.

Wardas, W., and Pyka, A. (1997). New visualizing agents for unsaturated higher fatty acids in TLC. *J. Planar Chromatogr.—Mod. TLC 10*: 63–67.

Watanabe, K., and Mizuta, M. (1995). Fluorometric detection of glycosphingolipids on thin-layer chromatographic plates. *J. Lipid Res. 36*: 1848–1855.

Weldon, P. J. (1996). Thin-layer chromatography of the skin secretions of vertebrates. In *Practical Thin-Layer Chromatography—A Multidisciplinary Approach*, B. Fried and J. Sherma (Eds.). CRC Press, Boca Raton, FL, pp. 105–130.

Wiesner, D. A., and Sweeley, C. C. (1995). Characterization of gangliosides by two-dimensional high-performance thin-layer chromatography. *Anal. Chim. Acta 311*: 57–62.

Wortmann, W., and Touchstone, J. C. (1973). Techniques for determination of specific activity of isotopic materials by thin layer chromatography. In *Quantitative Thin Layer Chromatography*, J. C. Touchstone (Ed.). Wiley, New York, pp. 23–44.

Wortman, W., Wortman, B., Schnabel, C., and Touchstone, J. C. (1974). Rapid determination of estriol of pregnancy by spectrodensitometry of thin layer chromatograms. *J. Chromatogr. Sci. 12*: 377–379.

Xu, G., Waki, H., Kon, K., and Ando, S. (1996). Thin-layer chromatography of phospholipids and their lyso forms: application to determination of extracts from rat hippocampal CA1 region. *Microchem. J. 53*: 29–33.

Zellmer, S., and Lasch, J. (1997). Individual variation of human plantar stratum corneum lipids, determined by automated multiple development of high-performance thin-layer chromatography plates. *J. Chromatogr., B: Biomed. Sci. Appl. 691*: 321–329.

16

Amino Acids

I. DEFINITION

Amino acids are organic compounds that contain the basic amine group, NH_2, and the acid carboxyl group, COOH. The general formula for amino acids is

$$\begin{array}{c} RCHCOOH \\ | \\ NH_2 \end{array}$$

For most amino acids the R represents a complex organic group; in glycine, the R represents H (hydrogen). Amino acids are usually known by their common rather than chemical names. For instance glycine, with the structural formula H_2N-CH_2-COOH, is known chemically as amino acetic acid. For a discussion of the structure of amino acids, consult Harborne (1984) and Bhushan and Martens (1996). In living aqueous systems, amino acids exist predominantly as dipolar ions referred to as "zwitterions" and represented by the formula

$$\begin{array}{c} RCHCOO^- \\ | \\ {}^+NH_3 \end{array}$$

Because of the ionic charge, separation of amino acids from inorganic salts may be difficult (Scott, 1969) and requires considerable sample cleanup as described in Chapter 4. Amino acids are soluble in water, but some are less soluble than others. Alcoholic 0.5 *M* HCl should be used to prepare solutions of amino acids that are only sparingly soluble in water.

Reference is usually made to the 20 amino acids found in proteins. These amino acids, along with the conventional 3- and 1-letter designations for each, are as follows: alanine (Ala, A); arginine (Arg, R); asparagine (Asn, N); aspartic acid (Asp, D); cysteine (Cys, C); glutamic acid (Glu, E); glutamine (Gln, Q); glycine (Gly, G); histidine (His, H); leucine (Leu, L); isoleucine (Ile, I); lysine (Lys, K); methionine (Met, M); phenylalanine (Phe, F); proline (Pro, P); serine (Ser, S); threonine (Thr, T); tryptophan (Trp, W); tyrosine (Tyr, Y); and valine (Val, V). According to Mant et al. (1992), the above-named amino acids are those most commonly determined by analytical techniques.

Numerous additional amino acids are found in animal and plant tissues, but these are generally not considered protein amino acids. Examples of such amino acids are γ-aminobutyric acid, hydroxyproline, ornithine, and pipecolic acid. These amino acids usually result from metabolic processes that occur in animals and plants. According to Harborne (1984), there may be more than 200 such nonprotein amino acids in plants.

II. FUNCTION

Amino acids are the building blocks of proteins; upon hydrolysis of plant and animal tissues, the 20 protein amino acids mentioned in Section I will usually be obtained. Amino acids are also found free (referred to as free-pool amino acids) in animal and plant tissues and in the blood of vertebrates and invertebrates. Qualitative and quantitative differences may exist in the free-pool amino acids of different species, and the significance of such variations is obscure. However, amino acid variations in species may be important in studies on the chemotaxonomy of organisms (Gilbertson and Schmid, 1975). By way of an example of differences in free-pool amino acids in related organisms, the most abundant free-pool amino acids in the rat tapeworm, *Hymenolepis diminuta*, are alanine and glutamic acid. By contrast, the most abundant free-pool amino acids in tapeworms of elasmobranchs (sharks and rays) are glycine and taurine (von Brand, 1973). It should be remembered, however, that a variety of analytical techniques are used to determine free-pool amino acids, and some of the interspecific variations reported in the literature may actually reflect differences in analytical procedures used.

From simple animals like protozoans, through more complex ones such as worms and insects, to the most highly evolved vertebrates, there appear to be 9 or 10 *essential* amino acids that must be supplied from an outside (exogenous) source (Prosser, 1973a). These essential amino acids are valine, leucine, isoleucine, lysine, arginine, methionine, threonine, phenylalanine, tryptophan, and histidine. Most animals appear to be able to synthesize the remaining protein amino acids required as the building blocks of proteins.

Amino acids play a role in osmoregulatory functions of animals. Usually in vertebrates the concentration of free-pool amino acids in the blood is relatively low. In invertebrates such as insects and molluscs, however, the concentration of free-pool amino acids in the hemolymph is relatively high and is related to an osmoregulatory function (Awapara, 1962).

Certain amino acids function as neurotransmitters, and perhaps the best-known amino acid in this role is γ-aminobutyric acid (GABA) (Prosser, 1973b). There is good evidence that GABA is a neuromuscular inhibitor in insects and crustaceans. Other amino acids, such as glycine and L-glutamate, have also been implicated as neurotransmitters (Prosser, 1973b).

In certain marine invertebrates and in parasitic worms, amino acids from the surrounding environment are taken in, either passively by diffusion or by active transport mechanisms. Conversely, some of these organisms also emit or excrete considerable amounts of amino acids through the general body surface or via the excretory system. For further discussion on this topic, consult von Brand (1973). By contrast, vertebrates excrete only minimal amounts of amino acids.

Although amino acids are generally considered as substances utilized for the synthesis of proteins and repair of body tissues, these compounds also serve as an energy source, particularly in times of starvation or deprivation. As discussed by Arme (1977) the amino acid proline is used as an energy reserve in blood-feeding insects and in certain blood-feeding protozoan parasites (hemoflagellates).

In plants, amino acids provide a major nitrogen source for the biosynthesis of numerous endogenous nitrogenous compounds. As mentioned in Section I, there are many nonprotein amino acids in plants, and the role of most of these substances is poorly understood (Harborne, 1984).

Haas et al. (1995) found that the stimulating snail-host cues for the chemoorientation of certain larval trematodes (i.e., echinostome cercariae) toward their hosts are small molecular amino acids, amino compounds, and peptides.

III. AMINO ACID TLC

TLC has been used to identify amino acids from plant and animal tissues and fluids. Heathcote (1979) described the use of various two-dimensional solvent systems with and without pretreatment of samples (for cleanup purposes) for the analysis of free-pool amino acids in bacteria and molds. Berry (1973) used various unidimensional solvent systems for TLC diagnosis of amino acidurias in human plasma. Hanes et al. (1961) used unidimensional solvent systems for the analysis of the 20 protein amino acids from plants. Harborne (1984) has discussed the use of various two-dimensional TLC solvent systems for the successful identification of amino acids in various plant samples. Bailey and Fried (1977) used various unidimensional solvent systems for the analysis of free-pool amino acids in a parasitic worm.

By careful use of various solvent systems, either in a unidimensional or double-dimensional mode, and of numerous specific chemical detection tests, TLC can be used for the identification of amino acids in plant and animal tissues and fluids.

It should be remembered that TLC studies of amino acids are an offshoot of similar studies using paper chromatography (PC). In fact, early PC solvent systems, such as those described by Levy and Chung (1953), can be used successfully for TLC on cellulose with little modification. Amino acid TLC is often used in conjunction with other analytical procedures, that is, TLC plus electrophoresis or TLC plus ion-exchange chromatography (Scott, 1969).

A perusal of the voluminous literature on the analysis of amino acids indicates that most separation procedures resort to automatic amino acid analyzer techniques. Teachers at small colleges or researchers on limited budgets may not have access to an automatic amino acid analyzer. Therefore, these workers must rely more heavily on planar chromatography procedures (both PC and TLC) for amino acid separations.

Studies by Heathcote (1979) indicate that quantitative densitometric TLC analyses of amino acids will provide results equivalent to that obtained from automated amino acid analysis procedures. In a study by Mack et al. (1979) on amino acids in mosquitoes, qualitative TLC procedures were first used to determine amino acid profiles in the tissue and hemolymph, and, subsequently, the automated amino acid analyzer was applied for quantification. The study of Mack et al. (1979) should be useful to those interested in qualitative and quantitative analysis of amino acids from biological samples.

Numerous publications are available on the quantitative determination of amino acids by TLC. The following studies are representative. Yamada et al. (1985) quantified the metabolites related to the tryptophan–NAD pathway by cellulose HPTLC using five multiple developments with four solvents and scanning at 254 nm. Lander and Treiber (1985) quantified L-alanine N-carboxyanhydride, derivatized with n-butylamine, developed on silica gel with ethyl acetate–ethanol–28% NH_4OH (80:20:4), and scanned at 210 nm. Lancaster and Kelly (1983) quantified S-alk(en)yl-L-cysteine sulfoxides in onion by silica gel TLC, detection with ninhydrin, and scanning. Chin and Koehler (1983) quantified histamine, tryptamine, phenethylamine, and tyramine in soy sauce by TLC-densitometry of dansyl derivatives. Monreal and McGill (1985) quantified free amino acids in soil using tritiated 1-fluoro-2,4-dinitrobenzene, two-dimensional TLC of the dinitrophenyl derivatives, and counting ^3H activity in the spots. Pastor-Anglada et al. (1984) quantified [^{14}C] dansyl amino acids in plasma by polyamide TLC and scintillation counting of scraped spots.

The following section provides a brief review of significant studies on the TLC of amino acids from 1983 to 1997.

Fater et al. (1983) used overpressurized TLC (OPTLC) to determine amino acids in cereals; very tight zones were formed in 75 min, and 10–20 samples were separated on a single plate. Abraham et al. (1983) also used OPTLC for the separation of reduced and oxidized forms of glutathione from amino acid mixtures.

Keller et al. (1984) separated cross-linking amino acids of elastin by two-dimensional silica gel TLC using butanol–acetic acid–water (4:1:1) and propanol–NH$_3$–water (8:1:11). Schwartz (1984) detected as little as 60 pmol of histidine phenylthiohydantoin by UV (366 nm) irradiation on a fluorescein-containing silica gel plate after development with ethanol–acetic acid (7:3). Henderson et al. (1985) reported that two-dimensional TLC was useful for screening aspartyl-glycosaminuria in the urine of children by heating ninhydrin-stained plates at 120°C.

Guenther et al. (1985) resolved enantiomers of *o*-alkyl amino acids by reversed-phase TLC on commercial Chiralplates developed with mixtures of methanol–water–acetonitrile. Marchelli et al. (1986) also separated amino acid and Dns–amino acid enantiomers on Chiralplates. Metrione (1986) reported that lissamine rhodamine B sulfonyl (laryl) derivatives of amino acids were detected by UV irradiation as fluorescent zones with a sensitivity of 400 fmol. Sarbu et al. (1987) noted that spraying with a 0.1% solution of 9-isothiocyanatoacridine spray with a mixture of derivatives in dichloromethane provided fluorescence detection of amino acids on silica gel R plates.

Bhushan and Ali (1987) tested amino acid separations on silica gel layers impregnated with various metal salts. Bhushan and Reddy (1989) reported the separation of phenylthiohydantoin (PTH) amino acids on silica gel with new mobile phases. Laskar and Basak (1988) described a new ninhydrin-based procedure that produced different colors and good sensitivity for amino acid detection. Bhushan and Reddy (1987) reviewed the TLC of PTH amino acids. Gankina et al. (1989) described a unidimensional multistep silica gel HPTLC method for the separation and identification of PTH and dansylamino acids. Bhushan et al. (1987) developed numerous solvent systems for effective separations of 2,4-dini-trophenyl-(DNP) amino acids. Bhushan (1988) reviewed the TLC resolution of enantiomeric amino acids and their derivatives. Kuhn et al. (1989) reported the amino acid enantiomer separation by TLC on cellulose of D- and L-tryptophan and methyltryptophan. Guenther (1988) determined TLC-separated enantiomers by densitometry.

Blackburn (1989) provided an update of his earlier handbook of amino acid chromatography and extended the number of tables devoted to the TLC of amino acids and amines to 56 (compared to 32 in the previous volume). The literature covered in the newer volume is mainly from 1981 to 1987. Some new methods of sample preparation, derivatization, and detection of these compounds are also described in this volume.

Grosvenor and Gray (1990) synthesized 2,4-dinitrophenylpyridium and used it as a colorimetric or fluorometric detection reagent for amino acids. Mueller and Eckert (1989) quantified glycine betaine in plants by use of two ion-exchange columns and silica gel TLC. Surgova et al. (1990) separated histamine from polyamines and determined it by spectrodensitometric TLC after dansyl derivatization.

Bhushan (1991) provided an extensive review on all aspects of the TLC of amino acids and their derivatives with references up to 1989. The procedures described in the review are based mainly on work done in Bhushan's laboratory, but contributions from other laboratories are also included. The methods described in this review provide an excellent starting point for the application of TLC to the analysis of amino acids and their derivatives.

Mustafa et al. (1992) used two-dimensional thin-layer chromatography on polyamide sheets to identify PTH–amino acids that have similar R_F values. The solvent systems used in this study were particularly effective for the separation of PTH–amino acids and provided a clear separation of all the PTH–amino acids studied with the exception of isoleucine/leucine.

Lepri et al. (1992) investigated the chromatographic behavior of racemic dinitropyridyl, dinitrophenyl, and dinitrobenzoyl amino acids on reversed-phase TLC plates developed with aqueous–organic mobile phases containing bovine serum albumin (BSA) as chiral agent. The method was simple, highly sensitive, and inexpensive, and the plates gave excellent resolution of most of the DL racemates with development time between 1 and 2 h. Varshney et al. (1992) described chromatographic separation of α-amino acids on antimony (V) phosphate–silica gel G plates in different aqueous, nonaqueous, and mixed solvent systems. They separated basic amino acids from neutral and acidic amino acids and their methods have proved useful for the quantitative separation of amino acids from drug samples.

Brink (1992) described a laboratory experiment for the qualitative analysis of amino acids present in egg lysozyme, involving hydrolysis with HCl, dansylation, and amino acid identification by TLC. The C-terminal amino acid was identified as leucine in a second experiment in which lysozyme was oxidized with acetic acid, hydrolyzed with carboxypeptidase A, and the cleaved acid dansylated and identified by TLC.

Pyka (1993) developed a new topological index for predicting the separation of D and L optical isomers of amino acids on chiral plates. LeFevre (1993) used reversed-phase TLC to separate dansyl–amino acid enantiomers. Das and

Sawant (1993) quantified amino acids in protein extracted from cane sugar after hydrolysis, derivatization with dansyl chloride, and silica gel HPTLC with a mobile phase of 5% EDTA–butanol–diethyl ether (5:10:35).

Bhushan and Parshad (1994) reported that amino acid separations on silica gel layers impregnated with transition metal ions were based on the complexing effect between the acids and the metals. Bhushan et al. (1994a) found that three new solvent systems, pyridine–benzene (2.5:20), methanol–carbon tetrachloride (1:20), and acetone–dichloromethane (0.3:8) improved the resolution and identification of 18 PTH amino acids compared to previously reported systems. Norfolk et al. (1994) compared separations of 18 amino acids on HPTLC silica gel, cellulose, and C-18 bonded silica gel layers; the ability to identify amino acids in the hemolymph and digestive gland–gonad complex of the medically important snail *Biomphalaria glabrata* was also studied; they also quantified by scanning densitometry alanine and aspartic acid from the hemolymph. Sinhababu et al. (1994) described a new spray reagent, *p*-dichloro-dicyanobenzoquinone, that detected amino acids with 0.1 to 1 μg detection limits and yielded various distinct colors that helped identify the acids. Jork and Ganz (1994) used TLC on chiral plates with a mobile phase of acetonitrile–methanol–water (4:1:1) and ninhydrin as the detection reagent for quality control of L-tryptophan. Clift et al. (1994) described an improved TLC technique for neonatal screening of amino acid disorders using dried blood spots. Mathur et al. (1994) resolved amino acid racemates on borate-gelled guaran-impregnated silica gel plates. Bhushan et al. (1994b) separated amino acid racemates on silica gel plates impregnated with a complex of copper and L-proline. Degtiar et al. (1994) used scanning densitometry to quantify L-lysine, L-threonine, L-homoserine in fermentation broths.

Siddiqi et al. (1995) separated alpha amino acids on tin(IV) selenoarsenate layers with a mobile phase of dimethyl sulfoxide (DMSO). Petrovic and Kastelan-Macan (1995) used scanning densitometry for the validation of quantitative amino acid analysis on mixed natural zeolite–microcrystalline cellulose layers. Ohtake et al. (1995) used cellulose TLC to analyze homocysteine from dried blood spots of patients with homocystinuria.

Bhushan et al. (1996) reported on comparative studies of HPLC and TLC for the separation of amino acids using Cu(II) ion and Bhushan and Parshad (1996) described separation by TLC of enantiomeric dansyl–amino acids using a macrocyclic antibiotic as a chiral selector. Malakhova et al. (1996) described video-densitometric methods for quantitative analysis by TLC of industrial amino acids. Huang et al. (1996) used cellulose TLC to separate certain aromatic amino acids into enantiomers using cyclodextrin mobile-phase additives. DeMeglio and Svanberg (1996) described a TLC procedure for the separation of proline and hydroxyprolene from biological samples. This technique is useful for studies on collagen metabolism, which often necessitates the distinction of these two amino acids. Bhushan and Martens (1996) extended and updated the earlier review of

Bhushan (1991) on amino acids and their derivatives; their new review includes references through 1994. Jain (1996) has reviewed studies on the applications of TLC to amino acid analysis of biological fluids and tissues. Such analyses are important in making diagnoses of inborn errors of amino acid metabolism. It was noted by Jain (1996) that TLC has proved useful to screen and quantify abnormal amounts of free amino acids in blood and urine samples. Shalaby (1996) in his chapter on TLC in food analysis has provided some protocols on amino acid analysis following the hydrolysis of protein samples. Fried and Haseeb (1996) in their chapter on TLC in parasitology have provided some information on the analysis of free pool amino acids from tissues of parasites and from the hemolymph of mosquitoes infected with *Plasmodium* (the malaria organism).

Steiner et al. (1998) used HPTLC to analyze amino acids in water conditioned by several medically important snails—*Biomphalaria glabrata, Helisoma trivolvis*, and *Lymnaea elodes*. Snail-conditioned water (SCW) provides information value (in the form of pheromones) to attract larval trematode parasites. The SCW samples were dried with air and reconstituted in 10% *n*-propanol and then applied to cellulose HPTLC plates and developed with *n*-propanol–water (7:3). Amino acids were detected with ninhydrin reagent and the resulting color sample zones were compared to known standards. The amino acids present in SCW (hR_F values and color reactions with ninhydrin given in parentheses) were as follows: an unknown (7, purple), aspartic acid (21, purple), serine (29, purple), alanine (40, purple), tryptophan (51, purple), valine (58, purple/orange), phenylalanine (64, light blue), and leucine (69, purple). The above amino acids were detected in the SCW of all the snails, except that phenylalanine was not detected in *Lymnaea elodes*.

IV. TLC EXPERIMENTS

TLC studies on amino acids can be performed on a variety of sorbents. Blackburn (1989) has provided 56 tables showing TLC data and separations of amino acids and amines using various combinations of layers, mobile phases, and detection reagents. Information on both free amino acids and amino acid derivatives was included.

As mentioned in Section III, separation techniques for amino acids designed originally for PC are generally applicable for TLC amino acid studies on cellulose. Solvent systems for the TLC of amino acids generally employ mixtures of alcohols, acids or bases, and water. TLC is advantageous compared to PC for amino acid analysis in that it is faster and provides more compact spots, leading to better sensitivity and resolution of compounds. Experiment 1 provides a simple introduction to the TLC analysis of amino acid standards on silica gel. Experiment 2 provides a similar experience with cellulose and amino acid standards. Experiment 3 uses a reversed-phase layer to separate amino acids. For TLC experiments on amino acids with animal and plant materials, see Fried and Sherma (1994).

V. DETAILED EXPERIMENTS

A. Experiment 1. TLC Separation of Amino Acids on Baker-flex Silica Gel IB2 Sheets

1. Introduction

This simple exercise provides an introduction to the TLC analysis of amino acids.

2. Preparation of Standards

Prepare individual standards of alanine, arginine, leucine, and valine (Sigma) as 1 μg/μl solutions in deionized water. Prepare each standard by dissolving 10 mg of an amino acid in 10 ml of water. Prepare a mixed standard of the four amino acids at 1 μg/μl for each. To obtain the mixed standard, weigh 10 mg of each amino acid and dissolve in 10 ml of water.

3. Layer

Silica gel IB2 sheets, used as received without pretreatment (Chapter 15, Experiment 1).

4. Application of Initial Zones

Draw an origin line 2 cm from the bottom of the sheet and a solvent front line 10 cm from the origin. Place 1 μl of the individual amino acids on separate origins. Place 1 μl of the mixed amino acid standard on a separate origin; use separate 1-μl Microcap micropipets for application of the standards.

5. Mobile Phase

n-Butanol–acetic acid–water (80:20:20).

6. Development

Develop the sheet for a distance of 10 cm from the origin in a saturated chamber (Chapter 15, Experiment 1) containing 100 ml of mobile phase. Because the mobile phase contains water, development time will be relatively long (approximately 2.5–3 h).

7. Detection

Incorporate 0.5% of ninhydrin (Shapiro et al., 1975) into the mobile phase prior to development. After development, remove the sheet from the tank, dry the sheet in air, and heat it in an oven at 90°C for 5 min.

8. Results and Discussion

The following separations (approximate R_F values included) will be achieved for both individual amino acids and the mixed standard: arginine, 0.09; alanine, 0.26; valine, 0.36; leucine, 0.48. The background color of the plate will be light yellow. The amino acid spots will be colored as follows: arginine, violet; alanine, rust; leucine and valine, purple-red.

B. Experiment 2. TLC Separation of Amino Acids on Baker-flex Cellulose Sheets

1. Introduction

The purpose of this experiment is to demonstrate TLC amino acid separations on cellulose.

2. Preparation of Standards

Prepare 1 µg/µl individual amino acid standards of alanine, valine, leucine, histidine, and glycine (Sigma) as described in Experiment 1 of this chapter. Prepare a mixed amino acid standard of 1 µg/µl of each amino acid as described in Experiment 1 of this chapter.

3. Layer

Baker-flex cellulose sheets, 20 × 20 cm, used as received.

4. Application of Initial Zones

Spot individual and mixed standards to the origins of the sheet as described in Experiment 1 of this chapter.

5. Mobile Phase

n-Butanol–acetic acid–water (60:15:25).

6. Development

Develop the sheet in 100 ml of the mobile phase in a saturated rectangular glass tank. Development time at room temperature will be 60 to 90 min.

7. Detection

Dissolve 0.5 g of ninhydrin in the mobile phase prior to development as described in Experiment 1. Following development, dry the plate in air and heat it at 90°C for 5 min. Amino acids will appear as various shades of purple.

8. Results and Discussion

The order of separation of the amino acids, along with approximate R_F values, will be as follows: histidine, 0.22; glycine, 0.33; alanine, 0.49; valine, 0.69; leucine, 0.82. Differences in the colors of amino acids detected by this procedure will be very noticeable and are as follows: histidine, blue-gray; glycine, purple-red; alanine, dark purple-violet; valine, purple-red; leucine, purple-red. The TLC system used as described is extremely sensitive and will achieve good results with applications of 0.1–0.5 µg per amino acid. The choice between silica gel and cellulose for the TLC of amino acids has been dictated mainly by personal preferences rather than clear-cut experimental advantages for one or the other layer.

C. Experiment 3. TLC Separation of Amino Acids on Whatman KC₁₈ Chemically Bonded Reversed-Phase Plates

1. Introduction

The purpose of this experiment is to demonstrate TLC amino acid separations on a reversed-phase plate as described in Sherma et al. (1983).

2. Preparation of Standards

Prepare a mixed standard of arginine, histidine, cysteine, serine, threonine, alanine, aspartic acid, valine, and tryptophan (Sigma) at 0.5 µg/µl for each amino acid in deionized water.

3. Layer

Whatman KC₁₈ chemically bonded reversed-phase plates, used as received without pretreatment.

4. Application of Initial Zones

Draw an origin line 2 cm from the bottom of the plate and a solvent front line 15 cm from the origin. Place 1 µl of the mixed amino acid standard at the origin.

5. Mobile Phase

n-Propanol–water (7:3).

6. Development

Develop the plate in a standard rectangular glass chamber that is lined with paper and preequilibrated with the mobile phase for at least 10 min prior to inserting the spotted layer. Development time will be approximately 3.5 h.

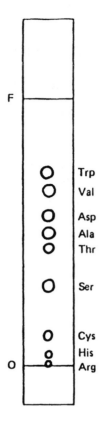

FIGURE 16.1 Separation of a nine-component amino acid mixture (0.5 µg each) on a Whatman KC$_{18}$ thin layer developed with *n*-propanol–water (7:3) in a saturated chamber. F, solvent front; O, origin; Arg, arginine; His, histidine; Cys, cysteine; Ser, serine; Thr, threonine; Ala, alanine; Asp, aspartic acid; Val, valine; Trp, tryptophan. [Reproduced with the permission of Marcel Dekker, Inc., from Sherma et al. (1983).]

7. Detection

Dry the plate in an oven at 100°C and then spray it with 0.1% ninhydrin in acetone; heat the plate again for 5 min or longer. Amino acids will be detected as colored spots on a white background.

8. Results and Discussion

The following separations (approximate R_F values included) will be achieved for the mixed standard: arginine, 0.0060; histidine, 0.063; cysteine, 0.11; serine, 0.30;

threonine, 0.47; alanine, 0.50; aspartic acid, 0.53; valine, 0.66; tryptophan, 0.74 (Figure 16.1).

Sherma et al. (1983) reported that their amino acid separations on C_{18} reversed-phase thin layers were similar to those previously achieved on conventional silica gel and cellulose by Sleckman and Sherma (1982). Sherma et al. (1983) concluded that it may be difficult to predict the relative separation behavior of numerous compounds including amino acids on the basis of chemical notions of "reversed"- or "normal"-phase chromatography.

REFERENCES

Abraham, M., Polyak, B., Szajani, B., and Boross, L. (1983). Analytical separation of reduced and oxidized forms of glutathione from amino acid mixtures by overpressured thin-layer chromatography. *J. Liq. Chromatogr. 6*: 2635–2645.

Arme, C. (1977). Amino acids in eight species of Monogenea. *Z. Parasitenk. 51*: 261–263.

Awapara, J. (1962). Free amino acids in invertebrates: A comparative study of their distribution and metabolism. In *Amino Acid Pools*, J. T. Holden (Ed.). Elsevier, Amsterdam, pp. 158–175.

Bailey, R. S., Jr., and Fried, B. (1977). Thin layer chromatographic analyses of amino acids in *Echinostoma revolutum* (Trematoda) adults. *Int. J. Parasitol. 7*: 497–499.

Berry, H. K. (1973). Detection of amino acid abnormalities by quantitative thin layer chromatography. In *Quantitative Thin Layer Chromatography*, J. C. Touchstone (Ed.). Wiley, New York, pp. 113–129.

Bhushan, R. (1988). Methods of TLC resolution of enantiomeric amino acids and their derivatives. *J. Liq. Chromatogr. 11*: 3049–3065.

Bhushan, R. (1991). Amino acids and their derivatives. In *Handbook of Thin-Layer Chromatography*, J. Sherma and B. Fried (Eds.). Marcel Dekker, New York, pp. 353–387.

Bhushan, R., and Ali, I. (1987). TLC resolution of amino acids in a new solvent and effect of alkaline earth metals. *J. Liq. Chromatogr. 10*: 3647–3652.

Bhushan, R., Ali, I., and Sharma, S. (1996). A comparative study of HPLC and TLC separation of amino acids using Cu(II) ion. *Biomed. Chromatogr. 10*: 37–39.

Bhushan, R., Chauhan, R. S., Ali, R., and Ali, I. (1987). A comparison of amino acids separation on zinc, cadmium, and mercury impregnated silica layers. *J. Liq. Chromatogr. 10*: 3653–3657.

Bhushan, R., Mahesh, V. K., and Varna, A. (1994a). Improved thin-layer chromatographic resolution of PTH amino acids with some new solvent systems. *Biomed. Chromatogr. 8*: 69–72.

Bushan, R., and Martens, J. (1996). Amino acids and their derivatives. In *Handbook of Thin-Layer Chromatography*, 2nd ed., J. Sherma and B. Fried (Eds.). Marcel Dekker, New York, pp. 389–425.

Bhushan, R., and Parshad, V. (1994). TLC of amino acids on thin silica gel layers impregnated with transition metal ions and their anions. *J. Planar Chromatogr.—Mod. TLC 7*: 480–483.

Bhushan, R., and Parshad, V. (1996). Thin-layer chromatographic separation of enantiomeric dansyl–amino acids using a macrocyclic antibiotic as a chiral selector. *J. Chromatogr., 736*: 235–238.

Bhushan, R., and Reddy, G. P. (1987). TLC of phenylthiohydantoins of amino acids: a review. *J. Liq. Chromatogr. 10*: 3497–3528.

Bhushan, R., and Reddy, K. R. N. (1989). TLC resolution of 18 PTH–amino acids using three solvent systems. *J. Planar Chromatogr.—Mod. TLC 2*: 79–82.

Bhushan, R., Reddy, G. P., and Joshi, S. (1994b). TLC resolution of DL amino acids on impregnated silica gel plates. *J. Planar Chromatogr.—Mod. TLC 7*: 126–128.

Blackburn, S. (1983). *CRC Handbook of Chromatography—Amino Acids and Amines*, Vol. 1. CRC Press, Boca Raton, FL.

Blackburn, S. (1989). *CRC Handbook of Chromatography—Amino Acids and Amines*, Vol. II. CRC Press, Boca Raton, FL.

Brink, A. (1992). Amino acid analysis in egg lysozyme. Methods of qualitative amino acid analysis and the determination of the C-terminal amino acid of a polypeptide. Part 2. *Prax. Naturwiss. Chem. 41*: 39–41.

Chin, K. D. H., and Koehler, P. E. (1983). Identification and estimation of histamine, tryptamine, phenethylamine, and tyramine in soy sauce by thin-layer chromatography of dansyl derivatives. *J. Food Sci. 48*: 1826–1828.

Clift, J., Hall, S. K., Carter, R. A., Denmeade, R., and Green, A. (1994). An improved thin-layer chromatography technique for neonatal screening for amino acid disorders using dried blood spots. *Screening 3*: 39–43.

Das, B., and Sawant, S. (1993). Quantitative HPTLC analysis of dansyl–amino acids. *J. Planar Chromatogr.—Mod. TLC 6*: 294–295.

Degtiar, W. G., Tyaglov, B. V., Degterev, E. V., Krylov, V. M., Malakhova, I. I., and Krasidov, V. D. (1994). Quantitative analysis of L-lysine, L-threonine, L-homoserine and cobalamins in fermentation broths. *J. Planar Chromatogr.—Mod. TLC 7*: 54–57.

DeMeglio, D. C., and Svanberg, G. K. (1996). A thin-layer chromatographic procedure for the separation of proline and hydroxyproline from biological samples. *J. Liq. Chromatogr. Relat. Technol. 19*: 969–975.

Fater, Z., Kemeny, G., Mincsovics, E., and Tyihak, E. (1983). Determination of amino acids in cereals by overpressured thin-layer chromatography. *Dev. Food Sci. 5A*: 483–488.

Fried, B., and Haseeb, M. A. (1996). Thin-layer chromatography in parasitology. In *Practical Thin-Layer Chromatography—A Multidisciplinary Approach*, B. Fried and J. Sherma (Eds.). CRC Press, Boca Raton, FL, pp. 51–70.

Fried, B., and Sherma, J. (1994). *Thin Layer Chromatography—Techniques and Applications*, 3rd ed., Marcel Dekker, New York, pp. 296–299.

Gankina, E. S., Malakhova, I. I., Kostyuk, I. O., and Belenkii, B. G. (1988). One-dimensional HPTLC of DNS- and PTH-amino acids. *J. Planar Chromatogr.—Mod. TLC 1*: 364–366.

Gilbertson, D. E., and Schmid, L. S. (1975). Free amino acids in the hemolymph of five species of fresh water snails. *Comp. Biochem. Biophys. 51B*: 201–203.

Grosvenor, P., and Gray, D. O. (1990). 2,4-Dinotrophenylpyridium chloride, a novel and versatile reagent for the detection of amino acids, primary and secondary amines, thiols, thiolactones, and carboxylic acids during planar chromatography. *J. Chromatogr. 504*: 456–463.

Guenther, K. (1988). Determination of TLC-separated amino acid enantiomers. *Analusis* *16*: 514–518.

Guenther, K., Martens, J., and Schickedanz, M. (1985). Resolution of optical isomers of thin-layer chromatography (TLC). Enantiomeric purity of L-dopa. *Fresenius' Z. Anal. Chem. 322*: 513–514.

Haas, W., Koerner, M., Hutterer, E., Wegner, M., and Haberl, B. (1995). Finding and recognition of the snail intermediate hosts by 3 species of echinostome cercariae. *Parasitology 110*: 113–142.

Hanes, C. S., Harris, C. K., Moscarella, M. A., and Tigrane, E. (1961). Stabilized chromatographic systems of high resolving power for amino acids. *Can. J. Biochem. Physiol. 39*: 163–190.

Harborne, J. B. (1984). *Phytochemical Methods—A Guide to Modern Techniques of Plant Analysis*, 2nd ed. Chapman & Hall, London.

Heathcote, J. G. (1979). Amino acids. In *Densitometry in Thin Layer Chromatography— Practice and Applications*, J. C. Touchstone and J. Sherma (Eds.). Wiley, New York, pp. 153–179.

Henderson, M. J., Allen, J. T., Holton, J. B., and Goodall, R. (1985). Extra heating of amino acids. *Clin. Chim. Acta 146*: 203–205.

Huang, M. B., Li, H. K., Li, G. L., Yan, C. T., and Wang, L. P. (1996). Planar chromatographic direct separation of some aromatic amino acids and aromatic amino alcohols into enantiomers using cyclodextrin mobile phase additives. *J. Chromatogr. A 742*: 289–294.

Jain, R. (1996). Thin-layer chromatography in clinical chemistry. In *Practical Thin-Layer Chromatography—A Multidisciplinary Approach*, B. Fried and J. Sherma (Eds.). CRC Press, Boca Raton, FL, pp. 131–152.

Jork, H., and Ganz, J. (1994). Opportunities and limitations of modern TLC/HPTLC in the quality control of L-tryptophan. *L-Tryptophan: Curr. Prospects Med. Drug Saf.* 338–350.

Keller, S., Ghosh, A. K., Turino, G. M., and Mandl, I. (1984). Separation of the crosslinking amino acids of elastin on thin-layer plates. *J. Chromatogr. 305*: 461–464.

Kuhn, A. O., Lederer, M., and Sinibaldi, M. (1989). Adsorption chromatography on cellulose. IV. Separation of D- and L-tryptophan on cellulose with aqueous solvents. *J. Chromatogr. 469*: 253–260.

Lancaster, J. E., and Kelly, K. E. (1983). Quantitative analysis of the S-alk(en)yl-L-cysteine sulfoxides in onion (*Allium cepa* L.). *J. Sci. Food Agric. 34*: 1229–1235.

Lander, R. J., and Treiber, L. R. (1985). Analysis of L-alanine N-carboxyanhydride by quantitative thin layer chromatography. In *Technical Applications of Thin Layer Chromatography, Proceedings of the Biennial Symposium on Thin Layer Chromatography*, 3rd ed. J. C. Touchstone and J. Sherma (Ed.). Wiley, New York, pp. 235–242.

Laskar, S., and Basak, B. (1988). Detection of amino acids on thin-layer plates. *J. Chromatogr. 436*: 341–343.

LeFevre, J. W. (1993). Reversed-phase thin-layer chromatographic separations of enantiomers of dansyl-amino acids using β-cyclodextrin as a mobile phase additive. *J. Chromatogr. 653*: 293–302.

Lepri, L., Coas, V., and Desideri, P. G. (1992). Planar chromatography of optical isomers with bovine serum albumin in the mobile phase. *J. Planar Chromatogr.—Mod. TLC 5*: 175–178.

Levy, A. L., and Chung, D. (1953). Two-dimensional chromatography of amino acids on buffered papers. *Anal. Chem. 25*: 396–399.

Mack, S. R., Samuels, S., and Vanderberg, J. P. (1979). Hemolymph of *Anopheles stephensi* from non-infected and *Plasmodium berghei*-infected mosquitoes. 2. Free amino acids. *J. Parasitol. 65*: 130–136.

Malakhova, I. I., Tyaglov, B. V., Degterev, E. V., Krasikov, V. D., and Degtiar, W. G. (1996). Video-densitometric methods for quantitative analysis by TLC of industrial amino acids. *J. Planar Chromatogr.—Mod. TLC 9*: 375–378.

Mant, C. T., Zhou, N. E., and Hodges, R. S. (1992). Amino acids and peptides. In *Chromatography. 5th Edition—Fundamentals and Applications of Chromatography and Related Differential Migration Methods—Part B: Applications*, E. Heftmann (Ed.). Elsevier, Amsterdam, pp. B 75–B 150.

Marchelli, R., Virgili, R., Armani, E., and Dossena, A. (1986). Enantiomeric separation of D,L-Dns-amino acids by one- and two-dimensional thin-layer chromatography. *J. Chromatogr. 355*: 354–357.

Mathur, V., Kanoongo, N., Mathur, R., Narang, C. K., and Mathur, N. K. (1994). Resolution of amino acid racemates on borate-gelled guaran-impregnated silica gel thin-layer chromatographic plates. *J. Chromatogr., 685*: 360–364.

Metrione, R. M. (1986). Separation of lissamine rhodamine B sulfonyl derivatives of amino acids by high-performance liquid chromatography and thin-layer chromatography. *J. Chromatogr. 363*: 337–344.

Monreal, C. M., and McGill, W. B. (1985). Centrifugal extraction and determination of free amino acids in soil solutions by TLC using tritiated 1-fluoro-2,4-dinitrobenzene. *Soil Biol. Biochem. 17*: 533–539.

Mueller, H., and Eckert, H. (1989). Simultaneous determination of monoethanolamine and glycine betaine in plants. *J. Chromatogr. 479*: 452–458.

Mustafa, G., Abbasi, A., and Zaidi, Z. H. (1992). Effective solvent systems for separation of PTH–amino acids on polyamide sheets. *J. Chem. Eng. 11*: 53–56.

Norfolk, E., Khan, S. H., Fried, B., and Sherma, J. (1994). Comparison of amino acid separations on high performance silica gel, cellulose, and C-18 reversed phase layers and application of HPTLC to the determination of amino acids in *Biomphalaria glabrata* snails. *J. Liq. Chromatogr. 17*: 1317–1326.

Ohtake, H., Hase, Y., Sakemoto, K., Oura, T., Wada, Y., and Kodama, H. (1995). A new method to detect homocysteine in dry blood spots using thin layer chromatography. *Screening 4*: 17–26.

Pastor-Anglada, M., Lopez-Tejero, D., and Remesar, X. (1984). Thin layer chromatography of 14C-dansyl amino acids for the quantification of plasma amino acid levels. *IRCS Med. Sci. 12*: 538–539.

Petrovic, M., and Kastelan-Macan, M. (1995). Validation of quantitative chromatographic analyses on laboratory-prepared thin layers. *J. Chromatogr. 704*: 173–178.

Prosser, C. L. (1973a). Nutrition. In *Comparative Animal Physiology*, 3rd ed., C. L. Prosser (Ed.). Saunders, Philadelphia, pp. 111–132.

Prosser, C. L. (1973b). Excitable membranes. In *Comparative Animal Physiology*, 3rd ed., C. L. Prosser (Ed.). Saunders, Philadelphia, pp. 457–504.

Pyka, A. (1993). A new optical topological index (I_{opt}) for predicting the separation of D and L optical isomers by TLC. Part III. *J. Planar Chromatogr.—Mod. TLC 6*: 282–288.

Sarbu, C., Marutoiu, C., Vlassa. M., and Liteanu, C. (1987). Direct fluorescent detection of chromatographically separated amino acids by means of 9-isothiocyanatoacridines. *Talanta 34*: 438–440.

Schwartz, T. W. (1984). Fluorescence detection of picomol amounts of the phenylthiohydantoin derivative of histidine in thin-layer chromatography. *J. Chromatogr. 299*: 513–515.

Scott, R. M. (1969). *Clinical Analysis by Thin-Layer Chromatography Techniques*. Ann Arbor–Humphrey Science Publishers, Ann Arbor, MI.

Shalaby, A. R. (1996). Thin-layer chromatography in food analysis. In *Practical Thin-Layer Chromatography—A Multidisciplinary Approach*. B. Fried and J. Sherma (Eds.). CRC Press, Boca Raton, FL, pp. 169–192.

Shapiro, I. L., Seidenberger, J. W., and Hoskin, S. (1975). A new centrifugal system for membrane filtration application to chromatography of amino acids in human urine, serum, and whole blood. *Anal. Lett. 8*: 71–89.

Sherma, J., Sleckman, B. P., and Armstrong, D. W. (1983). Chromatography of amino acids on reversed phase thin layer plates. *J. Liq. Chromatogr. 6*: 95–108.

Siddiqi, Z. M., and Rani, S. (1995). Studies of Tin(IV) selenoarsenate III: thin-layer chromatography of α-amino acids. *J. Planar Chromatogr.—Mod. TLC 8*: 141–143.

Sinhababu, A., Basak, B., and Laskar, S. (1994). Novel spray reagent for the identification of amino acids on thin-layer chromatography plates. *Anal. Proc. 31*: 65–66.

Sleckman, B. P., and Sherma, J. (1982). A comparison of amino acid separations on silica gel, cellulose, and ion exchange thin layers. *J. Liq. Chromatogr. 5*: 1051–1068.

Steiner, R. A., Fried, B., and Sherma, J. (1998). HPTLC determination of amino acids in snail conditioned water from *Biomphalaria glabrata*, two strains of *Helisoma trivolvis* and *Lymnaea elodes*. *Journal of Liquid Chromatography and Related Techniques 21*: 427–432.

Surgova, T. M., Sidorenko, M. V., Kofman, I. S., and Vinnitskii, V. B. (1990). Determination of histamine in the presence of polyamines by spectrodensitometric TLC. *J. Planar Chromatogr.—Mod. TLC 3*: 81–82.

Varshney, K. G., Khan, A. A., Maheshwari, S. M., and Gupta, U. (1992). Chromatographic separation of α-amino acids on antimony (V) phosphate–silica gel G plates from some synthetic mixtures and drug samples. *Bull. Chem. Soc. Jpn. 65*: 2773–2778.

von Brand, T. (1973). *Biochemistry of Parasites*, 2nd ed. Academic Press, New York.

Yamada, M., Minani, Y., Yuyama, S., and Suzuki, T. (1985). Determination of metabolites relating to the tryptophan–NAD pathway in rat urine by cellulose high performance thin-layer chromatography. *J. High Resolut. Chromatogr. Chromatogr. Commun. 8*: 69–72.

17

Carbohydrates

I. STRUCTURE

Carbohydrates are naturally occurring organic substances that contain mainly carbon, oxygen, and hydrogen. Some carbohydrates also contain additional elements such as nitrogen (glucosamines) or phosphorus (sugar phosphates). Carbohydrates are all relatively similar in chemical structure, are very polar, and show relatively high solubility in water and other polar solvents.

Monosaccharides are carbohydrates of low molecular weight consisting of a single basic sugar unit. Such naturally occurring carbohydrates with five carbon atoms per molecule are called pentoses, whereas others with six carbon atoms per molecule are called hexoses. The best-known natural pentoses are D-ribose, D-arabinose, and D-xylose. Of the naturally occurring hexoses, the best known are D-glucose, D-mannose, D-galactose, and D-fructose. Oligosaccharides are carbohydrates of intermediate molecular weight and consist of either two monosaccharide units linked by glycosidic bonding—for example, the disaccharides lactose, maltose, sucrose, and trehalose—or three monosaccharide units linked in a similar fashion—for example, the trisaccharide raffinose. Trisaccharides are often found in plants but rarely in animals. The best-known trisaccharide, raffinose, has been isolated from numerous plants.

Polysaccharides are carbohydrates of high molecular weight and consist of large aggregates of monosaccharide units joined via glycosidic bonding. Glycogen (animal starch), cellulose, and starch are the best-known naturally occurring polysaccharides. Thin-layer chromatography does not play an important role in the separation of polysaccharides. However, hydrolysis of polysaccharides will result in the formation of monosaccharides, which can be determined by TLC. For instance, starch, which is a polymer of D-glucose, can be hydrolyzed completely to glucose, and examination by TLC following hydrolysis should yield only one spot.

There are many different isomers of carbohydrates, but relatively few occur naturally. Hence, in the TLC analysis of sugars, the chromatographer is concerned with a limited number of biologically active carbohydrates. For a good discussion of the structure of carbohydrates of plants, consult Harborne (1984). For a similar discussion of the structure of carbohydrates of animals, consult Scott (1969).

II. FUNCTION

Sugars are manufactured as a result of photosynthesis and serve as the major source of energy storage in plants. The major free sugars in plants are glucose, fructose, and sucrose, along with lesser amounts of xylose, rhamnose, galactose, and arabinose (Harborne, 1984). The nonreducing disaccharide trehalose occurs mainly in fungi and yeast (Myrback, 1949). The major storage carbohydrate of plants is the polysaccharide starch, and another polysaccharide, cellulose, serves as the major constituent of the cell wall. Glycosides and nucleic acids also contain sugars. For instance, the sugar associated with ribose nucleic acid (RNA) is ribose. Sugars play a role in ecological interactions between insects and plants. For instance, during pollination, floral nectars help attract bees and other insects to flowers. Floral nectar contains mainly sucrose, glucose, and fructose (Harborne, 1984).

Animals ingest sugars from plants and store carbohydrates in the form of glycogen. The main sugar for animal respiration is the monosaccharide glucose (dextrose). However, some animals can utilize other sugars as an immediate source of energy. According to von Brand (1973), some parasitic protozoa use fructose and galactose in addition to glucose as a respiration source. One strain of crithidia (a trypanosomelike protozoan) apparently can utilize xylose as an energy source. Other parasitic protozoa may use mannose in preference to glucose as an energy source. According to von Brand (1973), parasitic worms (helminths) utilize a considerable number of sugars, that is, glucose, fructose, mannose, galactose, and maltose. Some parasitic worms utilize galactose as an energy source as readily as glucose.

Chromatographic and chemical procedures have been used to identify carbohydrates in a variety of invertebrate animals. The exact role of many of the sugars identified in invertebrates is poorly understood. Gross et al. (1958) identified glucosamine, galactose, glucose, mannose, fucose, and arabinose in spongin (skeletal material) of certain poriferans (sponges). Clark (1964) used paper chromatography (PC) to identify glucose and maltose in the body fluids of marine worms (polychaetes). Fairbairn (1958) used PC and chemical tests to identify glucose and trehalose in 71 species of invertebrates representing most of the major phyla. Fairbairn's study showed that trehalose is much more widespread in the animal kingdom than previously had been thought. According to Hochachka (1973), trehalose is the principal blood sugar in insects and may reach a concentration of 7% in the hemolymph. Anderton et al. (1993) found that hemolymph glucose and trehalose values were approximately the same (about 60 mg/dl for each sugar) in *Biomphalaria glabrata* snails maintained on a leaf lettuce–Tetramin diet. Interestingly, trehalose was thought to occur only in plants until Wyatt and Kalif (1956) isolated it from insect hemolymph. Trehalose is undoubtedly used as an energy reserve in many invertebrates. The gut of insects and other invertebrates contains the enzyme trehalase (Barnard, 1973), which hydrolyzes trehalose.

Gas chromatography techniques have been used for the quantitative analysis of sugars in invertebrates. Thus, Ueda and Sawada (1968) showed that the parasitic flatworm *Paragonimus* (lung fluke of man) contains about 70% glucose, 10% galactose, 8% mannose, and 6% ribose.

Simple sugars may occur in a polymerized state in animals, that is, in forms such as hexosamines and glucosamides. Moreover, cerebrosides that are complex lipids (glycolipids) may contain a variety of simple sugars. Antigenic polysaccharides occur in animals and contain proteins plus polysaccharides. Various sugars have been identified in these antigenic substances following acid hydrolysis, including rhamnose, fucose, galactose, xylose, hexosamine, and glucose.

Carbohydrates function in a variety of other ways in the animal kingdom. They may act as precursors for lipid synthesis as demonstrated in mosquitoes (Lakshmi and Subrahmanyam, 1975). In parasitic organisms (tapeworms) that lack a digestive system, leakage of sugars occurs through the tegument (von Brand, 1973). The absorption of sugars by passive diffusion and active transport has been well documented in tapeworms (von Brand, 1973). Whereas glycogen is usually synthesized from glucose in animals, at least in certain parasitic worms it can be synthesized from various other sugars such as fructose, sorbose, maltose, and saccharose (von Brand, 1973).

Homeostasis in mammals, under normal physiological conditions, is well known, and the concentration of blood glucose stays reasonably constant during this condition. In insects, this type of regulation apparently does not occur, and there is a rather wide fluctuation in sugar levels in the hemolymph (Florkin and Jeuniaux, 1974). Evidence for hemolymph glucose regulation in *Biomphalaria glabrata* snails in the same way as blood glucose regulation in vertebrates has been reported by Liebsch and Becker (1990). In parasitic infections—for example, malaria in the blood of vertebrates—serum sugar declines as a function of the infection (Sadun et al., 1966), presumably because the malaria parasite utilizes sugar from the blood. Mack et al. (1979) reported similar depletion of sugar in the hemolymph of mosquitoes infected with malaria. They suggested that this depletion may be related to direct utilization of sugar by the malaria parasite developing within the mosquito. Infection of snails by larval trematode parasites is characterized by decrease in the levels of many host tissue metabolites including hemolymph glucose (Thompson, 1985).

III. TLC OF CARBOHYDRATES

Thin-layer chromatography can be used as an analytical technique to detect sugars in biological samples. For a good discussion of sample preparation of biological specimens prior to carbohydrate TLC, see Churms (1981). Mack et al. (1979) have used TLC to detect carbohydrates in the hemolymph and whole-body homogenates of adult female *Anopheles* mosquitoes. Lato et al. (1968) used TLC to analyze carbohydrates in urine but indicated that their procedures were generally applicable to other body fluids. Fell (1990) and Anderton et al. (1993) used TLC to analyze hemolymph samples from insects and snails, respectively. These four papers should be consulted by those interested in performing carbohydrate TLC on biological samples.

Various sorbents are used in the TLC of sugars, the most popular ones being silica gel, cellulose, and kieselguhr (Ghebregzabher et al., 1976). Personal choice will often dictate which sorbent is used. Vomhof and Tucker (1965) have discussed the separation of simple sugars by cellulose TLC, whereas Menzies and Mount (1975) have mentioned the advantages of silica gel for use in carbohydrate TLC.

Cellulose TLC is often based on earlier procedures developed for paper chromatographic (PC) analyses of sugars. Separations on silica gel often call for impregnating plates with various inorganic ions such as borate, bisulfite, or phosphate (Supelco TLC Reporter, 1977; Jacin and Mishkin, 1965; Churms, 1981). Churms (1983) described carbohydrate TLC on cellulose, kieselguhr, and silica; she also discussed different reagents used to visualize carbohydrates separated by PC and TLC.

Precoated silica gel plates can be impregnated in the laboratory or can be purchased with impregnating agents already incorporated into the plate. A sugar separation on a precoated silica gel H layer impregnated with 0.3 M KH_2PO_4 during manufacture of the plate is described in the Supelco TLC Reporter (1977). Doner and Biller (1984) used amino–propyl-bonded silica high-performance thin-layer chromatographic plates impregnated with NaH_2PO_4 to separate sugars.

Silica gel plates impregnated with inorganic ions usually have certain advantages over unimpregnated silica gel or cellulose in the TLC of sugars. Separations on silica gel are faster than on cellulose and usually provide more compact spots, which can be important if quantitative TLC is to be used. A greater number of sugar-detection reagents can be used with silica gel than cellulose.

Numerous solvent systems have been described for the TLC of carbohydrates. For comprehensive listings, consult Ghebregzabher et al. (1976) or Scott (1969). Churms (1981) has provided tabular data showing various combinations of layers and solvent systems for TLC separations of carbohydrates.

Because sugars are polar compounds, most solvent systems employ water and other polar solvents. Hence, carbohydrate separations by TLC are relatively slow. Most TLC procedures for sugars use single-dimensional chromatography. However, Ghebregzabher et al. (1976) have described two-dimensional solvent systems for carbohydrate TLC. The enhanced resolution obtained with two-dimensional chromatography may compensate for the lengthiness involved in this procedure.

Metraux (1982) described a two-dimensional system for the separation of seven neutral and two acidic sugars from hydrolyzed plant cell wall polysaccharides. The sorbent was cellulose and plates were developed twice in the first direction with butanol–2-butenone–formic acid–water (8:6:3:3) and once in the second direction with phenol–water–formic acid (100:98:2). Spots were visualized with anisidine phthalate and heating at 100°C for 10 min.

Many detection procedures have been described for the visualization of sugars on TLC plates. For comprehensive listings of these procedures, consult Scott (1969), Ghebregzabher et al. (1976), and Churms (1981). Detection procedures are usually based on the reducing properties of some sugars, or the conversion of sugars to furfural derivatives (Mansfield, 1973). Additionally, nonspecific tests use sulfuric acid alone or in combination with other reagents that char sugars following heating of the plate. Detection by simple heating of plates has been reported (Klaus et al., 1991). Some detection procedures utilize enzymes such as glucose oxidase, trehalase, or invertase, which can be sprayed on the TLC plate and can provide more or less specific information about a particular sugar. For further information on the use of enzymes for sugar TLC detection procedures, consult Caldes and Prescott (1975).

Although numerous sugars exist, the number that occurs in biological samples is limited (Mansfield, 1973). According to Mansfield, the source of the sample should serve as a guide for the types of sugars to expect upon TLC. For instance, in the analysis of foods, glucose, fructose, sucrose, and maltose may be present, whereas in the analysis of clinical samples, one might find glucose, galactose, mannose, fructose, ribose, maltose, sucrose, lactose, glucosamines, and galactosamines.

Direct application of samples, particularly fluids, may be possible for the TLC of sugars, that is, without sample cleanup. However, various sample cleanup procedures may be required for some biological samples, particularly whole-body extracts of plant and animal tissues. Such cleanup procedures may include extraction, desalting, deproteinization, and other techniques described in Chapter 4 and in Churms (1981).

The following section provides a brief review of significant studies on the TLC of carbohydrates from 1983 to 1997.

Blom et al. (1983) desalted urine samples by ion-exchange chromatography and then determined oligosaccharides by TLC to diagnose lysosomal acid maltase deficiency. Parrado et al. (1983) separated mucopolysaccharides by continuous development TLC with propanol–NH$_4$OH–H$_2$O (40:60:5). Hirayama et al. (1984) resolved carbohydrates on silica gel sintered plates developed with butanol–acetone–water. Seller et al. (1984) determined reducing sugars, deoxy sugars, N-acetylamino sugars, and uronic acids by reaction with dansylhydrazine, two-dimensional TLC on polyamide sheets, and computer-controlled scanning fluorometry. Saamanen and Tammi (1984) determined unsaturated glycosaminoglycan disaccharides by TLC on cellulose with reflectance scanning at 232 nm. Ray et al. (1984) used a sulfosalicylic acid spray reagent (2% in 0.5 M H$_2$SO$_4$) to detect 0.2-μg amounts of various sugars on silica gel layers. Saito et al. (1985) determined the basic carbohydrate structure of gangliosides by the use of an HPTLC-immunostaining technique. Tugrul and Ozer (1985) determined 1–45 ng-amounts of monosaccharides in drug hydrolysates by the use of fluorodensitometry.

Vega et al. (1985) used TLC–densitometry to quantify sucrose, glucose, fructose, and xylose in natural products; compounds were detected with a diphenylamine–analine reagent and scanning was at 515 nm. Thieme et al. (1986) reported the R_F values for 16, 1-methoxybutyl D-malto-oligosaccharides. Rebers et al. (1986) analyzed amino sugars in polysaccharide hydrolysates by silica gel–TLC with an acetonitrile–acetic acid–ethanol–water mobile phase and detection with ninhydrin and silver nitrate. Lenkey et al. (1986) used TLC–densitometry to quantify sucrose and fructose in biological samples; the sugars were detected with thymol reagent. Tugrul and Ozer (1986) quantified monosaccharides in drug hydrolysates by fluorodensitometry after detection with aniline phthalate. Shimada et al. (1987) used TLC–densitometry to quantify unsaturated chondroitin sulfate disaccharides on silica gel after detection with a carbazol reagent. Suyama and Adachi (1987) detected sugars as opaque zones when silica plates were exposed to hexane vapors in a glass chamber.

Vajda and Pick (1988) separated mono-, di-, tri-, and oligosaccharides by forced-flow silica gel TLC. Doner (1988) used aminopropyl-bonded silica gel HPTLC plates to separate oligosaccharide homologs produced by the partial hydrolysis of cellulose, starch, chitin, and xylan. Patzch et al. (1988a, 1988b) reported the quantification of sugars from numerous samples using three different layers and 11 postchromatography derivatization methods. Brasseur et al. (1988) determined quencetin glycosides in plant extracts by separation on silica gel, detection with aminoethanol–diphenylborate, and densitometry at 533 nm. Shinomiya et al. (1989) detected unsaturated disaccharides from chondroitin as their dansylhydrazine derivatives.

Groesz and Braunsteiner (1989) used TLC–densitometry to quantify glucose, fructose, sucrose, and various fructo-oligosaccharides on silica gel 60 plates developed with a boric-acid–containing solvent; chromogenic detection was by use of thymolsulfuric acid, whereas fluorogenic detection was done after derivatization with zirconyl chloride. Klaus et al. (1989, 1990) separated 20 carbohydrates on amino-bonded plates and detected them with a sensitivity similar to that obtained with dip and spray reagents by heating and inspection under UV lights. Goso and Hotta (1990) analyzed mucin-derived oligosaccharides by TLC on aminopropyl-bonded silica gel plates developed with acetonitrile–10 mM triethyamine acetate (3:2). Torres et al. (1990) used TLC to separate six of the eight possible sucrose monostearate structural positional isomers and further differentiated them by use of a specific chromogenic detection reagent.

Pukl and Prosek (1990) used TLC–densitometry to quantify glucose, fructose, and saccharose; single or multiple developments were used with a solvent system consisting of acetonitrile–methanol–pH 5.5 phosphate buffer (85:5:15); sugars were detected with a diphenylamine–aniline reagent and scanning was at 440 and 515 nm. Fell (1990) used TLC–densitometry to quantify trehalose, glucose, and fructose at 125–2000 ng-levels in less than 1 µl samples of insect hemolymph.

Poukens-Renwart and Angenot (1991) used TLC–densitometry to quantify xylose, 3-O-methylglucose, and rhamnose in urine with arabinose as an internal standard; scanning was at 400 nm after the application of an aminobenzoic acid detection reagent in an automatic dipping chamber. Morcol and Velander (1991) used o-toluidine as detection reagent in the HPTLC determination of reducing sugars in protein hydrolysates with scanning at 295 nm. The sensitivity of the reagent for the sugars was in the range of 50–100 pmol.

Churms (1991) updated her earlier CRC volume on carbohydrates to include new material on the TLC of sugars, particularly in studies using HPTLC and bonded-phase plates.

Prosek et al. (1991) provided an extensive review of TLC for the analysis of carbohydrates. Considerable information on sample preparation of carbohydrates from numerous biological and nonbiological sources is given. Many chromatographic systems used routinely for the analysis of carbohydrates are discussed and evaluated. Detection methods involving the use of several derivatization procedures are considered. Finally, an extensive section concerned with the quantitative evaluation of sugars by densitometry or digital video scanning (DVS) is presented.

Klaus et al. (1991) presented a TLC method for the simultaneous analysis of uric acid, creatine, creatinine, and uric acid together with glucose in the urine and serum of patients with diabetes and other carbohydrate anomalies. The method requires no special sample preparation, and separations are done on amino-modified HPTLC precoated plates. Detection of all substances is achieved by simply heating the plates to give stable fluorescent derivatives.

Karlsson et al. (1991) used a specially designed TLC/fast-atom-bombardment mass spectrometry probe with continuous desorption and scanning over a moving TLC plate to resolve glycolipids with identical carbohydrate sequence into molecular species with differences in long-chain base and fatty acid. The technique was demonstrated for sulfatides with one or two sugar residues isolated from human kidney, GM3 ganglioside isolated from human malignant melanoma, and chemically modified gangliotetraosylceramide from mouse intestine.

Bosch-Reig et al. (1992) described a simple technique for the TLC separation and identification of 14 sugars from various biological samples and human milk. The following sugars were separated using monodimensional TLC along with two different mobile phases and two different detection reagents: L-fucose, D-galactose, D-glucose, lactose, N-acetylglucosamine, D-maltose, D-mannose, L-sorbose, fructose, D-xylose, glucuronic acid, N-acetyllactosamine, 3′ and 6′ sialyllactose, and maltodextrins (G$_2$–G$_8$).

Batisse et al. (1992) used TLC on silica gel to separate and quantify by spectrodensitometry the sugars in the cell wall of cherry. The following sugars were identified and quantified at nanomolecular levels: galactose, glucose, mannose, arabinose, xylose, rhamnose, and galacturonic, and glucuronic acids.

Churms (1992) has provided a review on the analysis of carbohydrates including recent studies on TLC. She discussed improvements in the separation of carbohydrates by the use of two-dimensional TLC on cellulose plates particularly for amino sugars as well as neutral and acidic monosaccharides. She noted that improved HPTLC and bonded-phase plates have been particularly helpful in separations of the higher oligosaccharides—i.e., malto-oligosaccharides, D-gluco-oligosaccharides, cellodextrins, oligogalacturonic acids, cyclodextrins, gangliosides, and others. She also discussed the use of new sensitive detection reagents for the planar chromatography of sugars.

Anderton et al. (1993) described a procedure for the analysis of sugars in the hemolymph and digestive gland–gonad complex (DGG) of *Biomphalaria glabrata* snails by quantitative HPTLC. They did a comparative study on a variety of chromatographic systems and detection reagents to determine the optimum conditions for the analysis. Preadsorbent silica gel plates (10 × 10 cm Merck HPTLC silica gel 60 CF$_{254}$) impregnated with sodium bisulfite and citrate buffer and developed with acetonitrile–water (85:15) provided the best resolution of the sugars and the tightest spots. α-Naphthol reagent gave dark blue-purple spots for the sugars, and a very slow rate of fading. R_F values for sugar standards in the above-mentioned chromatographic system were as follows: glucose, 0.40; trehalose, 0.17; mannose, 0.47; lactose, 0.16; fructose, 0.43; xylose, 0.60; sucrose, 0.26; maltose, 0.24; raffinose, 0.13; ribose, 0.66; and fucose, 0.63. The R_F values of the sugars in the snail hemolymph were as follows: trehalose, 0.17; glucose, 0.40; and unidentified, 0.09. The R_F values of the sugars in the DGG were: unidentified, 0.71; glucose, 0.39; trehalose, 0.17; unidentified, 0.10; and unidentified, 0.03. Szilagyi (1993) used overpressured development on silica gel layers impregnated with tricaprylmethylammonium chloride and methanol–water mobile phases to study the retention behavior of N-glycosides.

Yao et al. (1994) determined the glycosyl sequence of glycosides by enzymatic hydrolysis followed by the separation of the hydrolysate by TLC and direct analysis of the spots by fast atom bombardment–mass spectrometry (FABMS). Lodi et al. (1994) determined sugars in food and clinical samples without cleanup procedures using automated multiple development (AMD) with a polarity gradient based on acetonitrile–acetone–water on buffered amino layers and HPTLC–scanning densitometry for quantification. Robyt and Mukerjea (1994) used TLC-densitometry to determine nanogram amounts of maltodextrins and isomaltodextrins containing 1–20 glucose units with alpha-naphthol–sulfuric acid detection reagent. Perez et al. (1994) described a densitometric method using preadsorbent HPTLC silica gel 60 plates impregnated with sodium bisulfite and pH 4.8 citrate buffer, triple development with acetonitrile–water (85:15), and zone detection with alpha-naphthol reagent for the determination of sugars in the hemolymph and digestive gland–gonad complex of *Biomphalaria glabrata* snails infected with the larval trematode *Echinostoma caproni*.

Conaway et al. (1995) extended the above-mentioned study of Perez et al. (1994) to determine by densitometric–HPTLC sugars in *Helisoma trivolvis* snails infected with the larval trematode *Echinostoma trivolvis*. Zarzycki et al. (1995) examined the retention properties of alpha-, beta-, and gamma-cyclodextrins using water-wettable RP-C-18 layers and mobile phases composed of acetonitrile, methanol, ethanol, or propanol with water or methanol in acetonitrile. Esaiassen et al. (1995) analyzed n-acetylchitooligosaccharides by use of the Iatroscan TLC/flame ionization detection (FID) system. Brandolini et al. (1995) used automated multiple development (AMD)–HPTLC to monitor various carbohydrates—i.e., maltotetraose, maltotriose, maltose, glucose, and fructose—in different beers.

Pukl et al. (1996) extended and updated the earlier review of Prosek et al. (1991) on the TLC of carbohydrates. Abeling et al. (1996) used two unidimensional TLC systems to improve urinary screening methods for patients with oligosaccharide disorders. Sherma and Zulick (1996) described an HPTLC method for the simultaneous determination of fructose, glucose, and sucrose in sodas and iced teas. Kim and Sakano (1996) analyzed reducing sugars on TLC plates with modified Somogyri and Nelson reagents and with copper bicinchoninate. John et al. (1996) described methods for rapid quantification of sucrose and fuctan oligosaccharides by TLC and densitometry. Cesio et al. (1996) reported a fast and accurate TLC method to evaluate lactose intolerance by determining lactose in fecal samples. Jain (1996) described TLC methods useful in the clinical chemistry of sugar disorders, particularly those appropriate to analysis of blood and urine of patients with inherited and metabolic diseases. Fell (1996) provided extensive information on the TLC analysis of carbohydrates mainly relevant to entomology. However, the information provided on analytical techniques and approaches, solvent systems, visualization of sugars, and tables with R_F values of sugars in different mobile phases is transferable to disciplines other than entomology. Umesh et al. (1996) used HPTLC densitometry to analyze sugars in the hemolymph and digestive gland–gonad complex (DGG) of planorbid snails (*Biomphalaria glabrata* and *Helisoma trivolvis*) and found that the major sugars were glucose and maltose. The paper provides a good historical account of TLC studies of sugars in medically important snails and raises the unresolved question in regard to the occurrence of trehalose in planorbid snails.

Anderton et al. (1998) used quantitative HPTLC–densitometry to determine glucose in human serum. Analysis was done on preadsorbent silica gel plates with a mobile phase of acetonitrile–water (85:15) and a methanolic sulfuric acid detection reagent. The paper gives measurements of glucose obtained by HPTLC–densitometry before and after consumption of a sucrose solution. Anderton et al. (1998) also reported the R_F values of the sugar standards detected using the above-mentioned chromatographic system. These values were as follows: sucrose, 0.25; glucose, 0.31; trehalose, 0.18; mannose, 0.34; galactose, 0.27; xylose, 0.47; ribose, 0.48; lactose, 0.17; raffinose, 0.09; fructose, 0.33; melezitose, 0.15; and maltose, 0.19.

IV. TLC EXPERIMENTS

Experiment 1 uses cellulose for the separation of sugar standards, whereas Experiment 2 employs silica gel for the same purpose. For TLC experiments on sugars in plants and animals, see Fried and Sherma (1994).

V. DETAILED EXPERIMENTS

A. Experiment 1. TLC Separation of Sugars on Cellulose (Adapted from Instructions Supplied by J. T. Baker for Baker-flex Flexible TLC Sheets)

1. Introduction

This experiment is an introduction to TLC separation of sugars by multiple development on cellulose thin layers.

2. Preparation of Standards

Prepare 1-μg/μl individual standards of dextrose (glucose), galactose, lactose, and xylose (Sigma) by dissolving 10 mg of each sugar in 10 ml of deionized water. Prepare a mixed standard containing 1 μg/μl of each sugar. To do this, dissolve 10 mg of each sugar in a tube containing 10 ml of deionized water.

3. Layer

Baker-flex cellulose sheet, 20 × 20 cm, used as received.

4. Application of Initial Zones

Prepare an origin line 2 cm from the bottom of the plate and a solvent front line 10 cm from the origin. Spot 4 μl of each individual standard and 4 μl of the mixed standard on separate origins. Thoroughly dry the sheet in air prior to development.

5. Mobile Phase

n-Butanol–pyridine–water (60:40:30).

6. Development

Develop the sheet for a distance of 10 cm in 100 ml of the mobile phase in a tank that has been saturated with the mobile phase. Development at room temperature will take from 1.5 to 2 h. Remove the sheet and allow it to air dry in a fume hood for 1 h. Develop the sheet a second time for a distance of 10 cm in the same solvent. Remove the sheet from the tank and allow it to dry thoroughly prior to spraying.

7. Detection

Spray the sheet with an ethanolic solution of p-anisidine phthalate. To prepare the detection solution, dissolve 1.2 g of p-anisidine and 1.6 g of phthalic acid in 100 ml of ethanol. Spray the sheet using an aerosol sprayer and then heat the layer at 100°C for 10 min.

8. Results and Discussion

The order of separation and approximate R_F values should be as follows: lactose, 0.42; galactose, 0.58; dextrose (glucose), 0.63; xylose, 0.74. The separation should produce compact spots with brown to violet colors.

B. Experiment 2. TLC Separation of Sugars on Silica Gel

1. Introduction

This experiment, adapted from Lewis and Smith (1969), is an introduction to TLC separation of sugars on silica gel.

2. Preparation of Standards

Prepare 1-μg/μl individual standards of lactose, maltose, galactose, glucose, and mannose (Sigma) as described in Experiment 1. Prepare a mixed standard containing 1 μg/μl of each of the five sugars as described in Experiment 1.

3. Layer

Baker-flex silica gel IB 2 sheet, 20 × 20 cm. Impregnate the layer with 0.1 M sodium bisulfite (10.4 g of $NaHSO_3$ in 1 liter of deionized water) by dipping the sheet into a dipping tank or glass tray containing the bisulfite solution. Allow the sheet to air dry, and then activate it at 100°C for 30 min prior to spotting.

4. Application of Initial Zones

Prepare an origin line 2 cm from the bottom of the sheet and a solvent front line 10 cm from the origin. Spot 1 μl of each standard on separate origins.

5. Mobile Phase

Ethyl acetate–acetic acid–methanol–water (60:15:15:10).

6. Development

Develop the sheet for a distance of 10 cm in a tank saturated with 100 ml of the mobile phase. Remove the sheet and dry it thoroughly prior to spraying.

7. Detection

Spray the sheet with an α-naphthol–sulfuric acid visualizing reagent contained in an aerosol sprayer. To prepare the reagent, first add 5 g of α-naphthol to 33 ml of ethanol; then combine 21 ml of the ethanolic naphthol with 13 ml of H_2SO_4, 81 ml of ethanol, and 8 ml of H_2O. After the sheet is sprayed, heat it for about 5 min at 100°C.

8. Results and Discussion

The order of separation and approximate R_F values should be as follows: lactose, 0.14; maltose, 0.22; galactose, 0.32; glucose, 0.37; mannose, 0.42. The spots will be compact. This procedure is considerably faster and more sensitive than the separation using cellulose described in Experiment 1.

REFERENCES

Abeling, N. G. G. M., Rusch, H., and Van Gennip, A. H. (1996). Improved selectivity of urinary oligosaccharide screening using two one-dimensional TLC systems. *J. Inherited Metab. Dis. 19*: 260–262.

Anderton, C. A., Fried, B., and Sherma, J. (1993). HPTLC determination of sugars in the hemolymph and digestive gland–gonad complex of *Biomphalaria glabrata* snails. *J. Planar Chromatogr.—Mod. TLC 6*: 51–54.

Anderton, S. M., Layman, L. R., and J. Sherma (1998). Determination of blood glucose by quantitative TLC on preadsorbent silica gel plates. *J. Liq. Chromatogr. & Related Tech. 21*: 1045–1049.

Barnard E. A. (1973). Comparative biochemistry and physiology of digestion. In *Compar-*

ative Animal Physiology, 3rd ed., C. L. Prosser (Ed.). Saunders, Philadelphia, pp. 212–278.

Batisse, C., Daurade, M. H., and Bounias, M. (1992). TLC separation and spectro-densitometric quantitation of cell wall neutral sugars. *J. Planar Chromatogr.—Mod. TLC 5*: 131–133.

Blom, W., Luteyn, J. C., Kelholt-Dijkman, H. H., Huijmans, J. G. M., and Loonen, M. C. B. (1983). Thin-layer chromatography of oligosaccharides in urine as a rapid indication for the diagnosis of lysosomal acid maltase deficiency (Pompe's disease). *Clin. Chim. Acta 134*: 221–227.

Bosch-Reig, F., Marcot, M. J., and Minana, M. D. (1992). Separation and identification of sugars and maltodextrins by thin layer chromatography: application to biological fluids and human milk. *Talanta 39*: 1493–1498.

Brandolini, V., Menziani, E., Mazzotta, D., Cabras, P., Tosi, B., and Lodi, G. (1995). Use of AMD-HPTLC for carbohydrate monitoring in beers. *J. Food Compos. Anal. 8*: 336–343.

Brasseur, T. W., Wauters, J. N., and Angenot, L. (1988). Densitometric determination of quercetin heterosides in plant extracts. *J. Chromatogr. 437*: 260–264.

Caldes, G., and Prescott, B. (1975). A simple method for the detection and determination of trehalose by spot elution paper chromatography. *J. Chromatogr. 111*: 466–469.

Cesio, V., Heinzen, H., Servetto, C., Torres, M., and Moyna, P. (1996). Rapid determination of lactose intolerance. *Quim. Anal. (Barcelona) 15*: 140–143.

Conaway, C. A., Fried, B., and Sherma, J. (1995). High performance thin layer chromatographic analysis of sugars in *Helisoma trivolvis* (Pennsylvania strain) infected with larval *Echinostoma trivolvis* and uninfected *H. trivolvis* (Pennsylvania and Colorado strains). *J. Planar Chromatogr.—Mod. TLC 8*: 184–187.

Churms, S. C. (1981). *CRC Handbook of Chromatography—Carbohydrates*, Vol. 1. CRC Press, Boca Raton, FL.

Churms, S. C. (1983). Carbohydrates. In *Chromatography—Fundamentals and Applications of Chromatographic and Electrophoretic Methods—Part B: Applications*, E. Heftmann (Ed.). Elsevier, Amsterdam, pp. B224–B286.

Churms, S. C. (1991). *CRC Handbook of Chromatography—Carbohydrates, Vol. II.* CRC Press, Boca Raton, FL.

Churms, S. C. (1992). Carbohydrates. In *Chromatography—Fundamentals and Applications of Chromatography and Related Differential Migration Methods—Part B: Applications*, E. Heftmann (Ed.). Elsevier, Amsterdam, pp. B229–B292.

Clark, M. E. (1964). Biochemical studies on the coelomic fluid of *Nephtys hombergi* (Polychaeta: Nephtyidae), with observations on changes during different physiological states. *Biol. Bull. 127*: 63–84.

Doner, L. W. (1988). High-performance thin-layer chromatography of starch, cellulose, xylan, and chitin hydrolyzates. *Methods Enzymol. 160*: 176–180.

Doner, L. W., and Biller, L. M. (1984). High-performance thin-layer chromatographic separation of sugars: preparation and application of aminopropyl bonded-phase silica plates impregnated with monosodium phosphate. *J. Chromatogr. 287*: 391–398.

Esaiassen, M., Oeverboe, K., and Olsen, R. L. (1995). Analysis of N-acetyl-chitooligosaccharides by the Iatroscan TLC/FID system. *Carbohydr. Res. 273*: 77–81.

Fairbairn, D. (1958). Trehalose and glucose in helminths and other invertebrates. *Can. J. Zool. 36*: 787–795.

Fell, R. D. (1990). The qualitative and quantitative analysis of insect hemolymph sugars by high performance thin-layer chromatography. *Comp. Biochem. Physiol. 95A*: 539–544.

Fell, R. D. (1996). Thin-layer chromatography in studies on entomology. In *Practical Thin-Layer Chromatography—A Multidisciplinary Approach*, B. Fried and J. Sherma (Eds.). CRC Press, Boca Raton, FL, pp. 71–104.

Florkin, M., and Jeuniaux, C. (1974). Hemolymph: Composition. In *The Physiology of Insects*, 2nd ed., Vol. 5. M. Rockstein (Ed.). Academic Press, New York, pp. 255–307.

Fried, B., and J. Sherma (1994). *Thin-Layer Chromatography—Techniques and Applications*, 3rd ed. Marcel Dekker, New York, pp. 313–316.

Ghebregzabher, M., Rufini, S., Monalde, B., and Lato, M. (1976). Thin-layer chromatography of carbohydrates. *J. Chromatogr. 127*: 133–162.

Goso, Y., and Hotta, K. (1990). Analysis of mucin-derived oligosaccharides by medium-pressure gel-permeation and amino-plate thin-layer chromatography conducted in conjunction. *Anal. Biochem. 188*: 181–186.

Groesz, J., and Braunsteiner, W. (1989). Quantitative determination of glucose, fructose, and sucrose, and separation of fructo-oligosaccharides by means of TLC. *J. Planar Chromatogr.—Mod. TLC 2*: 420–423.

Gross, J., Dumsha, B., and Glazer, N. (1958). Comparative biochemistry of collagen. Some amino acids and carbohydrates. *Biochim. Biophys. Acta 30*: 293–297.

Harborne, J. B. (1984). *Phytochemical Methods—A Guide to Modern Techniques of Plant Analysis*, 2nd ed. Chapman & Hall, London.

Hirayama, H., Hiraki, K., Nishikawa, Y., and Fukazawa, S. (1984). Thin-layer chromatography of carbohydrates on silica gel sintered plates. *Kinki Daigaku Rikogakubu Kenkyu Hokoku 20*: 121–126; *Chem. Abstr.* 1984, 101,126298a.

Hochachka, P. W. (1973). Comparative intermediary metabolism. In *Comparative Animal Physiology*, 3rd ed., C. L. Prosser (Ed.). Saunders, Philadelphia, pp. 212–278.

Jacin, H., and Mishkin, A. R. (1965). Separation of carbohydrates on borate-impregnated silica gel G plates. *J. Chromatogr. 18*: 170–173.

Jain, R. (1996). Thin layer chromatography in clinical chemistry. In *Practical Thin-Layer Chromatography—A Multidisciplinary Approach*. B. Fried and J. Sherma (Eds.). CRC Press, Boca Raton, FL, pp. 131–152.

John, J. A. St., Bonnett, G. D., and Simpson, R. J. (1996). A method for rapid quantification of sucrose and fructan oligosaccharides suitable for enzyme and physiological studies. *New Phytol. 134*: 197–203.

Karlsson, K. A., Lanne, B., Pimlott, W., and Teneberg, S. (1991). The resolution into molecular species on desorption of glycolipids from thin-layer chromatograms, using combined thin-layer chromatography and fast-atom-bombardment mass spectrometry. *Carbohydr. Res. 221*: 49–61.

Kim, Y. K., and Sakano, Y. (1996). Analyses of reducing sugars on a thin-layer chromatographic plate with modified Somogyi and Nelson reagents, and with copper bicinchoninate. *Biosci. Biotechnol., Biochem. 60*: 594–597.

Klaus, R., Fischer, W., and Hauck, H. E. (1989). Use of a new adsorbent in the separation and detection of glucose and fructose by HPTLC. *Chromatographia 28*: 364–366.

Klaus, R., Fischer, W., and Hauck, H. E. (1990). Application of a thermal in situ reaction for fluorometric detection of carbohydrates on NH$_2$-layers. *Chromatographia 29*: 467–472.

Klaus, R., Fischer, W., and Hauck, H. E. (1991). Qualitative and quantitative analysis of uric acid, creatine, and creatinine together with carbohydrates in biological material by HPTLC. *Chromatographia 32*: 307–316.

Lakshmi, M. B., and Subrahmanyam, D. (1975). Trehalose of *Culex pipiens fatigans. Experientia 31*: 898–899.

Lato, M., Brunelli, B., and Ciuffini, G. (1968). Analysis of carbohydrates in biological fluids by means of thin layer chromatography. *J. Chromatogr. 36*: 191–197.

Lenkey, B., Csanyi, J., and Nanasi, P. (1986). A rapid determination of sucrose and fructose in biological samples by video densitometry. *J. Liq. Chromatogr. 9*: 1869–1875.

Lewis, B. A., and Smith, F. (1969). Sugars and derivatives. In *Thin Layer Chromatography*, 2nd ed., E. Stahl (Ed.). Springer-Verlag, New York, pp. 807–837.

Liebsch, M., and Becker, W. (1990). Comparative glucose tolerance studies in the freshwater snail *Biomphalaria glabrata*. Influence of starvation and infection with the trematode *Schistosoma mansoni. J. Comp. Physiol. B. Biochem. Syst. Environ. Physiol. 160*: 41–50.

Lodi, G., Betti, A., Brandolini, V., Menziani, E., and Tosi, B. (1994). Automated multiple development HPTLC analysis of sugars on hydrophilic layers: 1. Amino layers. *J. Planar Chromatogr.—Mod. TLC 7*: 29–33.

Mack, S. R., Samuels, S., and Vanderberg, J. P. (1979). Hemolymph of *Anopheles stephensi* from noninfected and *Plasmodium berghei*-infected mosquitoes. 3. Carbohydrates. *J. Parasitol. 65*: 217–221.

Mansfield, C. T. (1973). Use of thin-layer chromatography in determination of carbohydrates. In *Quantitative Thin Layer Chromatography*, J. C. Touchstone (Ed.). Wiley, New York, pp. 79–93.

Menzies, I. S., and Mount, J. N. (1975). Advantages of silica gel as a medium for rapid thin-layer chromatography of neutral sugars. *Med. Lab. Techno. 32*: 269–276.

Metraux, J. P. (1982). Thin-layer chromatography of neutral and acidic sugars from plant cell wall polysaccharides. *J. Chromatogr. 237*: 525–527.

Morcol, T., and Velander, W. H. (1991). An *o*-toluidine method for detection of carbohydrates in protein hydrolyzates. *Anal. Biochem. 195*: 153–159.

Myrback, K. (1949). Trehalose and trehalase. *Ergeb. Enzymforsch. 10*: 168–190.

Parrado, C., Alonso-Fernandez, J. R., Boveda, M. D., Fraga, J. M., and Pena, J. (1983). The use of continuous thin layer chromatography in the study of mucopolysaccharidoses. *J. Inherited Metab. Dis. 6*: 135–136.

Patzsch, K., Netz, S., and Funk, W. (1988a). Quantitative HPTLC (high-performance thin-layer chromatography) of sugars. Part 1: Separation and derivatization. *J. Planar Chromatogr.—Mod. TLC 1*: 39–45.

Patzsch, K., Netz, S., and Funk, W. (1988b). Quantitative HPTLC of sugars. Part 2: Determination in different matrixes. *J. Planar Chromatogr.—Mod. TLC 1*: 177–179.

Perez, M. K., Fried, B., and Sherma, J. (1994). High performance thin-layer chromatographic analysis of sugars in *Biomphalaria glabrata* (Gastropoda) infected with *Echinostoma caproni. J. Parasitol. 80*: 336–338.

Poukens-Renwart, P., and Angenot, L. (1991). A densitometric method for the determination of three clinically important monosaccharides in urine. *J. Planar Chromatogr.—Mod. TLC 4*: 77–79.

Prosek, M., Pukl, M., and Jamnik, K. (1991). Carbohydrates. In *Handbook of Thin-Layer Chromatography*, J. Sherma and B. Fried (Eds.). Marcel Dekker, New York, pp. 439–462.

Pukl, M., and Prosek, M. (1990). Rapid quantitative TLC analysis of sugars using an improved commonly used solvent system. *J. Planar Chromatogr.—Mod. TLC 3*: 173–176.

Pukl, M., Prosek, M., Golc-Wondra, A., and Jamnik, K. (1996). Carbohydrates. In *Handbook of Thin-Layer Chromatography*, 2nd ed., J. Sherma and B. Fried (Eds.). Marcel Dekker, New York, pp. 481–505.

Ray, B., Ghosal, P. K., Thakur, S., and Majumdar, S. G. (1984). Sulfosalicylic acid as spray reagent for the detection of sugars on thin-layer chromatograms. *J. Chromatogr. 315*: 401–403.

Rebers, P. A., Wessman, G. E., and Robyt, J. F. (1986). A thin-layer-chromatographic method for analysis of amino sugars in polysaccharide hydrolyzates. *Carbohydr. Res. 153*: 132–135.

Robyt, J. F., and Mukerjea, R. (1994). Separation and quantitative determination of nanogram quantities of maltodextrins and isomaltodextrins by thin-layer chromatography. *Carbohydr. Res. 251*: 187–202.

Saamanen, A. M., and Tammi, M. (1984). Determination of unsaturated glycosaminoglycan disaccharides by spectrophotometry on thin-layer chromatographic plates. *Anal. Biochem. 140*: 354–359.

Sadun, E. H., William, J. S., and Martin, L. K. (1966). Serum biochemical changes in malarial infections in men, chimpanzees, and mice. *Mil. Med. 131* (Suppl.): 1094–1106.

Saito, M., Kasai, N., and Yu, R. K. (1985). In situ immunological determination of basic carbohydrate structures of gangliosides on thin-layer plates. *Anal. Biochem. 148*: 54–58.

Scott, R. M. (1969). *Clinical Analysis by Thin-Layer Chromatography Techniques*. Ann Arbor–Humphrey Science Publishers, Ann Arbor, MI.

Seller, P., Thorn, W., and Neuhoff, V. (1984). Determination of sugars in the picomol range using dansylhydrazine. *Funkt. Biol. Med. 3*: 85–94.

Sherma, J., and Zulick, D. L. (1996). Determination of fructose, glucose, and sucrose in beverages by high-performance thin layer chromatography. *Acta Chromatogr. 6*: 7–12.

Shimada, E., Kudoh, N., Tomioka, M., and Matsumura, G. (1987). Separation of unsaturated chondroitin sulfate disaccharides in thin-layer chromatography on silica gel and their quantitative determination by densitometry. *Chem. Pharm. Bull. 35*: 1503–1508.

Shinomiya, K., Hoshi, Y., and Imanari, T. (1989). Sensitive detection of unsaturated disaccharides from chondroitin sulfates by thin-layer chromatography as their dansylhydrazine derivatives. *J. Chromatogr. 462*: 471–474.

Supelco TLC Reporter. (1977). *Carbohydrate Analysis by GC and TLC. TLC of Sugars*. Bulletin 774A, Supelco, Bellefonte, PA, pp. 4–5.

Suyama, K., and Adachi, S. (1987). A simple method for the visualization of the separated zones of sugars on silica-gel TLC plates without spray reagent. *J. Chromatogr. Sci.* *25*: 130–131.

Szilagyi, J., Kovacs-Hadady, K., and Kovacs, A. (1993). OPLC study of retention of aromatic N glycosides on layers impregnated with tricaprylmethylammonium chloride. *J. Planar Chromatogr.—Mod. TLC 6*: 212–215.

Thieme, R., Lay, H., Oser, A., Lehmann, J., Wrissenberg, S., and Boos, W. (1986). 3-Azi-1-methoxybutyl D-maltooligosaccharides specifically bind to the maltose/maltooligosaccharide-binding protein of *Escherichia coli* and can be used as photoaffinity labels. *Eur. J. Biochem. 160*: 83–91.

Thompson, S. N. (1985). Metabolic integration during the host associations of multicellular animal endoparasites. *Comp. Biochem. Physiol. 81B*: 21–42.

Torres, M., Dean, M. A., and Wagner, F. W. (1990). Chromatographic separations of sucrose monostearate structural isomers. *J. Chromatogr. 522*: 245–253.

Tugrul, L., and Ozer, A. (1985). TLC fluorodensitometric determination of monosaccharides in drug hydrolysates. Part II: Parameters of the TLC-fluorodensitometric monosaccharide analysis. *Acta Pharm. Turc. 27*: 30–32.

Tugrul, L., and Ozer, A. (1986). TLC–fluorodensitometric determination of monosaccharides in drug hydrolyzates. Part III: Application to various drug hydrolyzates. *Acta Pharm. Turc. 28*: 5–8.

Ueda, T., and Sawada, T. (1968). Fatty acid and sugar compositions of helminths. *Jap. J. Exp. Med. 38*: 145–147.

Umesh, A., Fried, B., and J. Sherma (1996). Analysis of sugars in the hemolymph and digestive gland–gonad complex (DGG) of *Biomphalaria glabrata* and *Helisoma trivolvis* (Colorado and Pennsylvania strains) maintained on restricted diets. *Veliger 39*: 354–361.

Vajda, J., and Pick, J. (1988). Separation of some mono-, di-, tri-, and oligosaccharides. *J. Planar Chromatogr.—Mod. TLC 1*: 347–348.

Vega, M., Valladeres, J., and Saelzer, R. (1985). Improved separation and quantitative analysis of sucrose, glucose, fructose and xylose in natural products by high-performance thin-layer chromatography (HPTLC). In *Proceedings of the International Symposium on Instrumental High Performance Thin-Layer Chromatography*, 3rd ed., R. E. Kaiser (Ed.). Institute for Chromatography, Bad Duerkeim, Germany, pp. 211–222.

Vomhof, D. W., and Tucker, T. C. (1965). The separation of simple sugars by cellulose thin layer chromatography. *J. Chromatogr. 17*: 300–306.

von Brand, T. (1973). *Biochemistry of Parasites*, 2nd ed. Academic Press, New York.

Wyatt, G. R., and Kalif, G. F. (1956). Trehalose in insects. *Fed. Proc. 15*: 388.

Yao, Z., Liu, G., Wen, H., He, W., Zhao, S., and Hu, W. (1994). Determination of the glycosyl sequence of glycosides by enzymic hydrolysis followed by TLC-FABMS. *J. Planar Chromatogr.—Mod. TLC 7*: 410–412.

Zarzycki, P. K., Nowakowska, J., Chmielewska, A., and Lamparczyk, H. (1995). Retention properties of cyclodextrins in reversed-phase HPTLC. *J. Planar Chromatogr.—Mod. TLC 8*: 227–231.

18

Natural Pigments

I. GENERAL

Kirchner (1978) has discussed the utility of thin-layer chromatography for the following natural pigments: chlorophylls, carotenoids, xanthophylls, flavonoids, anthocyanins, porphyrins, and bile. From the standpoint of practical TLC, the most important of these pigments are the chlorophylls, carotenoids, xanthophylls, and anthocyanins, and it is these pigments that are considered further in this chapter. For practical TLC of the flavonoids, consult Harborne (1984, 1992). For information on the TLC of porphyrins, see Doss (1972), Dolphin (1983), and Jacob (1992). Jain (1996) has provided useful information on the examination of porphyrins (in studies on clinical porphyrias) by TLC in clinical chemistry.

Practical TLC studies on bile pigments have been discussed by Scott (1969) and Dolphin (1983). Heirwegh et al. (1989) reviewed analytical procedures, including TLC, for the separation and identification of bile pigments including the biliverdins and bilirubins.

Isaksen (1991) provided an extensive review on the TLC of natural pigments. He stated that TLC is the most common method used to analyze natural pigments. Pigments considered by Isaksen include the flavonoids, anthocyanins, carotenoids, chlorophylls, chlorophyll derivatives, and porphyrins. Optimal chromatographic systems (i.e., sorbents and mobile phases for each pigment group) are considered in detail in the review. The Isaksen (1991) review was revised and updated by Andersen and Francis (1996).

Scheer (1988) provided information on the use of TLC for the analysis of chlorophylls, including tables of different sorbents used for the TLC analysis of chlorophylls and R_F values of chlorophylls and related pigments studied in various chromatographic systems. Gross (1991) has reviewed the salient literature on the TLC of chloroplast pigments and carotenoids from about 1963 to 1989.

Davies and Kost (1988) noted that TLC of carotenoids is one of the most modern methods of carotenoid separation. TLC of carotenoids has been done on a variety of layers using many different solvent systems. Davies and Kost (1988) provided an extensive list of tables on the TLC of carotenoids. Cikalo et al. (1992) used automated multiple development (AMD) techniques for the separation of complex plant carotenoid mixtures; they also evaluated the use of scanning densitometry for spectral identification and quantification of these carotenoids.

Harborne (1992) has reviewed the recent literature on the TLC of flavonoids and provided tables listing suitable sorbents and mobile phases for routine separations of these compounds.

Burnham and Kost (1988) noted that TLC of porphyrins is a useful technique with silica gel, polyamide (nylon), and talc often serving as the sorbents. With slight variations of the solvent systems, analytical TLC of porphyrins can often be converted to thick-layer or preparative-layer chromatography. Burnham and Kost (1988) provided numerous tables for the TLC separation of free acid porphyrins, porphyrin esters, and metalloporphyrins.

Jacob (1992) has reviewed studies on the TLC of porphyrins and noted that this technique is a versatile one for separating these compounds. The sorbent most often used is silica gel. TLC allows for the separation of most of the common classes of porphyrins—i.e., the porphyrin carboxylic acids and their esters, alkyl porphyrins, and porphyrin–metal complexes. TLC is also useful for the separation of the diagnostically important porphyrins from clinical specimens such as urine, feces, and blood. Routine methods for the separation by TLC of porphyrins from human specimens are now available. With various combinations of the mobile-phase benzene–ethyl acetate–methanol and silica gel sorbent, porphyrin methyl esters can be separated according to the number of carboxyl groups, with the R_F values being inversely proportional to that number (Jacob, 1992).

Lai et al. (1994) reported a simple technique to determine free porphyrins in urine and feces for the diagnoses of porphyria based on C-18 HPTLC with detection of fluorescent zones under long-wavelength UV light. Kus (1996) compared the TLC behavior of several crown ether derivatives of tetraphenylporphyrin on silica gel, alumina, and cellulose layers with different organic mobile phases.

II. FAT-SOLUBLE PIGMENTS

The chlorophylls, carotenoids, and xanthophylls are the best-known fat-soluble pigments and are often referred to as lipochromes. These pigments are found in plastids (chloroplasts or chromoplasts), within the cytoplasm of plant tissues. For further information, consult the Introduction of Experiment 1 and Goodwin (1952), Fox (1953), Strain and Sherma (1972), and Dolphin (1983). When lipochromes are found in animal tissues, they usually have accumulated there as a result of the ingestion of plant pigments. Animals generally store these pigments in particular organs (for example, the gonads) or in special cells (chromatophores) usually associated with the hypodermis or skin. Animals may alter the original plant pigment obtained by ingestion to produce a new chemical structure. The mechanisms involved in such pigment alterations are obscure.

Of the fat-soluble pigments, carotenoids have been best characterized. They are unsaturated hydrocarbons with branched chains of carbon atoms containing a cyclic structure at one or both ends. For the chemical structures of carotenoids, see Florey (1966).

Florey (1966) has discussed the role of carotenoids in the physiology of animals. These pigments often provide protective coloration for animals, particularly insects (adaptive coloration). Carotenoids are also involved in oxygen transfer and serve as electron acceptors for some invertebrates. Moreover, carotenoids are protective to organisms and serve as a radiation shield to filter out ultraviolet light. Carotenoids also play a major role as precursors for enzymes, vitamins, and hormones.

Chlorophylls are involved in photosynthetic reactions in plants. The structure and function of these pigments are discussed in the Introduction of Experiment 1.

Xanthophylls are often found in invertebrates along with carotenoids (Goodwin, 1952, 1972). The function of xanthophylls in animals is obscure.

III. WATER-SOLUBLE PIGMENTS

Flavonoids are the major water-soluble pigments in plants and are contained within the cell sap of plant tissues. The term "flavonoids" is derived from "flavus" (yellow) and refers to a diverse group of water-soluble pigments. Of all the flavonoids, the anthocyanins have the greatest importance from the standpoint of practical TLC. The anthocyanins are responsible for the variety of colors associated with autumnal (fall) foliage and with the variety of colors (mainly reds, scarlet, blues) associated with flowers and fruits. The anthocyanins yield anthocyanids and various sugars upon acid hydrolysis. During the TLC of anthocyanins, it is the anthocyanids that are detected.

Anthocyanins and anthocyanids are usually separated on microcrystalline cellulose. Andersen and Francis (1985) were able to separate flavonoids on cellulose in the mobile-phase HCl–formic acid–water (24.9:23.7:51.4). The R_F values obtained were 0.03 for delphinidin, 0.06 for cyanidin, 0.11 for pelargonidin, 0.26 for cyanidin 3-glucoside, 0.48 for cyanidin 3-rutinoside, and 0.95 for pelargoniden 3-sophoroside-5-glucoside.

Matysik and Benes (1991) determined the anthocyanin content in the petals of red poppy during flower development by six-stage polyzonal-stepwise gradient gel TLC–densitometry of the separated zones at 465 nm. Matysik (1992) determined anthocyanins in plant extracts by stepwise gradient elution with various combinations of the mobile-phase ethyl acetate–isopropanol–water–acetic acid on silica gel 60 plates. Densitograms were recorded at 465 or 580 nm.

Heimler and Pieroni (1991) separated the flavonoid and bioflavonoid glycosides of cypress leaves by HPTLC; the compounds were identified by in situ UV spectra and quantified by densitometry. Budzianowski (1991) described nine TLC systems for the separation of mixtures of flavonoid glucosides from their galactosidic analogs for homogeneity control of C-glycosylflavones. Heimler et al. (1988) described a two-dimensional reversed-phase HPTLC method for the separation of leaf flavonoids. Billeter et al. (1990) used TLC–densitometry to analyze flavonoids after derivatization with 2-aminoethyl diphenylborinate and polyethylene glycol 4000.

Heimler et al. (1992) determined the flavonoid content of olive leaves (*Olea europaea*) by using TLC and recording UV spectra directly on the silica layer. Three flavonoid glycosides, quercitin, flavonoid aglycon, and luteolin, along with chlorogenic acid were found in the olive leaves. Conde et al. (1992) used TLC to analyze flavonoids in the genus *Eucalyptus*. They described several solvent systems and spray reagents to identify various flavonoids from leaves, wood, and bark of species of eucalyptids. Two-dimensional chromatography on cellulose was the best separation method, although monodimensional chromatography on silica gel was also satisfactory for most of the flavonoids studied.

Burton et al. (1993) examined ipriflavone and its impurities by TLC with absorbance, fluorescence, and fluorescence quenching detection at 1 ng levels. Hadj-Mahammad and Meklati (1993) separated polyhydroxy- and polymethoxyflavones on silica gel developed with heptane-2-propanol (80:20) plus 5% acetic acid and on C-18 layers developed with methanol–water (80:20) plus 3% NaCl. Krawczyk et al. (1993) quantified flavonoids in rutin-containing buckwheat herb extract by silica gel TLC with scanning densitometry.

Bartolome (1994) reported polyamide TLC separation of flavanones from traces of other types of flavonoid compounds with crystallization of the flavonoid compounds before and after chromatography. Ficarra et al. (1994) determined the hydrophobic parameters of flavonoids by reversed-phase TLC. Nikolova-Damyanova et al. (1994) used TLC to quantify iridoid and flavonoid glucosides in species of *Linaria*.

Hiermann (1995) used different flavonoid patterns to distinguish *Epilobium angustifolium* from 11 other species of *Epilobium* by TLC on silica gel and polyamide and HPLC coupled with photodiode array detection. Rustan et al. (1995) developed TLC methods to determine flavonoids extracted from Mediterranean plants of the Labiatae group. Van Ginkel (1996) used TLC to identify alkaloids and flavonoids from the bark of *Uncaria tomentosa*.

IV. TLC OF PIGMENTS

Experiment 1 compares reversed-phase and adsorption TLC for the separation of fat-soluble chloroplast pigments from plant tissues. Experiment 2 uses silica gel TLC to examine lipophilic pigments in animal tissue. Experiment 3 uses cellulose TLC to examine water-soluble pigments (anthocyanins) in plant tissues.

V. DETAILED EXPERIMENTS

A. Experiment 1. Separation of Fat-Soluble Chloroplast Pigments from Plants by Adsorption and Reversed-Phase TLC

1. Introduction

The isolation and separation of fat-soluble leaf chloroplast pigments by adsorption chromatography on silica gel layers and reversed-phase partition chromatography on chemically bonded C_{18} layers are demonstrated in this experiment, based on the work of Strain and Sherma (1969) and Sherma and Latta (1978). The separability of these pigments reflects the capability of TLC to differentiate compounds with only small differences in chemical composition and molecular structure. The yellow carotenoid pigments fall into two principal groups, the carotenes and the xanthophylls. The most common carotenes, called α- and β-carotene, are isomeric, nonpolar, polyene hydrocarbons, $C_{40}H_{56}$, with 11 double bonds, all conjugated in β-carotene and only 10 conjugated in α-carotene. The xanthophylls are polar oxygen derivatives of carotenes. These derivatives contain from one to about six oxygen atoms, which occur as alcohol, keto, epoxy, or ester groups. The green chlorophylls are tetrapyrrolic compounds with an additional isocyclic ring and a central complexed magnesium atom. The two common chlorophylls, chlorophyll a, $C_{55}H_{72}O_5N_4Mg$, and b, $C_{55}H_{70}O_6N_4Mg$, differ only in the state of oxidation of a single group, a methyl, —CH_3, in a and an aldehyde, —CH=O, in b.

Chloroplast pigments are very susceptible to chemical change that can lead to the detection of artifacts during chromatography. These chemical alterations may be induced by changes in the plant material itself; they may occur during extraction, handling, and storage of pigments; and they may occur on active adsorbents. The procedures in this experiment, if closely followed, allow the separation and detection of the pigments actually present in the plant.

2. Extraction of Pigments from Leaves

Use fresh plant material as a source of chloroplast pigments and perform extraction as quickly as possible. Fresh spinach leaves obtained from a market and cocklebur (*Xanthium*) leaves grown in a greenhouse have been employed with success.

Place 2 g of tender, green leaves with large veins and petioles removed in a chilled blender with 60 ml of cold acetone or methanol and blend at high speed for 2 min. Centrifuge the mixture and pour the clear, green supernatant into a separatory funnel. Add 40 ml of cold petroleum ether (20–40°C) and 100 ml of saturated aqueous sodium chloride. Shake the mixture and allow the layers to separate. Discard the lower layer and wash the upper layer successively with two 100-ml portions of distilled water. Pour the washed upper, green layer out through the top of the funnel into a round-bottom flask and evaporate to dryness below 40°C under vacuum. A rotary evaporator is convenient, if available. If the extract is to be stored before use, keep under vacuum in a dark, cold place to avoid decomposition of pigments. Dissolve the residue in 1 ml of petroleum ether (60–110°C) to prepare the sample solution for TLC.

The xanthophylls and carotenes of plants are examined conveniently after removal of the chlorophylls by saponification with alkali. For this saponification, add 10 ml of 30% potassium hydroxide in methanol to the centrifuged acetone or methanol extract in the separatory funnel. After 30 min with occasional swirling, add 40 ml of cold petroleum ether (20–40°C)–diethyl ether (1:1) plus 100 ml of 10% aqueous sodium chloride solution. Wash the resultant upper golden-yellow layer with water and take to dryness as above. Prepare the sample solution by dissolving the residue in 1 ml of petroleum ether (60–110°C)–diethyl ether (1:1). This saponification procedure should not be employed with plant extracts containing xanthophylls that are decomposed by alkalies, as are fucoxanthin from diatoms and brown algae and peridinin from dinoflagellates.

3. Layers

Eastman Chromagram silica gel sheets No. 6061 for adsorption TLC. Whatman KC_{18} chemically bonded layers for reversed-phase TLC.

4. Initial Zones

Apply 3–5 μl of the final 1-ml extract solutions (saponified and unsaponified) to each layer on an origin line located 2 cm from the bottom edge.

5. Mobile Phases

Isooctane–acetone-diethyl ether (3:1:1) for silica gel layers. Methanol–acetone–water (20:4:3) for bonded-phase layers.

6. Development

Develop in separate rectangular glass tanks lined with Whatman No. 1 filter paper and equilibrated with the respective mobile phases for 15 min before inserting each layer. Wrap the tanks with aluminum foil to exclude light. Leave a small opening in the foil to observe the solvent front. Develop the silica gel for a distance of 15 cm past the origin and the bonded-phase layer for 10 cm. A total of 30–50 min will be required for the developments.

7. Detection and Identification

No visualization reagent is required because of the characteristic natural coloration of the pigments. Pour 6 M hydrochloric acid into a shallow tray and hold the layer over the escaping vapors (*Caution*: Work in a hood) to confirm the location of the neoxanthin zone (turns from yellow to blue-green) and violaxanthin zone (yellow to blue). The other leaf carotenoids remain yellow.

Further confirmation of identity is obtained by measuring absorption spectra of the separated zones in the visible region of the spectrum. For determination of the absorption spectra, the individual zones are removed (scraped) separately from the chromatogram, and each pigment is eluted. The colored solid from each zone is packed into a fresh tube for the elution of each pigment. Use ethanol for the elution and as the solvent for measurement of the absorption spectra of the carotenoids; use diethyl ether for the chlorophylls. Compare the spectra, with regard to shape and wavelengths of the absorption maxima, with recorded spectra for pure, authentic pigments (Strain, 1958; Strain et al., 1967). The λ_{max} for the leaf pigments are (in nanometers) as follows: β-carotene, 452, 480; lutein, 446, 474; violaxanthin, 442, 470; neoxanthin, 437, 466; chlorophyll a, 428.5, 660.5; and chlorophyll b, 452.5, 642.0. It may be necessary to chromatograph multiple spots or a streak of the extract to obtain sufficient pigments for spectroscopy.

8. Results and Discussion

As shown in Figure 18.1, leaves of seed-producing plants (spinach and cocklebur), spore-producing plants (ferns), and many green algae (e.g., *Chlorella pyrenoidosa* and *Ulva rigida*) yield the same pigments: carotenes, chlorophyll a, lutein, chlorophyll b, violaxanthin, and neoxanthin, plus a small amount of pheophytin a (chlorophyll a after loss of magnesium). On some layers of silica gel G (rather than on the Eastman sheet) developed with the same mobile phase, the same pigments have been observed but in a different sequence: The zone of lutein is behind both chlorophylls rather than between them as shown in Figure 18.1. This illustrates the importance of the layer used for TLC. The chromatogram of saponified extract is similar to that shown in Figure 18.1 except for the absence of the chlorophylls; the carotenoids are completely separated in the sequence, carotene (closest to the solvent front), lutein, violaxanthin, and neoxanthin.

Extracts of various other plants have been chromatographed in this same system, namely, other chlorophytes (e.g., *Chlorella vulgaris*, *Cladophora ovoidea*, and *Cladophora trichotoma*), siphonalean green algae (*Codium fragile* and *Codium setechellii*), blue-green algae (*Phormidium luridum*), *Euglena graci-*

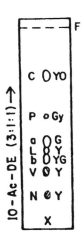

FIGURE 18.1 Chloroplast pigments separated from 3–5 µl of leaf extract by thin-layer chromatography on silica gel sheets. N = Neoxanthin; V = violaxanthin; L = lutein; b = chlorophyll *b*; a = chlorophyll *a*; C = carotene; F = solvent front; X = origin; Y = yellow; O = orange; G = green; /// = blue-green over HCl vapors, \\\ = blue over HCl vapors; Ac = acetone; IO = isooctane; DE = diethyl ether. [Reprinted with permission of the *Journal of Chemical Education*, Washington, DC, from Strain and Sherma (1969).]

lis, dinoflagellates (symbiotic algae from sea anemones), brown algae (*Sargassum* and *Cystophyllum*), and red algae (*Laurencia* and *Pterocladia*) (Strain and Sherma, 1969). The chromatograms revealed different characteristic pigment systems that are useful for classification of plants into major taxonomic groups.

Figure 18.2 shows the sequence of pigments obtained in the reversed-phase partition system. The chromatogram of the saponified extract should indicate the four carotenoids in the same locations, without the chlorophylls. As expected, the order of migration is reversed compared to the adsorption system illustrated in Figure 18.1. The chromatogram in Figure 18.2 results from spotting 5 µl of leaf extract solution. The application of a lower initial zone volume would improve the separations of the upper and bottom two zones. Alternatively, the use of a different mobile phase can emphasize these resolutions at the expense of the overall separation. Neoxanthin and violaxanthin were separated from each other and all other pigments by development for 35 min with methanol–acetone–water (2:2:1) (respective R_F values 0.23 and 0.18). Carotene (R_F 0.40) and chlorophyll *a* (R_F 0.53) were completely resolved in butanol–acetone (8:7).

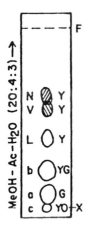

FIGURE 18.2 Choloroplast pigments from leaf extract separated on Whatman C_{18} reversed-phase layer. MeOH = methanol; for other abbreviations, see the legend of Figure 18.1. [Reprinted with permission of Elsevier Scientific Publishing Company, Amsterdam from Sherma and Latta (1978).]

In each case, spectra of pigments eluted from layers matched those from unaltered, authentic pigment standards prepared in the laboratory by repeated column and thin-layer chromatography of extracts from large masses of leaves. No artifacts except for a possible trace of pheophytin should be noted if instructions are carefully followed.

Other TLC studies on fat-soluble chloroplast pigments from plant tissues include the following. Suzuki et al. (1987) studied the retention of chlorophylls, pheophytins, and pheophorbides in C_{18} TLC and HPLC systems. Stauber and Jeffrey (1988) examined the photosynthetic pigments of 51 species of tropical and subtropical diatoms by normal- and reversed-phase TLC. Cserhati (1988) determined the lipophilicity of some photosynthetic pigments using reversed-phase TLC on various impregnated layers with acetone and ethanol as the mobile phase. Heimler et al. (1989) quantitatively determined chlorophylls in spruce needles by densitometry on cellulose layers.

Francis and Isaksen (1988) separated carotenoids on silica gel with tertiary alcohol–petroleum ether–developing solvents. In an earlier study, Isaksen and Francis (1986) had separated carotenoids on C_{18} layers developed with petroleum ether (40–60°C)–acetonitrile–methanol (1:1:2). Isaksen and Francis (1990) described a silver ion spray reagent that selectively discriminated carotenoids with β-end-groups as red zones. Bounias and Daurade (1989) reported that impregnation of silica gel layers with 125 mM D-glucose retards the oxidation of chlorophylls and preserves the integrity of carotenoids on the TLC plates.

Roy et al. (1991) extracted the pigments of green leafy vegetables in methanol–acetone–diethyl ether (1 : 1 : 1); the extracts were evaporated, reconstituted in the same solvent, and separated by TLC on silica gel G plates with a chloroform–petroleum ether–toluene–acetone (5 : 4 : 4 : 3) mobile phase. Pheophorbide *b*, mixed carotenoid, pheophorbide *a*, chlorophyll *b*, pheophytin *b*, chlorophyll *a*, pheophytin *a*, and carotene bands were obtained. This method can be used to detect artificial and natural colorants in food products.

Sherma et al. (1992) separated chloroplast pigments on silica gel and seven types of bonded silica gel plates. Spinach leaves were extracted in acetone, and the best overall separation was done on a C_{18} (Whatman) reversed-phase plate using the solvent system petroleum ether–acetonitrile–methanol (2: 4: 4). The R_F values and wavelengths (in nm) of maximum absorption of selected pigments on the C_{18} layers were as follows: β-carotene, 0.08, 455 nm; pheophytin, 0.24, not available; chlorophyll *a*, 0.36, 420 nm; and neoxanthin, 0.75, 440 nm.

Stability studies showed that the pigments degraded less on the C_{18} layers than on silica gel, and also when plates were wrapped in aluminum foil after development. Quantification of β-carotene, chlorophylls *a* and *b*, and lutein in the spinach leaves was done by scanning densitometry on the C_{18} layer at their wavelengths of maximum absorption. This article also provided extensive tabular data on the separation of chloroplast pigments in various chromatographic systems.

Johjima (1993) determined *cis*- and *trans*-carotenes of tangerine and yellow-tangerine tomatoes by TLC on mixed magnesia–alumina–cellulose–calcium sulfate (10:6:2:2) layers with stepwise gradient development from hexane–2-propanol–methanol (100:2:0.2) to pure hexane and scanning densitometry for quantification. Rosas-Romero et al. (1994) separated and quantified β-carotene, cantaxanthin, lutein, violaxanthin, and neoxanthin using rod TLC/flame ionization detection (FID) and a two-stage development technique. Schiedt (1995) reviewed procedures and materials for the TLC determination of carotenoids in biological extracts.

B. Experiment 2. Occurrence of Carotenoids and Xanthophylls in Snail Tissue

1. Introduction

Yellow pigments extracted from snails usually consist of carotenes and/or xanthophylls. These pigments have been demonstrated in the marine snail *Nassarius obsoletus* by Hoskin and Cheng (1975). Snails presumably obtain these pigments by feeding on plant tissues, and organisms with yellow coloration in their tissues should be suspected of containing these lipophilic pigments. The gonads and hepatopancreas (liver) of the snails should be particularly good organs to examine for pigments.

2. Preparation of Standards

Obtain β-carotene (Sigma) and prepare a 1-μg/μl solution in acetone. Obtain hen's egg yolk from a market and dissolve 0.1 ml of yolk in 1 ml of acetone; the predominant mixed pigment of egg yolk is xanthophyll (Romanoff and Romanoff, 1949). This preparation will provide a crude xanthophyll standard.

3. Preparation of Samples

Obtain snail material as described in Chapter 4. *Nassarius obsoletus* or *Biomphalaria glabrata* snails are preferable (see Fried and Sherma, 1990), but other marine and freshwater snails should also prove satisfactory. A particularly useful snail for this exercise is the medically important species *Biomphalaria glabrata*. This snail is available to investigators by contacting the Director, Biomedical Research Institute, 12111 Parklawn Drive, Rockville, MD 20852, USA. Extract 100 mg of tissue from the hepatopancreas (liver) or gonadal region of the snail in 2 ml of cold acetone in a test tube. Let the material stand for 5–10 min and then remove the supernatant with a Pasteur pipet. Pass the supernatant through a glass wool filter. Concentrate the filtrate under nitrogen (yellow-orange color) to about 100 μl. (*Caution*: Do not allow the sample to dry, or become exposed to air for long periods of time, because the pigments are susceptible to decomposition from heat, light, and autooxidation.) If the sample is to be stored, purge with nitrogen and store in a deep freeze at −20°C.

4. Layer

Baker-flex silica gel IB-2, 20 × 20 cm, used as received.

5. Mobile Phase

Acetone–hexane (30 : 70).

6. Application of Initial Zones

Spot 5-, 10-, and 20-μl samples on separate origins located 2 cm from the bottom of the sheet. Spot 1- and 5-μl aliquots of the β-carotene standard and 1 and 5 μl of the crude xanthophyll standard on separate origins.

7. Development

Develop the sheet for a distance of 10 cm above the origin in a saturated tank containing 100 ml of the mobile phase. No detection step is needed because pigments have natural color.

8. Results and Discussion

Snail tissue from *N. obsoletus* (Hoskin and Cheng, 1975) will show yellow-orange spots at R_F values of approximately 0.17 (xanthophylls) and 0.71 (carotenes). Other pigment spots may also appear. The R_F values of the β-carotene standard will be approximately 0.71, and the crude xanthophyll standard will have an approximate R_F value of 0.17.

If a spectrophotometer is available, the absorption maximum of the pure β-carotene standard dissolved in hexane should be determined. For further details, consult Hoskin and Cheng (1975). The absorbance maximum of the β-carotene should be about 455 nm. Elute the sample that has a chromatographic mobility similar to the β-carotene standard and prepare it for spectrophotometry. This material will show a spectrum similar to the β-carotene standard as a confirmation of identity.

Other recent studies on fat-soluble chloroplast pigments from animal tissues are as follows.

Fried et al. (1990) identified β-carotene in the digestive gland–gonad (DGG) complex of the freshwater snail *Helisoma trivolvis*. They used 10×20-cm high-performance (HP) silica gel plates (Whatman LHP-KDF) and the mobile-phase petroleum ether–diethyl ether–acetic acid (80:20:1). The pigment from the snail tissue had an R_F of 0.86, which was identical to that of an authentic β-carotene standard. A more complicated pattern of carotenoid pigments was obtained from the DGG of snails infected with the larval trematode parasite *Echinostoma trivolvis*. However, the identity of the pigments in the parasitized DGG, other than β-carotene, was not ascertained. In a later study (Fried et al., 1993), it was determined by TLC that the larval parasite stage (redia) isolated from the snail tissue had lutein in addition to β-carotene.

Sherma et al. (1992) used TLC to identify and quantify lutein and β-carotene from acetone extracts of snail bodies in two strains of *Helisoma trivolvis* (Colorado and Pennsylvania) snails and in the *Biomphalaria glabrata* snail. TLC of the snail extracts on C_{18} (Whatman) reversed-phase layers developed in petroleum ether–acetonitrile–methanol (2:4:4) showed a fast moving zone ($R_F = 0.08$) identical to a β-carotene standard. Scanning of in situ spectra confirmed pigment identities. The ratio of β-carotene to lutein was 1.1 in both strains of *Helisoma trivolvis*, but about 4.5 in *Biomphalaria glabrata*. The β-carotene to lutein ratio may be useful in the chemotaxonomy of planorbid snails.

For additional information on the TLC of carotenoids in animal parasites and insects, respectively, see Fried and Haseeb (1996) and Fell (1996).

C. Experiment 3. Separation of Anthocyanidins from Flowering Plants

1. Introduction

The purpose of this experiment is to demonstrate the use of cellulose TLC for the separation of anthocyanidins based on the works of Smith and Feinberg

(1965), Anditti and Dunn (1969), and Harborne (1984). Following acid hydrolysis, the anthocyanins yield anthocyanidins and sugar residues. Detailed descriptions of anthocyanidins are presented in Anditti and Dunn (1969) and in Harborne (1984).

2. Preparation of Standards

Anthocyanidin standards are not readily available, but marker solutions can be obtained by selecting particular plants with known anthocyanidins. For further details, consult Harborne (1984).

3. Preparation of Samples

Obtain flowering plants that have red or purple flowers or fruit. Plants in the following genera are readily available and provide a good source of anthocyanins: *Begonia, Hibiscus, Geranium, and Rhododendron.* Remove approximately 100 mg of petal or fruit material from a given plant species. Place the material in a test tube and add 0.5 ml of methanol–concentrated HCl (99:1). Use a glass rod to grind the plant material in the acidic methanol until the solution becomes heavily colored. Allow the material to stand for about 15 min and then remove the supernatant with a pipet. Concentrate the supernatant to approximately 100 μl under a stream of nitrogen gas. (*Caution*: Avoid use of air or strong light when handling anthocyanins because they are unstable).

4. Layer

Baker-flex cellulose or Avicel microcrystalline cellulose (Analtech) precoated layers, 20 × 20 cm, used as received.

5. Mobile Phase

n-Butanol–acetic acid–water (60:15:25).

6. Application of Initial Zones

Spot 5-, 10-, and 20-μl samples on separate origins, 2 cm from the bottom of the plate.

7. Development

Develop the plate for a distance of 10 cm above the origin in a saturated tank containing 100 ml of the mobile phase. No detection step is needed because pigments have natural color. Briefly dry the plate with a stream of nitrogen gas.

8. Results and Discussion

Look for red, magenta, and purple spots on the layer. Circle the spots immediately and record the colors. The color of spots may fade rapidly because anthocyanidins are unstable in air. It may be possible to distinguish up to six common anthocyanidins on the plate based on R_F values and color. R_F and colors from paper chromatography of the six most common anthocyanidin pigments separated in the *n*-butanol–acetic acid–water (60:15:25) solvent are as follows: delphiniden, 0.42, purple; petunidin, 0.52, purple; malviden, 0.58, purple; cyanidin, 0.68, magenta; peonidin, 0.71, magenta; pelargonidin, 0.80, red. Anthocyanidin separations by cellulose TLC are reportedly similar to those obtained in paper chromatography when identical solvents are used (Asen, 1965), so the above results should be approximated using the TLC experiment.

The TLC of anthocyanins along with other related phenolic compounds has been reviewed by Harborne (1992). Anderson and Francis (1985) described a method for separating anthocyanins and anthocyanidins from various plant sources by cellulose TLC using the mobile phase $HCl–HCOOH–H_2O$ (24.9 : 23.7 : 51.4).

Pothier (1996) provided useful information on flavonoids and anthocyanins in his review on TLC in the plant sciences. For additional information on isolation methods for anthocyanins and TLC of these compounds, including quantification by scanning densitometry, see Pothier (1996).

REFERENCES

Andersen, Ø. M., and Francis, G. W. (1985). Simultaneous analysis of anthocyanins and anthocyanidins on cellulose thin layers. *J. Chromatogr. 318*: 450–454.

Andersen, Ø. M., and Francis, G. W. (1996). Natural pigments. In *Handbook of Thin-Layer Chromatography*, 2nd ed., J. Sherma and B. Fried (Eds.). Marcel Dekker, New York, pp. 715–752.

Anditti, J., and Dunn, A. (1969). *Experimental Plant Physiology—Experiments in Cellular and Plant Physiology*. Holt, Rinehart and Winston, New York.

Asen, S. (1965). Preparative thin-layer chromatography of anthocyanins. *J. Chromatogr. 18*: 602–603.

Bartolome, E. R. (1994). Crystallization and thin layer chromatographic separation of flavanones from traces of other types of flavonoid compounds. *J. Planar Chromatogr.—Mod. TLC 7*: 70–72.

Billeter, M., Meier, B., and Sticher, O. (1990). Densitometric determination of flavonoids after derivatization. *J. Planar Chromatogr.—Mod. TLC 3*: 370–375.

Bounias, M., and Daurade, M. H. (1989). Quenching of the artifactual oxidation of chlorophyll pigments chromatographed on silica gel thin layers by impregnation of plates with glucose or thiosulfate. *Analusis 17*: 401–404.

Budzianowski, J. (1991). Separation of flavonoid glucosides from their galactosidic analogs by thin-layer chromatography. *J. Chromatogr. 540*: 469–474.

Burnham, B. F., and Kost, H-P. (1988). Porphyrins (exclusive of chlorophylls). In *CRC Handbook of Chromatography—Plant Pigments—Volume 1—Fat Soluble Pigments*, H-P. Kost (Ed.). CRC Press, Boca Raton, FL, pp. 189–231.

Burton, D. E., Bailey, D. L., and Lillie, C. H. (1993). Determination of ipriflavone and its impurities by thin-layer chromatography with absorbance and fluorescence detection. *J. Planar Chromatogr.—Mod. TLC 6*: 223–227.

Cikalo, M. J., Poole, S. K., and Poole, C. F. (1992). Instability of plant carotenoids to separation by multiple development thin layer chromatography. *J. Planar Chromatogr.—Mod. TLC 5*: 200–204.

Conde, E., Cadahia, E., and Garcia-Vallejo, M. C. (1992). Optimization of TLC for research on the flavonoids in wood and bark of species of the genus *Eucalyptus* L'Hertier. *Chromatographia 33*: 418–426.

Cserhati, T. (1988). Lipophilicity determination of some photosynthetic pigments by reversed-phase thin-layer chromatography. The effect of support and the organic phase in the eluent. *Chromatographia 25*: 908–914.

Davies, B. H., and Kost, H-P. (1988). Carotenoids. In *CRC Handbook of Chromatography—Plant Pigments—Volume 1—Fat Soluble Pigments*, H-P. Kost (Ed). CRC Press, Boca Raton, FL, pp. 3–183.

Dolphin, D. (1983). Porphyrins and related tetrapyrrolic substances. In *Chromatography—Fundamentals and Applications of Chromatographic and Electrophoretic Methods, Part B: Applications*, E. Heftmann (Ed.). Elsevier, Amsterdam, pp. B377–B406.

Doss, M. (1972). Thin-layer chromatography of porphyrins and complementary analytical methods. In *Progress in Thin-Layer Chromatography and Related Methods*, Vol. III. A. Niederwieser and G. Pataki (Eds.). Ann Arbor–Science Publishers, Ann Arbor, MI, pp.145–176.

Fell, R. D. (1996). Thin-layer chromatography in the study of entomology. In *Practical Thin-Layer Chromatography—A Multidisciplinary Approach*, B. Fried and J. Sherma (Eds.). CRC Press, Boca Raton, FL, pp. 72–104.

Ficarra, P., Ficcara, R., Constantino, D., Carulli, M., Tommasini, S., De Pasquale, A., and Calabro, M. L. (1994). Flavonoids: determination of hydrophobic parameters by reversed-phase high-performance thin-layer chromatography. *Boll. Chim. Farm. 133*: 221–227.

Florey, E. (1966). *An Introduction to General and Comparative Animal Physiology*. Saunders, Philadelphia.

Fox, D. L. (1953). *Animal Biochromes and Structural Colors*. Cambridge University Press, London.

Francis, G. W., and Isaksen, M. (1988). Thin-layer chromatography of carotenoids with tertiary alcohol-petroleum ether solutions as developing solvents. *J. Food Sci. 53*: 979–980.

Fried, B., Beers, K., and Sherma, J. (1993). Thin-layer chromatographic analysis of β-carotene and lutein in *Echinostoma trivolvis* (Trematoda) rediae. *J. Parasitol. 79*: 113–114.

Fried, B., and Haseeb, M. A. (1996). Thin-layer chromatography in parasitology. In *Practical Thin-Layer Chromatography—A Multidisciplinary Approach*, B. Fried and J. Sherma (Eds.). CRC Press, Boca Raton, FL, pp. 52–70.

Fried, B., Holender, E. S., Shetty, P. H., and Sherma, J. (1990). Effects of *Echinostoma trivolvis* (Trematoda) infection on neutral lipids, sterols, and carotenoids in *Helisoma trivolvis* (Gastropoda). *Comp. Biochem. Physiol. 97B*: 601–604.

Fried, B. and Sherma, J. (1990). Thin layer chromatography of lipids found in snails (Gastropoda: Mollusca). *J. Planar Chromatogr.—Mod. TLC 3*: 290–299.

van Ginkel, A. (1996). Identification of the alkaloids and flavonoids from *Uncaria tomentosa* bark by TLC in quality control. *Phytother. Res. 10* (Suppl. 1): S18–S19.

Goodwin, T. W. (1952). *The Comparative Biochemistry of the Carotenoids*. Chapman & Hall, London.

Goodwin, T. W. (1972). Pigments of Mollusca. In *Chemical Zoology*, M. Florkin and B. T. Scheer (Eds.). Academic Press, New York, pp.187–199.

Gross, J. (1991). *Pigments in Vegetables—Chlorophylls and Carotenoids*. Van Nostrand Reinhold, New York.

Hadj-Mahammed, M., and Meklati, B. Y. (1993). Polyhydroxy- and polymethoxy-flavones: behavior in high-performance thin-layer chromatography. *J. Planar Chromatogr.—Mod. TLC 6*: 242–244.

Harborne, J. B. (1984). *Pytochemical Methods—A Guide to Modern Techniques of Plant Analysis*, 2nd ed. Chapman & Hall, London.

Harborne, J. B. (1992). Phenolic Compounds. In *Chromatography, 5th Edition—Fundamentals and Applications of Chromatography and Related Differential Migration Methods, Part B: Applications*. E. Heftmann (Ed.). Elsevier, Amsterdam, pp. B363–B392.

Heimler, D., Michelozzi, M., and Boddi, V. (1989). Quantitative TLC determination of chlorophylls in spruce needles under mild pollution conditions. *Chromatographia 28*: 148–150.

Heimler, D., and Pieroni, A. (1991). High-performance quantitative thin-layer chromatography of flavonoid glycosides and biflavonoids of *Cupressus sempervirens* in relation to cypress canker. *Chromatographia 31*: 247–250.

Heimler, D., Pieroni, A., Tattini, M., and Cimato, A. (1992). Determination of flavonoids, flavonoid glycosides, and biflavonoids in *Olea europaea* L. leaves. *Chromatographia 33*: 369–373.

Heimler, D., Vidrich, V., and Buzzini, P. (1988). An HPTLC method for the separation and determination of leaf flavonoids. *Agrochimica 32*: 457–462.

Heirwegh, K. P. M., Fevery, J., and Blackaert, N. (1989). Chromatographic analysis and structure determination of biliverdins and bilirubins. *J. Chromatogr. 496*: 1–26.

Hiermann, A. (1995). Phytochemical characterization of *Epilobium angustifolium* and its differentiation from other *Epilobium* species by TLC and HPLC. *Sci. Pharm. 63*: 135–144.

Hoskin, G. P., and Cheng, T. C. (1975). Occurrence of carotenoids in *Himasthla quissetensis rediae* and the host, *Nassarius obsoletus*. *J. Parasitol. 61*: 381–382.

Isaksen, M. (1991). Natural pigments. In *Handbook of Thin-Layer Chromatography*, J. Sherma and B. Fried (Eds.). Marcel Dekker, New York, pp. 625–662.

Isaksen, M., and Francis, G. W. (1986). Reversed-phase thin-layer chromatography of carotenoids. *J. Chromatogr. 355*: 350–362.

Isaksen, M., and Francis, G. W. (1990). Silver ion spray reagent for the discrimination of β- and ε-end groups in carotenoids on thin-layer chromatograms. *Chromatographia 29*: 363–365.

Jacob, K. (1992). Porphyrins. In *Chromatography, 5th Edition. Fundamentals and Applications of Chromatography and Related Differential Migration Methods, Part B: Applications*. E. Heftmann (Ed.). Elsevier, Amsterdam, pp. B335–B362.

Jain, R. (1996). Thin layer chromatography in clinical chemistry. In *Practical Thin-Layer Chromatography—A Multidisciplinary Approach*, B. Fried and J. Sherma (Eds.). CRC Press, Boca Raton, FL, pp.132–152.

Johjima, T. (1993). Determination of cis-and trans-carotenes of tangerine and yellowish tangerine tomatoes by micro-thin-layer chromatography. *Engei Gakkai Zasshi 62*: 567–574.

Kirchner, J. G. (1978). *Thin Layer Chromatography*, 2nd ed., E. S. Perry (Ed.). *Techniques of Chemistry* Vol. XIV. Wiley, New York, pp.801–826.

Krawczyk, U., Petri, G., and Kery, A. (1993). Application of UV-visible spectrophotometry, TLC-densitometry and RP-HPLC for flavonoids determination in buckwheat herb extract. *Acta Pol. Pharm. 50*: 317–319.

Kus, P. (1996). Thin-layer chromatographic mobility of crown ether derivatives of mesotetraphenylporphyrin. *J. Planar Chromatogr.—Mod. TLC 9*: 435–438.

Lai, C. K., Lam, C. W., and Chan, Y. W. (1994). High-performance thin-layer chromatography of free porphyrins for diagnosis of porphyria. *Clin. Chem. 40*: 2026–2029.

Matysik, G. (1992). Thin-layer chromatography of anthocyanins with stepwise gradient elution. *J. Planar Chromatogr.—Mod. TLC 52*: 146–148.

Matysik, G., and Benes, M. (1991). Thin-layer chromatography and densitometry of anthocyanins in the petals of red poppy during development of the flowers. *Chromatographia 32*: 19–22.

Nikolova-Damyanova, B., Ilieva, E., Handjieva, N., and Bankova, V. (1994). Quantitative thin layer chromatography of iridoid and flavonoid glucosides in species of *Linaria. Phytochem. Anal. 5*: 38–40.

Pothier, J. (1996). Thin-layer chromatography in plant sciences. In *Practical Thin-Layer Chromatography—A Multidisciplinary Approach*, B. Fried and J. Sherma (Eds.). CRC Press, Boca Raton, FL, pp. 33–49.

Romanoff, A. L., and Romanoff, A. J. (1949). *The Avian Egg*. Wiley, New York.

Rosas-Romero, A. J., Herrera, J. C., Martinez De Aparicio, E., and Molina-Cuevas, E. A. (1994). Thin-layer chromatographic determination of β-carotene, cantaxanthin, lutein, violaxanthin and neoxanthin on Chromarods. *J. Chromatogr. 667*: 361–366.

Roy, A. K., Banerjee, A. K., and Chakrabarti, J. (1991). Separation of carotenoids, chlorophylls, and related pigments. *J. Inst. Chem. (India) 63*: 79.

Rustan, I., Michel, N., Sergent, M., Lesgards, G., and Luu, R. P. T. (1995). Adjustment of flavonoid analysis by thin-layer chromatography (TLC) using experimental research methodology: study of Labiatae plants. *Analusis 23*: 398–402.

Scheer, H. (1988). Chlorophylls. In *CRC Handbook of Chromatography—Plant Pigments—Volume I—Fat Soluble Pigments*, H-P. Korst (Ed.). CRC Press, Boca Raton, FL, pp. 235–309.

Schiedt, K. (1995). Thin-layer chromatography. *Carotenoids 1A*: 131–144.

Scott, R. M. (1969). *Clinical Analysis by Thin-Layer Chromatography Techniques*. Ann Arbor-Humphrey Science Publishers, Ann Arbor, MI.

Sherma, J., and Latta, M. (1978). Reversed phase thin layer chromatography of chloroplast pigments on chemically bonded C_{18} plates. *J. Chromatogr. 154*: 73–75.

Sherma, J., O'Hea, C. M., and Fried, B. (1992). Separation, identification, and quantification of chloroplast pigments by HPTLC with scanning densitometry. *J. Planar Chromatogr.—Mod. TLC 5*: 343–349.

Smith, O., and Feinberg, J. G. (1965). *Paper and Thin Layer Chromatography and Electrophoresis*. Shandon Scientific Co., London.

Stauber, J. L., and Jeffrey, S. W. (1988). Photosynthetic pigments in fifty-one species of marine diatoms. *J. Phycol. 24*: 158–172.

Strain, H. H. (1958). Significance of chloroplast pigments and of methods for their separation. *Ann. Priestley Lectures 32*: 1–80.

Strain, H. H., and Sherma, J. (1969). Modifications of solution chromatography illustrated with chloroplast pigments. *J. Chem. Educ. 46*: 476–483.

Strain, H. H., and Sherma, J. (1972). Investigations of the chloroplast pigments of higher plants, green algae, and brown algae and their influence upon the invention, modifications, and applications of Tswett's chromatographic method. *J. Chromatogr. 73*: 371–397.

Strain, H. H., Sherma, J., and Grandolfo, M. (1967). Alteration of chloroplast pigments by chromatography with siliceous adsorbents. *Anal. Chem. 39*: 926–932.

Suzuki, N., Saitoh, K., and Adachi, K. (1987). Reversed-phase high-performance thin-layer chromatography and column liquid chromatography of chlorophylls and their derivatives. *J. Chromatogr. 408*: 181–190.

19

Vitamins

I. GENERAL

Vitamin classification is initially based on fat solubility—i.e., A, D, E, and K—
versus water solubility—i.e., C, folic acid, nicotinic acid, and nicotinamide (B_3),
B_1, B_2, B_6, biotin, B_{12}, and pantothenic acid. DeLeenheer et al. (1985, 1992)
edited two valuable treatises on modern chromatographic analysis of vitamins.
Two tables have been compiled in this chapter with information on the chemical
terms for vitamins and their derivatives, vitamin function, sources of vitamins and
recent selected references containing TLC information. Table 19.1 summarizes
pertinent information on the fat-soluble vitamins and Table 19.2 does the same
for the water-soluble vitamins.

DeLeenheer et al. (1991) contributed a review on the TLC of lipophilic
vitamins. They noted that although the 1980s witnessed a proliferation of HPLC
methods for the determination of fat-soluble vitamins, the advent of newer
HPTLC–densitometric techniques may allow for a resurgence of interest in the
1990s for the analysis of lipophilic vitamins by TLC. Fried (1991) reviewed the
literature on the TLC of hydrophilic vitamins from about 1965 to 1988.

DeLeenheer and Lambert (1996) updated the references to 1994 and extended the coverage in DeLeenheer et al. (1991); they reemphasized the importance of new developments in HPTLC–densitometry in modern studies on the
analysis of lipophilic vitamins. Linnell (1996) stressed the importance of TLC
analysis of water-soluble vitamins, particularly in pharmaceutical preparations
and food products. He extended the earlier coverage by Fried (1991) and updated

TABLE 19.1 Summary of Information on the Fat-Soluble Vitamins

Vitamins	Chemical terms for vitamins and their derivatives	Functions	Sources	Selected references containing TLC information
A	Retinol; carotenoids; retinoic acid; 3-dehydroretinal	Essential for life, growth and maintenance; prevents xerophthalmia and night blindness	Egg yolk; cod liver oil; milk; butter; carrots	DeLeenheer et al. (1985); Furr er al. (1992)
D	Cholecalciferol; ergocalciferol; hydroxylated metabolites; 9,10-secosteroids	Calcium homeostasis	Milk; margarine	Jones et al. (1992)
E	Tocopherols; tocotrienols	Antioxidants; maintenance of red blood cell viability; possible role in electron transport	Oils; nuts; seeds	Nelis et al. (1985); Lang et al. (1992)
K	Phylloquinone (K_1); menaquinones (K_2); 2-methyl-1,4-naphthoquinone derivatives	Prevention of hemorrhagic diseases	Plant origins (K_1); putrified fish meal (K_2)	LeFevere et al. (1985); Lambert and DeLeenheer (1992)

the references to 1994. Dreassi et al. (1996) reviewed studies on TLC analysis of lipophilic and hydrophilic vitamins in pharmaceutical analysis, whereas Stahr (1996) has reviewed the work done on TLC of lipophilic vitamins relative to applications in veterinary toxicology.

II. TLC OF FAT-SOLUBLE VITAMINS

A. Vitamin A

Vitamin A is light sensitive and readily oxidized when exposed to air. To prevent isomerization and oxidation of this vitamin during sample preparation and TLC, work is best done in subdued light under a nitrogen atmosphere. TLC of this vitamin is usually done on silica gel plates activated at 120°C for 30–60 min before use. Some workers pretreat layers by spraying with an antioxidant to prevent degradation of substances on the plate. Mobile phases usually contain a large amount of an apolar solvent such as petroleum ether or hexane and a smaller amount of a more polar solvent such as diethyl ether or acetone. Spots can be detected under ultraviolet (UV) light, or by spraying plates with the Carr–Price reagent (antimony trichloride, see Chapter 8).

Specific examples of vitamin A separations by TLC include the following. Baczyk et al. (1971) separated vitamins A (palmitate), D_3, and E in pharmaceutical preparations using chloroform as the mobile phase; following development, the spots were located by UV light, scraped off the plate, and determined spectrophotometrically. Kahan (1967) used TLC to study vitamin A metabolites, and Targan et al. (1969) also used TLC on silica gel to fractionate and quantify β-carotene, retinol, retinal, retinyl esters, and retinoic acid in human serum after separation using a benzene–ethyl ether mobile phase. Fung et al. (1978) separated retinol, retinal, retinoic acid, and retinyl acetate standards on silica gel using acetone–light petroleum ether (18 : 82) as the mobile phase. Parizkova and Blattna (1980) used preparative TLC to analyze retinyl acetate degradation products. They employed various combinations of the mobile-phase hexane–diethyl ether to separate 14 substances.

Tables of R_F values of representative retinoids on silica gel and alumina have been given by John et al. (1965), Varma et al. (1964), and Fung et al. (1978). Tataruch (1984) used TLC for the quantitative analysis of retinol. Sliwick et al. (1990) used TLC and HPLC to compare the hydrophobicity of vitamin A derivatives.

TLC has been used less frequently in the 1990s for the analysis of retinoids because of the lability of these compounds to oxygen, and the increasing availability of HPLC equipment (Furr et al., 1992). TLC is still useful for rapid analysis of reaction products when performing synthetic modification of retinoids and for initial "scouting" analysis for the development of HPLC techniques. Furr et

TABLE 19.2 Summary of Information on the Water-Soluble Vitamins

Vitamins	Chemical terms for the vitamins and their derivatives	Functions	Sources	Selected references containing TLC information
C	Ascorbic acid; L-ascorbic acid	Prevention of scurvy; multiple roles in optimizing health and resistance to infectious diseases; numerous functions proposed	Fruit; fruit juices; spinach	Bui-Nguyen (1985); Parviainen and Nyyssönen (1992)
Folic acid	Folic acid (pteroylglutamic acid); 7,8-dihydrofolic acid; tetrahydrofolates	Folate coenzymes participate in one-carbon transfer reactions in cellular metabolism	Variety of foodstuffs; most animal tissues	McCormack and Newman (1985); Mullin and Duch (1992)
Nicotinic acid and nicotinamide (vitamin P.P) (also B$_3$)	Nicotinic acid (3-pyridine carboxylic acid or niacin); nicotinamide (nicotinic acid amide), niacinamide, 3-pyridine-carboxylic acid amide	Prevention of pellagra in humans and black tongue in dogs; metabolic functions of nicotinamide coenzymes	Yeast extracts; most food products, i.e., liver, heart, muscle meat	Hengen and deVries (1985); Shibata and Shimono (1992)
B$_1$	Thiamines, hydroxyethylthiamine	Antipolyneuritis factor in nerve functions; derivatives function as coenzymes in carbohydrate and amino acid metabolism	Liver; brain; heart; yeast	Kawasaki and Sanemori (1985); Kawasaki (1992)

B_2	Riboflavin; 7,8-dimethyl-10-(1'-d-ribityl) isoalloxazine; flavin mononucleotide (FMN); flavin-adenine dinucleotide (FAD)	Coenzymes in redox reactions; carry out 1- or 2-electron processes	Meat and other food products	Marletta and Light (1985); Nielsen (1992)
B_6	Pyridoxamine and related compounds	Deficiencies lead to reduction of coenzyme A in blood, increased excretion of vitamin C, reduced absorption of vitamin B_{12} and impaired antibody production	Meat; fish; cereals; fruit; milk; eggs	Ubbink (1992)
Biotin (vitamin H)	cis-Tetrahydro-2-oxothieno 3,4d imidazoline-4-valeric acid; biotin sulfoxides; biotin sulfone	Deficiencies lead to reduced carboxylase activities	Egg yolk; liver; milk; yeast	Frappier and Gaudry (1995); Gaudry and Ploux (1992)
B_{12}	Cyanocobalamin; cobalamins and corrinoids	Antipernicious anemia vitamin	Liver	Lindemans and Abels (1985); Lindemans (1992)
Pantothenic acid (vitamin B_5)	D(+)-pantothenic acid, D(+)-α,γ-dihydroxy-β,β-dimethylbutyryl-alanine	Coenzyme A (CoA, CoASH) the acyl transfer agent in two-carbon unit metabolism	Meat; liver; fish; yeast; cheeses; legumes	Velisek et al. (1992)

al. (1992) suggested that for initial "scouting" analysis (prior to HPLC), TLC on either silica gel or alumina plates with mobile phases of hexane–ether or hexane–acetone (4:1) gives adequate resolution of retinoids. According to Furr et al. (1992), on reversed-phase TLC plates, acetonitrile or methanol gives adequate mobility and resolution for retinyl acetate and retinol; less polar solvents (such as dichloroethane) mixed with acetonitrile give good resolution of long-chain retinyl esters, and acetonitrile–water or methanol–water mixtures are useful for the separation of polar retinoids.

Retinol and retinyl esters can be identified on TLC plates by their yellow-green fluorescence under long-wave (366-nm) UV light; most other retinoids are not fluorescent under UV and must be identified by nonspecific techniques such as iodine vapor or charring with sulfuric acid (Furr et al., 1992). Bargagna et al. (1991) developed TLC, HPTLC, and HPLC methods for identification of retinoic acids, retinol, and retinyl acetate in topical facial creams and solutions.

B. Vitamin D

Preliminary extraction of lipids is important prior to vitamin D analysis by TLC. Jones et al. (1992) provided a table of extraction procedures for vitamin D compounds. Care must be taken to avoid oxidizing vitamin D on the TLC plate. Some workers use 0°C and a nitrogen atmosphere during sample preparation and subsequent TLC of vitamin D.

Specific vitamin D studies include the following. Sklan et al. (1973) used TLC to separate vitamin D from cholesterol and in combination with GLC determined 25-hydroxyvitamin D_3 in chicken plasma. Vieth et al. (1978) used TLC to separate tritiated products in biochemical studies on kidney 25-OH-D hydroxylases. Mass spectrometric methods used to measure Vitamin D_3 and 25-OH-D_3 often utilize TLC as a prepurification step (Dueland et al., 1981). Czuczy and Morava (1982) extracted vitamin D from milk and determined it using TLC on silica gel with the mobile-phase benzene–ethanol (75:1). The chromatograms were visualized with $SbCl_3$ and the vitamin D level was determined fluorometrically at 550 nm.

HPTLC has been used for the separation of vitamin D metabolites using microparticulate silica gel as the stationary phase and a chloroform–ethyl acetate (1:1) mobile phase (Thierry-Palmer and Gray, 1983). They made a thorough study of HPTLC for the separation of a large variety of vitamin D metabolites. They used silica gel HPTLC–HLF plates (Analtech) and reported R_F values of the metabolites in the following three mobile phases: dichloromethane–isopropanol (90:10); chloroform–ethyl acetate (1:1); and hexane–isopropanol (85:15). Justova et al. (1984) used high-performance thin-layer chromatography (HPTLC) to determine 1,25-dihydroxyvitamin D_3 in human plasma. Jones et al. (1992) used TLC for the prepurification of saponified samples for gas–liquid chromatography (GLC). Wardas and Pyka (1995) evaluated various dyes—aniline blue, brilliant green, neutral red, thymol blue, phenol red and others—as detection reagents for vitamins D, A, and E after separation by adsorption and partition TLC.

C. Vitamin E

Considerable use is made of TLC in work with vitamin E. This vitamin, like A and D, is susceptible to light and oxidation during handling, sample preparation, and TLC. Preliminary extraction of lipids is necessary in the analysis of vitamin E from biological samples. TLC is used to separate tocopherols and tocotrienols occurring in nature (Nelis et al., 1985).

TLC is used to determine vitamin E in food products, animal tissues, and pharmaceutical preparations. Most tocopherols can be separated on silica gel plates using the following solvent systems: hexane–ethyl acetate (37:3), cyclohexane–diethyl ether (4:1), or dichloromethane. However, TLC does not resolve the positional isomers β- and α-tocopherol. Reversed-phase TLC on starch, cellulose, or talc layers impregnated with mineral oil, using the mobile-phase acetone–acetic acid (3:2), has been used to separate α-tocopherol from α-tocopherol acetate (Perisic-Janjic et al., 1976). Ruggeri et al. (1984) determined tocopherols in lipid extracts of the flagellate protozoan *Euglena gracilis* by TLC on silica gel GF plates with hexane–isopropyl ether (85:15) as the mobile phase, and by reversed-phase HPLC on C_{18} Micropak columns with methanol–water (95:5) as the mobile phase. When the results were compared, the TLC and HPLC systems were comparable in sensitivity, reproducibility, recovery, and ease of application.

Lang et al. (1992) provided a review of the TLC analysis of vitamin E including methods of sample preparation, appropriate stationary and mobile phases, methods of detection and quantification, and a useful table that summarizes TLC separation systems for vitamin E.

Lang et al. (1992) stated that although TLC has value as an inexpensive qualitative and semiquantitative assay for vitamin E, HPLC or GC is more suitable for quantitative analysis. However, given proper sample preparation, TLC is suitable for all sample types and for separating tocopherol and tocotrienol homologs. Although one-dimensional systems are usually satisfactory, additional resolution can be achieved by developing in a second direction. Quantification can be done by in situ densitometry. TLC is also useful for sample cleanup prior to other quantitative procedures such as GC or HPLC.

A technique that separates natural tocopherols uses silica gel G plates developed in the first dimension with chloroform and in the second dimension with hexane–isopropanol ether (80 : 20) (Chow et al., 1969). This system also separates most tocotrientols and naturally occurring tocophenyl esters (Lang et al.,

1992). Mueller-Mulot (1976) described a one-dimensional system that separates all tocopherols and tocotrienols but cannot resolve α-tocopherol from β-tocotrienol; the system uses three successive developments of a silica gel 60 plate with the mobile phase *n*-hexane–ethane acetate (92.5:7.5). Sliwick et al. (1993) separated tocopherol isomers and enantiomers by TLC and studied the relationship between R_F values and topological indexes.

D. Vitamin K

Considerable use has been made of TLC for vitamin K analyses, and numerous reviews are available (Sommer and Kofler, 1969; Bolliger and Konig, 1969; Dunphy and Brodie, 1971; LeFevere et al., 1985). Vitamin K is labile and should be protected from heat and light as discussed for the other fat-soluble vitamins. Adsorption, argentation, and partition modes of TLC have been used in vitamin K analysis (LeFevere et al., 1985). Numerous tables pertinent to vitamin K analysis by TLC have been provided by LeFevere et al. (1985) as follows: TLC properties of various naphthoquinones on silica gel G; R_F values for naphthoquinones on silica gel G impregnated with $AgNO_3$; R_F values of naphthoquinones for reversed-phase TLC; spray reagents for staining quinones and lipids; and applications of TLC to vitamin K analysis.

Specific applications of vitamin K studies by TLC include determination of K_1, K_3, and K_4 in pharmaceutical preparations using adsorption TLC on precoated alumina foil layers (Rittich et al., 1976) and use of partition rather than adsorption chromatography to separate K compounds in vitamin premixes on paraffin-impregnated cellulose layers (LeFevere et al., 1985). Baczyk et al. (1981) used silica gel H plates and a chloroform mobile phase to study the degradation of vitamins D_2 and $K_{1(20)}$ (phylloquinone) by ultraviolet light. Ersoy and Duden (1980) were able to separate $K_{1(20)}$ from α-tocopherol, β-carotene, and A and D vitamins in spinach using silica gel plates and the mobile-phase petroleum ether–benzene (6:1). Argentation TLC (5–20% $AgNO_3$-impregnated silica gel layers) can be used to separate the saturated phylloquinone from the unsaturated menaquinone K vitamin homologs. The mobile phase used for this separation is petroleum ether–chloroform–acetone (50:10:17) (Lichtenthaler et al., 1982).

Lambert and DeLeenheer (1992) contributed a review on the TLC analysis of K vitamins. They noted that although HPLC is usually the method of choice for analysis, there is some interest in HPTLC–densitometric analysis of the K vitamins. Their review covers isolation, extraction, and cleanup methods; uses of polar inorganic solvents; uses of modified silica (i.e., nonpolar bonded phases such as C_2, C_8, C_{18}, and phenyl); uses of mobile phases; and methods of detection and quantification of the K vitamins. Madden and Stahr (1993) assayed vitamin K in bovine liver by reversed-phase TLC with dichloromethane–methanol (7:3) mobile phase for separation, densitometric quantification, and confirmation by mass spectrometry.

III. TLC OF WATER-SOLUBLE VITAMINS

A. Vitamin C

Preparation of fruit juices and pharmaceutical samples prior to TLC presents no problems, but for samples containing protein, precipitation procedures must be used first (Bui-Nguyen, 1985). For instance, serum samples must be deproteinized by adding two volumes of methanol or acetonitrile for each volume of serum.

TLC has been used to determine ascorbic acid in foods, pharmaceutical preparations, and biological fluids. Paper chromatography was used to separate ascorbic and dehydroascorbic acid in plant extracts, and the spots were visualized using 2,5-dichlorophenol indophenol (Bui-Nguyen, 1985). This procedure is probably applicable to TLC on cellulose.

Numerous workers have used TLC to examine ascorbic acid in foods after purification of the ascorbic acid osazone derivative by column chromatography (Bui-Nguyen, 1985). TLC has been used to separate ascorbic and erythorbic hydrazone derivatives and also to measure the oxidized forms of ascorbic and erythorbic acids in meat and for the hydrazones of ascorbic and dehydroascorbic acid in blood plasma (Bui-Nguyen, 1985).

Brenner et al. (1964) separated ascorbic acid and erythorbic acid by TLC and detected the spots with 2,6-dichlorophenol indophenol or iodine. Illner (1984) separated ascorbic and isoascorbic acids by TLC on Silufol plates impregnated with metaphosphoric acid and glycerol. The mobile phase was water-methyl ethyl ketone and spots were detected by spraying with α,α'-dipyridyl iron (III) chloride. As little as 0.5% isoascorbic acid in the presence of ascorbic acid was separated.

Otsuka et al. (1986) used TLC to isolate and partially characterize degradation products of 2,3-diketo-L-gulonic acid (intermediates in the biosynthesis of ascorbic acid); the products were characterized by different spectrometric methods. Mandrou et al. (1988) devised a TLC procedure to determine ascorbic and dehydroascorbic acids in fruit juices and wine; the sample was reacted with 2,4-dinitrophenylhydrazine to form the osazones; the osazones were spotted on the TLC plate and quantified by in situ densitometry at 494 nm.

Parviainen and Nyyssonen (1992) reviewed the analytical techniques used to study vitamin C. They stated that although TLC was used in early analytical studies to increase selectivity in the analysis of ascorbic acid and related compounds, currently this method has little qualitative or quantitative use. They also suggested that HPTLC should prove valuable in the analysis of ascorbic acid and related compounds especially in the pharmaceutical, food, and animal sciences. Baranowska and Kadziolka (1996) used reversed-phase TLC to separate water and fat-soluble vitamins on RP-18 plates. A mixture of five water-soluble vitamins (C, nicotinic acid, nicotinamide, B_1, and rutin) was separated using water–methanol (5:4) and water–acetic acid (7:1) as mobile phases. A mixture of fat-soluble vitamins (A-acetate, E, E-acetate, and D_3) was separated using acetonitrile–benzene–chloroform (10:10:1) as mobile phase.

B. Folates

McCormack and Newman (1985) reviewed the TLC of folates, 4-amino derivatives of folates (methotrexate and related compounds), and unconjugated pteridines. Scott (1980) reviewed the separation and identification of naturally occurring folates (pteroylmonoglutamates and pteroates) by TLC. Although many solvent systems have been described, a single system (3% aqueous w/v NH_4Cl) containing 0.5% (v/v) mercaptoethanol (MET) as an antioxidant gives good separation of most of the naturally occurring pteroylmonoglutamates (McCormack and Newman, 1985). Scott (1980) presented a table showing TLC of naturally occurring folates on MN 300 UV_{254} cellulose. Separated folates can be detected on cellulose plates containing a fluorescent indicator by UV light at 254 or 366 nm. Reif et al. (1977) described a specific TLC procedure for the purification of folic acid standards.

Valerino et al. (1972) have shown that methotrexate (MTX) can be separated from its degradation product, 2,4-diamino-N-10-methyl pteroic acid (DAMPA), using DEAE cellulose plates and the mobile-phase 0.1 M ammonium bicarbonate at pH 8.8. According to Brown et al. (1973), MTX, DAMPA, and folates can be separated on cellulose plates using aqueous solutions containing mercaptoethanol. The mercaptoethanol was included to reduce the rate of oxidation of folate derivatives.

Unconjugated pteridines, which lack a *p*-aminobenzoylglutamic acid residue, are of interest because they can be formed by the oxidative transformation of reduced folate derivatives; moreover, pteridines in biological fluids are elevated under certain pathological conditions (McCormack and Newman, 1985).

Descimon and Barial (1966) used two-dimensional TLC on cellulose plates to separate pteridines present in the fruit fly, *Drosophila*. Wilson and Jacobsen (1977) modified the procedure of Descimon and Barial and obtained improved resolution and detection of the fluorescent substances found in *Drosophila*; Wilson and Jacobsen's two-dimensional TLC procedure involved an initial development in isopropanol–2% ammonium acetate (1 : 1) followed by development with 3% ammonium chloride, and then in situ fluorometric quantification.

For the preparation of samples containing folates, derivatives of folates, and unconjugated pteridenes in biological materials, consult McCormack and Newman (1985).

Mullin and Duch (1992) reviewed the analytical techniques used to study folic acid and reduced folate compounds. They noted that TLC has not had wide acceptance as a method for the separation of these compounds. TLC was supplanted first by ion-exchange chromatography and more recently by HPLC. Bhushan and Parshad (1994) separated vitamin B complex and folic acid on silica gel layers impregnated with transition metal ions; the compounds were detected by exposure of the layers to iodine vapor.

C. Nicotinic Acid and Nicotinamide (B₃)

TLC is used for the analysis of nicotinic acid and nicotinamides in pharmaceutical preparations, foods, and in biological materials (blood, urine, tissues).

In pharmaceutical preparations, nicotinic acid can be separated from nicotinamide by TLC on silica gel, aluminum oxide, and ion-exchange resins; detection of the spots can be made by examination of the plates under UV light or by specific detection reagents. For a table listing the R_F values of water-soluble vitamins, consult Hengen and deVries (1985). Nicotinic acid can be separated from nicotinamide and other water–soluble vitamins by TLC on silica gel using the mobile phase acetic acid–acetone–methanol–benzene (5:5:20:70). Expected R_F values of water-soluble vitamins in this system are as follows: B_2, 0.3; nicotinic acid, 0.75; nicotinamide, 0.55; B_6, 0.13; C, 0.27 (Hengen and deVries, 1985).

Prior to chromatographic analysis of foods, acid or alkaline pretreatment is used to liberate nicotinic acid (Hengen and deVries, 1985). Katsui (1972) described a method for determining nicotinic acid in coffee. After extraction with HCl and decoloration with lead carbonate, the solution was spotted on chromatographic paper and developed in a mobile phase of *n*-butanol–ammonia–water (100:2:16). Spots were visualized using cyanogen bromide and benzidine. This procedure is probably applicable to TLC on cellulose.

Analysis of nicotinic acid, nicotinamide, and its metabolites in biological materials, i.e., blood, plasma, urine, and tissues, is important in studies on biochemical pathways (Hengen and deVries, 1985). Pinder et al. (1971) described several paper and thin-layer chromatographic systems useful for the differentiation of nucleotides in tissues derived from nicotinamide and nicotinic acid. Hengen and deVries (1985) provided a table summarizing the R_F values of nicotinic acid, nicotinamide, and various intermediates of NAD+ and NADP+ synthesis for both paper and thin-layer chromatography. Haworth and Walmsley (1972) used two-dimensional TLC on silica gel for the identification of tryptophan metabolites in urine and resolved 32 compounds including nicotinic acid and nicotinamide. Kala et al. (1978) used silica gel TLC to examine urine for nicotinic acid and its metabolites after administration of nicotinyl alcohol. They used various mobile phases consisting of mixtures of chloroform–ethanol–25% ammonia.

Ropte and Aieloff (1985) determined nicotinamide in multivitamin tablets on silica gel plates using the mobile phase acetone–chloroform–butanol–25% ammonium hydroxide (30:30:40:5); quantification was by densitometric scanning of the plates at 262 nm. Ismaiel et al. (1985) also determined nicotinamide from multivitamin preparations following extraction in water. TLC was done on silica gel plates with the mobile phase chloroform–ethanol (60:25). Detection of spots was by exposure of the plate to cyanogen bromide and aniline vapor to give yellow spots, and quantification was by densitometric scanning at 468 nm. Sherma and Ervin (1986) used silica gel HPTLC plates to determine niacin and niacinamide from vitamin preparations; quantification was by densitometric scanning of fluorescence quenching.

Shibata and Shimono (1992) reviewed the analytical techniques used to study nicotinic acid and nicotinamide. They noted that TLC is generally not used for the separation of these compounds because of low sensitivity and low resolution. The method of choice for these compounds is HPLC and GC. However, TLC is still useful for the separation of labeled compounds. Shibata and Shimono (1992) presented a table showing the R_F values of nicotinic acid and nicotinamide and their metabolites using both paper and thin-layer chromatography. The table considers separations of these compounds using various stationary and mobile phases.

D. Vitamin B₁

Kawasaki and Sanemori (1985) reviewed liquid chromatographic studies on B₁ (thiamines) but did not consider TLC studies per se. Information on TLC of thiamines can be found in Levorato and Cima (1968), Ahrens and Korytnyk (1969), Ismaiel and Yassa (1973), and Bolliger and Konig (1969).

Thiamines can be extracted with water or aqueous alcohol, and extracts should be prepared under acid conditions at pH 4–6 to prevent decomposition. Thiamine can be separated from its esters, derivatives, and degradation products by TLC on cellulose using various solvent systems, such as isopropanol–water–trichloroacetic acid–25% ammonium hydroxide (71:9:20:0.3) or n-butanol–acetic acid–water (40:10:50). For a list of solvent systems and R_F values of thiamine and related compounds separated by TLC on cellulose and silica gel G, see Bolliger and Konig (1969). Thiamine and its esters and decomposition products can be detected on fluorescent layers using short-wave UV light or by spraying plates with the iodoplatinate or Dragendorff reagents (see Chapter 8).

Ziporin and Waring (1970) used two-dimensional TLC to separate thiamine and its metabolites and precursor compounds, including hydroxyethylthiamine, pyrimidine, thiazole fragments of thiamine, carboxylic acid, and sulfonic derivatives of N'-methylnicotinamide. Bican-Fister and Drazin (1973) used TLC to separate thiamine from riboflavin, pyridoxine, nicotinamide, and p-aminobenzoic acid in pharmaceutical preparations; the solvent systems used for this separation were various mixtures of glacial acetic acid–acetone–methanol–benzene (volume ratios not given). Ismaiel and Yassa (1973) used TLC to separate thiamine from calcium-pantothenate, choline, inositol, and cyanocobalamin in the mobile phase chloroform–ethanol–water (50:25:1). Ultraviolet detection of the compounds in both studies was carried out at 246 nm.

Kawasaki (1992) reviewed the use of TLC procedures for the analysis of thiamine and its metabolites and noted that this procedure was used mainly for the analysis of these compounds in pharmaceutical preparations.

Navas-Diaz et al. (1993) quantified low nanogram levels of thiamine, riboflavin, and niacin by fiber-optic fluorodensitometry after silica gel HPTLC with methanol–water (7:3) mobile phase. Perisic-Janjic et al. (1995) used TLC for the analysis of the B-complex vitamins in some commercial pharmaceutical B-complex vitamin products. The stationary phase used was a newly synthesized carbamide formaldehyde polymer (aminoplast) with mobile systems 1-butanol–water–acetone (25:9:5) and 1-butanol–methanol–benzene–water (20:10:10:8). Comparative studies examined the B-complex vitamins on cellulose thin layers.

E. Flavins

Marletta and Light (1985) reviewed TLC procedures for flavin determination in multivitamin preparations, foods, urine, and tissue samples. Flavins are light sensitive and, therefore, exposure to direct light or sunshine should be avoided during sample preparation and chromatography. If protected from light, riboflavin is relatively heat stable, thus facilitating sample preparation. Methods of sample preparation vary considerably depending on the source of flavins. Specific methods of sample preparation for multivitamins, foods, and biological samples have been listed by Marletta and Light (1985). They have also prepared an extensive table listing representative systems for TLC and paper chromatography of flavins.

The most widely used mobile phase for flavin separation by TLC is n-butanol–acetic acid–water (BAW). Thus, Yagi and Oishi (1971) separated mono-, di-, tri-, and tetrabutyrates of riboflavin on silica gel in BAW (4:1:5), and Hais et al. (1973) used the same mobile phase to separate riboflavin, 10-hydroxymethyl-flavinacetate, and lumichrome on cellulose.

During development, the TLC plates should be kept in the dark or in subdued light; this can be done by wrapping the tank in aluminum foil. The most sensitive method of visualizing flavins on a TLC plate is by their fluorescence under UV light.

Bochner and Ames (1982) used TLC to analyze cellular nucleotides (including flavin mononucleotide and flavin adenine dinucleotide) from ^{32}P-labeled cells of *Salmonella typhimurium*. They used two-dimensional chromatography with precoated polyethyleneimine cellulose TLC plates. The mobile phase in the first direction consisted of aqueous solutions of amine chlorides, i.e., tris-HCl at pH 8.0, and in the second direction the mobile phase consisted of a solution of saturated ammonium sulfate.

TLC can be used to differentiate riboflavin from FMN (flavin mononucleotide) and some flavin analogs using various mobile phases; riboflavin phosphates can be separated from riboflavin biphosphates, whereas the isomeric riboflavin phosphates co-migrate in all chromatographic systems studied (Nielsen et al., 1986). Bhushan and Ali (1987) used TLC to analyze riboflavin in different matrices and also to separate and identify vitamin B_2 from other water-soluble vitamins (B_1, B_6, and B_{12}). Thielemann (1974) used TLC to separate and identify vitamin B_2 from fat-soluble vitamins (A, D_2, and E) in multivitamin preparations.

Nielsen (1992) reviewed the chromatographic methods used for the analysis of flavins. He noted that HPTLC combined with densitometry has some advantages in the field of flavin analysis even when compared with analysis by HPLC. He stated that TLC is still a valuable tool for the isolation and purification of flavin compounds. Nielsen (1992) provided extensive tabular data on the determination of flavins and riboflavin by TLC.

F. Vitamin B$_6$

Vanderslice et al. (1985) reviewed liquid chromatographic studies on vitamin B_6 and provided useful information on sample preparation and extraction procedures prior to chromatography. They did not consider TLC studies in their review.

Bolliger and Konig (1969) reviewed TLC of vitamin B_6 compounds and provided a list of solvent systems and R_F values for the separation of these substances on silica gel G and cellulose. A representative TLC separation of B_6 compounds on silica gel with distilled water as the mobile phase yielded the following R_F values: pyridoxamene–2 HCl, 0.19; pyridoxalethylacetal–HCl, 0.36; pyridoxal, 0.15; pyridoxine–HCl, 0.52; pyridoxal-5′-phosphate, 0.64 (Bolliger and Konig, 1969).

Most B_6 compounds appear dark blue under UV light on silica gel layers containing a fluorescent indicator; B_6 compounds also yield blue spots when sprayed with 2,6-dichloroquinoechloroimide followed by subsequent treatment with ammonia vapor (Bolliger and Konig, 1969).

Sample preparation of B_6 compounds from pharmaceuticals usually presents no problems because these vitamins can be extracted with water, dilute acids, or methanol.

Bhushan and Ali (1987) used TLC to separate vitamin B_6 from the other water-soluble vitamins in multivitamin preparations; they used silica gel plates developed in butanol–benzene–acetic acid–water (8:7:5:3) or butanol–acetic acid–water (9:4:5).

Ubbink (1992) reviewed the chromatographic methods used to study vitamin B_6. He noted that with the advent of HPLC, TLC is rarely used as a quantitative technique in the analysis of this vitamin; however, it is still a convenient qualitative method to study the metabolism of vitamin B_6. Coburn and Mahuren (1987) examined vitamin B_6 metabolism in cats and found that although 70% of the ingested carbon-labeled pyridoxine appeared in the urine, only 2–3% of the excreted dose was 4-pyridoxic acid. Using TLC, it was determined that pyridoxine 3-sulfate and pyridoxal 3-sulfate were the major urinary metabolites of the ingested pyridoxine.

G. Biotin

Frappier and Gaudry (1985) reviewed TLC studies on biotin and presented a table of solvent systems frequently used for biotin separations. Although some work has been done on cellulose, most studies use silica gel layers. Solvent systems containing butanol are frequently used [i.e., butanol–water (1:1) or butanol–water–acetic acid (4:5:1)], and several detection reagents for biotin are available, including 1% potassium permanganate, 1% dimethylaminobenzaldehyde in HCl, iodine vapor, o-toluidine chloride, and p-dimethylaminocinnamaldehyde (p-DACA). Shimada et al. (1969) noted that reacting p-DACA in a sulfuric acid–methanol mixture with biotin yielded an intense red-orange color with maximal absorbance at 533 nm and that the reaction was specific for biotin.

Nuttall and Bush (1971) analyzed biotin in multivitamin preparations. The fat-soluble vitamins were first extracted, and the water-soluble materials were separated in three TLC systems; biotin was resolved using the mobile phase acetone–acetic acid–benzene–methanol (1:1:14:4) and detected by spraying the plate with o-toluidine–potassium iodide. Groningsson and Jansson (1979) determined biotin in the presence of other water-soluble vitamins using silica gel TLC and the mobile phase chloroform–methanol–formic acid (70:40:2); detection was by spraying with p-DACA.

Gaudry and Ploux (1992) reviewed TLC studies on biotin and noted that this technique is more convenient and gives better results than paper chromatography in the analysis of this vitamin from multivitamin mixtures. They noted that although silica gel is generally used as the support phase, cellulose powder has also been proposed; they provided tabular data on typical mobile phases used for the TLC analysis of biotin.

Gaudry and Ploux (1992) reported that another convenient visualization technique for TLC plates is bioautography. Using this technique, a TLC plate is inverted on the surface of an agar plate previously seeded with the bacteria *Saccharomyces cerevisiae* or *Lactobacillus plantarum*. This procedure allows for the diffusion of the chromatographed compounds, i.e., biotin and related derivatives.

H. Vitamin B₁₂

Lindemans and Abels (1985) reviewed TLC studies on vitamin B_{12} (cobalamin) and related corrinoids. Cobalamin is usually extracted from biological materials in hot 80% ethanol, but various other extraction procedures are available [reviewed in Lindemans and Abels (1985)].

Early paper chromatographic studies on plasma cobalamin extracts used Whatman 2 paper with *sec*-butanol–acetic acid–water (100:3:50) as the mobile phase (Lindstrand and Stahlberg, 1963). More recent studies on cobalamin separations have used silica and cellulose TLC. For instance, Firth et al. (1968) separated a large number of organocobalamins by cellulose TLC with four different solvent systems. Lindemans and Abels (1985) provided tabular data on TLC of vitamin B_{12} derivatives on cellulose.

The analysis of cobalamins from human plasma was facilitated by the introduction of silica gel–cellulose TLC and the use of a growth indicator for bioautography (Linnell et al., 1969) (see Chapter 8 for a discussion of bioautography). More recently, Linnell et al. (1970) designed a two-dimensional chromato-bioautographic method for the separation of the individual plasma cobalamins. Nexøe and Anderson (1977) used silica gel TLC for the determination of plasma cobalamins with *sec*-butanol–isopropanol–water–ammonia (30:45:25:2) as the developing solvent. Farquharson and Adams (1976) used silica gel TLC with *sec*-butanol–propanol–water–ammonia (7:4:3:1) as the mobile phase to determine B_{12} compounds in food.

Cobalamins have been examined on reversed-phase plates. A list of R_F values for cobalamins by reversed- versus normal-phase TLC has been presented by Lindemans and Abels (1985). The reversed-phase technique allows for the identification of unusual cobalamins and also has the advantage of having relatively short developing times (Fenton and Rosenberg, 1978).

Lindemans (1992) reviewed the use of TLC and paper chromatography for the analysis of cobalamins. His review contains extensive information on extraction procedures; bioautographic techniques for the analysis of cobalamins from the plasma of human subjects; tabular data on R_F values for cobalamins on reversed- and normal-phase TLC; and tabular data on various mobile phases used for the TLC analysis of vitamin B_{12} derivatives on cellulose.

I. Pantothenic Acid

Velisakek et al. (1992) reviewed the use of TLC and paper chromatography for the analysis of pantothenic acid (vitamin B_5). They presented considerable information on methods of sample handling and preparation, chromatographic systems (stationary and mobile phases), and detection methods. Tabular data were presented on detection methods, mobilities of pantothenic acid and coenzyme A in paper chromatographic systems, and mobilities of pantothenic acids and its derivatives in various TLC chromatographic systems.

Silica gel or cellulose TLC is often used for the determination of pantothenic acid in multivitamin preparations and mixed feeds. Ganshirt and Malzacher (1960) separated pantothenic acid on silica gel G plates with the mobile phase acetic acid–acetone–menthol–benzene (5:5:20:70). Thielemann (1974) used activated commercial silica gel sheets to separate pantothenic acid in multivitamin preparations. Puech et al. (1978) used silica gel GF_{254} layers to separate calcium pantothenates, pantolactone, and other degradation products of pantothenic acids from pharmaceutical preparations; the mobile phase used was chloroform–ethanol–acetic acid (10:7:3). Baczyk and Szczesniak (1975) separated the B-complex vitamins, including pantothenic acid, on cellulose $MN_{300}HR$ layers with *n*-butanol–acetic acid–water (8:1:11) as the mobile phase.

IV. DETAILED EXPERIMENTS

A. Experiment 1. TLC Separation of Fat-Soluble Vitamins on Baker-flex Aluminum Oxide IB-F Sheets

1. Introduction

This experiment provides an introduction to the TLC analysis of fat-soluble vitamins.

2. Preparation of Standards

Prepare individual standards (Sigma) of vitamin A palmitate, vitamin D_2 (calciferol), and vitamin E (dl-α-tocopherol) as 1-μg/μl solutions in benzene. Prepare each standard by dissolving 10 mg of the vitamin in 10 ml of benzene (benzene is a carcinogen and should be handled with extreme care). Prepare a mixed standard of the three vitamins by weighing 10 mg of each vitamin and dissolving in 10 ml of benzene to achieve a concentration of 1 μg/μl for each.

3. Layer

Aluminum oxide IB-F sheets (J. T. Baker), activated for 15 min in an oven at 110°C and then cooled in a desiccator containing phosphorus pentoxide.

4. Application of Initial Zones

Remove the sheet from the desiccator and draw an origin line 2 cm from the bottom of the sheet and a solvent front line 10 cm from the origin. Place 4 μl of the individual vitamin standards on separate origins. Place 4 µl of the mixed vitamin standard on a separate origin; use separate 4 µl Microcap micropipets for application of the standards. Place the sheet in a desiccator containing phosphorus pentoxide for 30 min.

5. Mobile Phase

Chloroform.

6. Development

To eliminate the possibility of oxidizing the vitamins, develop the spotted sheet immediately after it is removed from the desiccator. Develop the sheet for a distance of 10 cm from the origin in a saturated rectangular glass chamber containing 100 ml of mobile phase. The development time will be about 50 min.

7. Detection

Remove the sheet from the chamber and allow it to dry. Observe the sheet under 254 nm and 366 nm UV light. For permanent visualization, spray the sheet lightly with a solution of 1% vanillin in concentrated H_2SO_4. The vitamins will appear as brown-gray spots. The following separations (approximate R_F values included) will be achieved for both individual vitamins and the mixed standard: vitamin D_2, 0.18; vitamin E, 0.40; vitamin A palmitate, 0.67. A good separation is indicated by compact spots that do not tail. Some impurities may be noted in commercial vitamins.

B. Experiment 2. TLC Separation of Water-Soluble Vitamins on Silica Gel Plates

1. Introduction

This experiment, adapted from Bollinger and König (1959), provides an introduction to the TLC analysis of water-soluble vitamins.

2. Preparation of Standards

Prepare 5-µg/µl standards (Sigma) in deionized water of each of the following water-soluble vitamins: thiamine–HCl; cyanocobalamin (B_{12}); riboflavin (B_2); nicotinamide; pyridoxine–HCl (B_6); nicotinic acid, ascorbic acid. Prepare a mixed standard containing the seven water-soluble vitamins at respective concentrations of 6, 2, 1, 6, 6, 6, 6 µg/µl.

3. Layer

Silica gel GF_{254} (Merck), activated for 15 min in an oven at 110°C.

4. Application of Initial Zones

Draw an origin line 2 cm from the bottom of the plate and a solvent front line 15 cm from the origin. Spot 6 µl of thiamine HCl, 2 µl of cyanocobalamin, 1 µl of riboflavin, 6 µl of nicotinamide, 6 µl of pyridoxine–HCl, 6 µl of nicotinic acid, and 6 µl of ascorbic acid on separate origins. Spot 5 µl of the mixture on another origin.

5. Mobile Phase

Deionized water.

6. Development

Develop the plate for a distance of 15 cm from the origin in a rectangular glass chamber containing 100 ml of the mobile phase. Briefly dry the plate by warming it before visualization.

7. Detection

UV light, 254 nm.

8. Results and Discussion

The following separations (approximate R_F values included) will be achieved for both individual vitamins and the mixed standard: thiamine–HCl, 0.05; cyanocobalamin (B_{12}), 0.22; riboflavin (B_2), 0.40; nicotinamide, 0.49; pyridoxine–HCl (B_6), 0.52; nicotinic acid, 0.78; ascorbic acid, 0.96. The riboflavin will fluoresce yellow, whereas the other spots will appear dark. The riboflavin and pyridoxine–HCl spots may show some tailing.

REFERENCES

Ahrens, H., and Korytnyk, W. (1969). Pyridoxine chemistry. XXI. Thin-layer chromatography and thin-layer electrophoresis of compounds in the vitamin B_6 group. *Anal. Biochem. 30*: 413–420.

Baczyk, S., Bazanowska, K., Jakubowska, W., and Sobisz, I. (1971). Determination of the vitamins A, D_3 and E using thin-layer chromatography. *Fresenius' Z. Anal. Chem. 255*: 132–133.

Baczyk, S., Duczmal, L., Sobisz, I., and Swidzinska, K. (1981). Spectrophotometric determination of vitamins D_2 and K_1 in the presence of rutin. *Mikrochim. Acta 2*: 151–154.

Baczyk, S., and Szczesniak, L. (1975). Thin-layer chromatographic separation of selected B vitamins. *Acta Pol. Pharm. 32*: 347–350.

Baranowska, I., and Kadziolka, A. (1996). RPTLC and derivative spectrophotometry for the analysis of selected vitamins. *Acta Chromatogr. 6*: 61–71.

Bargagna, A., Mariani, E., and Dorato, S. (1991). TLC, HPTLC and HPLC determination of cis- and trans-retinoic acids, retinol and retinyl acetate in topically applied products. *Acta Technol. Legis Med.* 2: 75–86.

Bhushan, R., and Ali, I. (1987). TLC resolution of constituents of the vitamin-B complex. *Arch. Pharm.* 320: 1186–1187.

Bhushan, R., and Parshad, V. (1994). Separation of vitamin B complex and folic acid using TLC plates impregnated with some transition metal ions. *Biomed. Chromatogr.* 8: 196–198.

Bican-Fister, T., and Drazin, V. (1973). Quantitative analysis of water-soluble vitamins in multicomponent pharmaceutical forms. Determination of tablets and granules. *J. Chromatogr.* 77: 389–395.

Bochner, B. R., and Ames, B. N. (1982). Complete analysis of cellular nucleotides by two-dimensional thin layer chromatography. *J. Biol. Chem.* 257: 9759–9769.

Bolliger, H. R., and Konig, A. (1969). Vitamins, including carotenoids, chlorophylls, and biologically active quinones. In *Thin Layer Chromatography. A Laboratory Handbook*, E. Stahl (Ed.). Springer-Verlag, New York, pp. 259–311.

Brenner, C. S., Hinkley, D. F., Perkins, L. M., and Weber, S. (1964). Isomerization of the ascorbic acids. *J. Org. Chem.* 29: 2389–2392.

Brown, J. P., Davidson, G. E., and Scott, J. M. (1973). Thin-layer chromatography of pterpylglutamates and related compounds. *J. Chromatogr.* 79: 195–207.

Bui-Nguyen, M. H. (1985). Ascorbic acid and related compounds. In *Modern Chromatographic Analysis of the Vitamins*, A. P. DeLeenheer, W. E. Lambert, and M. G. M. DeRuyter (Eds.). Marcel Dekker, New York, pp. 267–301.

Chow, C. K., Draper, H. H., and Csallany, A. S. (1969). Assay of free and esterified tocopherols. *Anal. Biochem.* 32: 81–90.

Coburn, S. P., and Mahuren, J. D. (1987). Identification of pyridoxine 3-sulfate, pyridixal 3-sulfate, and N-methylpyridoxine as major urinary metabolites of vitamin B_6 in domestic cats. *J. Biol. Chem.* 262: 2642–2644.

Czuczy, P., and Morava, E. (1982). Thin-layer chromatographic analysis of vitamin D in human milk. *Anal. Chem. Symp. Ser.* 10: 483–487.

DeLeenheer, A. P., and Lambert, W. E. (1996). Lipophilic vitamins. In *Handbook of Thin-Layer Chromatography*, 2nd ed., J. Sherma and B. Fried (Eds.). Marcel Dekker, New York, pp. 1055–1077.

DeLeenheer, A. P., Lambert, W. E., and DeRuyter, M. G. M. (Eds.). (1985). *Modern Chromatographic Analysis of the Vitamins*. Marcel Dekker, New York.

DeLeenheer, A. P., Lambert, W. E., and Nelis, H. J. (1991). Lipophilic vitamins. In *Handbook of Thin-Layer Chromatography*. J. Sherma and B. Fried (Eds.). Marcel Dekker, New York, pp. 993–1019.

DeLeenheer, A. P., Lambert, W. E., and Nelis, H. J. (Eds.). (1992). *Modern Chromatographic Analysis of Vitamins*, 2nd ed. Marcel Dekker, New York.

Descimon, H., and Barial, M. (1966). Application de la chromatographie sur couche mince de cellulose a la seperation des pterines naturelles. *J. Chromatogr.* 25: 391–397.

Dreassi, E., Ceramelli, G., and Corti, P. (1996). Thin-layer chromatography in pharmaceutical analysis. In *Practical Thin-Layer Chromatography—A Multidisciplinary Approach*, B. Fried and J. Sherma (Eds.). CRC Press, Boca Raton, FL, pp. 231–247.

Dueland, S., Holmberg, I., Berg, T., and Pedersen, J. (1981). Uptake and 25-hydroxylation of vitamin D_3 by isolated rat liver cells. *J. Biol. Chem.* 256: 10430–10434.

Dunphy, P. J., and Brodie, A. F. (1971). Structure and function of quinones in respiratory metabolism. *Methods Enzymol. 18C*: 407–461.

Ersoy, L., and Duden, R. (1980). Reflectance photometric determination of alpha-tocopherol, vitamin K_1 and beta-carotene in spinach extracts. *Lebensm.-Wiss. Technol. 13*: 198–201.

Farquharson, J., and Adams, J. F. (1976). The forms of vitamin B_{12} in foods. *Br. J. Nutr. 36*: 127–136.

Fenton, W. A., and Rosenberg, L. E. (1978). Improved techniques for the extraction and chromatography of cobalamins. *Anal. Biochem. 90*: 119–125.

Firth, R. A., Hill, H. A. O., Pratt, J. M., and Thorp, R. G. (1968). Separation and identification of organocobalt derivatives of vitamin B_{12} on thin-layer cellulose. *Anal. Biochem. 23*: 429–432.

Frappier, F., and Gaudry, M. (1985). Biotin. In *Modern Chromatographic Analysis of the Vitamins*, A. P. DeLeenheer, W. E. Lambert, and M. G. M. DeRuyter (Eds.). Marcel Dekker, New York, pp. 477–496.

Fried, B. (1991). Hydrophilic vitamins. In *Handbook of Thin-Layer Chromatography*, J. Sherma and B. Fried (Eds.). Marcel Dekker, New York, pp. 987–992.

Fung, Y. K., Rhawan, G. R., and Sams, R. A. (1978). Separation of vitamin A compounds by thin layer chromatography. *J. Chromatogr. 147*: 528–531.

Furr, H. C., Barua, A. B., and Olson, J. A. (1992). Retinoids and carotenoids. In *Modern Chromatographic Analysis of Vitamins*, A. P. DeLeenheer, W. E. Lambert, and H. J. Nelis (Eds.). Marcel Dekker, New York, pp. 1–63.

Ganshirt, H., and Malzacher, A. (1960). Separation of several vitamins of the B group and of vitamin C by surface chromatography. *Naturwissenschaften 47*: 279–280.

Gaudry, M., and Ploux, O. (1992). Biotin. In *Modern Chromatographic Analysis of Vitamins*, A. P. DeLeenheer, W. E. Lambert, and H. J. Nellis (Eds.). Marcel Dekker, New York, pp. 441–467.

Groningsson, K., and Jansson, L. (1979). TLC determination of biotin in a lyophilized multivitamin preparation. *J. Pharm. Sci., 68*: 364–366.

Hais, I. M., Cerman, J., Lankasov, and Skranova, M. (1973). Recoveries and reproducibility in the thin-layer chromatographic fluorimetric determination of flavins by the elution method. VIII. Chromatography of riboflavin decomposition products. *J. Chromatogr. 66*: 311–319.

Haworth, C., and Walmsley, T. A. (1972). Study of the chromatographic behavior of tryptophan metabolites and related compounds by chromatography on thin layers of silica gel. I. Qualitative separation. *J. Chromatogr. 66*: 311–319.

Hengen, N., and deVries, J. X. (1985). Nicotinic acid and nicotinamide. In *Modern Chromatographic Analysis of the Vitamins*, A. P. DeLeenheer, W. E. Lambert, and M. G. M. DeRuyter (Eds.). Marcel Dekker, New York, pp. 477–496.

Illner, E. (1984). Separation of ascorbic and isoascorbic acids by thin-layer chromatography on Silufol-ready-made plates. *Pharmazie 39*: 864.

Ismaiel, S. A., Haney, W. G., and Abdel-Moety, E. M. (1985). Spectrodensitometric determination of nicotinamide in some multivitamin preparations. *Pharmazie 40*: 804–805.

Ismaiel, S. A., and Yassa, D. A. (1973). Determination of thiamine in pharmaceutical preparations by thin-layer chromatography. *Analyst 98*: 1–8.

John, K. V., Lakshmanan, M. R., Jungalwala, F. B., and Cama, H. R. (1965). Separation of vitamins A_1 and A_2 and allied compounds by thin-layer chromatography. *J. Chromatog. 18*: 53–56.

Jones, G., Trafford, D. J. H., Makin, H. L. J. M., and Hollis, B. W. (1992). Vitamin D: Cholecalciferol, ergocalciferol, and hydroxylated metabolites. In *Modern Chromatographic Analysis of Vitamins*, A. P. DeLeenheer, W. E. Lambert, and H. J. Nelis (Eds.). Marcel Dekker, New York, pp. 73–151.

Justova, V., Wildtova, Z., and Pacovsky, V. (1984). Determination of 1,25-dihydroxyvitamin D_3 in plasma using thin-layer chromatography and modified competitive protein binding assay. *J. Chromatogr. 290*: 107–112.

Kahan, J. (1967). Thin-layer chromatography of vitamin A metabolites in human serum and liver tissue. *J. Chromatogr. 30*: 506–513.

Kala, H., Danz, R., and Moldenhauer, H. (1978). In vivo testing of a 3-hydroxymethylpyridine hydrogen tartrate matrix tablet. Part 1. Determination of methods for metabolites of 3-hydroxymethylpyridines in urine. *Pharmazie 33*: 803–806.

Katsui, G. (1972). Pharmaceutical preparations. Fat-soluble vitamins and related compounds. *Yukagaku 21*: 719–724.

Kawasaki, T. (1992). Vitamin B_1: Thiamine. In *Modern Chromatographic Analysis of Vitamins*, A. P. DeLeenheer, W. E. Lambert, and H. J. Nellis (Eds.). Marcel Dekker, New York, pp. 319–354.

Kawasaki, T., and Sanemori, H. (1985). Vitamin B_1: Thiamines. In *Modern Chromatographic Analysis of the Vitamins*, A. P. DeLeenheer, W. E. Lambert, and M. G. M. DeRuyter (Eds.). Marcel Dekker, New York, pp. 385–412.

Lambert, W. E., and DeLeenheer, A. P. (1992). Vitamin K. In *Modern Chromatographic Analysis of Vitamins*, A. P. DeLeenheer, W. E. Lambert, and H. J. Nelis (Eds.). Marcel Dekker, New York, pp. 197–233.

Lang, J. K., Schillaci, M., and Irvin, B. (1992). Vitamin E. In *Modern Chromatographic Analysis of Vitamins*, A. P. DeLeenheer, W. E. Lambert, and H. J. Nelis (Eds.). Marcel Dekker, New York, pp. 153–195.

LeFevere, M. F. L., Claeys, A. E., and DeLeenheer, A. P. (1985). Vitamin K: Phylloquinone and menaguinones. In *Modern Chromatographic Analysis of the Vitamins*, A. P. DeLeenheer, W. E. Lambert, and M. G. M. DeRuyter (Eds.). Marcel Dekker, New York, pp. 201–265.

Levorato, C., and Cima, L. (1968). Thin-layer chromatography and determination of thiamine salts, phosphoric esters, disulfides, and their respective thiochromes. *J. Chromatogr. 32*: 771–773.

Lichtenthaler, H. K., Boerner, K., and Liljenberg, C. (1982). Separation of prenylquinones, prenylvitamins, and prenols on thin-layer plates impregnated with silver nitrate. *J. Chromatogr. 242*: 196–201.

Lindemans, J. (1992). Cobalamins. In *Modern Chromatographic Analysis of Vitamins*. A. P. DeLeenheer, W. E. Lambert, and H. J. Nellis (Eds.). Marcel Dekker, New York, pp. 469–513.

Lindemans, J., and Abels, J. (1985). Vitamin B_{12} and related corrinoids. In *Modern Chromatographic Analysis of the Vitamins*, A. P. DeLeenheer, W. E., Lambert, and M. G. M. DeRuyter (Eds.). Marcel Dekker, New York, pp. 497–540.

Lindstrand, K., and Stahlberg, K.-G. (1963). Vitamin B_{12} forms in human plasma. *Acta Med. Scand. 174*: 665–669.

Linnell, J. C. (1996). Hydrophilic vitamins. In *Handbook of Thin-Layer Chromatography*, 2nd ed, J. Sherma and B. Fried (Eds.). Marcel Dekker. New York. pp. 1047–1054.

Linnell, J. C., Hussein, H. A.-A., and Mathews, D. M. (1970). Two-dimensional chromatobioautographic method for complete separation of individual plasma cobalamins. *J. Clin. Pathol. 23*: 820–821.

Linnell, J. C., Mackenzie, H. M., Wilson, J., and Matthews, D. M. (1969). Patterns of plasma cobalamins in control subjects and in cases of vitamin B_{12} deficiency. *J. Clin. Pathol. 22*: 545–550.

Madden, U. A., and Stahr, H. M. (1993). Reverse phase thin layer chromatography assay of vitamin K in bovine liver. *J. Liq. Chromatogr. 16*: 2825–2834.

Mandrou, B., Charlot, C., and Tsobze, A. D. (1988). Determination of ascorbic acid and dehydroascorbic acid in wines and fruit juices. *Ann. Falsif. Expert. Chim. Toxicol. 81*: 323–332.

Marletta, M. A., and Light, D. R. (1985). Flavins. In *Modern Chromatographic Analysis of the Vitamins*, A. P. DeLeenheer, W. E. Lambert, and M. G. M. DeRuyter (Eds.). Marcel Dekker, New York, pp. 413–434.

McCormack, J. J., and Newman, R. A. (1985). Chromatographic studies of folic acid and related compounds. In *Modern Chromatographic Analysis of the Vitamins*, A. P. DeLeenheer, W. E. Lambert, and M. G. M. DeRuyter (Eds.). Marcel Dekker, New York, pp. 303–340.

Mueller-Mulot, W. (1976). Rapid method for the quantitative determination of individual tocopherols in oils and fats. *J. Amer. Oil. Chem. Soc. 53*: 732–736.

Mullin, R. J., and Duch, D. S. (1992). Folic acid. In *Modern Chromatographic Analysis of Vitamins*. A. P. DeLeenheer, W. E. Lambert, and H. J. Nellis (Eds.). Marcel Dekker, New York, pp. 261–283.

Navas Diaz, A., Guirado Paniagua, A., and Garcia Sanchez, F. (1993). Thin-layer chromatography and fiber-optic fluorometric quantitation of thiamine, riboflavin and niacin. *J. Chromatogr. 655*: 39–43.

Nelis, H. J., DeBevere, V. O. R. C., and DeLeenheer, A. P. (1985). Vitamin E: tocopherols and tocotrienols. In *Modern Chromatographic Analysis of the Vitamins*, A. P. DeLeenheer, W. E. Lambert, and M. G. M. DeRuyter (Eds.). Marcel Dekker, New York, pp. 129–200.

Nexøe, E., and Andersen, J. (1977). Unsaturated and cobalamin saturated transcobalamin I and II in normal human plasma. *Scand. J. Clin. Lab. Invest. 37*: 723–728.

Nielsen, P. (1992). Flavins. In *Modern Chromatographic Analysis of Vitamins*, A. P. DeLeenheer, W. E. Lambert, and H. J. Nellis (Eds.). Marcel Dekker, New York, pp. 355–398.

Nielsen, P., Rauschenbach, P., and Bacher, A. (1986). Preparation, properties, and separation by high-performance liquid chromatography of riboflavin phosphates. *Methods Enzymol. 122*: 209–220.

Nuttall, R. T., and Bush, B. (1971). Detection of ten components of a multi-vitamin preparation by chromatographic methods. *Analyst 96*: 875–878.

Otsuka, M., Kurata, K., and Arakawa, N. (1986). Isolation and characterization of an intermediate product in the degradation of 2,3-dikcto-L-gulonic acid. *Agric. Biol. Chem. 50*: 531–533.

Parizkova, H., and Blattna, J. (1980). Preparative thin-layer chromatography of the oxidative products of retinyl acetate. *J. Chromatogr. 191*: 301–306.

Parviainen, M. T., and Nyyssonen, K. (1992). Ascorbic acid. In *Modern Chromatographic Analysis of Vitamins*, A. P. DeLeenheer, W. E. Lambert, and H. J. Nellis (Eds.). Marcel Dekker, New York, pp. 235–260.

Perisic-Janjic, N., Petrovic, S., and Hadzic, P. (1976). Separation of fat-soluble vitamins by thin layer chromatography. *Chromatographia 9*: 130–132.

Perisic-Janjic, N. U., Popovic, M. R., and Djakovic, T. Lj. (1995). Quantitative determination of vitamin B complex constituents by fluorescence quenching after TLC separation. *Acta Chromatogr. 5*: 144–150.

Pinder, S., Clark, J. B., and Greenbaum, A. L. (1971). Assay of intermediates and enzymes involved in the synthesis of the nicotinamide nucleotides in mammalian tissues. *Methods Enzymol. 18B*: 20–46.

Puech, A., Monleaud, J., Jacob, M., and Lapiczak, G. (1978). Chromatographic study of vitamin B_5 and its degradation products. *Ann. Pharm. Fr. 36*: 191–195.

Reif, V. D., Reamer, J.-T., and Grady, L. T. (1977). Chromatographic assays for folic acid. *J. Pharm. Sci. 66*: 1112–1116.

Rittich, B., Simek, M., and Krska, M. (1976). Spectrophotometric determination of the group K vitamins in concentrates and pharmaceutical preparations after a chromatographic separation on Silufol. *Cesk. Farm. 25*: 64–66.

Ropte, D., and Aieloff, K. (1985). Densitometric determination of nicotinamide in multivitamin preparations after thin-layer chromatographic separation. *Pharmazie 40*: 793–794.

Ruggeri, B. A., Watkins, T. R., Gray, R. J. H., and Tomlins, R. I. (1984). Comparative analysis of tocopherols by thin-layer chromatography and high-performance liquid chromatography. *J. Chromatogr. 291*: 377–383.

Scott, J. M. (1980). Thin-layer chromatography of pteroylmonoglutamates and related compounds. *Methods Enzymol. 66*: 437–443.

Sherma, J., and Ervin, M. (1986). Quantification of niacin and niacinamide in vitamin preparations by densitometric thin layer chromatography. *J. Liq. Chromatogr. 9*: 3423–3431.

Shibata, K., and Shimono, T. (1992). Nicotinic acid and nicotinamide. In *Modern Chromatographic Analysis of Vitamins*, A. P. DeLeenheer, W. E. Lambert, and H. J. Nellis (Eds.). Marcel Dekker, New York, pp. 285–317.

Shimada, K., Nagase, Y., and Matsumoto, U. (1969). Labeling and application to microanalysis of biotin related compounds. 10. Microdetermination of *d*-biotin by isotope dilution. *Yakugaku Zasshi 89*: 436–440.

Sklan, D., Budowski, P., and Katz, M. (1973). Determination of 25-hydroxycholecalciferol by combined thin layer and gas chromatography. *Anal. Biochem. 56*: 606–609.

Sliwick, J., Kocjan, B., Labe, B., Kozera, A., and Zalejska, J. (1993). Chromatographic studies of tocopherols. *J. Planar Chromatogr.—Mod. TLC 6*: 492–494.

Sliwick, J., Podgorny, A., and Siwek, A. (1990). Chromatographic comparison of the hydrophobicity of vitamin A derivatives. *J. Planar. Chromatogr.—Mod. TLC 3*: 429–431.

Sommer, P., and Kofler, M. (1966). Physicochemical properties and methods of analysis of phylloquinones, menaquinones, ubiquinones, plastoquinones, menadione, and related compounds. *Vitam. Horm.* 24: 349–399.

Stahr, H. M. (1996). Planar chromatography applications in veterinary toxicology. In *Practical Thin-Layer Chromatography—A Multidisciplinary Approach*, B. Fried and J. Sherma (Eds.). CRC Press, Boca Raton, FL, pp. 249–264.

Targan, S. R., Merril, S., and Schwabe, A. D. (1969). Fractionation and quantification of β-carotene and vitamin A derivatives in human serum. *Clin. Chem.* 15: 479–486.

Tataruch, F. (1984). Method for the quantitative determination of retinol in blood serum by HPTLC. *Mikrochim. Acta 1*: 235–238.

Thielemann, H. (1974). One dimensional thin-layer chromatography of multivitamin preparations (Summavit 10 and Summavit drops) on activated foil UV 254. *Sci. Pharm.* 42: 145–149.

Thierry-Palmer, M., and Gray, T. K. (1983). Separation of the hydroxylated metabolites of vitamin D_3 by high-performance thin-layer chromatography. *J. Chromatogr.* 262: 460–463.

Ubbink, J. B. (1992). Vitamin B_6. In *Modern Chromatographic Analysis of Vitamins*, A. P. DeLeenheer, W. E. Lambert, and H. J. Nellis (Eds.). Marcel Dekker, New York, pp. 399–439.

Valerino, D. M., Johns, D. G., Zaharko, D. S., and Oliverio, V. T. (1972). Metabolism of methotrexate by intestinal flora. I. Identification and study of biological properties of the metabolite 4-amino-4-deoxy-n[10]-methylpteroic acid. *Biochem. Pharmacol.* 21: 821–831.

Vanderslice, J. T., Brownlee, S. G., Cortissoz, M. E., and Maire, C. E. (1985). Vitamin B_6 analysis: sample preparation, extraction procedures and chromatographic separations. In *Modern Chromatographic Analysis of the Vitamins*, A. P. DeLeenheer, W. E. Lambert, and M. G. M. DeRuyter (Eds.). Marcel Dekker, New York, pp. 435–476.

Varma, T. N. R., Panalaks, T., and Murray, T. K. (1964). Thin-layer chromatography of vitamin A and related compounds. *Anal. Chem.* 36: 1864–1865.

Velisakek, J., Davidek, J., and Davidek, T. (1992). Pantothenic acid. In *Modern Chromatographic Analysis of Vitamins*, A. P. DeLeenheer, W. E. Lambert, and H. J. Nellis (Eds.). Marcel Dekker, New York, pp. 515–560.

Vieth, R., Fraser, D., and Jones, G. (1978). Photography tank for continuous development of thin-layer chromatographic plates. *Anal. Chem.* 50: 2150–2152.

Wardas, W., and Pyka, A. (1995). New visualizing agents for fatty vitamins in TLC. *Chem. Anal. (Warsaw)* 40: 67–72.

Wilson, T. G., and Jacobsen, K. B. (1977). Isolation and characterization of pteridines from heads of *Drosophila melanogaster* by a modified thin-layer chromatography procedure. *Biochem. Genet.* 15: 307–319.

Yagi, K., and Oishi, N. (1971). Separation and determination of flavines using thin-layer chromatography. *J. Vitaminol. (Kyoto)* 17: 49–51.

Ziporin, Z. Z., and Waring, P. P. (1970). Thin-layer chromatography for the separation of thiamine, N′-methylnicotinamide, and related compounds. *Methods Enzymol.* 18A: 86–90.

20

Nucleic Acid Derivatives

I. PROPERTIES OF NUCLEIC ACIDS AND THEIR DERIVATIVES

Nucleotides consist of a pentose sugar joined to a nitrogen base (purine or pyrimidine) and phosphate group. Nucleic acids are polymerized from nucleotides and consist of ribonucleic acid (RNA) and deoxyribosenucleic acid (DNA). In DNA, the pentose sugar is D-deoxyribose and the bases are adenine, guanine, thymine, and cytosine; the pentose sugar in RNA is D-ribose and the bases are adenine, guanine, thymine, and cytosine. There also are other bases that occur infrequently in both RNA and DNA.

Most of the cell's DNA is in the nucleus, although DNA is also present in mitochondria and the chloroplasts of plants. In higher animals, nuclear DNA is bound to histones, which are basic proteins. The basic composition of nuclear DNA is characteristic of an organism and is expressed as the percentage of guanine and cytosine (G and C) bases. Base combinations vary in the animal kingdom. For instance, mammals have about 40% G and C, whereas invertebrates have lower values.

Cells contain more RNA than DNA, and the three types of RNA are ribosomal (rRNA), transfer (tRNA), and messenger (mRNA). rRNA constitutes about 75% of all the RNA, whereas tRNA accounts for 10–15% of the total and functions to transport amino acids in protein synthesis. mRNA consists of only 5–10% of the total and functions to carry information for protein synthesis from the nucleus to the cytoplasm.

TABLE 20.1   List of Important Nucleosides and Nucleotides

	Base	Nucleoside base and sugar	Nucleotide base and sugar[a] and phosphate
Purine	Adenine	Adenosine	Adenylic acid (AMP)
	Guanine	Guanosine	Guanylic acid (GMP)
	Hypoxanthine	Inosine	Inosinic acid (IMP)
Pyrimidine	Cytosine	Cytidine	Cytidylic acid (CMP)
	Thymine	Thymidine	Thymidine monophosphate
	Uracil	Uridine	Uridylic acid

[a]The sugar is ribose in RNA synthesis, and deoxyribose in DNA synthesis.
Source: Modified from Barrett (1981).

Nucleotides are the basic building blocks of nucleic acids, whereas nucleosides are nucleotides without the phosphate group (i.e., nucleoside = base + sugar, and nucleotide = base + sugar + phosphate).

The amount of nucleosides in cells is negligible; also deoxyribonucleotides occur at insignificant levels in tissues. Ribonucleotides are important cellular components and occur as 5'-monophosphates, di-, tri-, and cyclic 3',5'-phosphates. Nucleoside phosphates function as energy carriers and serve as precursors for the synthesis of nucleic acids. Cyclic nucleotides (i.e., cyclic 3',5'-AMP and cyclic-3',5'-GMP) also play a role in metabolic regulation.

Nucleic acids are broken down by nucleases, nucleotidases, and nucleosidases to give nucleotides, nucleosides, and purine and pyrimidine bases, respectively (see Table 20.1). Mangold (1969) and Cowling (1983) have reviewed the ways in which nucleic acids can be degraded. For instance, heating RNA or DNA at 100°C in 70% perchloric acid produces bases, whereas hydrolysis of RNA in HCl yields purine bases and pyrimidine nucleotides; alkaline hydrolysis of RNA yields both 3'-and 5'-ribonucleotides. The types of enzymes used play a role in the products obtained. For instance, RNase T_2 breaks down RNA to 3'-ribonucleotides, and further digestion with snake venom phosphodiesterase will produce nucleosides. DNA can be treated with pancreatic DNase I followed by snake venom phosphodiesterase to yield 5'-deoxyribonucleotides. An advantage of nuclease degradation is that products can be applied directly to thin layers without removal of reagents.

II.   TLC OF NUCLEIC ACID DERIVATIVES

A.   General

Mangold (1969) and Cowling (1983) have provided good reviews of this topic. TLC of nucleic acid derivatives can be done on various sorbents; cellulose, silica gel, ion-exchange layers, and numerous mobile phases have been used, including water and lithium chloride (LiCl). Nucleic acid derivatives are usually detected by examining chromatograms under short-wave ultraviolet (UV) light. Maps of the spots are often made following two-dimensional (2D) separation and visualization under the UV light.

B. Specifics

Numerous studies have been reported on the separation of nucleic acid derivatives using two-dimensional TLC on cellulose. The earlier literature has been reviewed by Mangold (1969). Significant 2D–TLC studies of nucleic acid derivatives on cellulose include the following. Randerath and Randerath (1964) used PEI-cellulose layers to separate 23 ribonucleotides. Di- and trinucleotides were separated along with some nucleotide sugars. Successive developments with increasing concentrations of LiCl were done in the first direction; the chromatogram was developed in the second direction with formic acid–sodium formate buffers by a stepwise development procedure (Figure 20.1). Raaen and Kraus (1968) used PEI-cellulose to resolve complex mixtures of nucleic acid bases, nucleotides,

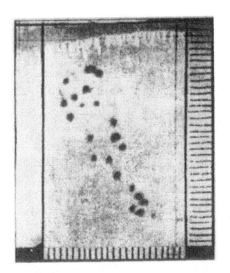
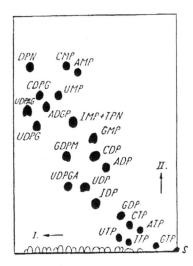

FIGURE 20.1 Separation of a mixture of mononucleotides by two-dimensional TLC on PEI cellulose. The chromatogram is on the left and the identified individual fractions are on the right. Consult Randerath and Randerath (1964) for the names of the individual components. [Reproduced with the permission of Elsevier Scientific Publishing Co. from Randerath and Randerath (1964).]

and nucleosides. Bases and nucleotides were separated from the nucleosides by developing with water in the first direction and a mobile phase consisting of butanol–methanol–water–ammonia (60:20:20:1) in the second direction. Randerath et al. (1972, 1980) introduced a sensitive radiochemical procedure for the analysis of nucleosides in which the *cis*-OH groups of the ribose sugar were oxidized with sodium periodate to give a mixture of nucleoside dialdehydes. Use of sodium [^{3}H] borohydride reduces the dialdehydes to trialcohols. Separation was on cellulose by development in the first direction with acetonitrile–ethyl acetate–*n*-butanol–2-propanol–6 *N* aqueous ammonia (40:30:10:20:27) and in the second direction with *t*-amyl alcohol–methyl ethyl ketone–water–formic acid (20:20:10:1). After development the products were visualized by low-temperature fluorography. Two-dimensional TLC on cellulose has also been used to screen disorders of purine and pyrimidine metabolism. Prior to TLC these substances must be isolated from urine by anion-exchange column chromatography (van Gennip et al., 1978). Figure 20.2 is a map of a two-dimensional chromatogram from van Gennip et al. (1978) showing the positions of various purines and pyrimidines.

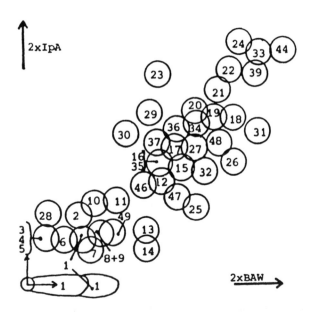

FIGURE 20.2 Map of a two-dimensional chromatogram showing the positions of various purines and pyrimidines. For the identity of these compounds, consult van Gennip et al. (1978). [Reproduced with the permission of Elsevier/ North Holland Biomedical Press from van Gennip et al. (1978).]

Bidimensional TLC studies have been done using silica gel as the sorbent. For instance, Lando et al. (1967) separated apurinic DNA oligonucleotides on silica gel with an isopropanol–water (75:25) mobile phase in the first direction and butanol–acetic acid–0.5% ammonia (90:30:40) in the second direction. Thymus DNA was separated into 21 spots, of which 16 were identified as nucleotides by UV spectrophotometry following enzymatic treatment. Issaq et al. (1977) used 2D–TLC on silica gel to separate a mixture of 19 adenine and uracil bases. Chloroform–methanol (90:10) was used for the first dimension and chloroform–propanol (90:30) for the second.

TLC has been combined with electrophoresis to facilitate nucleotide separations. Thus, Berquist (1965) mapped the enzymatic digests of RNA by separating 17 oligonucleotides and mononucleotides on a cellulose layer. Electrophoresis was first run at pH 2.5 and 50 V/cm, followed by two successive chromatographic developments at a right angle to the electrophoretic dimension. Diels and Wachter (1980) and Hashimoto et al. (1981) also combined gel electrophoresis and TLC in their RNA sequencing studies.

Unidimensional TLC of nucleic acid derivatives has been done on cellulose. Thus, Bij and Lederer (1983) used TLC on plastic sheets coated with cellulose to separate nucleic acid bases and nucleosides; water or aqueous solutions of $(NH_4)_2SO_4$ were used as mobile phases and the chromatography was done at room temperature using small glass jars covered with watch glasses. Detailed Experiment 1 of this chapter describes a unidimensional separation of nucleotides on PEI–cellulose.

Unidimensional TLC of nucleic acid derivatives has also been done on silica gel. Issaq et al. (1977) separated mixtures of alkylated guanines, adenines, uracils, and cytosines on silica gel $60F_{254}$, in various chloroform–alcohol solvent systems. The addition of 1 ml of ammonium hydroxide to the mobile phase prevented streaking and resulted in nondistorted developed spots. Figueira and Ribeiro (1983) separated adenine, adenosine, inosine, hypoxanthine, 2-AMP, 5-AMP, ADP, ATP, cAMP, and dibutyryl cAMP also on silica gel $60F_{254}$, and Sleckman et al. (1985) used conventional and high-performance silica gel to separate four representative nucleotides and their corresponding nucleosides. Detailed Experiment 2 of this chapter describes a unidimensional separation of nucleotides and nucleosides on silica gel.

Krauss and Schmidt (1985) separated adenosine-5'-phosphosulfate, adenosine-5'-phosphoramidate, and cyclic adenosine-5'-monophosphate nucleotides on silica gel GF plates with a mobile phase consisting of 2-propanol–25% ammonia solution–ethanol (6:3:1); detection of the nucleotides was by fluorescence quenching. Sarbu and Marutoiu (1985) were able to detect 10-ng amounts of purines on silica gel R with 1% uranyl acetate and UV radiation at 254 and 366 nm. Feldberg and Reppucci (1987) separated anomeric purine nucleosides on Chiralplates with methanol–water–acetonitrile (50:50:30) mobile phase.

Popovic and Perisic-Janjic (1988) separated purine-derivative drugs on starch and cellulose layers and quantified the derivatives by fluorodensitometry. Lepri et al. (1989a) examined the chromatographic behavior of purines, pyrimidines, and nucleosides on untreated and detergent-impregnated silanized silica gel plates. Rosier and Van Peteghem (1988) detected oxidatively modified 2'-deoxyguanosine-3'-monophosphate by using ^{32}P postlabeling and anion-exchange PEI–cellulose TLC.

Lepri et al. (1989b) examined the chromatographic behavior of 13 cyclic nucleotides and eight nicotinamide adenine coenzymes on untreated and detergent impregnated silanized silica, and on silica gel–NH$_2$ plates developed with aqueous–organic solutions at different pH values and ionic strengths. Keith et al. (1990) used two-dimensional TLC on cellulose plates to separate and quantify three adenosine derivatives obtained by venom phosphodiesterase digestion of poly (ADP-ribose). Jacks and Jones (1990) used HPTLC and an image analyzer detection system to assay the activity of 2'3'-cyclic nucleotide phosphohydrolase.

Van Gennip et al. (1991) contributed a chapter on the application of TLC and HPTLC for the detection of aberrant purine and pyrimidine metabolism in man. The authors provided extensive information on sample preparation of purines and pyrimidines, particularly from body fluids, blood, and tissue cells. The chapter is extensively illustrated with maps of two-dimensional chromatograms showing the positions of various purines and pyrimidines from normal and metabolically impaired subjects. Cserhati (1991) presented information on the retention behavior of some synthetic nucleotides, mainly 5-substituted deoxyuridine derivatives, on cyano (CN), diol, and amino (NH$_2$) precoated HPTLC plates. Brown (1991) provided a review with many references on numerous methods, including TLC, useful for the separation and identification of purines, pyrimidines, nucleosides, and nucleotides.

Sawinsky et al. (1992) described the separation of 5'-mononucleotides by overpressured layer chromatography (OPLC) on HPTLC silica gel 60F$_{254}$ plates with the mobile phase n-butanol–acetone–acetic acid–5% aqueous ammonia–water (9:5:2:2:3). After development, the plates were dried in cold air and the spots visualized under shortwave UV light (254 nm). The R_F values of the mononucleotides were as follows: 5'-AMP, 0.26; 5'-GMP, 0.38; 5'-IMP, 0.47; 5'-UMP, 0.51; 5'-CMP, 0.74. Steinberg and Cajigas (1992) provided detailed information on a two-dimensional thin-layer chromatographic technique to measure DNA deoxynucleotide modification. The TLC separation retains normal deoxynucleotide retention values and allows rapid visual assessment of DNA quality. The assay uses commercially available enzymes and is carried out in two dimensions on standard polyethyleneimine (PEI)–cellulose (Sigma) sheets. The assay can assess the maintenance of DNA quality and integrity. The monophosphate separation is easily carried out by two-dimensional PEI-cellulose TLC using solvent A (0.1 M acetic acid, pH 3.5 with NaOH) in the first dimension and solvent B [5.6M

$(NH_4)_2SO_4$, 0.12 M Na_2 EDTA, and 0.035 M $(NH_4)HSO_4$, pH 4] in the second dimension.

Savela and Hemminki (1993) used butanol extraction and nuclease P1-enhanced ^{32}P-postlabeling and thin layer radiochromatography to analyze cigarette smoke–induced DNA adducts in human granulocytes and lymphocytes. Turteltaub et al. (1993) coupled multidimensional TLC with accelerator mass spectrometry (MS) to study DNA adduction with heterocyclic amines.

Srinivas and Srinivasulu (1994) separated mixtures of purines and pyrimidines on layers of the zeolite heulandite used as the adsorbent; the R_F values were given for adenine, guanine, thymine, cytosine, uracil, adenosine, and hypoxanthine. Foersti et al. (1994) compared TLC- and HPLC- ^{32}P-postlabeling assay for cisplatin–DNA adducts, and showed that one of the standards, platinated d (TGG), could be analyzed with recovery of 31% for both methods when the amount of adduct was 0.5–100 fmol; while for platinated DNA, TLC had measurement sensitivity of 3.2 fmol and recovery of 28% compared to respective values of 8–40 fmol and 16% for HPLC.

Keith (1995) presented reference maps of mobilities of modified ribonucleotides and nucleosides separated by two-dimensional TLC on cellulose. Paw and Misztal (1995) isolated cytosine, uracil, and their arabinosides from human plasma by solid-phase extraction and Adsobex minicolumns and separated them by horizontal and ascending TLC in various mobile phases.

DeLong (1996) applied TLC to analyze phosphoromonothioate and phosphorodithioate oligonucleotides radiolabeled with either ^{32}P or [14]C. TLC coupled with radioactivity was compared to polyacrylamide gel electrophoresis (PAGE) and HPLC. TLC proved to be a rapid and sensitive alternative to the above-mentioned methods and was particularly suited for the analysis of the chemically modified oligonucleotides.

Steinberg et al. (1996) provided a definitive chapter on the TLC of nucleic acids and their derivatives including an extensive review of the pertinent literature. They stressed the need for a uniform approach to the TLC analysis of nucleic acids because such an approach would be helpful in making data comparisons. Their review included new technical information that they used successfully in their laboratory relative to the TLC of nucleic acids. Their procedures are especially useful in situations where expensive equipment is not available. They also stressed greater use of data quantification by methods described in the review. This chapter should be read by any chromatographer who intends to use TLC as a primary method of analysis for nucleic acids and their derivatives.

Koskinen et al. (1997) isolated DNA from rat livers treated with tamoxifen and analyzed the samples by the ^{32}P-postlabeling method. Sample analysis was done both by reversed-phase HPLC with ^{32}P online detection and by TLC on polyethyleneimine plates followed by autoradiography. The HPLC method resolved five adduct peaks, whereas TLC only resolved two groups of adduct spots.

It was concluded that HPLC provided better resolution and ease of quantification than TLC for these analyses. Kotynski et al. (1997) determined the TLC detection limits of phosphorothioate analogs of nucleotides using UV, iodine, HCl vapors, the iodine–azide reagent, and the molybdate reagent. The iodine–azide reagent proved most useful for the selective TLC detection of thiophosphoryl nucleotides. Paw and Misztal (1997) extracted various antimetabolites of pyrimidine bases (i.e., cytarabine, ftorafur, 6-azauridine, 5-fluorouracil, and trifluorothymidine) and separated them by silica gel–TLC in horizontal chambers; the compounds were visualized by UV light with detection limits in the 200–300 ng range.

III. DETAILED EXPERIMENTS

A. Experiment 1. TLC Separation of Monophosphate Nucleotides on Baker-flex Cellulose PEI-F Sheets

1. Introduction

This exercise provides an introduction to the TLC analysis of nucleotides.

2. Preparation of Standards

Prepare individual standards (Sigma) of adenosine-5'-phosphoric acid, cytidine 5'-phosphate (sodium salt), guanosine-5'-phosphate (sodium salt), and uridine-5'-phosphate (disodium salt) as 1-μg/μl solutions in deionized water. Prepare each standard by dissolving 10 mg of the nucleotide in 10 ml of water. Prepare a mixed standard of the four nucleotides at 1 μg/μl for each. To obtain the mixed standard, weigh 10 mg of each nucleotide and dissolve in 10 ml of water to achieve a concentration of 1 μg/μl for each.

3. Layer

Cellulose PEI-F sheets, used as received without pretreatment.

4. Application of Initial Zones

Draw an origin line 2 cm from the bottom of the sheet and a solvent front line 10 cm from the origin. Place 4 μl of the individual nucleotide standards on separate origins. Place 4 μl of the mixed nucleotide standard on a separate origin; use 4-μl Microcap micropipets for application of the standards. Air dry the sheet for 30 min to allow the water to evaporate from the sample spots.

5. Mobile Phase

1.0 M solution of lithium chloride (LiCl).

6. Development

Develop the sheet for a distance of 10 cm from the origin in a rectangular glass chamber containing 100 ml of the mobile phase. It is not necessary to equilibrate the tank with the mobile phase for this ion-exchange separation which is not affected by the vapor pressure of the mobile phase. Development time will be about 1 h.

7. Detection

After development, remove the sheet from the tank and allow it to air dry at room temperature. Locate the spots by observing the sheet under the short-wave UV light. Although cellulose PEI fluoresces strongly at 366 nm and weakly at 254 nm, the nucleotides are observed only at 254 nm.

8. Results and Discussion

The following separations (approximate R_F values included) will be achieved for both individual nucleotides and the mixed standard: guanosine-5'-phosphate (sodium salt), 0.34; adenosine-5'-phosphoric acid, 0.48; cytidine-5'-phosphate (sodium salt), 0.60; and uridine-5-phosphate (disodium salt), 0.74. The compounds will appear as dark spots against a bright fluorescent background at 254 nm.

B. Experiment 2. TLC Separation of Nucleotides and Nucleosides on Silica Gel

1. Introduction

This exercise from the work of Sleckman et al. (1984) provides for the separation of four representative nucleotides and their corresponding nucleosides on silica gel.

2. Preparation of Standards

Obtain the following nucleotide standards (Sigma): adenosine-5-monophosphate; uridine-5-monophosphate; cytidine-5-monophosphate; guanosine-5-monophosphate. Also, obtain the following nucleoside standards (Sigma): adenosine; uridine; cytidine; guanosine. Prepare individual standards at a concentration of 3μg/μl.

3. Layer

Whatman LK5DF, 20 × 20 cm, preadsorbent silica gel plates, used as received with no pretreatment.

4. Application of Initial Zones

Apply 5 μl (15 μg) of each standard to the preadsorbent zones of the plate with either a Microcap micropipet or a Drummond digital microdispenser.

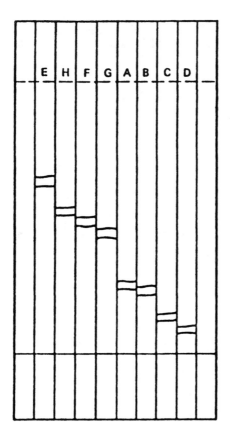

FIGURE 20.3 Separation of nucleotides and nucleosides on a silica gel layer. Compounds: E, adenosine; H, guanosine; F, uridine; G, cytidine; A, adenosine-5-monophosphate; B, uridine-5-monophosphate; C, cystidine-5-monophosphate; D, guanosine-5-monophosphate. [Reproduced with the permission of John Wiley and Sons, Inc., from Sleckman et al. (1985).]

5. Mobile Phase

Isopropanol–water–ammonia (7 : 2 : 1).

6. Development

Develop the plate in a rectangular glass tank lined with paper and allowed to equilibrate with the mobile phase for 10 min prior to inserting the layer. Allow the plate to develop for a distance of 13.5 cm past the origin (sorbent–preadsorbent interface). Development time will be approximately 3.5 h.

7. Detection

Visualize by fluorescence quenching under a short-wave UV lamp. The layer contains a fluorescent indicator that is activated at 254 nm.

8. Results and Discussion

The results of the separation are shown in Figure 20.3. The approximate R_F values are as follows: D, guanosine-5-monophosphate, 0.09; C, cystidine-5-monophosphate, 0.13; B, uridine-5-monophosphate, 0.23; A, adenosine-5-monophosphate, 0.25; G, cytidine, 0.44; F, uridine, 0.48; H, guanosine, 0.53; E, adenosine, 0.63. This system should allow the separation of six of the eight compounds. Some overlap will be noted between (A) adenosine-5-monophosphate and (B) uridine-5-monophosphate.

REFERENCES

Barrett, J. (1981). *Biochemistry of Parasitic Helminths*. University Park Press, Baltimore.

Berquist, P. L. (1965). A thin layer method for the mapping of enzymatic digest of ribonucleic acids. *J. Chromatogr. 19*: 615–618.

Bij, K. E., and Lederer, M. (1983). Thin-layer chromatography of some nucleobases and nucleosides on cellulose layers. *J. Chromatogr. 268*: 311–314.

Brown, E. G. (1991). Purines, pyrimidines, nucleosides, and nucleotides. *Methods Plant Biochem. 5*: 53–90.

Cowling, G. J. (1983). Nucleic acids. In *Chromatography—Fundamentals and Applications of Chromatographic and Electrophoretic Methods, Part B: Applications*, E. Heftmann (Ed.), Elsevier, Amsterdam, pp. B345–B375.

Cserhati, T. (1991). Retention behavior of some synthetic nucleosides on cyano (CN), diol, and amino (NH_2) precoated high-performance thin-layer chromatographic plates. *J. Chromatogr. Sci. 29*: 210–216.

DeLong, R. K. (1996). Thin-layer chromatography of radiolabeled oligonucleotide analogs: a rapid and sensitive purity assay. *Nucleosides Nucleotides 15*: 1433–1437.

Diels, L., and de Wachter, R. (1980). An RNA sequencing method based on a two-dimensional combination of gel electrophoresis and thin layer chromatography. *Arch. Int. Physiol. Biochim. 88*: B26–B27.

Feldberg, R. S., and Reppucci, L. M. (1987). Rapid separation of anomeric purine nucleosides by thin-layer chromatography on a chiral stationary phase. *J. Chromatogr. 410*: 226–229.

Figueira, M. A., and Ribeiro, J. A. (1983). Separation of adenosine and its derivatives by thin-layer chromatography on silica gel. *J. Chromatogr. 325*: 317–322.

Foersti, A., Staffas, J., and Hemminki, K. (1994). Comparison of TLC- and HPLC- ^{32}P-postlabelling assay for cisplatin-DNA adducts. *Carcinogenesis 15*: 2829–2834.

Hashimoto, S., Wold, W. S. M., Brackmann, K. H., and Green, M. (1981). Nucleotide sequences of 5′-termini of adenovirus 2 early transforming region Ela and Elb messenger ribonucleic acids. *Biochemistry 20*: 6640–6647.

Issaq, H. J., Barr, E. W., and Zielinski, W. L. (1977). Separation of alkylated guanines, adenines, uracils and cytosines by thin-layer chromatography. *J. Chromatogr. 131*: 265–273.

Jacks, A. S., and Jones, M. A. (1990). 2′,3′-Cyclic nucleotide phosphohydrolase activity determined using an image analyzer detection system. *Anal. Biochem. 184*: 321–324.

Keith, G. (1995). Mobilities of modified ribonucleotides on two-dimensional cellulose thin-layer chromatography. *Biochimie 77*: 142–144.

Keith, G., Desgres, J., and De Murcia, G. (1990). Use of two-dimensional thin-layer chromatography for the components study of polyadenosine diphosphate ribose. *Anal. Biochem. 191*: 309–313.

Koskinen, M., Rajaniemi, H., and Hemminki, K. (1997). Analysis of tamoxifen-induced DNA adducts by ^{32}P-postlabelling assay using different chromatographic techniques. *J. Chromatogr. Biomed. Sci. Appl. 691*: 155–160.

Kotynski, A., Kudzin, Z. H., Okruszek, A., Krajewska, D., Olesiak, M., and Sierzchala, A. (1997). Iodine-azide reagent in detection of thiophosphoryl nucleotides in thin-layer chromatography systems. *J. Chromatogr. 773*: 285–290.

Krauss, F., and Schmidt, A. (1985). Simple separation of adenosine-5′-phosphosulfate, cyclic adenosine-5′-phosphoramidate, and cyclic adenosine-5′-monophosphate nucleotides involved in adenosine-5′-phosphosulfate degradation. *J. Chromatogr. 348*: 296–298.

Lando, D., de Rudder, J., and Privat de Garilhe, M. (1967). Séparation des oligonucleotides du DNA apurinique par chromatographie bidimensionnelle sur couches minces de gel de silice. *J. Chromatogr. 30*: 143–148.

Lepri, L., Coas, V., Desideri, P. G., Pettini, L., and Checchini, L. (1989a). Planar chromatography of cyclic nucleotides on untreated and detergent-impregnated silanized silica plates. *J. Planar Chromatogr.—Mod. TLC 2*: 456–460.

Lepri, L., Coas, V., Desideri, P. G., and Checchini, L. (1989b). Planar chromatography of closely related steroids and nucleotides on untreated and detergent-impregnated silanized silica plates. *J. Planar Chromatogr.—Mod. TLC 2*: 214–218.

Mangold, H. K. (1969). Nucleic acids and nucleotides. In *Thin Layer Chromatography*, 2nd ed., E. Stahl (Ed.). Springer-Verlag, Berlin, pp. 786–807.

Popovic, M., and Perisic-Janjic, N. (1988). Separation and fluorodensitometric determination of some purine derivative drugs by thin-layer chromatography on starch and cellulose. *Chromatographia 26*: 244–246.

Paw, B., and Misztal, G. (1995). Chromatographic analysis (TLC) of uracil arabinoside, cytosine arabinoside, uracil, and cytosine in human plasma. *Acta Pol. Pharm. 52*: 455–457.

Paw, B., and Misztal, G. (1997). Thin-layer chromatographic analysis of antimetabolites of pyrimidine bases in human plasma. *Chem. Anal. 42*: 37–40.

Raaen, H. P., and Kraus, F. E. (1968). Resolution of complex mixtures of nucleic acid bases, nucleosides and nucleotides by two-dimensional thin layer chromatography on polyethyleneimine-cellulose. *J. Chromatogr. 35*: 531–537.

Randerath, E., and Randerath, K. (1964). Resolution of complex nucleotide mixtures by two-dimensional anion-exchange thin layer chromatography. *J. Chromatogr. 16*: 126–129.

Randerath, E., Gupta, R. C., and Randerath, E. (1980). Tritium and phosphorus-32 derivative methods for base composition and sequence analysis of RNA. *Methods Enzymol. 65*: 638–680.

Randerath, E., Yu, C.-T., and Randerath, K. (1972). Base analysis of ribopolynucleotides by chemical tritium labeling. Methodological study with model nucleosides and purified tRNA species. *Anal. Biochem. 48*: 172–198.

Rosier, J., and Van Peteghem, C. (1988). Detection of oxidatively modified 2'- deoxyguanosine-3'-monophosphate, using phosphorus-32 postlabeling and anion-exchange thin-layer chromatography. *J. Chromatogr. 434*: 222–227.

Sarbu, C., and Marutoiu, C. (1985). A new detection method for purines in thin-layer chromatography. *Chromatographia 20*: 683–684.

Savela, K., and Hemminki, K. (1993). Analysis of cigarette-smoke–induced DNA adducts by butanol extraction and nuclease P1-enhanced ^{32}P-postlabeling in human lymphocytes and granulocytes. *Environ. Health Perspect. 101* (Suppl. 3): 145–150.

Sawinsky, J., Halasz, A., and Tyihak, E. (1992). Separation of 5'-mononucleotides by overpressured layer chromatography (OPLC). *J. Planar Chromatogr.—Mod. TLC 5*: 390–391.

Sleckman, B. P., Touchstone, J. C., and Sherma, J. (1985). Thin layer chromatography of nucleotides and nucleosides. In *Techniques and Applications of Thin Layer Chromatography*, J. C. Touchstone and J. Sherma (Eds.), Wiley, New York, pp. 331–341.

Srinivas, B., and Srinivasulu, K. (1994). Separation of some purines and pyrimidines by thin-layer chromatography using heulandite as an adsorbent. *J. Indian Chem. Soc. 71*: 711–712.

Steinberg, J. J., and Cajigas, A. (1992). Enzymatic shot-gun 5'-phosphorylation and 3'-sister phosphate exchange: a two-dimensional thin-layer chromatographic technique to measure DNA deoxynucleotide modification. *J. Chromatogr. 574*: 41–55.

Steinberg, J. J., Cajigas, A., and Oliver Jr., G. W. (1996). Nucleic acids and their derivatives. In *Handbook of Thin-Layer Chromatography*, 2nd ed., J. Sherma and B. Fried (Eds.). Marcel Dekker, New York, pp. 921–969.

Turteltaub, K. W., Vogel, J. S., Frantz, C. E., and Fultz, E. Studies on DNA adduction with heterocyclic amines by accelerator mass spectrometry: a new technique for tracing isotope-labeled DNA adduction. *IARC Sci. Publ. 124*: 293–301.

van Gennip, A. H., Abeling, N. G. G. M., and DeKorte, D. (1991). Application of TLC and HPTLC for the detection of aberrant purine and pyrimidine metabolism in man. In *Handbook of Thin Layer Chromatography*, J. Sherma and B. Fried (Eds.). Marcel Dekker, New York, pp. 863–906.

van Gennip, A. H., van Noordenburg-Huistra, D. Y., DeBrec, P. K., and Wadman, S. K. (1978). Two-dimensional thin-layer chromatography for the screening of disorders of purine and pyrimidine metabolism. *Clin. Chim. Acta 86*: 7–20.

21

Steroids and Terpinoids

I. STEROIDS: GENERAL

Scott (1969) considered steroids as lipids usually extracted with organic solvents and varying considerably in polarity according to the kind and degree of substitution present in the molecule. He classified steroids into a number of groups based on function as follows: sterols and steroid alcohols, usually with double bonds; sex hormones, steroids produced mainly in the testis (androgens) or ovary (estrogens); adrenocortical hormones, steroids produced in the cortex of the adrenal gland; bile acids, steroids usually bonded to taurine or glycine and functioning as emulsion-stabilizing agents in the intestine; sapogenins, plant products with a steroid bonded to carbohydrates; cardiac glycosides, plant products similar to sapogenins and used as heart stimulants; and vitamin D, a steroid precursor of the active vitamin.

Heftmann (1983) reviewed chromatographic techniques for the analysis of sterols, vitamin D, sapogenins and alkaloids, estrane derivatives, androstane derivatives, corticosteroids, miscellaneous hormones, bile acids, ecdysteroids, and lactone derivatives.

Touchstone (1986) provided extensive information on the chromatographic analyses of the aforementioned steroids and outlined methods for their extraction and sample preparation. He also provided extensive tabular data on numerous steroids separated by GC, HPLC, and TLC.

There are numerous classes and types of steroids, both natural and synthetic, and steroids show variable characteristics: Some are hydrophilic and others lipophilic; some are stable and others unstable. Touchstone (1986) emphasized the importance of sample handling and preparation in steroid analysis. Because steroids are frequently isolated from blood or urine, the first step often involves liquid–liquid partition or separating steroids from water-soluble components by shaking the sample with an organic solvent. Modern cleanup methods (see Chapter 4) using columns that have been prepared commercially are of value in the extraction and sample preparation of steroids. Specific examples of modern cleanup methods for each class of steroid are given in Lamparczyk (1992).

Szepesi and Gazdag (1991) contributed a chapter on the TLC of steroids including information on sample preparation, stationary-phase and mobile-phase systems useful for the separation of steroidal pharmaceuticals; they provided detailed methods of detection and quantification of steroids. The authors compiled extensive tables listing the methods used for identification and purity testing of different steroids described in recent editions of the *United States Pharmacopoeia* and the *British Pharmacopoeia*. They also compiled a table of selected practical applications, including stationary and mobile phases, methods of detection and quantification, and references to some selected TLC systems applied to the investigation of steroids from 1980 to 1988. Szepesi and Gazdag (1996) updated their review to include coverage through 1994. Dreassi et al. (1996) have reviewed the application of TLC to steroids as relevant to pharmaceutical analysis. Jain (1996) has provided some information on the analysis of steroid hormones in his review on TLC in clinical chemistry.

Lamparczyk (1992) provided extensive coverage of the chromatographic analyses of steroids and updated information in Touchstone (1986). The book includes information on steroid nomenclature, chromatographic methods used in steroid analyses, and sample preparation. Although not restricted to TLC, the book provides extensive information on all aspects of TLC related to androgens, bile acids, ecdysteroids, estrogens, pregnanes, and corticoids.

Numerous in situ detection procedures (see Chapter 8) are available for steroids following their separation by TLC. These include the usual nondestructive procedures (i.e., UV light, iodine, and fluorescein), and destructive procedures (i.e., H_2SO_4, phosphomolybdic acid). There are numerous destructive methods for locating steroids in situ and most have been described in Touchstone (1986). Probably the most universal one is spraying the plate with 10% sulfuric acid in methanol and heating for 10 min at 90°C. According to Lamparczyk (1992), most steroids can be visualized using 10 g of copper sulfate and 5 ml *o*-phosphoric acid dissolved in 95 ml of H_2O for normal phase TLC and the same reagents dissolved in 95 ml of methanol for reversed-phase TLC. Specific spray reagents that locate particular functional groups are also available to detect certain steroids. For instance, the Liebermann–Burchard reaction with acetic anhydride–sulfuric acid for the detection of Δ^5-3-hydroxysterols, phosphomolybdic acid for reducing steroids, and chloramine-T–sulfuric acid for detecting various steroids at levels of 0.2–5 μg (Scott and Sawyer, 1975; Bajaj and Ahuja, 1979). An extensive list of detection reagents for steroids has been provided by Touchstone (1986).

This chapter examines representative TLC methods for androstanes (androgens), estrane derivatives (estrogens), corticosteroids (corticoids and pregnanes), bile acids, and ecdysteroids. Sterols have been considered in Chapter 15 with lipids, vitamin D in Chapter 19 with vitamins, and sapogenins and glycosides with alkaloids in Chapter 22 on pharmaceuticals.

II. TLC OF ANDROGENS

The androgens are C_{19} steroids produced mainly in the testis but also in the adrenal gland and ovary. Representative androgens include testosterone, dihydrotestosterone, and synthetic anabolic steroids such as trenbolone (Oehrle et al., 1975). Androgens are the most abundant of the neutral steroids and can be difficult to separate because they have relatively few functional groups and low polarity.

Lisboa and Hoffman (1975) separated numerous androstanes and androstenes by TLC on silica gel $60F_{254}$ using mobile phases of cyclohexane ethyl acetate (1:1) and *n*-hexane–ethyl acetate (3:1). Petrovic and Petrovic (1976) used two-dimensional TLC on silica gel G to separate various androgen derivatives; the mobile phase in the first direction was benzene–acetone and in the second direction, benzene–ethyl acetate. Spots were detected by spraying the plates with 50% H_2SO_4 in methanol, and then heating them for 10–15 min at 100°C. Similarly, Kerebel et al. (1977) separated testosterone and related C_{19} steroids on silica gel $60F_{254}$ with a mobile phase of chloroform–ethyl acetate (4:1). Spots were visualized by spraying the plates with H_2SO_4–methanol (7:3) and then heating them at 120°C for 15 min. Puah et al. (1978) used TLC as a purification step in the radioimmunoassay of testosterone and 5α-dihydrotestosterone conjugates in human urine and serum. Tinschert and Trager (1978) separated 13 steroids including 4-androstenes and 5-androstenes on silica gel and 1,2-propanediol-impregnated cellulose. Sunde et al. (1979) used silica gel TLC to separate the epimeric 3-hydroxyandrogens. They used multiple development with methylene chloride–ethyl acetate (9:1) as a mobile phase and observed that androsterone is more polar than epiandrosterone. O'Shannessy (1984) used silver nitrate–impregnated silica gel plates to separate the steroids 5α-androstenediol-5α-androstanediol and dehydroepiandrosterone–5α-dihydrostestosterone. The TLC method was based on the addition of bromine across the double bond of the steroids while on the plate. The steroids were detected by spraying the plates with ethanol–H_2SO_4 (1:1) and heating them at 110°C for 10–20 min.

Brown et al. (1988) separated 12 androgens on silica gel 60 plates using a single development in the mobile phase methylene chloride–methanol–water (225:15:1.5). Better separation of these androgens was obtained on silica gel plates developed twice in the same direction using the mobile phase light petroleum–diethyl ether–acetic acid (48:50:2). Van Look et al. (1989) separated 12 related testosterone compounds on 10 × 10-cm silica gel plates in a mobile phase of either chloroform–acetone (9:1) or cyclohexane–ethyl acetate–ethanol (77.5:20:2.5).

Androgens may be present in meat as additives to improve weight gain in cattle. Garcia et al. (1991) described a method for screening meat for residues of hormonal growth promoters. Five grams of meat were extracted in acetonitrile–diethyl ether–HCl (1:1:1). The extract was evaporated, dissolved in chloroform, dried, and fractionated on a silica gel column eluted with hexane–ethyl acetate (1:1). The eluate was evaporated, dissolved in chloroform, and chromatographed on a silica gel 60 plate, first with chloroform, and then, after drying, in the same direction with hexane–ethyl acetate (95:5). The impurities were scraped off the bottom of the plate, and the plate was developed again with cyclohexane–ethyl acetate (1:1) or benzene-ethyl acetate (7:1). Plates were sprayed with 20% ethanolic-H_2SO_4, and anabolic hormones were detected by spot colors under daylight and 366-nm ultraviolet light. Diethylstilbestrol, progesterone, testosterone, trenbolone, and zeranol could be detected, and testosterone, trenbolone, and zeranol could be quantified by densitometry at 265, 366, and 313 nm, respectively.

Srinivas and Srinivasulu (1993) found that heulandate layers developed with carbon disulfide–pyridine (1:1) were effective for the separation of steroid hormones,—dehydroepiandrosterone could easily be separated from mixtures of cholesterol and estradiol benzoate or testosterone phenylpropionate, and testosterone could be separated from mixtures of cholesterol and testosterone phenylpropionate or estradiol benzoate. Likewise, estradiol could easily be separated from mixtures of cholesterol and estradiol benzoate. Daeseleire et al. (1994) applied HPTLC and GC/MS for the detection of anabolic steroids used as growth promotors in illicit cattle fattening within the European market. Agrawal et al. (1995) developed a sensitive, reliable, and rapid silica gel TLC method for the separation, identification, and quantification of stereospecific androgen metabolites.

III. TLC OF ESTROGENS

TLC has been used for the analysis of natural and synthetic estrogens in pharmaceutical preparations. Considerable TLC literature is available on separations involving the three major human estrogens, estrone, estriol, and estradiol-17β.

Estrogenic steroids, particularly the 2-hydroxy derivatives, are labile, and methods are needed for their protection during chromatography. A good method of protection is one in which thin layers of silica are impregnated with a solution of ascorbic acid. The ascorbic acid stabilizes the estrogens and does not interfere with chromatography. For details of this method and for procedures on extraction and sample preparation of estrogens, consult Touchstone (1986).

Estrogens are present both free and conjugated with glucuronic and sulfuric acids. Conjugated steroids can be hydrolyzed by acid or enzymes. Acid hydrolysis of estrogen conjugates yields free estrone, estriol, and estradiol. Estrogens may be present in meat as additives to improve weight gain in cattle, and TLC methods are available for their analysis (Wortberg et al., 1978).

TLC is used for the analysis of both free and conjugated estrogens. Rojkowski and Broadheat (1972) separated estrogens and their derivatives on silica gel GF$_{254}$ (Merck) with a mobile phase of benzene–ethyl acetate (50:10); the spots were detected by heating the plates at 70°C for 30 min after spraying them with 50% aqueous sulfuric acid. They also reported the mobilities of estrone and estradiol-17β in various mobile phases. Penzes and Oertel (1972) reported the R_F values of free estrogens and dansyl derivatives on silica gel G in various mobile phases; detection was by fluorescence. Additional representative TLC separations of estrogens include separation of estrones and their derivatives (Petrovic and Petrovic, 1976); separation of estradiol and diethylstibestrol and their iodinated derivatives (Mansinger et al., 1977); and separation of estrogen conjugates on reversed-phase C$_{18}$ (Merck) plates (Rao et al., 1978). Tsang (1984) used two solvent systems for multiple development in the same direction to give good separation of estrone sulfate, 17-estradiol 3-sulfate, and 17-estradiol 17-sulfate on silica gel 60F$_{254}$ plates. Solvent system 1 consisted of ethanol–chloroform–ammonia (45:90:1.8), and solvent system 2 contained ethanol–ethyl acetate–ammonia (45:90:1.8).

Watkins and Smith (1986) separated various conjugated estrogens using four different chromatographic systems. Good separations of these compounds were achieved when C$_{18}$ layers and a mobile phase of absolute methanol–0.5% tetramethyl ammonium chloride (1:1) were used. The R_F values of the conjugated estrogens in this system were as follows: estrone-3-sulfate, 0.03; estradiol-17β-glucuronide, 0.11; estrone-b-glucuronide, 0.24; estiol-16a-glucuronide, 0.32; estriol-3-sulfate, 0.48; and estriol-3-glucuronide, 0.66.

Van Look et al. (1989) separated seven estrogens on 10 × 10-cm silica gel 60 plates with a mobile phase of cyclohexane–ethyl acetate–ethanol (77.5:20:2.5). The R_F values of the separated compounds were as follows: ethynylestradiol, 0.23; estradiol, 0.18; estradiol benzoate, 0.34; estradiol valerate, 0.42; estradiol phenylpropionate, 0.40; trenbolone acetate, 0.32; and estradiol cypionate, 0.47.

Lamparczyk et al. (1989) separated three estrogens on HPTLC–C$_{18}$, Merck (10 × 10-cm) plates using absolute alcohols as the mobile phases. R_F values with ethanol were 0.83 for estriol, 0.77 for estradiol, and 0.74 for estrone; values for the corresponding estrogens using butanol as the mobile phase were 0.92, 0.85, and 0.83, respectively.

Experiment 1 in this chapter is concerned with the TLC separation of estrogens.

IV. TLC OF CORTICOIDS AND PREGNANES

Corticoids are steroids from the cortex of the adrenal gland and other natural or synthetic compounds having similar activity. They are C_{21} substances differing mainly in the number and location of the oxygen function. Representative corticoids are cortisone, cortisol, tetrahydrocortisone, corticosterone, and aldosterone. Pregnanes are C_{21} steroids present in pregnancy urine. A representative pregnane, 5β-pregnane-3α, 20α-diol (pregnanediol), is a metabolite of progesterone. A related compound, 3,2α-preganedione, is present in the pregnancy urine of mares. Some pregnanes can be isolated from the cortex of the adrenal gland. Corticoids are usually extracted from plasma, urine, or rat adrenal glands. For details on the sample preparation of corticoids and pregnanes, consult Touchstone (1986).

TLC can be used to separate numerous corticoids and other C_{21} steroids without derivatization. Sample preparation prior to TLC is very important, particularly if the corticoid is from a biological matrix.

Smith and Hall (1974) listed the R_F values of 19 corticoids separated by TLC on silica gel in six different mobile phases. Gleispach (1977) used TLC to resolve 10 natural and synthetic corticoids on silica gel and listed R_F values of these steroids following their separation in 22 different mobile phases. Byrne et al. (1977) separated 17 natural and synthetic corticoids by TLC on silica gel 60 in a mobile phase of either methylene chloride–dioxane–water (120:30:50) or methylene chloride–dioxane–water (100:50:50). Bailey et al. (1978) listed R_F values for numerous corticoid acetates separated by TLC on silica gel GF$_{255}$ in ethylene dichloride–methanol–water (95:5:0.2). Herkner et al. (1978) separated numerous corticoids and related steroids by TLC on silica gel G in the mobile phase cyclohexane–ethyl acetate (20:80) and obtained the following R_F values: aldosterone, 0.08; cortisol, 0.23; corticosterone, 0.24; cortisone, 0.28; 11-deoxycortisol, 0.48; deoxycorticosterone, 0.55; testosterone, 0.61; 17α-hydroxy-pregnenolone, 0.75; androstenedione, 0.76; dehydropiandrosterone, 0.77; 17α-hydroxy-progesterone, 0.80; pregnenolone, 0.83; and progesterone, 0.84.

Gleispach (1977) separated closely related pregnanes on silica gel GF$_{254}$ developed five times in petroleum ether–diethyl ether (40:60). Hara and Mibe (1975) gave R_F values of numerous pregnanes separated in 11 mobile phases on both silica gel and alumina layers. Yamaguchi and Hayashi (1976) separated 15 pregnanes and androgens on silica gel GF$_{254}$ using either the mobile phase chloroform–methanol (9 : 1) or ethyl acetate–benzene (1 : 1); spots were detected by fluorescence quenching.

Brown et al. (1988) reported R_F values for nine common corticosteroids separated on silica gel 60 plates with the mobile phase methylene chloride–methanol–water (225:15:1.5). Van Look et al. (1989) listed the R_F values of six common pregnanes separated on silica gel 60 plates (10 × 10 cm) with the mobile phase chloroform–acetone (9:1). Singh et al. (1989) chromatographed 14 different pregnanes and corticoids on silica gel plates using seven different mobile phases. R_F values of all of these steroids were given in five of the seven solvent systems. Armstrong et al. (1988) used reversed-phase C_{18}, $KC_{18}F$ Whatman (20 × 20-cm) plates to chromatograph numerous isomeric pregnenes. The mobile phase was acetonitrile–water (30:70 or 35:65) plus 0.3 M hydroxypropyl-β-cyclodextrin. The R_F values of the isomeric pregnenes were reported.

Al-Habet and Rogers (1989) separated eight pregnanes and corticoids on silica gel plates (10 × 20 cm) using the mobile phase chloroform–ethanol–water (90:10:2). The R_F values of these steroids were as follows: progesterone, 0.77; cortisone, 0.45; prednisone, 0.43; dexamethasone, 0.32; cortisol, 0.30; methylprednisolone, 0.27; prednisolone, 0.25; and methylprednisolone sodium succinate, 0.07. Gaignage et al. (1989) separated dexamethasone and its derivatives on silica gel G-60 plates (Merck, 20 × 20 cm) using three different mobile phases. R_F values of the dexamethasone derivatives were reported for each mobile phase. Most of the derivatives were separated in the mobile phase chloroform–ethanol (9:1).

Touchstone and Fang (1991) described methods for fluorometric labeling of tetrahydroprogesterones. 7-Diethylamino coumarin-3-carbohydrazide was used to label the ketone group of the tetrahydroprogesterones to form fluorescent derivatives that could be detected with high sensitivity. The isomeric 3-hydroxypregnanes separated readily by HPTLC after derivatization. This separation was not possible using the underivatized isomers.

Hoebus et al. (1993) used silica gel TLC with various mobile phases to identify corticosteroid hormones available on the European market. Detection was by fluorescence after the plates were sprayed with alcoholic sulfuric acid and heated. Datta and Das (1994) identified and quantified corticosteroids and their esters in pharmaceutical preparations of creams and ointments. The preparations were dissolved in chloroform, centrifuged to remove water and insoluble material, and silica gel TLC with hexane mobile phase was used to wash out base ingredients followed by chloroform–ethyl acetate (1:1) for free steroids or (2:1) for esters and spectrodensitometry at 240 nm; recoveries were 99 ± 1% and the agreement of results with the official methods used was excellent.

V. TLC OF BILE ACIDS

TLC has been used to separate free bile acids and their derivatives. With the use of commercial reversed-phase layers, convenient methods are now available for the resolution of conjugated bile acids (Touchstone, 1986).

Goswami and Frey (1974) reported R_F values for 10 bile acids separated on silica gel in 15 mobile phases. Kim and Kritchevsky (1976) used seven different mobile phases to separate lithocholic, deoxycholic, chenodeoxycholic, and cholic acids and their glycine and taurine conjugates on silica gel G layers. Ikawa and Goto (1975) devised two-dimensional thin-layer chromatography methods for the separation of bile acids and salts from the human gallbladder. Ikawa (1976) purified methyl esters of bile acids by preparative TLC on silica gel-sintered plates.

Rocchi and Salivioli (1976) separated various bile acids on silica gel G with a mobile phase consisting of either isoamyl acetate–proprionic acid–propanol–water (4:3:2:1) or butanol–acetic acid–water (10:1:1). Jirsa and Soldatova (1978) described a specific color reaction for the detection of chenodeoxycholic acid on TLC plates. When plates were sprayed with acetic anhydride containing 10% H_2SO_4, chenodeoxycholic acid gave a specific red color; deoxycholic and cholic acids gave yellow spots and lithocholic acid was not colored. Taylor et al. (1979) reviewed the TLC procedures used to separate free and conjugated forms of lithocholic acid from bile mixtures. They described a novel chromatographic separation of lithocholic acid from a mixture of four unconjugated bile acids. The system consisted of Gelman ITLC-type silica gel chromatography sheets. Seprachrom miniature chromatography chambers, an isooctane–isopropyl ether–acetic acid (2:1:0.4) solvent system, and a 5% H_2SO_4 in methanol detection spray. Raedsch et al. (1979) separated individual monosulfated primary bile acid conjugates by reversed-phase partition TLC on octadecyl-bonded silica gel plates. The solvent system was acetonitrile containing calcium carbonate. They reported excellent separation of the 3- and 7-monosulfated glycine conjugates as well as 3- and 7-monosulfated taurine conjugates of cholic and chenodeoxycholic acids.

Touchstone et al. (1980) separated conjugated bile acids on reversed-phase layers (Whatman LKC_{18}) using the mobile phase chloroform–0.3 M $CaCl_2$ (1:1). Spots were detected by spraying the plates with 10% H_2SO_4 in ethanol and then heating the layers at 170°C for 2 min and observing fluorescence. The R_F values of the acids were as follows: taurocholic acid, 0.38; glycocholic acid, 0.31; taurochenodeoxycholic acid, 0.26; taurodeoxycholic acid, 0.23; glycochenodeoxycholic acid, 0.20; glycodeoxycholic acid, 0.17; and glycolithocholic acid, 0.10. Detailed Experiment 2 in this chapter is a modification of the Touchstone et al. (1980) study.

Szepesi et al. (1982) designed a TLC method for the isolation and determination of chenodeoxycholic acid and related compounds using silica gel 60 plates with the mobile phase chloroform–ethyl acetate–acetic acid–2 methoxyethanol (9:9:2:1). Van den Ende et al. (1983) combined TLC and in situ fluorometry to determine 0.05–0.1 µg of free bile acids and bile acids conjugated with glycine or taurine.

During the 1980s, sophisticated detection systems were used for the TLC analysis of bile acids including direct scanning fluorometry and liquid secondary–ion mass spectrometry. As an example of the latter technique (use of direct positive and negative ion detection), Dunphy and Busch (1990) separated cholic, chenodeoxycholic, and lithocholic acids on aluminum sheets precoated with silica gel and the mobile phase isooctane–ethyl acetate–acetic acid (10:10:2).

Iida et al. (1986) separated numerous 5α and 5β bile acid methyl esters using both normal- and reversed-phase TLC. Merck (20 × 20 cm) silica gel 60 plates were used for the normal-phase TLC and good separations were achieved in the mobile phase hexane–ethyl acetate (83:17); for the reversed-phase separations, RP-18 Merck (10 × 10 cm) plates were used, and the plates were developed in a methanol–acetonitrile–water–formic acid (47.5:47.5:5.5:0.5) solvent system.

Busch (1987) described an instrument for the direct two-dimensional analysis of developed TLC plates with a liquid-state secondary-ion mass spectrometry (LSIMS) source. According to Dunphy and Busch (1990), this method has proved effective in providing qualitative and semiquantitative information about mixtures of bile acids.

Lamparczyk (1992) provided detailed methods for the sample preparation and TLC determination of fecal bile acids (cholic, chenodeoxycholic, deoxycholic, lithocholic, and ursideoxycholic acids). After the sample is cleaned up by liquid–liquid extraction, from 2 to 20 μl of purified fecal bile solution (no more than 20 μg of bile acids) is spotted in the preadsorbent zone of a 20 × 20-cm silica gel plate. The plate is developed in isooctane–2-propanol–acetic acid (30:10:1) for 40 min, dried, and developed again with isooctane–ethyl acetate–acetic acid (10:10:2) for 65 min. For quantitative analysis, the plate is dipped for 2 sec into a 0.2%, 2,7-dichlorofluorscein ethanol solution. Bile acid fluorescence is measured with a TLC scanner and the results calculated by the peak-height method. Rivas-Nass (1994) evaluated the influence of temperature, ionic strength, mobile-phase pH, and addition of modifiers to the solvent system on the silica gel TLC of bile acids.

VI. TLC OF ECDYSTEROIDS

The ecdysteroids comprise a group of polyhydroxylated steroids related in structure to ecdysone (Lamparczyk, 1992). They are typically polar, involatile, and thermally unstable. For a list of the trivial and systematic names of the more common ecdysteroids, see Lamparczyk (1992). Ecdysteroids are the molting hormones of arthropods, mainly crustaceans and insects. Particular glands in both insects and crustaceans are responsible for the production of ecdysteroid hormones. There are about 20 ecdysteroids (zooecdysteroids) known mainly from insects and crustaceans. A greater array of ecdysteroids (over 100) have been isolated from plants (phytoecdysteroids) and presumably some of these compounds provide defense mechanisms against phytophagous insects (Lamparczyk, 1992). Sample preparation is very important in ecdysteroid analysis because these compounds may be unstable, present in low concentrations, and may be closely related to other compounds from the source. Numerous procedures for the extraction and sample preparation of both phytoecdysones and zooecdysones have been given by Touchstone (1986).

Morgan and Poole (1976) reviewed the early work on insect ecdysteroids, including methods of extraction of ecdysones from insect tissue as well as solvent systems useful for TLC studies on silica gel. Ruh and Black (1976) separated ecdysone from 20-hydroxyecdysone on Whatman LK5F silica gel plates with the mobile phase chloroform–methanol (4:1); the R_F values for ecdysone and 20-hydroxyecdysone were 0.68 and 0.60, respectively. Spots were detected by fluorescence quenching or by spraying the plates with either vanillin–H_2SO_4 or anisaldehyde. Koolman et al. (1979) separated ecdysone metabolites from the blowfly, *Calliphora*, by TLC on silica gel with chloroform–methanol (80:20) mobile phase. The R_F values of the steroids were: ecdysone, 0.27; ecdysterone, 0.21; inokesterone, 0.19; 26-hydroxyecdysone, 0.15; and 20,26-dihydroxyecdysone, 0.11.

Wilson et al. (1981) reported that the recovery of ecdysones could be improved by the use of reversed-phase TLC plates (C_2 C_8 and C_{18}) Dinan et al. (1981) separated ecdysteroids on silica gel GF_{254} with continuous development for 2 h in the mobile phase chloroform–ethanol (9:2); the R_F values of the separated compounds were as follows: 3-dehydroecdysone, 0.88; 3-dehydro-20-hydroxyecdysone, 0.77; 3 epiecdysone, 0.53; ecdysone, 0.45, 3-epi-20 hydroxyecdysone, 0.34; and 20-hydroxyecdysone, 0.28. Wilson et al. (1981) gave the R_F values of numerous ecdysteroids separated on three types (C_2, C_8, and C_{18}) of reversed-phase TLC plates; mobile phases consisted of either 2-propanol–water or acetonitrile–water.

Wilson et al. (1982) found that C_{12}-bonded plates and paraffin-coated silica gel plates enhanced the resolution and recovery of ecdysones. Wilson (1985) found that C_{12} and C_{18} nonhydrophobic TLC plates gave better resolution than the more hydrophobic C_{18} reversed-phase HPTLC plates. Wilson and Lafont (1986) used TLC to examine the metabolites of [^{3}H] 20-hydroxyecdysone in the feces of *Locusta migratoria* (locusts) and noted that the metabolites could be resolved using two-dimensional chromatography with two solvent systems.

Dinan (1987) gave R_F values of 27 different ecdysone acyl esters and their 2,3-acetonide derivatives following separation on HPTLC silica gel 60 plates (Merck, 10 × 10 cm) in chloroform–ethanol (9:1) mobile phase; the R_F values of 22 of these compounds were obtained following separation on HPTLC RP-18 plates (Merck) in the mobile phase absolute methanol. Wilson and Lewis (1987) separated ecdysteroids on silica gel HPTLC plates (Merck), a single development in the mobile phase dichloromethane–methanol (80:20). The R_F values of the ecdysteroids were: makisterone A, 0.23; ecdysone, 0.26; polypodine B, 0.27; ponasterone C, 0.40; ponasterone A, 0.44; and polypodine B 2-cinnamate, 0.73. The authors also used multiple development with methylene chloride–methanol (92.5 : 7.5) to enhance the separation of some of these ecdysteroids. Whiting and Dinan (1988) separated ecdysteroid conjugates from extracts of the eggs of *Acheta domesticus* (crickets) by TLC. They also used this approach for the preliminary identification of ecdysteroid conjugates that were then purified by HPLC and identified by mass spectrometry. Wilson et al. (1988) observed that the presence of labeled nonpolar fractions are more obvious after radioscanning of developed plates than if samples are run on HPLC columns used for the separation of the polar metabolites. Wilps and Zoller (1989) used TLC to separate ecdysteroids from ovarian tissue and hemolymph of *Phormia terraenovae* (blowflies) in studies on ecdysteroid production during development.

Wilson (1992) described the use of various boronic acids for the derivatization and separation of ecdysteroids by normal- and reversed-phase TLC. The derivatives were prepared from butyl-, phenyl-, and aminophenylboronic acids by over-spotting a methanolic solution of the reagent on to the applied ecdysteroids at the origin of the TLC plate. Butylboronic acid was also used as a mobile-phase additive to modify chromatographic behavior. A 20,22-diol group was an essential structural requirement for the formation of a cyclic boronate of an ecdysteroid. These procedures allowed for the selective separations of such compounds as ponasterone A and 2-deoxyecdysone. Pis and Harmatha (1992) also described the use of phenylboronic acid as a versatile derivatization agent for TLC studies of ecysteroids.

Fell (1996) has provided useful information on analytical techniques and approaches for the separation of ecdysteroids by TLC. He noted that ecdysteroid separation can be achieved on silica gel TLC plates, HPTLC plates, RP-TLC and RP-HPTLC plates. Plates with a fluorescent indicator (F_{254}) allow for ecdysteroid detection visually by UVC (fluorescence quenching). Paraffin-coated plates can be used (by developing plates in dichloromethane containing 5 to 7.5% paraffin oil) to improve the separation and recovery of ecdysteroids. For details on the preparation of these plates, see Fell (1996). For a practical approach to the analysis of ecdysteroids by TLC including methods of sample preparation, choice of chromatographic systems (i.e., combination of plates and solvent systems) and methods of detection, see Fell (1996).

VII. TERPINOIDS: GENERAL

Croteau and Ronald (1983) and Coscia (1984) reviewed chromatographic studies on terpinoids. Those authors considered terpinoids as ubiquitous, structurally diverse, lipophilic compounds. Terpinoids exist in various polymeric forms of the unsaturated branch-chain pentane, isoprenol. Terpenes are characterized as hemiprenes (C_5), monoterpenes (C_{10}), sesquiterpenes (C_{15}), diterpenes (C_{20}), and triterpenes (C_{30}). A representative hemiprene (hemiterpene) is mevalonic acid, a precursor in the biosynthesis of isoprenoids. Monoterpenes contain numerous

essential oils, such as terpineol, used as a clearing agent in the histological preparation of biological specimens. Sesquiterpenes contain numerous plant hormones
such as abscissic acid that are structurally similar to juvenile hormones in insects.
The diterpenes contain gibberellins, which are active as plant hormones.
Triterpenes consist of substances such as squalane and squalene, precursors involved in sterol synthesis. Tetraterpenes consist of carotenoids and chloroplast
pigments. They are considered in Chapter 18. Vitamins D, E, and K are terpinoids
and are considered in Chapter 19.

VIII. TLC OF TERPINOIDS

TLC of terpinoids has been reviewed by Croteau and Ronald (1983) and Coscia
(1984). The latter author has provided 36 tables of TLC separations of terpinoids,
reagents used to detect terpinoids on TLC plates, and detailed methods for sample
preparation of the diverse groups of terpinoids.

Hemiterpenes, monoterpenes, sesquiterpenes, diterpenes, and triterpenes
are all suitable for separation by TLC. Representative separations are as follows:
Cardemil et al. (1974) separated hemiterpene alcohols on silica gel impregnated
with 25% $AgNO_3$ and ethyl acetate as the mobile phase. Battaile et al. (1961)
separated monoterpenes on silica gel using various combinations of hexane–ethyl
acetate as developing solvents. Croteau and Ronald (1983) discussed the application of silica gel and silica gel layers impregnated with $AgNO_3$ to separate sesquiterpenes. Robinson and West (1970) used argentation TLC on silica gel to separate diterpenes in hexane–benzene (7:3). Argentation TLC was also used to
separate triterpenes (Ikan, 1965).

Terpinoids can be detected on TLC plates using iodine, concentrated
H_2SO_4, phosphomolybdic acid, and fluorescein dyes. As mentioned previously,
Coscia (1984) has listed extensive detection procedures for terpinoids.

Koops et al. (1991) used TLC as a cleanup step prior to the identification
of various terpinoids by gas chromatography–mass spectrometry from seedlings
of *Euphorbia lathyris*. The plant latex consisted mainly of triterpene alcohols,
whereas the epicuticular wax contained mainly triterpene ketones.

Kohli and Kumar (1991) reported that silica gel impregnated with chloro-
platinic acid (H_2PtCl_6) was superior to plates impregnated with silver nitrate
($AgNO_3$) for the separation of certain terpinoids by TLC.

Nikolova-Damyanova et al. (1992) described methods for the analysis of
terpinoids by TLC from extracts of various algae and invertebrates from the Black
Sea.

Fell (1996) noted that the primary use of TLC for terpinoids in entomologi-
cal studies has been for work on juvenile hormone biosynthesis and metabolism.
Because juvenile hormones are sesquiterpinoid compounds, his review was limited mainly to the TLC analysis of those compounds. He presented several exam-

ples of the separation of sesquiterpinoids by TLC on silica gel and gave details on mobile phases, detection reagents, and quantification of these compounds.

IX. DETAILED EXPERIMENTS

A. Experiment 1. Separation of Estrogens on Whatman HPTLC Silica Gel Plates

1. Introduction

This experiment (in two parts, A and B) is an introduction to the TLC separation of steroids on high-performance TLC plates. It is adapted from Whatman's TLC Technical Series, Volume 2. In A, estriol, estradiol, and estrone are separated. In B, estriol, estrone, 17-hydroxy-progesterone, pregnenolone, and progesterone are separated.

2. Preparation of Standards

Prepare solutions containing 0.01–0.1 mg/ml of each compound (Sigma) (estriol, estradiol, estrone, 17-hydroxyprogesterone, pregnenolone, and progesterone) in chloroform or chloroform–methanol (9:1).

3. Layer

Whatman LHP-K silica gel, 10 × 10 cm, used as received.

4. Application of Initial Zones

In A, spot 5 µl of estriol, estradiol, and estrone on separate origins in the preadsorbent area of a plate. Also, spot a mixture of the three steroids on a separate origin. In B, spot 5 µl of estriol, estrone, 17-hydroxyprogesterone, pregnenolone, and progesterone on separate origins in the preadsorbent area of another plate; also, spot a mixture of the five steroids on a separate origin.

5. Mobile Phase

In A, use methylene chloride–methanol (92:8). In B, use chloroform–acetone (90 : 10).

6. Development

Develop the plates 6–8 cm in paper-lined, equilibrated (10 min) chambers. Time of development should be 8–17 min.

7. Detection

Spray the plates with concentrated ethanol–H_2SO_4 (92 : 8) and heat them at 110°C for 5 min; examine the plates under long-wave ultraviolet (UV) light. The sensitivity of detection is 0.01 µg.

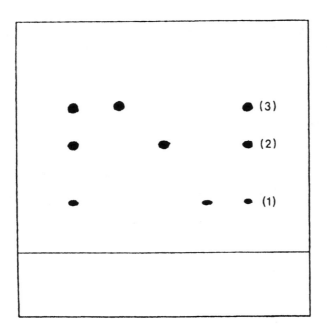

FIGURE 21.1   Separation of estriol (1), estradiol (2), and estrone (3) on a Whatman LHP-K plate in methylene chloride–methanol (92 : 8). Photograph reproduced from Whatman's TLC Technical Series, Volume 2 with permission.

8.   Results and Discussion

The approximate R_F values of the compounds in A should be as follows: estriol, 0.25; estradiol, 0.51; and estrone, 0.68 (Figure 21.1). In B, the approximate R_F values of the compounds should be as follows: estriol, 0.03; estrone, 0.57; 17-hydroxyprogesterone, 0.38; pregnenolone, 0.45; and progesterone, 0.65 (Figure 21.2).

B.   Experiment 2. Separation of Conjugated Bile Acids on Whatman KC$_{18}$ Reversed-Phase Plates

1.   Introduction

This experiment, adapted from Whatman's TLC Technical Series, Volume 1, and Touchstone et al. (1980), is an introduction to TLC separation of bile acids on reversed-phase TLC plates.

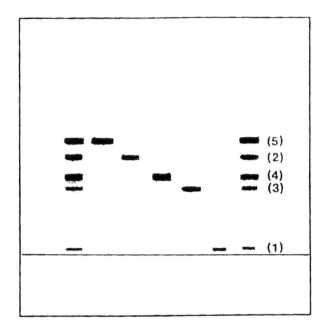

FIGURE 21.2 Separation of estriol (1), estrone (2), 17-hydroxyprogesterone (3), pregnenolone (4), and progesterone (5), on a Whatman LHP-K plate in chloroform–acetone (90 : 10). Photograph reproduced from Whatman's TLC Technical Series, Volume 2 with permission.

2. Preparation of Standards

Prepare 100 ng/µl solutions of taurocholic acid, glycocholic acid, taurochenodeoxycholic acid, taurodeoxycholic acid, glycochenodeoxycholic acid, glycodeoxycholic acid, and glycolithocholic acid (Sigma) in methanol.

3. Layer

Whatman KC$_{18}$F reversed-phase plate, 20 × 20 cm, used as received.

4. Application of Initial Zones

Apply 2.5 µl of each bile acid to separate origins of the plate.

5. Mobile Phase

Ethanol–0.3 M calcium chloride (pH 3.0)–dimethyl–sulfoxide (25 : 25 : 2).

6. Development

Develop to a distance of 17 cm from the origin in a paper-lined chamber; approximate time of development is 3 h.

7. Detection

Dry the plate at 170°C, and spray it lightly with ethanol–H_2SO_4 (90 : 10); heat the plate in an oven at 180°C for 2 min to produce fluorescence.

8. Results and Discussion

The approximate R_F values and order of migration should be as follows: glycolithocholic acid, 0.10; glycodeoxycholic acid, 0.17; glycochenodeoxycholic acid, 0.20; taurodeoxycholic acid, 0.23; taurochenodeoxycholic acid, 0.26; glycocholic acid, 0.31; and taurocholic acid, 0.38.

REFERENCES

Agrawal, A. K., Pampori, N. A., and Shapiro, B. H. (1995). Thin-layer chromatographic separation of regioselective and stereospecific androgen metabolites. *Anal. Biochem.* 224: 455–457.

Al-Habet, M. H., and Rogers, H. J. (1989). Two chromatographic methods for the determination of corticosteroids in human biological fluids: pharmacokinetic applications. *J. Pharm. Sci.* 78: 660–666.

Armstrong, D. W., Faulkner, J. R., Jr., and Han, S. M. (1988). Use of hydroxypropyl- and hydroxyethyl-derivatized beta-cyclodextrins for the thin-layer chromatographic separation of enatiomers and diastereomers. *J. Chromatogr.* 452: 323–330.

Bailey, K., By, A. W., and Lodge, B. A. (1978). Description and chromatographic investigation of Mexican medication for arthritis and asthma, including an unusual corticosteroid. *J. Chromatogr.* 166: 299–304.

Bajaj, K. L., and Ahuja, J. L. (1979). General spray reagent for the detection of steroids on thin-layer plates. *J. Chromatogr.* 172: 417–419.

Battaile, J., Dunning, R. L., and Loomis, W. D. (1961). Biosynthesis of terpenes. I. Chromatography of peppermint oil terpenes. *Biochim. Biophys. Acta* 51: 538–544.

Brown, J. W., Carballeira, A., and Fishman, L. M. (1988). Rapid two-dimensional thin-layer chromatographic system for the separation of multiple steroids of secretory and neuroendocrine interest. *J. Chromatogr.* 439: 441–447.

Busch, K. L. (1987). Direct analysis of thin-layer chromatograms and electrophoretograms by secondary ion mass spectrometry. *Trends Anal. Chem.* 6: 95–100.

Byrne, J. A., Broun, I. K., Chaubal, M. G., and Malone, M. H. (1977). Thin-layer chromatographic screening procedure for some drugs used in treatment of arthritis: "Arthritic cures" prescribed in Mexico. *J. Chromatogr.* 137: 489–492.

Cardemil, E., Vicuna, J. R., Jabalquinto, A. M., and Cori, O. (1974). Separation of isoprenoid alcohols and aldehydes by thin-layer chromatography. *Anal. Biochem.* 59: 636–638.

Coscia, C. J. (Ed.). (1984). *CRC Handbook of Chromatography—Terpinoids*, Vol. 1. CRC Press, Boca Raton, FL, p. 183.

Croteau, R., and Ronald, R. C. (1983). Terpinoids. In *Chromatography; Fundamentals and Applications of Chromatographic and Electrophoretic Methods, Part B*: Applications, E. Heftmann (Ed.). Elsevier, Amsterdam, pp. B149–B189.

Daeseleire, E., Vanoosthuyze, K., and Van Peteghem, C. (1994). Application of high-performance thin-layer chromatography and gas chromatography mass spectrometry to the detection of new anabolic steroids used as growth promoters in cattle farming. *J. Chromatogr. 674*: 247–253.

Datta, K., and Das, S. K. (1994). Identification and quantitation of corticosteroids and their esters in pharmaceutical preparations of creams and ointments by thin-layer chromatography and densitometry. *J. AOAC Int. 77*: 1435–1438.

Dinan, L. (1987). Thin-layer chromatography of ecdysone acyl esters and their 2,3-acetonide derivatives. *J. Chromatogr. 411*: 379–392.

Dinan, L. N., Donnakey, P. L., Rees, H. H., and Goodwin, T. W. (1981). High-performance liquid chromatography of ecdysteroids and their 3-EPI, 3-dehydro and 26-hydroxy derivatives. *J. Chromatogr. 205*: 139–145.

Dreassi, E., Ceramelli, G., and Corti, P. (1996). Thin-layer chromatography in pharmaceutical analysis. In *Practical Thin-Layer Chromatography—A Multidisciplinary Approach*, B. Fried and J. Sherma (Eds.). CRC Press, Boca Raton, FL, pp. 231–247.

Dunphy, J. C., and Busch, K. L. (1990). Thin-layer chromatography/liquid secondary-ion mass spectrometry in the determination of major bile salts in dog bile. *Talanta 37*: 471–480.

Fell (1996). Thin-layer chromatography in the study of entomology. In *Practical Thin-Layer Chromatography—A Multidisciplinary Approach*, B. Fried and J. Sherma (Eds.). CRC Press, Boca Raton, FL, pp. 71–104.

Gaignage, P., Logray, G., Marlier, M., and Dreze, P. (1989). Applications of chromatographic and spectrometric techniques to the study of dexamethasone related compounds. *Chromatographia 28*: 623–630.

Garcia, G., Saelzer, R., and Vega, M. (1991). Screening of meat for residues of hormonal anabolic growth promoters. *J. Planar Chromatogr.—Mod. TLC 4*: 223–225.

Gleispach, H. (1977). Dunnschicht-Chromatographische Probleme bei der Cortisalbestimmung mittels Protein-bindingstechnik. *Chromatographia 10*: 40–44.

Goswami, S. K., and Frey, C. (1974). A novel method for the separation and identification of bile acids and phospholipids of bile on thin-layer chromatograms. *J. Chromatogr. 89*: 87–91.

Hara, S., and Mibe, K. (1975). Systematic analysis of steroids. Solvent selectivity of the mobile phase in thin-layer chromatography in relation to the mobility and the structure of steroidal pharmaceuticals. *Chem. Pharm. Bull. 23*: 2850–2859.

Heftmann, E. (1983). Steroids. In *Chromatography—Fundamentals and Application of Chromatographic and Electrophoretic Methods, Part B: Applications*, E. Heftmann (Ed.). Elsevier, Amsterdam, pp. B191–B222.

Herkner, K., Nowotny, P., and Waldhausel, W. (1978). Die Bestimmung von Aldosteron in Horn und Plasma. *J. Chromatogr. 146*: 273–281.

Hoebus, J., Daneels, E., Roets, E., and Hoogmartens, J. (1993). Identification of corticosteroid hormones by thin-layer chromatography. *J. Planar Chromatogr.—Mod. TLC 6*: 269–273.

Iida, T., Momose, T., Shinohara, T., Goto, J., Nambara, T., and Chang, F. C. (1986). Separation of allo bile acid stereoisomers by thin-layer and high-performance liquid chromatography. *J. Chromatogr. 366*: 396–402.

Ikan, R. (1965). Thin-layer chromatography of tetra-cyclic triterpenes on silica impregnated with silver nitrate. *J. Chromatogr. 17*: 591–593.

Ikawa, S. (1976). Purification of methylated bile acids recovered from thin-layer chromatograms on silica gel-sintered plates. *J. Chromatogr. 117*: 227–231.

Ikawa, S., and Goto, M. (1975). Two-dimensional thin-layer chromatography of bile acids. *J. Chromatogr. 114*: 237–243.

Jain, R. (1996). Thin-layer chromatography in clinical chemistry. In *Practical Thin-Layer Chromatography—A Multidisciplinary Approach*, B. Fried and J. Sherma (Eds.). CRC Press, Boca Raton, FL, pp. 131–152.

Jirsa, M., and Soldatova, H. (1978). A specific reaction for chenodeoxycholic acid. *J. Chromatogr. 157*: 449–450.

Kerebel, A., Moffin, R. F., Berthou, F. L., Picart, D., Bardou, L. G., and Floch, H. H. (1977). Analysis of $C_{19}O_3$ steroids by thin-layer and gas-liquid chromatography and mass spectrometry. *J. Chromatogr. 140*: 229–244.

Kim, H. K., and Kritchevsky, D. (1976). Quantitation of bile acids and bile salts by a thin-layer chromatographic charring method. *J. Chromatogr. 117*: 222–226.

Kohli, J. C., and Kumar, S. (1991). Novel procedure for the thin-layer chromatography of terpinoids on platinum ion-silica gel layers. *Natl. Acad. Sci. Lett. (India). 14*: 325–329.

Koolman, J., Reum, L., and Karlson, P. (1979). 26-Hydroxyecdysone, 20,26-dihydroxyecdysone, and inokesterone detected as metabolites of ecdysone in the blowfly *Calliphora vicina* by radiotracer experiments. *Z. Physiol. Chem. 360*: 1351–1355.

Koops, A. J., Baas, W. J., and Groeneveld, H. W. (1991). The composition of phytosterols, latex triterpenols, and wax triterpinoids in the seedling of *Euphorbia lathyris* L. *Plant Sci. 74*: 185–192.

Lamparczyk, H. (1992). *CRC Handbook of Chromatography—Analysis and Characterization of Steroids*. CRC Press, Boca Raton, FL.

Lamparczyk, H., Ochocka, R. Jr., and Zarzycki, P. (1989). Evidence for electrostatic interactions of steroids in thin-layer chromatography. *Chromatographia 27*: 565–568.

Lisboa, B. L., and Hoffman, U. (1975). Application of over-run thin-layer and gas-liquid chromatography in the separation of closely related $C_{19}O_2$ steroids. *J. Chromatogr. 115*: 177–182.

Mansinger, D., Marcus, C. S., Tarle, M., Casanora, J., and Wolf, W. (1977). Preparation and high-performance liquid chromatography of iodinated diethylstilbestrols and some related steroids. *J. Chromatogr. 130*: 129–138.

Morgan, D. W., and Poole, C. F. (1976). The extraction and determination of ecdysones in arthropods. *Adv. Insect. Physiol. 12*: 17–62.

Nikolova-Damyanova, B., Stefanov, K., Seizova, K., and Popov, S. (1992). Extraction and rapid identification of low molecular weight compounds from marine organisms. *Comp. Biochem. Physiol. 103B*: 733–736.

Oehrle, K. -L., Vogt, K., and Hoffmann, B. (1975). Determination of trenbolone and trenbolone acetate by thin-layer chromatography in combination with a fluorescence colour reaction. *J. Chromatogr. 114*: 244–246.

O'Shannessy, P. J. (1984). Separation of the steroidal pairs 5α-androstenediol-5α-androstanediol and dehydroepiandrosterone-5α-dihydrotestosterone by thin-layer chromatography. *J. Chromatogr. 295*: 308–311.

Penzes, L. P., and Oertel, G. W. (1972). Determination of steroids by densitometry of derivatives. III. Micro-assay of estrogens as dansyl derivatives. *J. Chromatogr. 74*: 359–365.

Petrovic, J. A., and Petrovic, S. M. (1976). Separation of some androgens and estrogens by thin-layer chromatography. *J. Chromatogr. 119*: 625–629.

Pis, J., and Harmatha, J. (1992). Phenylboronic acid as a versatile derivatization agent for chromatography of ecdysteroids. *J. Chromatogr. 596*: 271–275.

Puah, C. M., Kjeld, J. M., and Joplin, G. F. (1978). A radioimmunochromatographic scanning method for the analysis of testosterone conjugates in urine and serum. *J. Chromatogr. 145*: 247–255.

Raedsch, R., Hofmann, A. F., and Tserng, K. (1979). Separation of individual sulfated bile acid conjugates as calcium complexes using reversed-phase partition thin-layer chromatography. *J. Lipid Res. 20*: 796–800.

Rivas-Nass, A., and Muellner, S. (1994). The influence of critical parameters on the TLC separation of bile acids. *J. Planar Chromatogr.—Mod. TLC 7*: 278–285.

Rao, P. N., Purdy, R. H., Williams, M. C., Moore, P. H., Jr., Goldzreker, J. W., and Layne, D. S. (1978). Metabolites of estradiol-17β in bovine liver: Identification of the 17-β-D-glucopyranoside of estradiol-17α. *J. Steroid Biochem. 10*: 179–185.

Robinson, D. R., and West, C. A. (1970). Biosynthesis of cyclic diterpenes in extracts from seedlings of *Ricinus communis* L. I. Identification of diterpene hydrocarbons formed from mevalonate. *Biochemistry 9*: 70–79.

Rocchi, E., and Salivioli, G. F. (1976). Determination of biliary salts by thin-layer chromatography associated with direct fluorimetry. *Minerva Gastroenterol. 22*: 130–134.

Rojkowski, K. M., and Broadheat, G. D. (1972). The monochloroacetylation of oestrogens prior to gas-liquid chromatography with electron capture detection. *J. Chromatogr. 69*: 373–376.

Ruh, M. F., and Black, C. (1976). Separation and detection of α- and β-ecdysone using thin-layer chromatography. *J. Chromatogr. 116*: 480–481.

Scott, R. M. (1969). *Clinical Analysis by Thin-Layer Chromatography Techniques*. Ann Arbor-Humphrey Science Publishers, Ann Arbor, MI.

Scott, R. M., and Sawyer, R. T. (1975). Detection of steroids with molybdovanodophosphoric acids on thin layer chromatograms. *Microchem. J. 20*: 309–312.

Singh, A. K., Gordon, B., Hewetson, D., Granley, K., Ashraf, M., Mishra, U., and Dombrovskis, D. (1989). Screening of steroids in horse urine and plasma by using electron impact and chemical ionization gas chromatography—mass spectrometry. *J. Chromatogr. 479*: 233–242.

Smith, P., and Hall, C. J. (1974). Solvents for the adsorption chromatography of adrenocortical steroids. *J. Chromatogr. 101*: 202–205.

Srinivas, B., and Srinivasulu, K. (1993). Thin-layer chromatographic studies of some steroid hormones using heulandite, a zeolite as an adsorbent. *J. Indian Chem. Soc. 70*: 853–854.

Sunde, A., Stenstad, P., and Eik-Nes, K. B. (1979). Separation of epimeric 3-hydroxyandrostanes and 3-hydroxyandrostenes by thin layer chromatography on silica gel. *J. Chromatogr. 175*: 219–221.

Szepesi, G., Dudas, K., Pap, A., Vegh, Z., Mincsovics, E., and Tylhak, T. (1982). Quantitative analysis of chenodeoxycholic acid and related compounds by a densitometric thin-layer chromatographic method. *J. Chromatogr. 237*: 137–143.

Szepesi, G., and Gazdag, M. (1991). Steroids. In *Handbook of Thin Layer Chromatography*, J. Sherma and B. Fried (Eds.). Marcel Dekker, New York, pp. 907–938.

Szepesi, G., and Gazdag, M. (1996). Steroids. In *Handbook of Thin-Layer Chromatography*, J. Sherma and B. Fried (Eds.). Marcel Dekker, New York, pp. 971–999.

Taylor, W. A., Blass, K. G., and Ho, C. S. (1979). Novel thin-layer chromatographic separation and spectrofluorometric quantitation of lithocholic acid. *J. Chromatogr. 168*: 501–507.

Tinschert, W., and Träger, L. (1978). Application of over-run thin-layer and gas-liquid chromatography in the separation of closely related $C_{19}O_2$ steroids. *J. Chromatogr. 115*: 177–182.

Tomsova, Z. (1991). Thin-layer chromatography of urinary 17-oxosteroids using dansylhydrozine as a prelabeling reagent. *J. Chromatogr. 570*: 396–398.

Touchstone, J. C. (Ed.). (1986). *CRC Handbook of Chromatography—Steroids*. CRC Press, Boca Raton, FL.

Touchstone, J. C., and Fang, X. (1991). Fluorometric labeling of tetrahydroprogesterones. *Steroids 56*: 601–605.

Touchstone, J. C., Levitt, R. E., Levin, S. S., and Soloway, R. D. (1980). Separation of conjugated bile acids by reverse phase thin layer chromatography. *Lipids 15*: 386–387.

Tsang, C. P. W. (1984). Thin-layer chromatographic separation of estrogen sulfates. *J. Chromatogr. 294*: 517–518.

Van den Ende, A., Raedecker, C. E., and Mariuhu, W. M. (1983). Microanalysis of free and conjugated bile acids by thin-layer chromatography and in situ spectrofluorometry. *Anal. Biochem. 134*: 153–162.

Van Look, L., Deschuytere, P., and Van Pateghem, C. (1989). Thin-layer chromatographic method for the detection of anabolics in fatty tissues. *J. Chromatogr. 489*: 213–218.

Watkins, T. R., and Smith, A. (1986). Reversed-phase thin-layer chromatography of estrogen glucuronides. *J. Chromatogr. 374*: 221–222.

Whiting, P., and Dinan, L. (1988). The occurrence of apolar ecdysteroid conjugates in newly-laid eggs of the house cricket, *Acheta domesticus*. *J. Insect Physiol. 34*: 625–631.

Wilson, I. D. (1992). The use of boronic acids for the normal and reversed phase TLC of ecdysteroids. *J. Planar Chromatogr.—Mod. TLC 5*: 316–318.

Wilson, I. D. (1985). Thin-layer chromatography of ecdysteroids. *J. Chromatogr. 318*: 373–377.

Wilson, I. D., Bielby, C. R., and Morgan, E. D. (1982). Studies on the reversed-phase thin-layer chromatography of ecdysteroids on C_{12} bonded and paraffin coated silica. *J. Chromatogr. 242*: 202–206.

Wilson, I. D., and Lafont, R. (1986). Thin-layer chromatography and high performance thin-layer chromatography of [3H] metabolites of 20-hydroxyecedysone. *Insect Biochem. 16*: 33–40.

Wilson, I. D., and Lewis, S. (1987). Separation of ecdysteroids by high-performance thin-layer chromatography using automated multiple development. *J. Chromatogr. 408*: 445–448.

Wilson, I. D., Morgan, E. D., Robinson, P., Lafont, R., and Blais, C. (1988). A comparison of radio–thin layer and radio–high performance liquid chromatography for ecdysteroid metabolism studies. *J. Insect Physiol. 34*: 707–711.

Wilson, I. D., Scalia, S., and Morgan, E. D. (1981). Reversed-phase thin-layer chromatography for the separation and analysis of ecdysteroids. *J. Chromatogr. 212*: 211–219.

Wilps, H., and Zoller, T. (1989). Origin of ecdysteroids in females of the blowfly *Phormia terraenovae* and their relation to reproduction and metabolism. *J. Insect. Physiol. 35*: 709–717.

Wortberg, C., Woller, R., and Chulamorakot, T. (1978). Detection of estrogen-like compounds by thin-layer chromatography. *J. Chromatogr. 156*: 205–210.

Yamaguchi, Y., and Hayashi, C. (1976). Determination of urinary pregnanetrial using thin-layer chromatography and 3α-hydroxy steroid-dehydrogenase. *Jap. J. Clin. Chem. 5*: 140–143.

22

Pharmaceuticals

I. TLC: GENERAL

An examination of the most recent issue of Sherma's (1998) biennial review on planar chromatography shows that more TLC application papers on pharmaceutical drugs and alkaloids than on any other compound type have appeared from 1995 to 1997. The next largest number of entries for TLC application papers is on lipids.

The literature on TLC of pharmaceuticals is voluminous, and significant reviews include Macek (1972), Auterhoff (1977), Gilpin (1979), Gonnet and Marichy (1980), and Pachaly (1983). Quantitative thin-layer chromatography in pharmaceutical analysis has been reviewed by Munson (1984), and Wagner et al. (1984) contributed an excellent thin-layer chromatography atlas on plant drugs. A main feature of the atlas is a collection of 165 striking color photographs of TLC plates illustrating the separations considered in the text. The book provides details on sample preparation, sorbents, solvent systems, and detection reagents for the TLC of numerous drugs derived from plant materials.

Wilson (1991) has contributed a review with 34 references on recent advances in TLC applications to drug metabolism studies. TLC continues to be widely used in drug metabolism studies, especially where radioisotopes are employed. Recent advances in the field have led to increased chromatographic performance and to instrument automation of many of the steps in the TLC process. The advances have included the introduction of much improved HPTLC plates together with a wide range of bonded silica phases, providing a great range of chromatographic selectivities. Additionally, because it is possible to obtain mass spectra directly from TLC plates, without the need for time-consuming preparative scale isolations of unknowns, drug metabolites can be more rapidly identified.

Ng (1991) has provided a review on TLC and HPTLC analysis of the more popular pharmaceuticals and illicit drugs covering the literature from about 1975 to 1988. This reference includes information on sample preparation, layers, mobile phases, detection reagents, and structures of popular human and animal drugs. Numerous examples are provided for qualitative and quantitative TLC analysis of pharmaceutical drugs in various dosage forms. The review of Ng (1991) was extended and updated by Szepesi and Nyiredy (1996) to include references through 1995.

Gough (1991) has edited a book on the analysis of drugs of abuse using various analytical procedures. Several chapters include information on the use of TLC in drug analysis. The applications section of the first chapter includes a well-organized compendium of TLC systems which are of value for the analysis of solid dosage forms and natural products such as opium and cannabis.

Roberson (1991) provided a brief review with 29 references on the use of TLC in the analysis of drugs of abuse. Although the examples used are a small fraction of those available in the literature, they serve to show the versatility of TLC in the analysis of drugs.

Ojanpera (1992) contributed a review with 55 references on toxicological drug screening by TLC. The article covers computerized system evaluation, automated sample application, instrumental modes of development, and computerized identification by use of R_F values for numerous compounds.

Touchstone (1992) has provided information on methods of sample preparation, layers, mobile phases, and detection reagents for pharmaceuticals, particularly drugs of abuse such as amphetamines, barbiturates, benzodiazepines, cannabinoids, cocaine, methaqualone, opiates, and anorexicants.

Renger (1993) discussed the role of planar chromatography as a tool in pharmaceutical analysis. Advances in instrumentation, layers, and methodology have enhanced the status of modern planar chromatography and allowed for improvement in reliability, accuracy, and reproducibility. Therefore, planar chromatographic methods offer a simple, realistic, and economical alternative to the other chromatographic techniques used in pharmaceutical analysis.

Extensive use has been made of reversed-phase thin-layer chromatography in pharmaceutical separations. Siouffi et al. (1979) considered the properties and operating conditions of reversed-phase layers for drug separations. Likewise, Brinkman and deVries (1979, 1980) discussed the use of reversed-phase thin layers for the separation of selected pharmaceuticals. Numerous drug separations on Whatman KC_{18} reversed-phase layers have been described in the Whatman TLC Technical Series, Volume 1.

Brinkman and deVries (1979, 1980) also described TLC applications of drugs on high-performance silica gel plates. Numerous separations of pharmaceuticals on Whatman HPTLC plates and a useful list of references of drugs separated by HPTLC have been presented in the Whatman TLC Technical Series, Volume 2.

Some comparative information is available on drug separations done by TLC versus HPLC. Thus, Egli and Keller (1984) used silica gel TLC and reversed-phase TLC on C_{18}-bonded layers to separate 28 drugs in a variety of widely used mobile phases. Their TLC separations compared favorably with similar studies done using HPLC.

Bonicamp (1985) described a simple TLC experiment that involves the separation and identification of some commonly used drugs. She used the Toxi-Lab A TLC system, which screens a variety of basic drugs in less than 1 h.

Fishbein (1983) reviewed TLC studies on certain classes of pharmaceuticals and on drugs of abuse and abused drugs. He considered the significant TLC studies on barbiturates, tricyclic antidepressants, benzodiazepines, antiarrhythmic drugs, phenacetin and acetaminophen, and several miscellaneous drugs. This chapter considers the aforementioned drugs along with several others.

Dreassi et al. (1996) reviewed TLC in pharmaceutical analysis and included applications on alkaloids, sulfur-containing compounds, and oxygen- and nitrogen-containing compounds (including the more commonly used β-blockers). Jain (1996) discussed uses of TLC for the analysis of drugs in clinical chemistry. Ojanpera (1996) reviewed thin-layer chromatography in forensic toxicology. She described TLC analysis of major drugs of abuse—amphetamines, cannabinoids, cocaine and metabolites, lysergids, and opiates. She also described dedicated TLC methods for substances such as alkaloids, benzodiazepines, β-adrenergic blocking drugs, phenothiazine, sulfa drugs, and sympathomimetics: Wilson (1996) reviewed TLC in drug analysis of biological fluids. He described the capabilities, advantages and disadvantages of various TLC techniques. He also gave an example of the use of HPTLC with scanning densitometry for quantification of antipyrine in human plasma. The use of TLC–mass spectrometry and TLC–tandem mass spectrometry in the identification of the compounds separated by TLC was discussed.

A number of experiments have been described in detail on pharmaceuticals in the Machery-Nagel (MN) "TLC Applications" catalog. Because of the availability of this catalog from MN (see Chapter 23 in this book for details), the experiments are not described herein. The following applications of TLC experiments are recommended (the application number of the experiment as listed in the MN catalog is given in parentheses): antiarrhythmic drugs (28), antihypertensive drugs and β-blockers (31), barbiturates (33–36), cannabinoids (45), cocaine metabolites (47), and sulphonamides (150).

II. TLC OF BARBITURATES

Barbiturates are used as sedatives. According to Fishbein (1983), more than 50 of the numerous barbiturates synthesized are marketed for clinical use. Among the best known are the barbitals, buta-, pento-, seco-, and phenobarbital (see Chapter 10, Figure 10.6). There is widespread abuse of barbiturates, and techniques are needed to detect these sedatives in human blood samples, particularly in individuals who have attempted suicide. Phenobarbital is also used as an anticonvulsant in the treatment of epileptics. Blood levels in patients must be monitored to determine the balance between the therapeutic versus toxic levels of the drug.

Barbiturates have been separated in the Whatman Applications Laboratory (TLC Technical Services, Volume 1) on KC_{18} reversed-phase plates with the mobile phase methanol–0.5 M NaCl (70:30). Spots were detected using diphenylcarbazone–mercuric sulfate with a sensitivity of about 1 μg. The R_F values of the various barbiturates were as follows: secobarbital, 0.34; pentobarbital, 0.38; amobarbital, 0.38; mephobarbital, 0.41; allylisobutyl barbituric acid, 0.45; butabarbital, 0.48; and phenobarbital, 0.55.

Kovacs-Hadady (1995) studied the retention behavior of barbiturates by overpressurized liquid chromatography (OPLC) on silica gel layers impregnated with paraffin or dodecyltrimethylammonium bromide and developed with mixtures of methanol and water.

III. TLC OF TRICYCLIC ANTIDEPRESSANTS

The tricyclic antidepressants are among the most widely prescribed drugs for the treatment of mental disorders and therapy for depression (Fishbein, 1983). Widely used antidepressants are amitriptyline and its demethylated metabolite, nortriptyline. Other tricyclics include doxepin, clomipramine, and dibenzepine.

Po and Irwin (1979) used TLC to separate numerous tricyclic neuroleptic tranquilizers. Samples were dissolved in ethyl acetate, and the mobile phase consisted of mixtures of different n-alcohols with water. Circular development in a Camag U chamber was used, and spots were detected by fluorescence quenching. The R_F values of some of the better known drugs developed in methanol–water (90:10) were amitriptyline, 0.17; clopenthixol, 0.40; doxepin, 0.42; nortriptyline, 0.37; and promazine, 0.14. Shirke et al. (1994) determined amitriptyline and chlordiazepoxide in combined dosage forms using ethyl acetate–methanol–diethylamine (9.5:0.5:0.05) mobile phase and scanning at 245 nm.

IV. TLC OF BENZODIAZEPINES

Benzodiazepine-derivative tranquilizers are widely prescribed in the United States. Chlordiazepoxide (Librium) is a commonly prescribed antianxiety (tranquilizer) drug. A related drug, diazepam (Valium), is used in therapy, and its concentration, along with the metabolite nitrazepam, must be monitored in body fluids. A TLC procedure for determining nitrazepam in human urine has been described (Inoue and Niwaguchi, 1985).

TLC methods for the separation of various benzodiazepine tranquilizers have been reported. Thus, Okumura and Nagaoka (1979) separated various benzodiazepines in methanol–water (2:1 or 3:1), and spots were detected by fluorescence quenching. Tranquilizers and their metabolites have been chromatographed on Whatman HP-KF plates (Whatman Technical Series, Volume 2). Samples prepared in methanol were separated in the mobile phase ethyl acetate–methanol (95:5); spots were detected by fluorescence quenching under 254-nm ultraviolet (UV) light; the compounds along with their R_F values were n-desmethycheordiazepoxide, 0.11; chlordiazepoxide, 0.21; demoxepam, 0.38; oxazepam, 0.45; and diazepam, 0.60.

White et al. (1991a) used high-performance thin-layer chromatography to determine chlordiazepoxide, demoxepam, and 2-amino-5-chlorobenzophenone in tablets. The HPTLC silica gel $60F_{254}$ plates were predeveloped with ethyl acetate–NH_4OH and the mobile phase consisted of acetonitrile–NH_4OH (99.9:0.1). Plates were quantified by scanning densitometry in the absorbance mode with a Hg light source at 254 nm; recoveries of compounds were 92–100%. White et al. (1991b) analyzed diazepam (Valium) and related compounds in bulk and tablets by HPTLC–densitometry using silica gel $60F_{254}$ plates and scanning at 254 nm. Drug recovery was 94–115%.

Patil and Shingare (1993) reported on TLC detection of certain benzodiazepines. Otsubo et al. (1995) detected benzodiazepines and zopiclone in human serum at 0.1–0.4 μg/ml by HPTLC R_F values and spot colors in three systems. Sarbu and Cimpan (1996) developed a quantitative TLC method using plates with a fluorescent indicator to determine certain 1,4-benzodiazepines in pharmaceuticals. They provided calibration curves for the quantitative analysis of diazepam, oxazepam, and chlordiazepoxide.

V. TLC OF THEOPHYLLINE

Theophylline is a bronchodilator used in the treatment of asthmatics. It can be monitored by TLC following the direct application of urine from asthmatic patients to Whatman LHP-KF plates (Whatman Technical Series, Volume 2). Theophylline was also determined following the direct application of 20 μl of serum to silica gel plates; the mobile phase was chloroform–isooctane–methanol (70:20:10), and the drug was detected by fluorescence quenching under 254-nm UV light (Heilweil and Touchstone, 1981). Shinde et al. (1994) used HPTLC to determine theophylline and etofylline in dosage forms using chloroform–methanol (9:1) mobile phase and scanning at 278 nm.

VI. TLC OF SULFONAMIDES

Sulfonamides are used for the prevention and treatment of human infectious diseases and also as growth promoters in swine. It is necessary to monitor pork products for drug residues of sulfonamides. Various sulfonamides were separated on Whatman LKC$_{18}$F layers in a methanol–0.5 M NaCl (50:50) mobile phase; spots were detected with UV light (366 nm) after spraying with fluorescamine reagent followed by 0.5% triethanolamine in chloroform to stabilize fluorescence. The R_F values of the sulfonamides were as follows: sulfabromomethazine, 0.19; sulfadimethoxine, 0.39; sulfamethazine, 0.59; sulfathiazole, 0.76; sulfaguanidine, 0.90 (Whatman TLC Technical Series, Volume 3).

Thomas et al. (1981) quantified sulfamethazine in swine tissues at 0.1 ppm by densitometry of the fluorescamine derivative on preadsorbent silica gel layers. Tissue was extracted with ethyl acetate, and the drug was back extracted into methylene chloride, and concentrated. Samples and standards were developed with chloroform–tertiary butanol (80:20).

Martin et al. (1993) described a silica gel–TLC method for the determination of sulfonamide residues in eggs. The sample was homogenized, extracted with methylene chloride/acetone (1:1), and purified by column chromatography. The sulfonamides were detected with fluorescamine and quantified by densitometric scanning.

Agababa et al. (1996) developed a simple method for TLC determination of sulfamethoxazole, trimethoprim, and impurities of sulfanilamide and sulfanilic acid using a chloroform–n-heptane–ethanol (3:3:3) mobile phase. The zones were quantified by scanning densitometry in the reflectance/absorbance mode at 260 nm. Abjean (1997) described a method for multiclass and multiresidue qualitative detection of chloramphenicol, nitrofuran, and sulfonamide residues in animal muscle. The drugs were extracted from 1 gram of tissue with 2 ml of ethanol acetate and purified by silica solid-phase extraction. After elution of the cartridge, the collection solution was evaporated; the residue was dissolved in methanol and chromatographed on silica gel 60–HPTLC plates.

VII. TLC OF ASPIRIN, PHENACETIN, AND CAFFEINE

TLC is useful for the determination of aspirin, phenacetin, and caffeine in the analgesic tablet, APC. A sample from a methanolic solution of an APC tablet was applied to a Whatman KC$_{18}$ reversed-phase plate. Spots were developed for 15 cm in an unequilibrated tank (about 70 min), and detected by fluorescence quenching; the R_F values were phenacetin, 0.26; caffeine, 0.47; and aspirin, 0.74 (Sherma and Beim, 1978). Detailed Experiment 2 in this chapter is concerned with TLC-densitometric analysis of aspirin and caffeine in analgesic tablets.

Sherma and Rolfe (1992) described the analysis of analgesic tablets (Excedrin Extra Strength) containing aspirin, acetaminophen, and caffeine by high-performance thin-layer chromatography (HPTLC) on silica gel layers. The experiment also introduced students to the use of HPTLC for screening and/or quantitative analysis of drugs. Safety precautions for conducting this type of an experiment were given.

Spell and Stewart (1994) used HPTLC to determine *p*-aminosalicylic acid and its major impurity, *m*-aminophenol, in bulk and pharmaceuticals with chloroform–ethyl acetate–acetic acid–methanol (46:29:21:4) mobile phase. Agababa et al. (1995) developed an HPTLC method for the determination of acetylsalicylic acid and the impurity salicylic acid in tablets. Argekar and Kunjir (1996) described a method for simultaneous determination of aspirin and dipyridamole in tablets by HPTLC. McLaughlin and Sherma (1996) described a method for determining salicylic acid in anti-acne pharmaceuticals involving separation on a preadsorbent HPTLC–silica gel plate with fluorescent phosphor; compound detection was by fluorescence quenching and quantification was by densitometric scanning. Salicylic acid was detected and quantified on the plate at levels as low as 1 μg.

VIII. TLC OF ALKALOIDS

Many alkaloids are derivatives of tertiary amines and, as the name implies, are alkaline in nature. Numerous drugs of abuse as well as abused drugs are alkaloids. An extensive list of alkaloids, their sample preparation, solvent systems, and detection reagents have been given by Wagner et al. (1984).

Various alkaloids prepared in methanol were separated on KC_{18} reversed-phase layers in a methanol–0.5 M NaCl (65:35) mobile phase (Whatman, TLC Technical Series, Volume 1). Spots were detected with the iodoplatinate reagent diluted with ethanol (2:1) just prior to use. The R_F values of the separated alkaloids were as follows: Valium, 0.16; methadone, 0.27; quinine, 0.37; phencyclidine, 0.49; meperidine, 0.52; Talwin, 0.57; cocaine, 0.58; codeine, 0.72; and morphine, 0.72.

Detailed Experiment 1 in this chapter is concerned with the separation of numerous alkaloid drugs from urine.

Pothier et al. (1991) described the separation of numerous alkaloids in plant extracts by overpressured layer chromatography on alumina plates with only a single solvent, ethyl acetate, as the mobile phase. The method was applied with success to plant extracts and to the determination of hR_F values of 81 authentic samples of natural alkaloids or their derivatives.

Alemany et al. (1993) resolved a mixture of cannabinoids in human plasma by extraction with a C-18 Sep Pak cartridge, derivatization with dansyl chloride and scanning at 340 nm. Jain (1993) studied interferences of adulterants on TLC screening of morphine, diazepam, and phenobarbitol and found that morphine was the most sensitive to all of the adulterants used.

Bieganowska and Petruczynik (1994) studied the separation of 15 alkaloids on unmodified silica gel in reversed-phase ion-association systems. Tombesi et al. (1994) purified 15 alkaloids (including atropine, cocaine, codeine, and quinine) by TLC decomposition of their picrates using silica gel and alumina sorbents; identification was based on melting points and UV-visible spectra.

Lemire and Busch (1994) separated alkaloids derived from *Sanguinaria canadensis* by one- and two-dimensional TLC and capillary zone electrophoresis (CZE). The higher loadings possible by TLC allowed easier identification of mixture components compared to CZE, and reproducibility of R_F and mass spectral data was excellent. Bailey (1994) described a new TLC method for detecting 5 µg amounts of cocaethylene and cocaine using solid-phase extraction (SPE) of urine buffered to pH 9.3, silica gel TLC with hexane–toluene–diethylamine (13:4:1) mobile phase and iodoplatinate spray reagent. Parvais et al. (1994) detected pyrrolizidine alkaloids in borage (*Borago officinalis*) seed oil on silica gel plates developed with dichloromethane–methanol–25% aqueous ammonia solution (79:20:1) with detection by Mattock, Dragendorff, and iodoplatinate reagents.

Mink et al. (1995) used diffuse reflectance Fourier transform infrared spectroscopy (DRIFT) as an in situ detection method for the qualitative and quantitative analysis of heroin, cocaine, and codeine separated by TLC. A detection limit of 2 µg was attained on microcrystalline cellulose layers with a Bio-Rad Digilab type interferometer. A reliable analysis was made with 10 to 15 µg drug per spot. Tam et al. (1995) determined the indole alkaloids cantharanthine and vindoline in *Cantharanthus roseus* by scanning at 280 and 310 nm, respectively, after separating them by triple development in petroleum ether–diethyl ether–acetone–ethanol (70:10:20:1) mobile phase. Munro and White (1995) evaluated diazonium salts as visualization reagents for amphetamines (50 ng sensitivity). Lillsunde and Korte (1995) reviewed TLC screening and GC/MS confirmation techniques for drugs of abuse and cited 128 references.

Kamble et al. (1996) described a specific and sensitive chromogenic reagent consisting of a 1:1 mixture of a 1% aqueous solution of cupric chloride and a 1% aqueous solution of potassium ferricyanide for the detection of heroin (diacetylmorphine). Heroin gave a dark brown spot after spraying with the reagent. The reagent did not react positively to the following adulterants: barbiturates, benzodiazepines, methaqualone, and caffeine. The detection reagent was sensitive to about 1 µg of heroin. Gupta and Verma (1996) described a TLC–densitometric method for the determination of major opium alkaloids (i.e., morphine, codeine, thebaine, papaverine, and narcotine) in poppy straw samples. The analysis combined the separation of compounds on silica gel 60 F_{254} plates with spot detection by spraying the plates with Dragendorff reagent.

IX. TLC OF CARDIAC GLYCOSIDES AND ANTIARRHYTHMIC DRUGS

TLC is useful for monitoring drugs used in the treatment of cardiac disorders. Drugs such as lidocaine and digitalis and its derivatives are used to regulate heartbeat, thus preventing cardiac arrhythmias. Malikin et al. (1984) have reported on the use of HPTLC to monitor the serum of patients treated with various drugs used to control cardiac arrhythmias.

Bloch (1980) separated digoxin and related digitalis glycosides on KC_{18} reversed-phase plates. Samples in chloroform–methanol (2:1) were separated in a methanol–water (7:3) mobile phase and detected by spraying with chloramine T-trichloroacetic acid reagent; fluorescent spots were viewed under 366-nm UV light. The R_F values of the separated compounds were digoxigenin mono-digitoxoside, 0.72; digoxigenun bis-digitoxoside, 0.65; digoxin, 0.60; gitoxin, 0.45; and digitoxin, 0.35.

Tivert and Backman (1993) separated enantiomers of the β-blocking drugs metoprolol, propranolol, and alprenolol on diol layers with dichloromethane and a chiral counterion, N-benzoylcarbonylglycyl-L-proline, as mobile-phase additive. Ponder and Stewart (1994) developed an HPTLC method for the determination of digoxin, digoxigenin, bisdigitoxoside, and gitoxin in digoxin drug substance and tablets on C-18 reversed-phase plates using water–methanol–ethyl acetate (25:24:1) mobile phase and absorbance and fluorescence scanning.

X. DETAILED EXPERIMENTS

A. Experiment 1. Separation of Drugs and Their Determination in Urine by TLC

1. Introduction

This experiment describes a procedure for the separation and detection of selected drugs and their identification in urine using TLC, as devised in the Whatman Laboratories. The procedure involves extraction of the drugs from urine; application of concentrated extracts to a Whatman precoated, preadsorbent plate; development of the chromatograms; and visualization and identification of the drugs and/or metabolites based on R_F and color. Further information and details on the identification of many more drugs and metabolites than are included in this experiment are available from Whatman, Bulletin No. 502, TLC Toxicology System, and in the monograph by Stahl (1973).

2. Preparation of Standards

Obtain standards of morphine, codeine, quinine, dextroamphetamine, methamphetamine, methadone, and cocaine from a commercial source such as Applied

Science Laboratories. Purchase of drug standards in the United States usually requires that qualified researchers and analysts obtain a Drug Enforcement Agency (DEA) license, although dilute solutions of some drugs can be purchased without a license. Requirements can be ascertained by contacting a regional DEA office. Prepare standards as 1-mg/ml individual and mixture solutions in methanol, and dilute 1:20 to a level of 1 µg/20 µl of each drug.

3. Preparation of Samples

Store urine specimens under refrigeration. This experiment can be tested by adding a known amount of drug to "blank" urine known not to contain that chemical. However, results will not be exactly the same as in actual urine samples, where the parent drug may be conjugated, most likely with glucuronic acid, or metabolized.

Check the urine specimen with pH paper and adjust the pH within a range of 8.3–10.3. Pour 10 ml of specimen into a 6-oz (175-ml) bottle and add 1.0 ml of pH 9.3 $NH_4OH–NH_4Cl$ buffer and 22 ml of chloroform–propan-2-ol (isopropanol) (3:1). Prepare the buffer by adding concentrated ammonia (ammonium hydroxide) to a saturated aqueous ammonium chloride solution until the required pH is obtained as measured by a pH meter. A larger volume of urine will yield more drug for identification, but the same ratio of urine to solvent (10:22) must be maintained; 10 ml is a minimum sample that should be employed.

Shake the bottle gently for 3 min and separate the phases by pouring the mixture through a folded cone of Whatman phase-separating paper (Grade 1PS). Collect the water-free solvent in a test tube, add one drop of propan-2-ol–acetic acid (1:1), and evaporate to dryness. The acetic acid converts free bases to acetate salts so that they are not lost by volatilization. Dissolve the residue in 50 µl of methanol to prepare the spotting solution for TLC.

4. Layer

Whatman Linear-K5D silica gel plate.

5. Application of Initial Zones

By means of Microcap micropipets, apply 20 µl of each individual drug and the seven-component drug mixture solutions, representing 1 µg of each compound. Apply 20 µl of the concentrated urine extract prepared as described in Section 3 with a Microcap micropipet or Drummond Dialamatic microdispenser (Chapter 5, Figure 5.2). Dispense the solutions in the center of the preadsorbent area of the channels in a downward, brushlike motion. Apply no sample closer than 0.5 cm to the bottom edge of the plate so that the mobile phase is not contacted. After all samples have been applied, dry the plate thoroughly with a stream of warm air not exceeding 80°C.

6. Mobile Phase

Ethyl acetate–methanol–concentrated ammonium hydroxide (29:6.5:1.5).

7. Development

Saturate a glass rectangular tank with mobile phase by covering the side of the chamber facing the plate with blotter-weight paper (Whatman 3 or 3 MM), pouring in the mobile phase, and covering the tank. Allow to stand 5 min (only) and then insert the plate and cover quickly. Allow the mobile phase to ascend for 12 cm beyond the preadsorbent–silica gel interface.

8. Detection

Heat the developed plate at 110–120°C for 5 min. Spray the plate while still hot with Marquis Reagent and view immediately under 254- and 366-nm ultraviolet light. Prepare the reagent by mixing 1 ml of reagent-grade formaldehyde with 100 ml of concentrated sulfuric acid; prepare fresh weekly.

9. Results and Discussion

The drugs appear on the chromatogram with the following approximate R_F values (measured from the layer interface) and colors:

Morphine	0.07, charcoal gray in visible light
Codeine	0.18, charcoal gray in visible light
Quinine	0.18, brilliant fluorescent blue under UV light
Methamphetamine	0.20, brick red fluorescence
Dextroamphetamine	0.36, brick red fluorescence
Methadone	0.65, red fluorescence
Cocaine	0.72, red fluorescence

In analysis of actual urine specimens, it must be realized that many drugs, such as cocaine, are more or less rapidly metabolized in the body into one or more metabolites. For detection of the parent drug, it is essential to obtain the urine specimen used for analysis within a short time after ingestion of the drug.

Along with color and R_F value, the presence and position of metabolites can be a vital aid for certain identification of drugs present. For example, quinine forms two metabolites detected at R_F 0.10 and 0.15 as blue fluorescent bands by the method described above.

Other solvent systems and detection reagents are recommended for the identification and confirmation of drugs and their metabolites. These include ethyl acetate–methanol–concentrated ammonium hydroxide (90:7:3) with detection by fluorescamine, diphenylcarbazone, and iodoplatinate reagents for D-amphetamine, barbiturates, and acidic drugs (e.g., phenobarbital, Dilantin, Mellaril, Valium, chlorpromazine, Lomotil, and Talwin); and ethyl acetate–methanol–concentrated ammonium hydroxide–water (85:13.5:0.5:1.0) with detection by ninhydrin and iodoplatinate for alkaloids and amphetamines (e.g., methadone, quinine, morphine, Darvon, Librium, Valium, and amitriptyline). Details are available from Whatman for these analyses. Cross-referencing among the three chromatographic systems is an added method for confirmation of the presence of a particular drug residue.

The procedures described permit detection of substances in amounts as low as 1 µg on the TLC plate, or in concentration levels as low as 200 ng/ml. Many other drugs besides the seven chosen for this experiment can be separated and detected by the Whatman system.

B. Experiment 2. Quantitative Determination of Aspirin and Caffeine in Analgesic Tablets

1. Introduction

The purpose of this experiment is to analyze analgesic tablets containing aspirin and caffeine by scanning of quenched zones on preadsorbent high-performance reversed-phase C_{18}–F plates. The procedure is based on the paper by Sherma et al. (1985).

2. Preparation of Standards

Prepare appropriate stock and working standard solutions of aspirin and caffeine in methanol to bracket the amounts of ingredients contained in the sample solution. For analysis, for example, the aspirin stock solution should contain approximately 80 mg/ml and caffeine approximately 5.5 mg/ml. Working standards are mixtures of about 1.0 ml of each solution, 1.5 ml of each solution, and 2.0 ml of each solution made up to volume with methanol in 25-ml volumetric flasks.

3. Preparation of Samples

Place a tablet in a 100-ml volumetric flask and crush with a flattened glass rod. Add 50 ml of methanol and a magnetic stirbar and stir for 10 min. Fill the solution to the line with methanol. Inert ingredients will remain undissolved and settle to the bottom of the flask.

4. Layer

Whatman 20 × 20-cm chemically bonded C_{18} RP plate with fluorescent phosphor, preadsorbent strip, and 19 channels. Prewash by development with methylene chloride–methanol (1:1) and air dry.

5. Application of Initial Zones

Apply 10 µl of each of the three standard aspirin and caffeine mixtures and 10 µl of sample solution in duplicate by streaking onto preadsorbent areas of adjacent channels using Drummond microcaps.

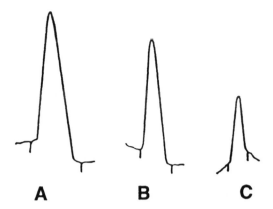

FIGURE 22.1 Densitometer scans of (A) aspirin (34.8 μg), (B) caffeine (30.1 μg), and (C) phenacetin (0.50 μg) at 254 nm on phosphor-containing silica gel HPTLC plates. The attenuation setting of the integrator/recorder was ×6 in each case.

6. Mobile Phase

Methanol–0.5 M NaCl in 1% aqueous acetic acid (2:3).

7. Development

Develop for a distance of 12 cm beyond the preadsorbent C_{18} junction in a paper-lined, vapor-saturated chamber.

8. Detection

Air dry the developed plate for 25 min and view under 254-nm UV light. Zones will appear dark against a bright fluorescent background.

9. Quantification

Scan the aspirin and caffeine samples and standards with a densitometer such as the Kontes Model 800 using a 254-nm light beam, the length of which matches the width of the channels. Plot the average areas of the duplicate standards versus micrograms spotted for aspirin and caffeine and interpolate the micrograms of each component represented by the average area of the sample spot from the standard curve. The micrograms spotted for 10 μl of the three standards specified above are 32, 48, and 64 for aspirin and 2.2, 3.3, and 4.4 for caffeine. After determining the micrograms of aspirin and caffeine in the sample zones from the calibration curves, the values are multiplied by the dilution factor 100,000/10 to convert to milligrams per tablet.

10. Results and Discussion

Results obtained are usually within ± 10% of the label values for the tablet ingredients. Respective R_F values for aspirin, caffeine, and phenacetin are 0.65, 0.29, and 0.10, so products containing any combination of these compounds can be analyzed. The amount of phenacetin in the spotted sample aliquot must be bracketed with standards as described above for aspirin and caffeine. Figure 22.1 illustrates typical densitometry scans for the three compounds.

REFERENCES

Abjean, J. P. (1997). Planar chromatography for the multiclass, multiresidue screening of chloramphenicol, nitrofuran, and sulfonamide residues in pork and beef. *J. AOAC Int. 80*: 737–740.

Agababa, D., Lazarevic, R., Zivanov-Stakic, D., and Vladimirov, S. (1995). HPTLC assay of acetylsalicylic acid and the impurity salicylic acid in pharmaceuticals. *J. Planar Chromatogr.—Mod. TLC 8*: 393–394.

Agababa, D., Radiovic, A., Vladimirov, S., and Zivanov-Stakic, D. (1996). Simultaneous TLC determination of co-trimoxazole and impurities of sulfanilamide and sulfanilic acid in pharmaceuticals. *J. Chromatogr. Sci. 34*: 460–464.

Alemany, G., Gamundi, A., Nicolau, M. C., and Saro, D. (1993). A simple method for plasma cannabinoid separation and quantification. *Biomed. Chromatogr. 7*: 273–274.

Argekar, A. P., and Kunjir, S. S. (1996). Simultaneous determination of aspirin and dipyridamole in pharmaceutical preparations by high performance thin layer chromatography. *J. Planar Chromatogr.—Mod. TLC 9*: 65–67.

Auterhoff, H. (1977). *Thin Layer Chromatograms and IR Spectra. Identification of Drugs*, 3rd ed. Wissenschaftliche Verlagsgesellschaft, Stuttgart.

Bailey, D. N. (1994). Thin-layer chromatographic detection of cocaethylene in human urine. *Am. J. Clin. Pathol. 101*: 342–345.

Bieganowska, M. L., and Petruczynik, A. (1994). Thin-layer reversed-phase chromatography of some alkaloids in ion-association systems. *Chem. Anal. (Warsaw) 39*: 139–147.

Bloch, D. E. (1980). Reverse phase thin layer chromatographic procedure for identification of digoxin and related fluorescing materials. *J. Assoc. Off. Anal. Chem. 63*: 707–708.

Bonicamp, J. M. (1985). Separation and identification of commonly used drugs: A thin layer chromatography experiment for freshman chemistry. *J. Chem. Educ. 62*: 160–161.

Brinkman, U. A. T., and de Vries, G. (1979). The use of chemically bonded stationary phases in high-performance thin-layer chromatography (HPTLC). *HRC&CC J. High Resolut. Chromatogr. Chromatogr. Commun. 2*: 79–81.

Brinkman, U. A. T., and deVries, G. (1980). The use of chemically bonded phases in high-performance thin layer chromatography. *J. Chromatogr. 192*: 331–340.

Dreassi, I., Ceramelli, G., and Corti, P. (1996). Thin-layer chromatography in pharmaceutical analysis. In *Practical Thin-Layer Chromatography—A Multidisciplinary Approach*, B. Fried and J. Sherma (Eds.). CRC Press, Boca Raton, FL, pp.231–247.

Egli, R. A., and Keller, S. (1984). Comparison of silica-gel and reversed-phase thin-layer chromatography and liquid chromatography in the testing of drugs. *J. Chromatogr. 291*: 249–256.

Fishbein, L. (1983). Pharmaceuticals. In *Chromatography—Fundamentals and Applications of Chromatographic and Electrophonetic Methods, Part B: Applications*, E. Heftmann (Ed.). Elsevier, Amsterdam, pp. B287–N330.

Gilpin, R. K. (1979). Pharmaceuticals and related drugs. *Anal. Chem. 51*: 275R–287R.

Gonnet, C., and Marichy, M. (1980). Chromatographic analysis of pharmaceutical products of toxicological interest in chemically bonded thin layers. Influence of the organic modifier nature on selectivity. *J. Liquid Chromatogr. 3*: 1901–1912.

Gough, T. A. (Ed.). (1991). *The Analysis of Drugs of Abuse*. Wiley, Chichester/New York.

Gupta, M. M., and Verma, R. K. (1996). Combined thin-layer chromatography–densitometry method for the quantitative estimation of major alkaloids in poppy straw samples. *Indian J. Pharm. Sci. 58*: 161–163.

Heilweil, E., and Touchstone, J. C. (1981). Theophylline analysis by direct application of serum to thin layer chromatograms. *J. Chromatogr. Sci. 19*: 594–597.

Inoue, T., and Niwaguchi, T. (1985). Determination of nitrazepam and its main metabolites in urine by thin-layer chromatography and direct densitometry. *J. Chromatogr. 339*: 163–169.

Jain, R. (1993). Interference of adulterants in thin layer chromatography method for drugs of abuse. *Indian J. Pharmacol. 25*: 240–242.

Jain, R. (1996). Thin-layer chromatography in clinical chemistry. In *Practical Thin-Layer Chromatography—A Multidisciplinary Approach*, B. Fried and J. Sherma (Eds.). CRC Press, Boca Raton, FL, pp.131–152.

Kamble, V. W., Garad, M. V., and Dongre, V. G. (1996). A new chromogenic spray reagent for the detection and identification of heroin (diacetylmorphine): part 1. *J. Planar Chromatogr.—Mod. TLC 9*: 280–281.

Kovacs-Hadady, K. (1995). Study of the retention behavior of barbiturates by overpressured layer chromatography on silica gel layers impregnated with dodecyltrimethylammonium bromide. *J. Planar Chromatogr.—Mod. TLC 8*: 47–51.

Lemire, S. W., and Busch, K. L. (1994). Chromatographic separation of the constituents derived from *Sanguinaria canadensis*: thin-layer chromatography and capillary zone electrophoresis. *J. Planar Chromatogr.—Mod. TLC 7*: 221–228.

Lillsunde, P., and Korte, T. (1995). Thin-layer chromatographic screening and gas chromatographic/mass spectrometric confirmation in the analysis of abused drugs. In *Analysis and Addiction of Misused Drugs*, J. A. Adamovics (Ed.), Marcel Dekker, New York, pp.221–265.

Macek, K. (Ed.). (1972). *Pharmaceutical Applications of Thin-Layer and Paper Chromatography*. Elsevier, Amsterdam.

Malikin, G., Lam, S., and Karmen, A. (1984). Therapeutic drug monitoring by high performance thin-layer chromatography. *Chromatographia 18*: 253–259.

Martin, E., Duret, M., and Vogel, J. (1993). Determination of sulfonamide residues in eggs. *Mitt. Geb. Lebensmittelunters. Hyg. 84*: 274–280.

McLaughlin, J. R., and Sherma, J. (1996). Quantitative HPTLC determination of salicylic acid in topical acne medications. *J. Liq. Chromatogr. Relat. Technol. 19*: 17–21.

Mink, J., Horvath, E., Kristof, J., Gal, T., and Veress, T. (1995). Direct analysis of thin-layer chromatographic spots of narcotics by means of diffuse reflectance Fourier transform infrared spectroscopy. *Mikrochim. Acta 119*: 129–135.

Munro, C., and White, P. (1995). Evaluation of diazonium salts as visualization reagents for the thin layer chromatographic characterization of amphetamines. *Sci. Justice 35*: 37–44.

Munson, J. W. (1984). Quantitative thin layer chromatography in pharmaceutical analysis. *Drugs Pharm. Sci. 11*: 155–178.

Ng, L. L. (1991). Pharmaceuticals and drugs. In *Handbook of Thin Layer Chromatography*, J. Sherma and B. Fried (Eds.). Marcel Dekker, New York, pp.717–755.

Ojanpera, I. (1992). Toxicological drug screening by thin-layer chromatography. *Trends Anal. Chem. 11*: 222–229.

Ojanpera, I. (1996). Thin-layer chromatography in forensic toxicology. In *Practical Thin-Layer Chromatography—A Multidisciplinary Approach*, B. Fried and J. Sherma (Eds.). CRC Press, Boca Raton, FL, pp. 193–230.

Okumura, T., and Nagaoka, T. (1979). Separation of benzodiazepine minor tranquilizers by means of reversed phase thin layer chromatography. *Bunseki Kagabu 28*: 584–587.

Otsubo, K., Seto, H., Futagami, K., and Oishi, R. (1995). Rapid and sensitive detection of benzodiazepines and zopiclone in serum using high-performance thin-layer chromatography. *J. Chromatogr., B: Biomed. Appl. 669*: 408–412.

Pachaly, P. (1983). *Thin Layer Chromatography in the Pharmacy: Rapid and Simple Identification of Common Pharmaceutical Drugs, Extracts and Tinctures*, 2nd ed. Wissenschaftliche Verlagsgesellschaft, Stuttgart.

Patil, V. B., and Shingare, M. S. (1993). Thin layer chromatographic detection of certain benzodiazepines. *J. Planar Chromatogr.—Mod. TLC 6*: 497–498.

Parvais, O., Vander Stricht, B., Vanhaelen-Fastre, R., and Vanhaelen, M. (1994). TLC detection of pyrrolizidine alkaloids in oil extracted from the seeds of *Borago officinalis*. *J. Planar Chromatogr.—Mod. TLC 7*: 80–82.

Po, L. W., and Irwin, W. J. (1979). The identification of tricyclic neuroleptics and sulfonamides by high performance thin layer chromatography. *HRC&CC, J. High Resolut. Chromatogr. Chromatogr. Commun. 2*: 623–627.

Ponder, G. W., and Stewart, J. T. (1994). High-performance thin-layer chromatographic determination of digoxin and related compounds, digoxigenin bisdigitoxoside and gitoxin, in digoxin drug substance and tablets. *J. Chromatogr. 659*: 177–183.

Pothier, J., Galand, N., and Viel, C. (1991). Separation of alkaloids in plant extracts by overpressured layer chromatography with ethyl acetate as mobile phase. *J. Planar Chromatogr.—Mod. TLC 4*: 392–396.

Renger, B. (1993). Quantitative planar chromatography as a tool in pharmaceutical analysis. *J. AOAC Int. 76*: 7–13.

Roberson, J. (1991). The use of thin-layer chromatography in the analysis of drugs of abuse. In *Analysis of Drugs of Abuse*, T. A. Gough (Ed.). Wiley, Chichester, pp. 3–22.

Sarbu, C., and Cimpan, G. (1996). Determination of 1,4-benzodiazepines by quantitative TLC. A regression study employing the ladder of power. *J. Planar Chromatogr.—Mod. TLC 9*: 126–128.

Sherma, J. (1998). Planar chromatography. *Anal. Chem. 70*: 7R–26R.

Sherma, J., and Beim, M. (1978). Determination of caffeine in APC tablets by densitometry on chemically bonded C_{18} reversed phase thin layers. *HRC&CC, J. High Resolut. Chromatogr. Chromatogr. Commun. 1*: 308–309.

Sherma, J., and Rolfe, C. D. (1992). The analysis of analgesic tablets containing aspirin, acetaminophen, and caffeine by silica gel HPTLC: a student laboratory experiment. *J. Planar Chromatogr.—Mod. TLC 5*: 197–199.

Sherma, J., Stellmacher, S., and White, T. J. (1985). Analysis of analgesic tablets by quantitative high performance reversed phase TLC. *J. Liq. Chromatogr. 8*: 2961–2967.

Shinde, V. M., Tendolkar, N. M., and Desai, B. S. (1994). Simultaneous determination of theophylline and etofylline in pharmaceutical dosage forms by quantitative TLC. *J. Planar Chromatogr.—Mod. TLC 7*: 133–136.

Shirke, P. P., Patel, M. K., Tamhane, V. A., Tirodkar, V. B., and Sethi, P. D. (1994). Determination of amitriptyline hydrochloride and chlordiazepoxide in combined dosage forms by high performance thin layer chromatography. *East Pharm. 37*: 179–180.

Siouffi, A. M., Wawrzynowicz, F., Bressolle, F., and Guiochon, G. (1979). Problems and applications of reversed-phase thin-layer chromatography. *J. Chromatogr. 186*: 563–574.

Spell, J. C., and Stewart, J. T. (1994). Quantitative analysis of *p*-aminosalicylic acid and its major impurity *m*-aminophenol in *p*-aminosalicylic acid drug substance by HPTLC and scanning densitometry. *J. Planar Chromatogr.—Mod. TLC 7*: 472–473.

Stahl, E. (1973). *Drug Analysis by Chromatography and Microscopy. A Practical Supplement to Pharmacopoeias.* Ann Arbor Science Publishers, Ann Arbor, MI.

Szepesi, G., and Nyiredy, S. (1996). Pharmaceuticals and drugs. In *Handbook of Thin-Layer Chromatography*, 2nd ed., J. Sherma and B. Fried (Eds.). Marcel Dekker, New York, pp. 819–876.

Tam, M. N., Nikolova-Damyanova, B., and Pyuskyulev, B. (1995). Quantitative thin layer chromatography of indole alkaloids. II. Catharanthine and vindoline. *J. Liq. Chromatogr. 18*: 849–858.

Thomas, M. H., Soroka, K. E., Simpson, R. M., and Epstein, R. L. (1981). Determination of sulfamethazine in swine tissues by quantitative thin-layer chromatography. *J. Agric. Food Chem. 29*: 621–624.

Tivert, A. M., and Backman, A. (1993). Separation of the enantiomers of β-blocking drugs by TLC with a chiral mobile phase additive. *J. Planar Chromatogr.—Mod. TLC 6*: 216–219.

Tombesi, O. L., Maldoni, B. E., Bartolome, E. R., Haurie, H. M., and Faraoni, M. B. (1994). Purification of alkaloids by thin layer chromatographic decomposition of their picrates. *J. Planar Chromatogr.—Mod. TLC 7*: 77–79.

Touchstone, J. C. (1992). *Practice of Thin Layer Chromatography*, 3rd ed. Wiley, New York.

Wagner, H., Bladt, S., and Zgainski, E. M. (1984). *Plant Drug Analysis: A Thin Layer Chromatography Atlas.* Springer Verlag, Berlin.

White, D. J., Stewart, J. T., and Honigberg, I. L. (1991a). Quantitative analysis for chlordiazepoxide hydrochloride and related compounds in drug substance and tablet dosage form by HPTLC and scanning densitometry. *J. Planar Chromatogr.—Mod. TLC 4*: 330–332.

White, D. J., Stewart, J. T., Honigberg, I. L., and Irwin, L. (1991b). Quantitative analysis of diazepam and related compounds in drug substance and tablet dosage form by HPTLC and scanning densitometry. *J. Planar Chromatogr.—Mod. TLC 4*: 411–415.

Wilson, I. D. (1991). Recent advances in thin-layer chromatography application to drug metabolism studies. *J. Biopharm. Sci. 2*: 173–189.

Wilson, I. D. (1996). Thin-layer chromatography: A neglected technique. *Ther. Drug Monit. 18*: 484–492.

23

Miscellaneous Applications

I. INTRODUCTION

This chapter provides thin-layer chromatography information on three disparate subjects that may be of interest to separation scientists. Background information is presented in the introduction section of each experiment. The experiments in this chapter are concerned with the TLC separation of organic acids (Experiment 1), the TLC separation of penicillins (Experiment 2), and the TLC experiments of pesticides (Experiments 3 and 4).

II. DETAILED EXPERIMENTS

A. Experiment 1. Separation of Organic Acids by TLC

1. Introduction

Organic acids are readily separated by TLC, and those acids of most interest to the biologist are the Krebs-cycle intermediates such as citric, malic, succinic, and lactic. These compounds are referred to as carboxylic acids because of the presence of the carboxyl (CO_2H) group. Malic and succinic acids contain two carboxyl groups (dicarboxylic acids), whereas citric acid is a tricarboxylic compound. Numerous studies are available on paper chromatographic determinations of dicarboxylic and tricarboxylic acids from plants (Harborne, 1984). Many of these studies can be modified for the TLC separation of organic acids on cellulose layers.

Plants accumulate considerable amounts of organic acids, particularly in the cell sap of their tissues. Invertebrates also accumulate organic acids as a result of aerobic and anaerobic fermentative processes (von Brand, 1973) and may excrete these acids, particularly as lactic and succinic acids. For the most part, the details of the excretion of organic acids in both free-living and parasitic invertebrates have not been elucidated; for further discussion, see von Brand (1973).

Organic acids are soluble in water, colorless, and mainly nonvolatile. They can be detected by their reactions with acid–base indicators such as bromcresol green or bromthymol blue. Although most separations of organic acids have been achieved by paper chromatography (PC), some literature is available on TLC separations by cellulose, silica gel, and polyamide layers (Harborne, 1984).

Most separation studies have used plant tissues which are extracted with warm ethanol or methanol. The extracts are then filtered and the supernatant used directly for PC or TLC analysis. Harborne (1984) has discussed the use of cellulose, silica gel, and polyamide layers for the separation of plant organic acids by TLC. Cellulose layers may be used with similar solvents as in paper chromatography. Silica gel plates should be developed with solvent systems such as methanol-5 M NH$_4$OH (4:1) or benzene–methanol–acetic acid (79:14:7). The latter solvent system will separate maleic ($R_F = 0.07$) from fumaric acid ($R_F = 0.23$). Polyamide layers can be used to separate the acids of the tricarboxylic acid cycle with a solvent system of di-isopropyl ether–petroleum ether–carbon tetrachloride–formic acid–water (5:2:20:8:1). Less information on the TLC of organic acids from animal tissues is available, but some information can be found in Rasmussen (1967), Passera et al. (1964), Bleiweiss et al. (1967), Lupton (1975), and Hanai (1982).

The "TLC Applications" catalogue of Machery-Nagel (MN) (available from MN by writing to Machery-Nagel, GmbH & Co., KG, P.O. Box 101352, D-52313, Duren, Germany) describes separation of organic acids on silica gel glass plates (SIL G-25) dried 1 h at 105°C prior to use. Separation was based on the application of 1 µl aliquots of various organic acids developed by ascending chromatography in a saturated chamber for a distance of 15 cm from the origin (1.5 h) with a mobile phase of n-pentyl formiate–chloroform–formic acid (20:70:10). The acids were detected by dipping the plates in an ethanolic solution of bromocresol green (gives yellow spots on a blue background). The R_F values for various organic acids were as follows; adipic (0.36), glutaric (0.41), succinic (0.28), malonic (0.20), fumaric (0.37), maleic (0.30), mandelic (0.44), α-hydroxybuturic (0.38), lactic (0.24), glycolic (0.16), malic (0.08), and citric (0.04).

Shalaby (1996) discussed applications of TLC to determine organic acids in food preservatives (i.e., benzoic, citric, and lactic acids) in foods and soft drinks. He also described an experiment for separating 10 to 20 µl samples of 0.5% aqueous solutions of tartaric, malic, lactic, and succinic acids on cellulose TLC plates by ascending chromatography using various mobile phases including n-butanol–formic acid (95:5). Detection was achieved by spraying the plates with acridine solution (250 mg of acridine in 200 ml of ethanol) and viewing them under UV light after drying.

Tyman (1996) provided a good discussion of applications of TLC to the analysis of carboxylic acids used as preservatives or as medicinal agents. Of particular interest is mention of the use of TLC for the simultaneous detection of benzoic, citric, and lactic acids in soft drinks, benzoic acid in textile fibers and food preparations, and hydroxybenzoic acid in wines. A variety of techniques and layers (including polyamide and silanized silica gel) have been used in these separations.

The experiment described here shows the utility of cellulose TLC for the separation of organic acids. Modifications of this design should result in successful separations of organic acids found in plant and animal tissues.

2. Preparation of Standards

Obtain the following substances: DL-lactic acid (85%); DL-malic acid; malonic acid–disodium salt monohydrate; sodium succinate hexahydrate; citric acid monohydrate. Except for the lactic acid, all of these substances are solids at room temperature. Prepare each solid as 5-µg/µl standards in deionized water by dissolving 10 mg of each in 2 ml of deionized water. Prepare lactic acid as a 6-µg/ µl standard as follows: Weigh a test tube and then add enough lactic acid with a pipet to increase the weight of the tube by 12 mg. Add 2 ml of deionized water to the tube. This will produce a lactic acid solution of 6 µg/µl.

3. Layer

Precoated Avicel microcrystalline cellulose plates, 20 × 20 cm (Analtech), used as received.

4. Application of Initial Zones

With a Microcap micropipet or a 10-µl digital microdispenser, spot 2 and 4 µl of each standard at separate origins 2 cm from the bottom of the plate.

5. Mobile Phase

n–Butanol–formic acid–water (40:10:50).

6. Development

Develop in a tank saturated with the mobile phase for a distance of 12 cm past the origin.

7. Detection

Remove the plate from the tank and dry it thoroughly with a gentle stream of air. Spray the plate with a solution of 40 mg of bromthymol blue in 100 ml of 0.010 M (0.40 g/L) sodium hydroxide. The organic acids will appear as blue spots on a slightly yellow background without heating. The sensitivity of this detection reagent is about 10–20 μg for the organic acids used in this study.

8. Results and Discussion

Based on earlier results from paper chromatography, the approximate R_F values should be as follows: citric acid, 0.37; malic acid, 0.45; malonic acid, 0.63; and lactic acid, 0.67.

Additional organic acids can also be separated in this system, e.g., tartaric with a probable R_F of 0.22, succinic with an approximate R_F of 0.74, and fumaric with an approximate R_F of 0.89. Tricarboxylic acids can also be detected by reaction with pyridine–acetic anhydride (7:3) where citric acid gives a yellow color and other acids give either brown or brown-yellow colors (Harborne, 1984).

Additional organic acid applications have been reviewed in Sherma and Fried (1982, 1984), Hanai (1982) and Sherma (1986a, 1988a, 1992a). Significant studies on carboxylic acids and related compounds include the following. Carboxylic acids were separated by TLC on silica gel using the mobile phase isopropyl ether–formic acid–H_2O (90:7:3); spots were detected by spraying with methanolic–$CuSO_4 \cdot 5H_2O$ and quantified by reflectance densitometry at 690 nm (Gaberc-Porekar and Socic, 1979). Two-dimensional TLC on silica gel–cellulose mixed layers was used to separate lactic acid and citric acid cycle intermediates (Riley and Mix, 1980). Aconitic, citric, isocitric, itaconic, malic, maleic, glyoxylic, hippuric, and α-ketoglutamic acids, sodium pyruvate, and leucine were separated by TLC on silica gel and the spots were detected by spraying the plates with 0.8 M dicyclohexylcarbodiimide in butanol (Chen, 1982). Dicarboxylic acids were separated on silica gel layers with a chloroform–methanol–25% NH_4OH solvent system and detected by spraying with a solution of acridine or its derivatives (Sarbu et al., 1983). Dicarboxylic and polycarboxylic acids were separated on cellulose layers and detected with a $AgNO_3$–vanillin reagent (Radowitz et al., 1984). Carboxylic acids in wine and juice were quantified by HPTLC separation; detection was with a xylose–aniline reagent and densitometry was at 546 nm (Lin and Tanner, 1985). TLC was used to separate citric acid from trichloracetic acids on plates coated with $CaSO_4$ containing ZnO (Rathore et al., 1985).

B. Experiment 2. Separation of Penicillins by Reversed-Phase TLC

1. Introduction

Penicillins are usually separated by TLC on silica gel, but reversed-phase layers are also used. Because of the possibility of nucleophilic reactions of β-lactam molecules, which can happen after spotting penicillins on silica gel layers, the use of reversed-phase plates has certain advantages (Kreuzig, 1991).

The purpose of this experiment, adapted from Whatman TLC Technical Series, Volume 1, is to describe procedures for the separation of penicillins on Whatman reversed-phase C_{18} layers.

The penicillins are one of numerous classes of antibiotics. Wagman and Weinstein (1983) described antibiotics as substances, produced by microorganisms, that are capable of inhibiting the growth of microorganisms. They reviewed TLC procedures useful for the identification and differentiation of numerous antibiotics and considered the following classes: aminoglycosides, macrolides, penicillins, cephalosporins, tetracyclines, polypeptides, and polyenic antifungals.

Probably the best known classes are the penicillins and tetracyclines, and these are considered further in this chapter. For more detailed information on the TLC of antibiotics, see Sherma and Fried (1982, 1984), Wagman and Weinstein (1984), Sherma (1986a, 1990, 1992a, 1996), and Kreuzig (1991, 1996).

2. Preparation of Standards

Obtain the following individual penicillin standards (Sigma) and prepare 1-mg/ml solutions of each in methanol: ampicillin, penicillin G, methicillin, oxacillin, and cloxacillin.

3. Layers

Whatman KC_{18}, 20 × 20-cm layer, used as received without pretreatment.

4. Application of Initial Zones

Apply 5 μl each of the individual standards and mixtures to separate origins located 2–2.5 cm from the bottom edge of the plate using Microcap micropipets or a 10-μl digital microdispenser.

5. Mobile Phase

Methanol–0.1 M K_2HPO_4 (pH 8.7) (55:45).

6. Development

Develop for a distance of 11 cm from the origin in a paper-lined glass chamber that has been preequilibrated with the mobile phase for 10 min. Development time is about 1 h.

7. Detection

Detect spots with iodine vapor. The sensitivity of detection is 1 µg.

8. Results and Discussion

The approximate R_F values of the separated compounds are: ampicillin, 0.74; penicillin G, 0.58; methicillin, 0.65; oxacillin, 0.46; and cloxacillin, 0.42.

This exercise can also be done on Whatman $LKC_{18}F$, 20 × 20-cm preadsorbent plates (Whatman TLC Technical Series, Volume 3). The six individual penicillins are prepared as 1-mg/ml solutions in methanol, and 5 µl of each sample is applied to separate origins on the plate. The mobile phase is methanol–H_2O–1% $NaHCO_3$ (45:55:0.8), and the plate is developed for a distance of 10 cm from the origin in a glass tank that has been preequilibrated with the mobile phase for 10 min. Time of development is about 70 min. The spots are detected with iodine vapor and the approximate R_F values of the separated penicillins are as follow: cloxacillin, 0.18; oxacillin, 0.22; penicillin G, 0.34; methicillin, 0.41; ampicillin, 0.66; and 6-aminopenicillinic acid, 0.82.

Kreuzig (1996) described a TLC separation by ascending chromatography for 5 cm for the control of penicillin V fermentations using Merck HPTLC silica gel plates and a mobile phase of toluene–ethyl acetate–acetic acid (40:40:20); the dried plates were scanned at 268 nm and the R_F values for 4-hydroxy-penicillin V, penicillin V, and phenoxyacetic acid were 0.34, 0.50, and 0.60, respectively. Kreuzig (1996) reemphasized the fact that because of widespread use of tetracyclines in veterinary medicine (mainly as feed additives) and in agriculture (e.g., in honeybee culture to prevent foul brood of honeybees), numerous TLC procedures for tetracycline analysis are available.

Numerous authors [reviewed in Wagman and Weinstein (1983)] have used silica gel with different solvent mixtures to characterize a variety of penicillins. Hendrickx et al. (1984) described the separations of 18 penicillins on silica gel and silanized silica gel using 35 mobile phases. Khier et al. (1984) separated penicillin derivatives from closely related degradation products by reversed-phase TLC using the mobile phase methanol–aqueous KH_2PO_4–methyl cyanide (8:4:1).

Kreuzig (1991) has contributed a valuable review on TLC and HPTLC analysis of various antibiotics covering the literature from 1980 to 1990. This reference includes extensive information on sample preparation, layers, mobile phases, detection reagents, and in situ quantification of penicillins, cephalosporins, aminoglycosides, macrolides, tetracyclines, and miscellaneous other antibiotics. Kreuzig (1996) extended his earlier coverage of antibiotics in Kreuzig (1991) to include significant literature references through 1994.

Sherma (1996) cited eight references on applications of TLC in antibiotics based on examining the significant literature in analytical chemistry from about

1993 to 1995. These references included TLC studies mainly on the following antibiotics: polyethers, tetracyclines and oxytetracyclines, neomycin sulfate, minocycline, oxolinic acid, erythromycin, tylosin, and macrolides.

As mentioned in the Introduction (Section 1), TLC has been widely used for the separation and determination of tetracyclines. Thus, Gatsonis and Ageloudis (1982) separated tetracycline, oxytetracycline, and chlortetracycline by TLC on kieselguhr plates prepared from a slurry with 5% EDTA solution; acetone–ethyl acetate–water (20:10:3) was used for development, and the plates were scanned at 400 nm for tetracycline, and 363 nm for oxytetracyclines and chlortetracyclines. Oka et al. (1983) separated various tetracyclines on precoated silica gel plates using the mobile phase chloroform–methanol–5% disodium EDTA (65:20:5), and the absorbance of the spots was recorded by densitometry. Oka et al. (1984a, 1984b) described the use of reversed-phase thin layers and the detection reagent Fast Violet B for the analysis of tetracycline antibiotics.

Residual tetracyclines in animal tissues were determined by solid-phase extraction on a C_{18} SPE column followed by TLC on an EDTA-treated silica gel plate and UV densitometry (Ikai et al., 1987). Seven tetracyclines were separated on silica gel layers impregnated with buffered EDTA and a mobile phase of dichloromethane–methanol–water (59:35:6) by Naidong et al. (1989). Residues of tetracyclines along with aminoglycosides, macrolides, and chloramphenicol were screened in poultry meat by cellulose and silica gel TLC following extraction and cleanup on C_{18} SPE columns (Vega et al., 1990).

C. Experiment 3. Thin-Layer Chromatography of Insecticides

1. Introduction

The purpose of Experiments 3 and 4 is to present methods for the TLC determination of pesticides. Experiment 3 describes procedures for the separation and detection of organochlorine (OCl), organophosphorus or organophosphate (OP), and N-methylcarbamate insecticides and metabolites (Sherma and Bloomer, 1977; Sherma et al., 1977, 1978; Sherma, 1978). Experiment 4 describes the quantitative TLC determination of three classes of herbicides after isolation from water by conventional separatory funnel extraction or solid-phase extraction (SPE) (Sherma, 1986c; Sherma and Boymel, 1983).

Pesticides may be classified according to either their use (e.g., insecticides, herbicides, fungicides, fumigants) or their chemical type. Organohalides, organophosphates, and N-methylcarbamates are among the most important types of insecticides, and these classes are studied in this experiment. The general techniques of TLC analysis are similar in all cases. Organochlorine pesticides are not acutely toxic, but they are persistent in the environment and are stored in fatty tissues. Organophosphorus compounds are acutely toxic but are generally nonpersistent and not stored in fatty tissues. The carbamate insecticides are relatively low in acute toxicity and are nonpersistent. The pharmacological action of both OP and carbamate insecticides is based on their cholinesterase-inhibiting properties.

The chemistry of different classes of pesticides has been reviewed by Benson (1969, 1971), Benson and Jones (1967), and Mel'nikov (1971). The chemical structures, trade and common names, and chemical, physical, analytical, toxicity, and usage data are contained in the *Pesticide Dictionary* (1992) and the *Agrochemicals Handbook* (1991).

The TLC of all classes of insecticides, herbicides, fungicides, and related industrial chemicals (PCBs, chlorinated dioxins and furans, etc.) has been reviewed by Sherma (1973, 1980, 1986b, 1991a, 1991b, and 1992b) and by Follweiler and Sherma (1984). These references include extensive information on sample preparation, layers, mobile phases, detection reagents, qualitative identification, and quantification by scanning densitometry for many pesticides in a wide variety of sample matrices. Fodor-Csorba (1996) has contributed an extensive review on the uses of TLC for the analysis of pesticides. The review contains 12 tables, eight figures, and 183 references.

Sherma (1996) reviewed the literature on procedures and applications of TLC/HPTLC for studies on pesticides (i.e., insecticides, herbicides and fungicides), based on coverage of the significant literature in analytical chemistry from about 1993–1995. In a more recent review (Sherma, 1997), he discussed the advances in the applications of TLC/HPTLC for the separation, detection, and determination of pesticides and related compounds from the 1994–1996 literature. Analyses were covered for various samples—food and environmental, residues of pesticides (including insecticides), herbicides, and fungicides belonging to different chemical classes. He also included references on formulation analysis and the use of radio–TLC for studies on pesticide metabolism. Representative chromatographic systems for practical separations and analyses of various pesticides were given.

Bladek (1996) reviewed the uses of TLC in the analysis of pesticides in the environment. Shalaby (1996) discussed the uses of TLC for the analysis of pesticides from fatty and nonfatty foods. Stahr (1996) described the uses of TLC in veterinary toxicology for pesticide analysis. In addition to coverage of the usual insecticides, he provided considerable information on the TLC of rodenticides and avicides (particularly anticoagulants such as warfarin and alkaloids such as strychnine).

The Machery-Nagel catalog on "TLC Applications" (address for obtaining this catalog was given earlier in this chapter) lists 15 experiments (applications 92–106) for TLC studies on pesticides and herbicides. These applications are recommended for perusal by chromatographers interested in this area of TLC.

2. Preparation of Standards

Obtain organochlorine and carbamate pesticide standards from a commercial source such as Polyscience Corp. or ChemService (Westchester, PA). Obtain standards of phorate and its metabolites from the manufacturer, American Cyanamid Co., Princeton, NJ. Prepare individual and mixture solutions at a level of 1 μg/μl by dissolving 25 mg of each compound in 25 ml of solvent. For the study of chlorinated pesticides, prepare individual solutions and a mixture of heptachlor, endosulfan (Thiodan), heptachlor epoxide, lindane, dieldrin, and dicofol (Kelthane) in hexane. For organophosphate pesticides, use phorate and its metabolites phoratoxon, phorate sulfoxide, and phoratoxon sulfoxide in ethyl acetate. For carbamates, use 3-keto carbofuran (a metabolite of carbofuran), carbaryl, and metalkamate (Bux) in ethyl acetate.

3. Layers

Silica gel G precoated 20 × 20-cm plates (Analtech or Brinkmann) for organocholines and silica gel K-5 or K-6 plates (Whatman) for organophosphates and carbamates, used as received without pretreatment. Plates with preadsorbent spotting area can be used.

4. Application of Initial Zones

Apply 1μl each of the individual pesticides and mixtures to separate origins located 2–2.5 cm from the bottom edge of the plate using Microcap micropipets.

5. Mobile Phases

n-Hexane–acetone (99:1) for organochlorines, hexane–acetone–chloroform (65:30:5) for organophosphates, and hexane–acetone–chloroform (75:15:5) for carbamates.

6. Development

Use rectangular glass N-tanks in each case. Develop organochlorines in a paper-lined, saturated tank until the mobile phase reaches the top of the plate. Develop OP pesticides and carbamates in unsaturated tanks for a distance of 10 cm above the origin.

7. Detection

Chromatograms are dried in a fume hood for 15 min. Support the plate containing the organochlorine spots almost vertically in the fume hood and spray the layer thoroughly with silver nitrate detection reagent. Begin at the top of the layer and gradually move down with a horizontal sweeping motion, wetting the gel completely. Prepare the reagent as follows: Dissolve 20 g of reagent-grade $AgNO_3$ in 100 ml of distilled water and store in a dark bottle; mix 4 ml of this solution with 100 ml of redistilled acetone plus 10 ml of distilled water plus 6 ml of concentrated ammonium hydroxide plus 8 drops of 2-phenoxyethanol. After spraying, leave the plate in the fume hood for 30 min and then place the plate under a UV light source such as a germicidal lamp. Irradiate until the pesticides become visible as purplish-black spots on a white or canary yellow background. The optimum irradiation time and lamp-to-layer distance must be determined to produce the maximum spot-to-background contrast.

Detect phorate insecticide and its metabolites by dipping the plate for 5 sec in a 5% solution of magnesium chloride hexahydrate in methanol and then air drying. Next dip for 5 sec into a 0.3% solution of N,2,6-trichlorobenzoquinoneimine in hexane (stored under refrigeration in an amber bottle and prepared fresh weekly). Heat for 25 min at 100°C to produce orange to red-orange spots.

Detect carbamates by dipping the dried chromatogram into methanolic potassium hydroxide solution (56.0 g/L) for 5 sec, air drying, and then dipping for 5 sec in a solution of 50 mg of p-nitrobenzenediazonium fluoborate in 180 ml of acetone plus 20 ml of diethylene glycol. After air drying again, 3-ketocarbofuran and matalkamate will appear as pink spots and carbaryl deep blue.

8. Results and Discussion

One microgram of each compound is readily visualized with the described detection reagents, and sensitivity as low as 100 ng is achieved for most compounds with proper technique. The standard solutions can be diluted 1:10 to produce a concentration level of 100 ng/µl and 1–10 µl applied for chromatography if spots are too large and poorly resolved at the 1-µg level. Preadsorbent plates produce compact, band-shaped zones.

The chlorinated compounds should be completely separated in the order listed in Section 2, with heptachlor nearest the solvent front and dicofol closest to the origin. Other common organochlorine insecticides that can be detected and identified by this TLC procedure include aldrin, DDE, DDT, TDE (DDD), endrin, BHC, daconil, methoxychlor, captan, toxaphene, and chlordane. Alternative mobile phases that should be tested for any particular mixture not separated by hexane–acetone (99:1) include hexane (recommended for aldrin, heptachlor, p,p'-DDE, o,p'- and p,p'-DDT and p,p'-TDE); benzene–hexane (1:1); and hexane–methanol (99:1) (applicable to more polar pesticides).

Approximate R_F values and zone colors and shapes of the phorate compounds are as follows: phorate (0.91, red-orange, elliptical); phoratoxon (0.59, orange, elliptical); phorate sulfoxide (0.34, red-orange, elliptical); and phoratoxon sulfoxide (0.12, orange, round). Other phorate metabolites that can be studied in this experiment include phorate sulfone and phoratoxon sulfone.

Metalkamate will give an approximate R_F value of 0.59, carbaryl 0.48, and 3-keto carbofuran 0.25. Other N-methyl carbamates that are detectable in the chromatographic system described are carbofuran (R_F 0.47, pink), propoxur

(0.46, blue), aminocarb (0.52, yellow), and mexacarbate (0.58, yellow). The TCQ reagent used to detect the phorate compounds will detect *N*-methyl carbamates if initial dipping in 3% potassium hydroxide replaces the magnesium chloride dip.

The detection procedures described here are sufficiently sensitive, stable, and linear to allow densitometry of residues in various sample substrates. These procedures are fully described in the references in subsection 1.

D. Experiment 4. Quantitative Determination of Herbicides in Water

1. Introduction

The purpose of this experiment is to demonstrate the determination of herbicide residues in water. It is applicable to the analysis of spiked control (pesticide-free) water, tap water, or environmental water samples from creeks, lakes, ponds, or rivers. Herbicides are extracted by use of a separatory funnel or SPE on a C_{18} microcolumn, and quantification is carried out by scanning densitometry after TLC separation and zone detection.

Herbicides are the most widely used type of pesticides, and triazines, *N*-arylcarbamates, and chlorophenoxy acids are among the most important classes of herbicides. Herbicides are used to control unwanted grasses and weeds in a variety of crops, such as cotton, alfalfa, sunflower, sorghum, rice, and a variety of fruits and vegetables. The following experiment involves the analysis of atrazine as an example of the triazines, 2,4-D for the chlorophenoxy acid herbicides, and chlorpropham for the carbamates.

2. Preparation of Standards

Obtain atrazine, 2,4-D, and chlorpropham analytical standards from a commercial source such as PolyScience (Niles, IL). Prepare 10.0-mg/ml stock solutions in ethyl acetate (atrazine, chlorpropham) or diethyl ether–hexane (6 : 4 v/v) (2,4-D). Prepare TLC standards and spiking solutions as needed by appropriate dilution of these stock solutions.

3. Preparation of Samples

Actual environmental water samples or control (blank) samples spiked with the pesticides can be analyzed.

Analyze a 250-ml water sample containing 10 ppb of atrazine by passing through a 75-ml reservoir into a C_{18} extraction column (6 ml, J. T. Baker, Phillipsburg, NJ) held in a Baker-12 vacuum manifold, at a flow rate of 5 ml/min. (Prewash the column before use with 12 ml of methanol followed by 12 ml of deionized water.) Wash the column with 10 ml of deionized water (discard) and dry the column by drawing vacuum for 5 min. Remove the column from the manifold and elute the atrazine into a 4-ml screw-top vial by passing three 1-ml portions of methanol through the column with gentle pressure from a pipet bulb. Evaporate the eluate just to dryness under a gentle stream of nitrogen with the vial in a 40°C water bath. Dissolve the residue in 100 μl of methylene chloride to prepare the solution for TLC analysis.

Analyze a 250-ml sample of water containing 10 ppb of 2,4-D in the same way, except acidify the sample initially to pH 2 with concentrated HCl, precondition the column with acidified water, and pass 10 ml of acidified water before eluting the 2,4-D with methanol.

Analyze a 1000-ml sample of water containing 500 ppb of chlorpropham by extraction in a separatory funnel with 100-, 50-, and 50-ml portions of methylene chloride after adding 50 ml of saturated NaCl solution. Concentrate the combined extracts on a rotary evaporator to a volume of several milliliters, transfer quantitatively to a calibrated vial, and evaporate just to dryness with nitrogen gas. Dissolve the residue in 200 μl of ethyl acetate to prepare the solution for TLC analysis.

4. Layer

Use analtech silica gel GF plates (20 × 20 cm) impregnated by dipping into ammoniacal silver nitrate solution for atrazine and 2,4-D. Prepare the dip solution by mixing 216 ml of fresh silver nitrate stock solution (17 g of $AgNO_3$ dissolved in 25 ml of deionized water, made up to 1 L with acetone), 20 ml of deionized water, and 6 ml of 6 M ammonium hydroxide. Dip for several seconds.

Whatman $KC_{18}DF$ chemically bonded reversed-phase plates (20 × 20 cm) for chlorpropham.

Channeled plates with preadsorbent area facilitate scanning and sample application. If preadsorbent plates are used, do not impregnate the preadsorbent area with silver nitrate detection reagent. Preclean plates by development to the top in chloroform–methanol (1:1).

5. Application of Initial Zones

Spot duplicate 20-μl aliquots of the atrazine or 2,4-D sample solution (equivalent to 500 ng for 100% recovery from a 10-ppb solution) along with 300 ng, 500 ng, and 700 ng of standard (3 μl, 5 μl, and 7 μl of a 100 ng/μl standard solution). Spot duplicate 1-μl aliquots of the chlorpropham sample solution (equivalent to 625 ng for 100% recovery from a 500-ppb solution) along with 250 ng, 500 ng, and 1000 ng of standard.

6. Mobile Phase

Chloroform–acetone (9:1) for atrazine, hexane–glacial acetic acid–diethyl ether (72:30:18) for 2,4-D, and methanol–acetonitrile–tetrahydrofuran–water (50:15:8:27) for chlorpropham.

7. Development

Develop plates in a paper-lined glass chamber presaturated with solvent for 10 min for a distance of 12 cm above the origin.

8. Detection

Detect atrazine and 2,4-D as black spots on a white to light gray background by exposure of the air-dried plate to UV light from a Hanovia 679A 176W germicidal lamp (or equivalent) for 10 min.

Detect chlorpropham as a red-purple spot on a white background by spraying the air-dried plate lightly and evenly with 6 N ethanolic hydrochloric acid (60 ml of concentrated HCl + 50 ml of absolute ethanol). Cover the wet but not soaked layer with a clean glass plate and heat for 10 min in an oven set at 180°C. Cool the plate and spray with 1% sodium nitrite solution (1.0 g of $NaNO_2$ in 20 ml of distilled water diluted to 100 ml with 2 N ethanolic HCl). Air dry the layer and spray with fresh 1% N-(1-naphthyl)ethylenediamine dihydrochloride solution (1.0 g of reagent dissolved in 10 ml of H_2O, and diluted to 100 ml with ethanol) until visibly saturated.

9. Quantification

Scan the spots at the wavelength of maximum absorption, as determined by measurement of the in situ spectrum of each compound after detection if a densitometer that allows spectral scanning is available. With the Kontes Model 800 fiber-optics densitometer, the white phosphor (400-nm peak, 300-nm bandwidth) was used. Calculate a calibration equation from the areas and weights of the series of standards and calculate the weight of pesticide in the sample from the average areas of the duplicate sample aliquots using this equation. If analyzing a spiked sample, calculate recovery by comparison to the theoretical amount. If analyzing a real sample, calculate concentration of the pesticide by multiplying the calculated sample weight times the ratio of the reconstitution volume to the spotted volume and divide by the weight of sample.

10. Results and Discussion

The approximate R_F values in the respective TLC systems specified above are: atrazine, 0.64; 2,4-D, 0.54; and chlorpropham, 0.32.

The silver nitrate detection reagent functions by reduction of Ag^+ to metallic silver by the UV light at the location of the chlorinated herbicides on the layer. Chlorpropham is detected by hydrolysis with HCl to a primary amine, followed by diazotization with $NaNO_2$ and coupling with the Bratton–Marshall reagent, N-(1-naphthyl)ethylenediamine.

Recoveries of pesticides from spiked water samples using the described procedures were in the range 78–100%. Extracts from water samples containing impurities that are co-extracted and detected may require an additional cleanup step involving techniques such as solvent partitioning (Sherma and Koropcheck, 1977) or alumina column chromatography (Sherma and Boymel, 1983; Sherma and Miller, 1980).

The quantification procedures described are suitable for water samples with the pesticide concentrations specified. For other concentrations, adjustments must be made in the sample volume, reconstitution volume, and/or aliquot volume spotted so that the area of the sample spot is bracketed by the areas of at least two standards.

SPE is also an efficient procedure for recovery of organochlorine (Sherma, 1988b) and organophosphorus (Sherma and Bretschneider, 1990) pesticides from water and can be combined with the TLC procedures described in Experiment 3 for quantification of these compounds in water. SPE and TLC have also been used to quantify the benzoylurea insecticide diflubenzuron in water (Sherma and Rolfe, 1993).

REFERENCES

Agrochemicals Handbook, 3rd Edition. (1991). The Royal Society of Chemistry, CRC Press, Boca Raton, FL.

Benson, W. R. (1969). Chemistry of pesticides, *Ann. N.Y. Acad. Sci. 160*: 7–29.

Benson, W. R. (1971). Literature of pesticide chemistry. II. *J. Assoc. Off. Anal. Chem. 54*: 192–201.

Benson, W. R., and Jones, H. A. (1967). Literature of pesticide chemistry. *J. Assoc. Off. Anal. Chem. 50*: 22–28.

Bladek, J. (1996). Thin-layer chromatography in environmental analysis. In *Practical Thin-Layer Chromatography—A Multidisciplinary Approach*, B. Fried and J. Sherma (Eds.). CRC Press, Boca Raton, FL, pp. 153–168.

Bleiweiss, A. S., Reeves, H. C., and Ajl, S. J. (1967). Rapid separation of some common intermediates of microbial metabolism by thin layer chromatography. *Anal. Biochem. 20*: 335–338.

Chen, S. C. (1982). Dicyclohexycarbodiimide as a simple and specific fluorgenic spray reagent for some di- and tricarboxylic acids. *J. Chromatogr. 238*: 480–482.

Fodor-Csorba, K. (1996). Pesticides. In *Handbook of Thin-Layer Chromatography*, 2nd ed., J. Sherma and B. Fried (Eds.). Marcel Dekker, New York, pp. 753–817.

Follweiler, J. M., and Sherma, J. (1984). *Handbook of Chromatography—Pesticides*, Vol. 1. CRC Press, Boca Raton, FL.

Gaberc-Porekar, V., and Socic, H. (1979). New chromogenic reagent for carboxycylic acids on thin-layer plates. *J. Chromatogr. 178*: 307–310.

Gatsonis, C. D., and Ageloudis, C. A. (1982). Quantitative determination of antibiotics by means of densitometry on TLC. Part 2: Determination of tetracyclines. *Pharmazie 37*: 649–650.

Hanai, T. (1982). *Handbook of Chromatography: Phenols and Organic Acids*. CRC Press, Boca Raton, FL.

Harborne, J. B. (1984). *Phytochemical Methods—A Guide to Modern Techniques of Plant Analyses*, 2nd ed. Chapman & Hall, London.

Hendrickx, S., Roets, E., Hoogmartens, J., and Vanderhaeghe, H. (1984). Identification of penicillins by thin-layer chromatography. *J. Chromatogr. 291*: 211–218.

Ikai, Y., Oka, H., Kawamura, N., Yamada, M., Harada, K., and Suzuki, M. (1987). Improvement of chemical analysis of antibiotics. XIII. Systematic simultaneous analysis of residual tetracyclines in animal tissues using thin-layer and high performance liquid chromatography. *J. Chromatogr. 411*: 313–323.

Khier, A. A., Blaschke, G., and El Sadek, M. (1984). Spectrodensitometric determination of some penicillin derivatives. *Anal. Lett. 17*: 1667–1675.

Kreuzig, F. (1991). Antibiotics. In *Handbook of Thin-Layer Chromatography*, J. Sherma and B. Fried (Eds.). Marcel Dekker, New York, pp. 407–437.

Kreuzig, F. (1996). Antibiotics. In *Handbook of Thin-Layer Chromatography*, 2nd ed., J. Sherma and B. Fried (Eds.). Marcel Dekker, New York, pp. 445–480.

Lin, L., and Tanner, H. (1985). Quantitative HPTLC analysis of carboxylic acids in wine and juice. *HRC&CC, J. High Resolut. Chromatogr. Chromatogr. Commun. 8*: 126–131.

Lupton, C. J. (1975). Qualitative analysis of carboxylic acids by partition thin-layer chromatography. *J. Chromatogr. 104*: 223–224.

Mel'nikov, N. N. (1971). Chemistry of pesticides. *Residue Rev. 36*: 1–480.

Naidong, W., Cachet, T., Roets, E., and Hoogmartens, J. (1989). Identification of tetracyclines by TLC. *J. Planar Chromatogr.—Mod. TLC 2*: 424–429.

Oka, H., Uno, K., Harada, K. I., Kaneyama, Y., and Suzuki, M. (1983). Improvement of the chemical analysis of antibiotics. I. Simple method for the analysis of tetracyclines on silica gel high-performance thin-layer plates. *J. Chromatogr. 260*: 457–462.

Oka, H., Uno, K., Harada, K., and Suzuki, M. (1984a). Improvement of chemical analysis of antibiotics. III. Simple method for the analysis of tetracyclines on reversed phase thin layer plates. *J. Chromatogr. 284*: 227–234.

Oka, H., Uno, K., Harada, K., Hoyashi, M., and Suzuki, M. (1984b). Improvement of chemical analysis of antibiotics. VI. Detection reagents for tetracyclines in thin-layer chromatography. *J. Chromatogr. 295*: 129–239.

Passera, C., Pedrotti, A., and Ferrari, G. (1964). Thin-layer chromatography of carboxylic acids and ketoacids of biological interest. *J. Chromatogr. 14*: 289–291.

Pesticide Dictionary. (1992). In *Farm Chemicals Handbook*. Meister Publishing Co., Willoughby, OH.

Radowitz, W., Mueller, H., Mannsfeld, M., and Fiebig, W. (1984). Thin-layer chromatography of di- and polycarboxylic acids. *Z. Chem. 24*: 267–269.

Rasmussen, H. (1967). Separation and detection of carboxylic acids by thin-layer chromatography. *J. Chromatogr. 26*: 512–514.

Rathore, H. S., Kumari, K., and Agrawal, M. (1985). Quantitative separation of citric acid from trichloroacetic acid on plates coated with calcium sulfate containing zinc oxide. *J. Liq. Chromatogr. 8*: 1299–1317.

Riley, R. T., and Mix, M. X. (1980). Thin-layer separation of citric acid cycle intermediates, lactic acid, and the amino acid taurine. *J. Chromatogr. 189*: 286–288.

Sarbu, C., Horn, M., and Marutoui, C. (1983). Thin-layer chromatographic detection method for dicarboxylic acids. *J. Chromatogr. 281*: 345–347.

Shalaby, A. R. (1996). Thin-layer chromatography in food analysis. In *Practical Thin-Layer Chromatography—A Multidisciplinary Approach*, B. Fried and J. Sherma (Eds). CRC Press, Boca Raton, Fl., pp. 169–192.

Sherma, J. (1973). Thin layer chromatography: Recent advances. In *Analytical Methods for Pesticides and Plant Growth Regulators*, Vol. VII, G. Zweig and J. Sherma (Eds.). Academic Press, New York, Chap. 1.

Sherma, J. (1978). Determination of pesticides in water by densitometry on preadsorbent TLC plates. *Am Lab. 10(10)*: 105–109.

Sherma, J. (1980). Quantitative TLC. In *Analytical Methods for Pesticides and Plant Growth Regulators*, Vol. XI, G. Zweig and J. Sherma (Eds.). Academic Press, New York, Chap. 3.

Sherma, J. (1986a). Thin-layer and paper chromatography. *Anal. Chem. 58*: 69R–81R.

Sherma, J. (1986b). Thin layer chromatography. In *Analytical Methods for Pesticides and Plant Growth Regulators*, Vol. XIV, G. Zweig and J. Sherma (Eds.). Academic Press, New York, Chap. 1.

Sherma, J. (1986c). Determination of triazine and chlorophenoxy herbicides in natural water samples by solid phase extraction and quantitative thin layer chromatography. *J. Liq. Chromatogr. 9*: 3433–3438.

Sherma, J. (1988a). Thin-layer and paper chromatography. *Anal. Chem. 60*: 74R–86R.

Sherma, J. (1988b). Determination of organochlorine insecticides in waters by quantitative TLC and C-18 solid phase extraction. *J. Liq. Chromatogr. 11*: 2121–2130.

Sherma, J. (1990). Planar chromatography. *Anal. Chem. 62*: 371R–381R.

Sherma, J. (1991a). Thin layer chromatography of pesticides. *J. Planar Chromatogr.—Mod. TLC 4*: 7–14.

Sherma, J. (1991b). Pesticides. *Anal. Chem. 63*: 118R–130R (and previous reviews by this author in biennial *Applications Reviews* issues published in odd-numbered years beginning in 1981).

Sherma, J. (1992a). Planar chromatography. *Anal. Chem. 64*: 134R–147R.

Sherma, J. (1992b). Pesticides. In *Chromatography*, 5th ed., E. Heftmann (Ed.). Elsevier, Amsterdam, Chap. 23 in Part B.

Sherma, J. (1996). Planar chromatography. *Anal. Chem. 68*: 1R–19R.

Sherma, J. (1997). Review: determination of pesticides by thin-layer chromatography. *J. Planar Chromatogr.—Mod. TLC 10*: 80–89.

Sherma, J., and Bloomer, J. (1977). Separation, detection, and densitometric determination of chlorinated insecticides on silica gel and aluminum hydroxide papers. *J. Chromatogr. 135*: 235–240.

Sherma, J., and Boymel, J. (1983). Densitometric quantification of urea, carbamate, and anilide herbicides on C_{18} reversed phase thin layers using Bratton–Marshall reagent. *J. Liq. Chromatogr. 6*: 1183–1192.

Sherma, J., and Bretschneider, W. (1990). Determination of organophosphorus insecticides in water by C-18 solid phase extraction and quantitative TLC. *J. Liq. Chromatogr. 13*: 1983–1989.

Sherma, J., and Fried, B. (1982). Paper and thin-layer chromatography. *Anal. Chem. 54*: 45R–57R.

Sherma, J., and Fried, B. (1984). Paper and thin-layer chromatography. *Anal. Chem. 56*: 48R–63R.

Sherma, J., and Koropchack, J. (1977). Determination of chlorophenoxy herbicides by densitometry on thin layer chromatograms. *Anal. Chim. Acta 91*: 259–266.

Sherma, J., and Miller, N. T. (1980). Quantitation of the triazine herbicides atrazine and simazine in water by thin layer chromatography with densitometry. *J. Liq. Chromatogr. 3*: 901–910.

Sherma, J., and Rolfe, C. (1993). Determination of diflubenzuron residues in water by solid phase extraction and quantitative high performance thin layer chromatography. *J. Chromatogr. 643*: 337–339.

Sherma, J., Klopping, K. E., and Getz, M. E. (1977). Determination of pesticide residues by quantitative thin layer densitometry on microscope slides. *Am. Lab. 9(12)*: 66–73.

Sherma, J., Kovalchick, A. J., and Mack, R. (1978). Quantitative determination of carbaryl in apples, lettuce, and water by densitometry of thin layer chromatograms. *J. Assoc. Off. Anal. Chem. 61*: 616–620.

Stahr, H. M. (1996). Planar chromatography applications in veterinary toxicology. In *Practical Thin-Layer Chromatography—A Multidisciplinary Approach*, B. Fried and J. Sherma (Eds.). CRC Press, Boca Raton, FL, pp. 250–264.

Tyman, J. P. (1996). Phenols, aromatic carboxylic acids, and indoles. In *Handbook of Thin-Layer Chromatography*, 2nd ed., J. Sherma and B. Fried (Eds.). Marcel Dekker, New York, pp. 877–920.

Vega, M., Geshe, E., Garcia, G., and Saelzer, R. (1990). Screening of antibiotic residues in poultry meat. *J. Planar Chromatogr.—Mod. TLC 3*: 437–438.

von Brand, T. (1973). *Biochemistry of Parasites*, 2nd ed. Academic Press, New York.

Wagman, G. H., and Weinstein, M. J. (1983). Antibiotics. In *Chromatography—Fundamentals and Applications of Chromatography and Electrophoretic Methods, Part B: Applications*, E. Heftmann (ed.). Elsevier, Amsterdam, pp. B331–B344.

Wagman, G. H., and Weinstein, M. (1984). *Chromatography of Antibiotics*. Elsevier, Amsterdam, p.510.

Directory of Manufacturers and Sources of Standards, Sample Preparation Supplies, and TLC Instruments, Plates, and Reagents

Earlier editions of this book contained a rather extensive international directory of manufacturers and suppliers of standards, sample preparation supplies, and instruments and products for thin-layer chromatography. This type of information is now readily available on the internet world wide web through browsers such as Netscape, making the inclusion of a printed directory less important. Therefore, only a limited list of the major manufacturers and suppliers of analytical standards, sample preparation products, and TLC products and instruments, many of which we have dealt with in our research, is given below.

The following are three useful sources of information on instruments and supplies for TLC:

1. *Analytical Chemistry Lab Guide* ('97–'98) is available in the form of a 356-page magazine as well as on-line at the following American Chemical Society web address: <http://pubs.acs.org>. To find information on TLC, click on "Lab Guide 1997–98 Edition" under the "Electronic Directories" heading, then click on "Instrument and Laboratory Supplies" under the "Search by" heading. Type "tlc" in the blank "Instrument/Supply" box and click on "Submit Query," and a list of TLC supplies will be provided. Clicking on a product will display a list of suppliers, and clicking on a company name will display detailed information about the company, including address, telephone, fax, e-mail, and internet addresses, and its products. Options for use of the Lab Guide include

browsing or searching the following categories: instrument and laboratory supplies, laboratory reagents and standards, services, OEM supplies and suppliers, and company directory.

2. *American Laboratory Buyers' Guide Edition* is available as a 272 page magazine (dated January 1998). Information is available on-line at <http://www.iscpubs.com>. At this web site, click on "Publications" followed by "North American Publications" and "Buyers Guide—1998" or "European Publications" and "Buyers Guide—International Laboratory Internet Directory." Type "thin layer chromatography" as the search parameter, click on "start" to view a list of companies and products, and click on any of these entries for product and ordering information.

3. *Buyers' Guide 1997–1998, LC-GC Magazine*, volume 15, number 8, pp. 681–796, August, 1997. The guide contains a listing of corporate web sites and internet addresses but is not itself available on the internet world wide web.

LIST OF COMPANIES

Advanced Separation Technologies Inc. (Astec), 37 Leslie Court, P.O. Box 297, Whippany, NJ 07981

Aldrich Chemical Co. Inc., 1001 West Paul Avenue, Box 355, Milwaukee, WI 53201

Alltech-Applied Science, 2701 Carolean Industrial Drive, State College, PA 16801–7544.

American Type Culture Collection, 12301 Parklawn Drive, Rockville, MD 20852

Analtech Inc., 75 Blue Hen Road, P.O. Box 7558, Newark, DE 19714

Anspec/Ohio, 12 West Shelby Boulevard, Suite 4, Columbus, OH 43085

J.T. Baker Inc., 222 Red School Lane, Phillipsburg, NJ 08865

Bioscan/RSS, Inc., 4590 MacArthur Boulevard, N.W., Washington, DC 20007

Bodman Industries, P.O. Box 2421, Aston, PA 19014

Brinkmann Instruments, Inc., 1 Cantiague Road, P.O. Box 1019, Westbury, NY 11590–0207

Camag Scientific Inc., 515 Cornelius Harnett Drive, Wilmington, NC 28401

Carolina Biological Supply Co., Burlington, NC 27215

Champlain Biological Service, Glen Gardner, NJ 08826

Desaga GmbH, ln den Ziegelwiesen 1–7, Postfach 12 80, D-69153, Wiesloch, Germany

Drummond Scientific Co., 500 Parkway, Broomall, PA 19008

EG&G Berthold, Laboratorium Prof. Dr. Berthold GmbH & Co. KG, Postfach 100163, D-75312, Bad Wildbad, Germany

EM Science, 480 Democrat Road, P.O. Box 70, Gibbstown, NJ 08027–1297

Fisher Scientific Co., 711 Forbes Avenue, Pittsburgh, PA 15219–4785

Fotodyne Inc., 950 Walnut Ridge Drive, Hartland, WI 53029

Gelman Sciences, 600 South Wagner Road, Ann Arbor, MI 48103–9019

Grand Isle Biological Company (GIBCO), 3175 Staley Road, P.O. Box 68, Grand Isle, NY 14072

Harrison Research Inc., 840 Moana Court, Palo Alto, CA 94306

Hitachi Instruments, 3100 North First Street, San Jose, CA 95134

Kimble/Kontes, 1022 Spruce Street, P.O. Box 729, Vineland, NJ 08360

Macherey-Nagel GmbH & Co. KG, P.O. Box 101352, Neumann-Neander Strasse, D-52348, Dueren, Germany

Matreya Inc., 500 Tressler Street, P.O. Box I, Pleasant Gap, PA 16823

Merck KGaA, LPRO Chrom, Frankfurter Strasse, 64271 Darmstadt, Germany

Molecular Dynamics, 928 East Arques Avenue, Sunnyvale, CA 94086–4520.

Nu-Check Prep Inc., P.O. Box 295, Elysian, MN 56208

Packard Instrument Co., 800 Research Way, Meriden, CT 06450–3271

PolyScience, 6600 West Touhy Avenue, Niles, IL 60714

Presque Isle Cultures, P.O. Box 8191, Presque Isle, PA 16505

Regis Technologies Inc., 8210 Austin Avenue, Morton Grove, IL 60053–0519

Schleicher & Schuell Inc., 10 Optical Avenue, P.O. Box 2012, Keene, NH 03431

Sea Life Supply (formerly Peninsula Marine Biologic), 740 Tioga Avenue, Sand City, CA 93955

Scanalytics-CSPI, 40 Linnell Circle, Billerica, MA 01821

Serdary Research Laboratories Inc., 1643 Kathryn Drive, London, ON, N6G 2R7, Canada

Shimadzu Scientific Instruments, 7102 Riverwood Drive, Columbia, MD 21046

Sigma Chemical Co., P.O. Box 14508, St. Louis, MO 63178

Steraloids Inc., P.O. Box 310, Wilton, NH 03086

Supelco Inc., Supelco Park, Bellefonte, PA 16823

United States Biochemical Corporation (USBC), P.O. Box 22400, Cleveland, OH 44122

Ward's Natural Science Establishment, Inc., 5100 West Henrietta Road, P.O. Box 92912, Rochester, NY 14692–9012

Whatman Inc., 9 Bridewell Place, Clifton, NJ 17014

Woods Hole Biological Co., Supply Department, Marine Biological Laboratory, Woods Hole, MA 02543

Glossary

ABSORPTION. The interaction of certain wavelengths of radiation with a substance so that they are not transmitted, or the "taking-in" of a gas or liquid by a solid both on its surface and throughout its structure.

ACCURACY. The agreement between an experimental result (a single measurement or the mean of several replicate measurements) and the true or theoretical value.

ACTIVATION. The process of heating an adsorbent layer to drive off moisture and convert it to its most attractive or retentive state.

ACTIVITY. The surface property of a solid (adsorbent) that causes a particular degree of surface interaction and resultant migration for different substances, as influenced by the size of the available surface and the surface energy.

ACTIVITY GRADES (Brockmann activity grades). A standard grading system for the activity (adsorptivity) of alumina based on deactivation with water. Grade I is anhydrous alumina and has the highest activity. Grades II, III, IV, and V contain 3%, 6%, 10%, and 15% (by weight) of water, respectively. Activity grades have been used to define silica gel as well as alumina.

ADSORPTION. The attraction between the surface atoms of a solid and an external molecule by intermolecular forces such as hydrogen bonds, London forces, electrostatic forces, and charge-transfer forces. The adsorbent is the stationary phase for adsorption TLC. *Adsorption chromatography* involves separation of substances, based on their polarity, in a solid–liquid system involving competing interaction between adsorption on the surface and dissolving in the mobile phase.

ALUMINA (ALUMINUM OXIDE). A naturally basic polar adsorbent with a variety of pore sizes that is used for chromatography after thermal dehydration at a specified temperature to control its activity grade.

AMINO PHASE. Chemically bonded silica gel with a NH_2 functional group and short-chain n-alkyl (e.g., n-propyl) spacer group. Aminopropyl silica gel is polar and can function as a weakly basic ion exchanger or in straight-or reversed-phase modes, depending on the mobile phase.

ANALYTE. A solute that is to be identified or, more often, quantitatively determined by thin-layer chromatography or other method.

ANALYTICAL TLC. Thin-layer chromatography performed on 100–250-μm layers for the purpose of separation, identification, or quantification of substances.

ANTICIRCULAR TLC. Development of a layer from an outer circle of initial zones toward the center.

ARGENTATION TLC. Thin-layer chromatography using a layer, usually silica gel, impregnated with silver nitrate for the purpose of improving the separation of certain compounds.

ASCENDING DEVELOPMENT. The usual mode of thin-layer chromatography development in which the mobile phase moves upward by capillary action of the sorbent, as opposed to circular, descending, or horizontal development. The process is referred to as one-dimensional ascending development when chromatography is carried out in only one direction.

AUTOMATED MULTIPLE DEVELOPMENT (AMD). Linear, ascending, multiple developments are carried out in an automated instrument. The development distance is increased and the strength of the mobile phase is reduced (stepwise gradient elution) for each step. For silica gel, the mobile phase composition changes from more polar to less polar. AMD leads to zones with reduced diffusion and increased resolution because of a concentration effect.

AZEOTROPE (AZEOTROPIC MIXTURE). A liquid mixture of two or more substances that behaves like a single substance in that the vapor produced by partial evaporation of the liquid has the same composition as the liquid.

BAND. A chromatographic zone, usually in the shape of a horizontal line rather than a round spot.

BINDER. Any chemical added to a sorbent to improve the stability or hardness of the layer.

BONDED PHASE. A stationary phase chemically bonded to (as opposed to mechanically deposited on) a support material.

CAPACITY FACTOR (k'). A measure of sample retention by a layer:

$$k' = \frac{1 - R_F}{R_F}$$

CELLULOSE. Organic sorbents used for separations of hydrophilic compounds such as amino acids and carbohydrates by normal-phase partition chromatography. Native, fibrous and microcrystalline (rod-shaped) celluloses are available commercially and differ in separation properties.

CENTRIFUGAL LAYER CHROMATOGRAPHY. Analytical or preparative chromatography in which solvent is driven through the layer by centrifugal force in circular, anticircular, or linear modes. Centrifugal layer chromatography (CLC), which is also termed *rotation planar chromatography* (RPC), improves separation and reduces separation times compared to normal development but involves complex instruments.

CHAMBER. The tank, jar, or vessel of other type in which thin-layer chromatographic development is carried out. A *normal chamber* is a closed, glass vessel used for linear, ascending one- or two-dimensional development with (saturated) or without (unsaturated) mobile-phase–soaked filter paper liner.

CHAMBER SATURATION. Equilibration of the chamber or tank with mobile-phase vapor before and during development of the plate. In normal chambers, equilibration is achieved by placing a mobile-phase–wetted piece of filter paper against the back and side walls of the closed chamber 10 min before inserting the plate for development.

CHIRAL PHASE. A layer used for separation of optically active compounds (enantiomers). The most widely used chiral phase is prepared from C-18 (octadecyl) chemically bonded silica gel impregnated with a Cu(II) salt and an optically active enantiomerically pure hydroxyproline derivative and operates with a ligand-exchange mechanism.

CHROMATOGRAM. The series of zones on or in the layer after development (the developed TLC plate), or the record of the separation presented as a densitometer scan, photograph, photocopy, or video image.

CHROMATOGRAPHIC SOLVENT. Solvent or mixture of solvents used as the mobile phase.

CHROMATOGRAPHIC SYSTEM. The combination of the solvent, sorbent, and the sample mixture. The interactions of the chromatographic system determine the selectivity of the separation.

CHROMATOGRAPHY. A method of analysis in which the flow of mobile phase promotes the separation of substances by differential migration from a narrow initial zone in a porous, sorptive, stationary medium.

CHROMATOSTRIPS. An early term used for narrow thin-layer plates.

CHROMOGENIC. A reagent or reaction causing a solute to become colored. Chromogenic reagents are also termed *dyeing reagents* or *staining reagents*.

CLEANUP. Removal of interfering impurities from the analyte during sample preparation prior to TLC by a method such as extraction or column chromatography.

COCHROMATOGRAPHY. Development of a mixture prepared by adding a known standard to a sample thought to consist of or contain that substance. Formation of two zones from the mixture indicates the nonequivalence of the standard and sample, whereas the inability to separate the mixture is one piece of evidence for confirmation of identity.

COLUMN CHROMATOGRAPHY. Chromatography carried out by passing a liquid (or gaseous) mobile phase through a stationary phase packed in a column (*see* Open-Column Chromatography and HPLC).

CONDITIONING. Uptake of the vapor phase by the layer during chamber saturation prior to development of the layer.

CONJUGATE. The combined form in which drugs or pesticides can be found in plant or human samples, e.g., glucuronide or sulfate. Conjugates usually require hydrolysis by acid, base, or enzymes to free the analyte prior to TLC analysis.

CONTINUOUS DEVELOPMENT. Development of a layer for a greater distance than the actual plate length; also called overrun development.

CYANO PHASE. Silica gel chemically bonded with a CN functional group through a hydrophobic spacer. The layer has polarity between amino and dimethylsiloxane reversed-phase plates and can function with normal- or reversed-phase mechanisms, depending on the mobile phase.

DEACTIVATION. Adding moisture to an adsorbent layer to lower its retention of polar solutes.

DEMIXING. Separation of mobile-phase components during development, leading to the formation of one or more solvent fronts along the layer. Mobile phase demixing occurs mostly during TLC in an unsaturated normal chamber.

DENSITOMETER. Also called a *scanner* or *chromatogram spectrophotometer*; the instrument that measures absorbance or fluorescence for quantitative analysis or records spectra for qualitative analysis of zones. Commercial densitometers are of the mechanical slit-scanning or video-recording type.

DENSITOMETRY. Quantification of a zone directly on the layer with an instrument that measures color, absorption of ultraviolet light, fluorescence, or radioactivity.

DEPROTEINIZATION. A sample preparation procedure involving removal of protein, e.g., by precipitation with a reagent.

DERIVATIZATION. Reaction of solutes before chromatography or directly on the layer for the purpose of facilitating separation or detection.

DESALTING. A sample preparation procedure involving removal of salts by some procedure such as ion exchange or dialysis.

DESCENDING CHROMATOGRAPHY. Chromatography in which the mobile phase moves downward through the layer.

DESTRUCTIVE DETECTION. A detection method that irreversibly changes the chemical nature of the solute, e.g., sulfuric acid charring.

DETECTION. The process of locating a separated substance on a chromatogram, whether by physical methods, chemical methods, or biological methods (visualization). Detection is based on color, fluorescence, or UV absorption that is intrinsic to the separated compounds or induced by application of a suitable reagent, usually postchromatography, by dipping or spraying.

DEVELOPMENT. The movement of the mobile phase through the layer to form the chromatogram. This term does *not* mean detection of the zones.

DEVELOPMENT SOLVENT. The mobile phase.

DIATOMACEOUS EARTH (KIESELGUHR). Naturally occurring amorphous silicic acid, formed from the skeletons of microscopic marine organisms (fossils). It has a small surface area of low activity and a large pore diameter and is used mainly as a support in partition TLC and for preparing preadsorbent zones.

DIOL PHASE. Silica gel chemically bonded with a DIOL (alcoholic hydroxy) functional group through a hydrophobic spacer. The layer is polar and has a mechanism similar to unmodified silica gel.

DISPLACEMENT TLC. Mode of TLC in which the mobile phase contains a "displacer component" that has a higher affinity for the stationary phase than any of the solutes to be separated. This is in contrast to the usual elution mode of TLC, in which the mobile phase has a smaller affinity than the analytes for the sorbent. In displacement chromatography, development leads to formation of a "displacement train" of zones, all moving with the same velocity in reverse order of affinity for the sorbent and having higher concentration than in linear elution chromatography. Displacement TLC has been used to scout for optimum displacers and determine solute distribution for displacement column HPLC, which is used mostly for preparative separations. [Kalasz, H., Kerecsen, L., and Nagy, J. (1984). Conditions and parameters dominating displacement TLC. *J. Chromatogr. 316*: 95–104.]

DISTRIBUTION CONSTANT (K). *See* Partition Coefficient.

DROP CHROMATOGRAPHY. Application of a drop of mixture solution to a layer, followed by applications of drops of a solvent on top of the sample to develop it into concentric rings.

EDGE EFFECT. The greater migration (higher R_F value) of a substance located at the side edges of a plate compared to the center. It is remedied by lining the tank with mobile-phase wetted filter paper to improve chamber saturation.

EFFICIENCY. The narrowness of a peak compared with the length of time the component is in contact with the stationary phase. Efficiency is measured by plate number, N, or height equivalent to a theoretical plate, H.

ELUENT. A solvent used to remove or elute a substance away from the sorbent during recovery for preparative or quantitative TLC. Also, the mobile phase used to perform column chromatography, but an improper term for the mobile phase used for development in thin-layer chromatography.

ELUOTROPIC SERIES. A series of solvents arranged in order of increasing ability to displace a solute, i.e., increasing polarity in adsorption chromatography.

ELUTION. The removal of a solute from a sorbent by passages of a suitable solvent. The eluent is the solvent and the effluent or eluate is the liquid flowing from the sorbent (the eluent ± the solute). *Elution development* is defined as the development of a small amount of sample through a column or flat bed of sorbent with a liquid less strongly sorbed than the sample. The terms *elution development* and *chromatography* are generally used synonymously; the term *displacement development* is used when the developing solvent is more strongly sorbed that the sample, and the term *frontal analysis* is used when the sample is passed continuously through the sorbent.

ENRICHMENT. The concentration effect obtained during sample preparation. If a compound contained in 1 liter of water is collected on a solid-phase extraction column and is subsequently eluted with 2 ml of a solvent, an enrichment factor of 500 is obtained.

FLAT-BED (or PLANAR) CHROMATOGRAPHY. Common term for PC and TLC.

FLORISIL. Magnesium silicate adsorbent.

FLUORESCENCE. Emission of visible radiation by molecules excited by UV radiation. Substances that fluoresce are characterized by rigid structures and pi-electron systems.

FLUORESCENCE INDICATOR. A fluorescing substance added to a layer to allow detection of colorless, UV-absorbing substances under UV light as dark spots on a green or blue background. The indicators are usually manganese-activated zinc silicate or acid-stable alkaline earth tungstate (254 nm) or an optical brightener (366 nm). The detection method is termed *fluorescence quenching*.

FORCED-FLOW PLANAR CHROMATOGRAPHY (FFPC). Planar chromatography in which solvent migrates through the layer under the influence of external pressure (overpressured layer chromatography, OPLC) or centrifugal force (rotation planar chromatography, RPC).

FRONT. The visible boundary at the junction of the mobile-phase wetted layer and the "dry" layer.

FLUOROGENIC. A reagent or reaction causing a solute to become fluorescent.

GC. Gas chromatography, most often gas-liquid partition chromatography (GLC).

GEL FILTRATION (GEL PERMEATION). Exclusion chromatography, using a polymer gel layer. Gel filtration uses a Dextran-type gel to separate water-soluble species. Gel permeation uses a styrene-divinylbenzene-type gel to separate organic species.

GRADIENT. The change in layer or mobile-phase composition made to improve separations.

H (HETP). *See* Plate Number.

HARD LAYER. An abrasion-resistant sorbent layer bound to the backing by an organic polymer.

HIGH-PRESSURE PLANAR LIQUID CHROMATOGRAPHY (HPPLC). Automated circular separation under high pressure.

HOMOLOG. A member of a homologous series, which is a related succession of compounds each containing one or more carbon atoms and two more hydrogen atoms than the one before it in the series. The alcohols methanol, ethanol, and propanol are homologs.

HORIZONTAL LINEAR DEVELOPMENT. Development in a commercial chamber from one end or from both ends to the center.

HPLC. High-performance column liquid chromatography, performed under high pressure with narrow-bore columns containing small-diameter packings and continuous-flow detectors.

HPTLC. High-performance thin-layer chromatography, performed on layers with small, uniform, densely packed sorbent particles. HPTLC layers are more homogeneous, thinner, and smaller than layers for TLC, leading to faster, higher resolution separations with improved detection sensitivity.

hR_F. $100 \times R_F$.

HYDROPHILIC. Substances that are soluble in water or other polar solutions.

HYDROPHOBIC. Substances that are soluble in nonpolar hydrocarbon solvents and insoluble in water.

IMPREGNATED LAYER. A layer containing a liquid or solid chemical or mixture of chemicals added to aid separation or detection of solutes.

IN SITU. Occurring "in place," directly on the layer.

ION EXCHANGE. A process whereby ions of the same charge sign replace one another in a given phase. In chromatography, the term usually refers to systems in which the stationary phase is made up of an ionic polymer. This can be a synthetic resin, a cellulose or Sephadex-type exchanger, or various types of inorganic materials.

IR. Infrared.

ISOMER. One of two or more compounds having the same molecular weight and formula, but differing with respect to the arrangement or configuration of the atoms. Examples of isomers are ortho(1,2)-, meta(1,3)-, and para(1,4)-dichlorobenzene.

LIPOPHILIC. *See* Hydrophobic.

MASS SPECTROMETRY. A method for structure determination based on ionization of the compound and the sorting out of ions in a mass analyzer.

MASS TRANSFER. Movement of a solute between the stationary and mobile phases.

METABOLITE. The product(s) resulting from one or more chemical reactions that change or break down a compound such as a drug or pesticide in a biological system (e.g., human or animal body, or plant) or the environment.

MICROGRAM (μg). 10^{-6} g or 1000 ng.

MIGRATION. Travel of a solute through the layer in the direction of the mobile-phase flow.

MOBILE PHASE. The moving liquid phase used for development, often called the development solvent or solvent.

MULTIPLE DEVELOPMENT. Repeated development of the chromatogram in the same direction with one mobile phase or different mobile phases, each one for the same or different distances. Improves the separation of mixtures of substances with R_F values below 0.5.

MULTIMODAL SEPARATION. A separation involving two distinct techniques, such as GC and TLC or HPLC and TLC. TLC is normally the second of the two techniques used. The coupling of two separation techniques is also called "multidimensional chromatography."

N-CHAMBER. A square or rectangular tank or chamber, or a cylinder, with (saturated) or without (unsaturated) a paper liner and covered with a top, for development of TLC or HPTLC plates.

NANOGRAM (ng). 10^{-9} g or 0.001 μg.

nm. Nanometer or 10^{-9} m.

NMR (or NUCLEAR MAGNETIC RESONANCE SPECTROMETRY). A method for structure determination of organic compounds based on the magnetic properties of different nuclei.

NONDESTRUCTIVE DETECTION. Detection of a substance on a chromatogram by a process that will not permanently change the chemical nature of the substances being detected.

NORMAL-PHASE CHROMATOGRAPHY. Adsorption or partition TLC in which the stationary phase is polar in relation to the mobile phase. Also referred to as *straight phase chromatography.*

OPEN-COLUMN CHROMATOGRAPHY. Liquid-column chromatography performed in the classical manner in a relatively large bore, usually glass column under gravity or low-pressure flow.

OPLC. Overpressured or overpressure layer chromatography, in which mobile phase is forced by a pump through a sealed analytical or preparative layer.

ORIGIN. The location of the applied sample; also, the starting point for chromatographic development of the applied sample.

OVERRUN DEVELOPMENT. Continuous development beyond the top edge of the layer.

PARTICLE SIZE. The size (diameter) of the particles comprising the layer. Smaller particles (mean and distribution or range) provide better efficiency and resolution.

PARTITION COEFFICIENT OR RATIO (K_d). The ratio of concentrations of solute in each phase: $K_d = C_s/C_m$, where C_s and C_m are the concentrations in the stationary and mobile phases, respectively.

PARTITION TLC. Separation of solutes based on differential distribution between a stationary liquid supported on the layer and the liquid mobile phase. In normal-phase partition, the stationary liquid is more polar than the mobile phase, whereas in reversed-phase partition the stationary liquid is the less polar. The term is also used for TLC on bonded phases in some cases.

PC. Paper chromatography.

PEI CELLULOSE. Microcrystalline cellulose modified with polyethyleneimine; acts as an anion exchanger.

PICOGRAM (pg). 10^{-12} g or 10^{-3} ng.

PLANAR (or FLAT BED) CHROMATOGRAPHY (PC). Common term for thin-layer or paper chromatography.

PLATE HEIGHT (HETP or H). Indicates the diffusion or broadening of the zones during migration. The length of the layer divided by the number of plates (N). A small plate height results in better resolution.

PLATE NUMBER (N). $N = 16\,(d_R/W)^2$, where d_R is the distance of migration of the spot and W is its width (size in the direction of development). N can be measured using the spots on the plate or their densitometric scans.

PMD. Programmed multiple development. The repeated development of a TLC plate with the same mobile phase or different mobile phases in the same direction for gradually increasing distances, using an automated commercial instrument. Also termed automated multiple development (AMD).

POLARITY. The effect of the combined interactions between a functional group and a layer or mobile phase, i.e., dispersion, dipole, and/or hydrogen bonding forces, is designated as "polarity." In chromatographic systems, benzene is described as a more polar solvent than toluene, although only the latter has a true dipole moment, the classical requirement for polarity. In chromatographic behavior, hydrocarbons and halogens are of low polarity, esters and ketones have intermediate polarity, and alcohols and amines are highly polar. The polarity of a silica gel surface is reduced by impregnation or chemical bonding with a hydrocarbon.

POLYAMIDE. Separations are based on formation of hydrogen bonds between the functional groups of the sample (phenols, amino acid derivatives, heterocyclic nitrogen compounds, carboxylic and sulfonic acids) and the carbonyl oxygen of the amide group, as well as a general reversed-phase partition mechanism (water is the weakest solvent). Available as two polycondensation products, polyundecanamide (PA 11) and polycaprolactam (PA 11).

PORES. The open channels within the sorbent structure. The *pore diameter* is the mean size of the opening, and the *pore volume* is the internal volume of the porous particles filled with mobile phase.

POSTCHROMATOGRAPHIC. Occurring after TLC separation, as in spray or dip application of detection reagents.

PREADSORBENT (CONCENTRATING) ZONE. A strip of inactive sorbent (silicon dioxide or diatomaceous earth) situated below, and adjacent to, the active sorbent that provides the separation. The preadsorbent zone simplifies and speeds manual sample application and forms narrow, band-shaped initial zones at the interface of the layers.

PRECHROMATOGRAPHIC. Occurring before TLC separation, as in derivatization of spotted samples at the origin or *preimpregnation* of the plate with a reagent to promote separation or detection before application of initial zones.

PRECISION. The agreement or lack of scatter among replicate determinations or analyses, without regard to the true answer.

PRECOATED PLATES. Commercial plates or sheets sold with the layer already formed and ready for use.

PRELOADING. Sorption of gaseous molecules by the layer prior to development with the mobile phase; also called layer conditioning or preadsorption.

PREPARATIVE TLC. Thin-layer chromatography of larger quantities of material on thick layers for the purpose of isolating separated substances for further analysis or use. The term is sometimes abbreviated PLC (preparative layer chromatography).

R_F. The ratio of the distance of migration of the center of a zone divided by the distance of migration of the solvent front, both measured from the origin.

R_M. A value used in relating chromatographic behavior to chemical structure:

$$R_M = \log \left(\frac{1}{R_F} - 1 \right)$$

R_X. The same as R_F except that the distance of solvent front migration is replaced by the distance of migration of some reference compound X.

RADIAL (CIRCULAR) DEVELOPMENT. Development of a layer in such a manner as to form circular or arc-shaped zones. Some workers differentiate circular and radial TLC by the use of "circular" for the case where one initial zone is developed into circular zones and "radial" for development of a series of initial zones, spotted in a circular pattern, into arcs. The layer is usually arranged horizontally during radial development.

RADIOCHROMATOGRAPHY. Chromatography of radioactively labeled substances. The resultant chromatogram is a *radiochromatogram* or *radiogram*.

RELATIVE HUMIDITY. The percentage of water saturation of the ambient air or chamber atmosphere at a given temperature. A relative humidity of 40–65% is normally used for TLC, and it is important to control humidity for consistent results. Control of humidity is achieved by placing a saturated salt solution or a sulfuric acid solution of specified concentration inside the chamber.

RESOLUTION. The ability to separate two zones. The mathematical description of resolution is

$$R = \frac{2(R_{F_1} - R_{F_2})}{W_1 + W_2}$$

where R_{F_1} and R_{F_2} are the respective R_F values of any two zones and W_1 and W_2 are their respective zone lengths (in the direction of development). Migration distances can be used instead of R_F values in this equation. Resolution is the result of the combined contributions of efficiency (zone compactness), selectivity (separation of zone centers), and capacity (average R_F value of the pair of substances to be separated).

REVERSE- or REVERSED-PHASE CHROMATOGRAPHY. Liquid-liquid partition TLC in which the stationary phase is nonpolar compared to the mobile phase. The layer can be impregnated or bonded in nature.

ROD TLC. TLC carried out on sintered glass rods, usually in conjunction with a scanning flame ionization detector.

SANDWICH CHAMBER (S-CHAMBER). A developing chamber formed from the plate itself, a spacer, and a layered or nonlayered cover plate that stands in a trough containing the mobile phase, or some other type of small-volume chamber in which vapor-phase saturation occurs quickly.

SATURATED. The condition of a chamber that is lined with paper and equilibrated with mobile-phase vapors before beginning chromatographic development.

SECONDARY FRONT. An additional solvent or mobile-phase front, below the primary front, that occurs because the mobile-phase components demix.

SELECTIVITY. The ability of a chromatographic system to produce different R_F values for the components of a mixture, i.e., to separate the zone centers.

SENSITIVITY. The ability to detect or measure a small mass of analyte.

SEPARATION NUMBER. A measure of the separating power of a TLC system: The number of substances completely separated (resolution $= 1$) between $R_F = 0$ and $R_F = 1$ by a homogeneous mobile phase (no solvent gradients in the direction of development).

SOFT LAYER. A sorbent layer prepared without binder or with gypsum binder (*see* Hard Layer).

SOLID-PHASE EXTRACTION. An alternative to traditional separatory funnel extraction in which an analyte is extracted from a liquid sample by use of a solid packed in a small column, cartridge, or disk. The sample is forced through the solid, which is an adsorbent or bonded phase, with the aid of vacuum or pressure; the analyte is retained on the solid, and it is subsequently eluted with a small volume of a strong solvent, usually resulting in significant enrichment.

SOLUTE. A general term for the compounds or ions being chromatographed.

SOLVENT FRONT. The farthest point of movement of the mobile phase during development.

SOLVENT STRENGTH (ε^0). A measure of the polarity of a solvent for liquid-solid adsorption chromatography. It is based on the free energy of adsorption onto a standard surface. Values for common solvents range from 0.00 (pentane) to 0.95 (methanol).

SORPTION. A general term for the attraction between a layer and a solute, without specification of the type of physical mechanism (i.e., adsorption, partition, ion exchange) or mixed mechanism involved. Sorbent is a related general term referring to the layer itself.

SPOT. Used synonymously with zone, but usually meant to indicate a round or elliptical shape.

SPOT CAPACITY. Same as SEPARATION NUMBER.

STATIONARY PHASE. The solid sorbent layer, with or without any impregnation agent, preloaded vapor molecules, or immobilized mobile-phase component(s).

STEPWISE DEVELOPMENT. Development using a mobile phase whose composition is changed using discontinuous, stepped gradients, in contrast to continuously variable gradient elution.

STREAK. An initial zone in the shape of a narrow horizontal line at the origin.

STREAKING. *see* Tailing.

SUPERCRITICAL FLUID EXTRACTION (SFE). A technique for extraction of analytes from sample matrices using a dense gas. Carbon dioxide, which becomes a supercritical fluid when used above its critical pressure (1070 psi) and temperature (31°C), is the most widely applied extraction medium for SFE because it is nontoxic, nonflammable, and facilitates extractions at low temperatures in a nonoxidizing environment.

SURFACE AREA. The total or specific surface area includes the surface area of the sorbent particles plus the surface area of the pores, usually measured in m^2/g. Surface area is directly related to the degree of interaction between the sample and sorbent.

TAILING. Formation of a zone with an elongated rear portion, often leading to incomplete resolution.

THROUGHPUT. A term used mostly in the context of "sample throughput," which indicates the number of samples that can be analyzed by a particular method in a given period of time. TLC has high sample throughput because of the ability to apply multiple samples to a single plate. "Solvent throughput" is a term that designates the amount of solvent passing through the layer. For example, there is higher solvent throughput with an unsaturated N-chamber compared to a saturated N-chamber because of the increased solvent that passes through the layer to replace the solvent that has evaporated from it.

TLC. Thin-layer chromatography.

TLG. Thin-layer gel chromatography, in which separations are based mainly on solute sizes.

TRAILING. *See* Tailing.

TWIN-TROUGH (DOUBLE-TROUGH) CHAMBER. A glass chamber for linear ascending development with a glass ridge on the bottom dividing it into two sections for holding mobile phase or other solutions. Facilitates vapor pre-equilibration of the layer and minimizes solvent usage.

TWO-DIMENSIONAL DEVELOPMENT. Successive development of a chromatogram with the same solvent or different solvents in directions at a 90° angle to each other.

UNSATURATED. The condition of a chamber that has the mobile phase and plate added together so that equilibration with the vapors is occurring during chromatographic development.

UV. Ultraviolet.

VISUALIZATION. Detection of the zones on a chromatogram.

ZONE. The area of distribution on the layer containing the individual solutes or mixture before, during, or after chromatography. The initial zone is the applied sample prior to development. Band, zone, and spot are often used more or less interchangeably, but spot usually denotes a round zone and band a flat, horizontally elongated zone.

Index